# Modern Acoustics and Signal Processing



For further volumes:
http://www.springer.com/series/3754

**Series Preface for Modern Acoustics and Signal Processing**

In the popular mind, the term "acoustics" refers to the properties of a room or other environment—the acoustics of a room are good or the acoustics are bad. But as understood in the professional acoustical societies of the world, such as the highly influential Acoustical Society of America, the concept of acoustics is much broader. Of course, it is concerned with the acoustical properties of concert halls, classrooms, offices, and factories—a topic generally known as architectural acoustics, but it is also concerned with vibrations and waves too high or too low to be audible. Acousticians employ ultrasound in probing the properties of materials, or in medicine for imaging, diagnosis, therapy, and surgery. Acoustics includes infrasound—the wind-driven motions of skyscrapers, the vibrations of the earth, and the macroscopic dynamics of the sun.

Acoustics studies the interaction of waves with structures, from the detection of submarines in the sea to the buffeting of spacecraft. The scope of acoustics ranges from the electronic recording of rock and roll and the control of noise in our environments to the inhomogeneous distribution of matter in the cosmos.

Acoustics extends to the production and reception of speech and to the songs of humans and animals. It is in music, from the generation of sounds by musical instruments to the emotional response of listeners. Along this path, acoustics encounters the complex processing in the auditory nervous system, its anatomy, genetics, and physiology—perception and behavior of living things.

Acoustics is a practical science, and modern acoustics is so tightly coupled to digital signal processing that the two fields have become inseparable. Signal processing is not only an indispensable tool for synthesis and analysis, it informs many of our most fundamental models about how acoustical communication systems work.

Given the importance of acoustics to modern science, industry, and human welfare Springer presents this series of scientific literature, entitled Modern Acoustics and Signal Processing. This series of monographs and reference books is intended to cover all areas of today's acoustics as an interdisciplinary field. We expect that scientists, engineers, and graduate students will find the books in this series useful in their research, teaching, and studies.

July 2012

William M. Hartmann
Series Editor-in-Chief

**The ASA Press**

The ASA Press imprint represents a collaboration between the Acoustical Society of America and Springer dedicated to encouraging the publication of important new books in acoustics. Published titles are intended to reflect the full range of research in acoustics. ASA Press books can include all types of books published by Springer and may appear in any appropriate Springer book series.



ASA Press

## Acoustical Society of America

The mission of the **Acoustical Society of America** (www.acousticalsociety.org) is to increase and diffuse the knowledge of acoustics and promote its practical applications. The ASA is recognized as the world's premier international scientific society in acoustics, and counts among its more than 7,000 members professionals in the fields of bioacoustics, engineering, architecture, speech, music, oceanography, signal processing, sound and vibration, and noise control.

Since its first meeting in 1929, The Acoustical Society of America has enjoyed a healthy growth in membership and in stature. The present membership of approximately 7500 includes leaders in acoustics in the United States of America and other countries. The Society has attracted members from various fields related to sound including engineering, physics, oceanography, life sciences, noise and noise control, architectural acoustics; psychological and physiological acoustics; applied acoustics; music and musical instruments; speech communication; ultrasonics, radiation, and scattering; mechanical vibrations and shock; underwater sound; aeroacoustics; macrosonics; acoustical signal processing; bioacoustics; and many more topics.

To assure adequate attention to these separate fields and to new ones that may develop, the Society establishes technical committees and technical groups charged with keeping abreast of developments and needs of the membership in their specialized fields. This diversity and the opportunity it provides for interchange of knowledge and points of view has become one of the strengths of the Society.

The Society's publishing program has historically included the *Journal of the Acoustical Society of America*, the magazine *Acoustics Today*, a newsletter, and various books authored by its members across the many topical areas of acoustics. In addition, ASA members are involved in the development of acoustical standards concerned with terminology, measurement procedures, and criteria for determining the effects of noise and vibration.

Jens Blauert
Editor

# The Technology of Binaural Listening

ASA Press

Springer

*Editor*
Jens Blauert
Fak. Elektrotechnik
LS Allgm.Elektrotechn.+Akustik
Univ. Bochum
Bochum
Germany

ISBN 978-3-642-37761-7 ISBN 978-3-642-37762-4 (eBook)
DOI 10.1007/978-3-642-37762-4
Springer Heidelberg New York Dordrecht London

Library of Congress Control Number: 2013938218



Printed on acid-free paper

Springer is part of Springer Science+Business Media (www.springer.com)

# Preface

For more than six decades, scientists have developed binaural models of the human auditory system. Some of them, for instance, the so-called coincidence model and the so-called equalization-and-cancellation model, have indeed been quite successful in explaining basic functions of binaural localization and signal detection.

Recently, advances in digital signal processing and the availability of suitable hardware have prepared the ground for applying binaural models in the context of modern technology in many stimulating ways. Recognizing this situation, in 2009 15 laboratories in Europe and the US have founded the research group AABBA, with the aim of setting up "*Aural Assessment by Means of Binaural Algorithms.*" Now, after its first 4-year term of activity, the group presents their most relevant results in the form of this book.

All chapters underwent the same rigorous review process as known from international archival journals. In other words, each chapter was reviewed by three members of the AABBA group plus by at least two anonymous external reviewers.

Yet, unlike usual journal papers, the chapters have been written with the intention to provide, besides their scientifically new content, illustrative introductions to their respective areas. In addition, there is an introductory chapter on binaural modeling at large and a chapter that gives an outlook to the future of these models. The volume further provides a MATLAB toolbox that enables readers to construct binaural models of their own, tailored to their specific demands. These features make the book suitable for teaching and a versatile source of references.

Current members of AABBA are research labs at Universities/Technical Universities in Helsinki, Boston, Cardiff, Oldenburg, Lyon, Troy NY (Rensselaer), Bochum, Berlin, Copenhagen (Lyngby), Dresden, Eindhoven, Munich, Paris (Pierre et Marie Curie), Patras, Rostock, Toulouse, and at the Austrian Academy's Acoustics Research Institute in Vienna. Altogether, 50 coauthors participated in writing this volume.

It is hoped that the book, which is indeed the first and only of its kind so far, will pave the way to many further advanced applications of binaural models in technology in the near future. AABBA intends to stay part of this exciting development and has just started on a second 4-years term of joint R&D activity.

Bochum, March 2013 Jens Blauert

# Contents

**An Introduction to Binaural Processing** . . . . . . . . . . . . . . . . . . . . . . 1
A. Kohlrausch, J. Braasch, D. Kolossa and J. Blauert

**The Auditory Modeling Toolbox** . . . . . . . . . . . . . . . . . . . . . . . . . . . 33
P. L. Søndergaard and P. Majdak

**Trends in Acquisition of Individual Head-Related Transfer Functions** . . . . . . . . . . . . . . . . . . . . . . . . . . . . . . . . . . . . . . 57
G. Enzner, Chr. Antweiler and S. Spors

**Assessment of Sagittal-Plane Sound-Localization Performance in Spatial-Audio Applications** . . . . . . . . . . . . . . . . . . . . . . . . . . . . . 93
R. Baumgartner, P. Majdak and B. Laback

**Modeling Horizontal Localization of Complex Sounds in the Impaired and Aided Impaired Auditory System** . . . . . . . . . . . . 121
N. Le Goff, J. M. Buchholz and T. Dau

**Binaural Scene Analysis with Multidimensional Statistical Filters** . . . . 145
C. Spille, B. T. Meyer, M. Dietz and V. Hohmann

**Extracting Sound-Source-Distance Information from Binaural Signals** . . . . . . . . . . . . . . . . . . . . . . . . . . . . . . . . . . 171
E. Georganti, T. May, S. van de Par and J. Mourjopoulos

**A Binaural Model that Analyses Acoustic Spaces and Stereophonic Reproduction Systems by Utilizing Head Rotations** . . . . 201
J. Braasch, S. Clapp, A. Parks, T. Pastore and N. Xiang

**Binaural Systems in Robotics** . . . . . . . . . . . . . . . . . . . . . . . . . . . . . 225
S. Argentieri, A. Portello, M. Bernard, P. Danès and B. Gas

**Binaural Assessment of Multichannel Reproduction** . . . . . . . . . . . . . . 255
H. Wierstorf, A. Raake and S. Spors

**Optimization of Binaural Algorithms for Maximum Predicted Speech Intelligibility** . . . . . . . . . . . . . . . . . . . . . . . . . . . . . . . . 279
A. Schlesinger and Ch. Luther

**Modeling Sound Localization with Cochlear Implants** . . . . . . . . . . . . 309
M. Nicoletti, Chr. Wirtz and W. Hemmert

**Binaural Assessment of Parametrically Coded Spatial Audio Signals** . . . . . . . . . . . . . . . . . . . . . . . . . . . . . . . . . . . . . . 333
M. Takanen, O. Santala and V. Pulkki

**Binaural Dereverberation** . . . . . . . . . . . . . . . . . . . . . . . . . . . . 359
A. Tsilfidis, A. Westermann, J. M. Buchholz, E. Georganti and J. Mourjopoulos

**Binaural Localization and Detection of Speakers in Complex Acoustic Scenes** . . . . . . . . . . . . . . . . . . . . . . . . . . . . . . . . . 397
T. May, S. van de Par and A. Kohlrausch

**Predicting Binaural Speech Intelligibility in Architectural Acoustics** . . . . . . . . . . . . . . . . . . . . . . . . . . . . . . . . . . . . . . . 427
J. F. Culling, M. Lavandier and S. Jelfs

**Assessment of Binaural–Proprioceptive Interaction in Human-Machine Interfaces** . . . . . . . . . . . . . . . . . . . . . . . . . . 449
M. Stamm and M. E. Altinsoy

**Further Challenges and the Road Ahead** . . . . . . . . . . . . . . . . . . . . . . 477
J. Blauert, D. Kolossa, K. Obermayer and K. Adiloğlu

**Index** . . . . . . . . . . . . . . . . . . . . . . . . . . . . . . . . . . . . . . . . 503

# Contributors

**Kamil Adiloğlu** Neural Information Processing Group, Institut für Softwaretechnik und Theoretische Informatik, Technische Universität Berlin, Berlin, Germany; now with HörTech gGmbH, Oldenburg, Oldenburg, Germany

**M. Ercan Altinsoy** Chair of Communication Acoustics, Technische Universität Dresden, Dresden, Germany

**Christiane Antweiler** Institute of Communication Systems and Data Processing, RWTH Aachen University, Aachen, Germany

**Sylvain Argentieri** Institute for Intelligent Systems and Robotics, Pierre and Marie Curie University (Univ Paris 06), National Centre for Scientific Research (CNRS), Paris, France

**Robert Baumgartner** Acoustics Research Institute, Austrian Academy of Sciences, Vienna, Austria

**Mathieu Bernard** Institute for Intelligent Systems and Robotics, Pierre and Marie Curie University (Univ Paris 06), National Centre for Scientific Research (CNRS), Paris, France

**Jens Blauert** Institute of Communication Acoustics, Ruhr-Universität Bochum, Bochum, Germany

**Jonas Braasch** Graduate Program in Architectural Acoustics, Center of Cognition, Communication and Culture, Rensselaer Polytechnic Institute, Troy, NY, USA

**Jörg M. Buchholz** National Acoustic Laboratories, Australian Hearing, Department of Linguistics, Macquarie University, Sydney, Australia

**Samuel Clapp** Graduate Program in Architectural Acoustics, Center of Cognition, Communication and Culture, Rensselaer Polytechnic Institute, Troy, NY, USA

**John Culling** School of Psychology, Cardiff University, Cardiff, UK

**Patrick Danès** Laboratory for Analysis and Architecture of Systems, Université Paul Sabatier (Univ. Toulouse), National Centre for Scientific Research (CNRS), Toulouse, France

**Torsten Dau** Department of Electrical Engineering, Centre for Applied Hearing Research, Technical University of Denmark, Kgs. Lyngby, Denmark

**Mathias Dietz** Department of Medical Physics and Acoustics, Carl-von-Ossietzky Universität, Oldenburg, Germany

**Gerald Enzner** Institute of Communication Acoustics, Ruhr-Universität Bochum, Bochum, Germany

**Bruno Gas** Institute for Intelligent Systems and Robotics, Pierre and Marie Curie University (Univ Paris 06), National Centre for Scientific Research (CNRS), Paris, France

**Eleftheria Georganti** Audio and Acoustic Technology Group, Electrical and Computer Engineering Department, University of Patras, Rio, Patras, Greece

**Werner Hemmert** Bio-Inspired Information Processing, Institute of Medical Engineering, Technische Universität München, Garching, Germany

**Volker Hohmann** Department of Medical Physics and Acoustics, Carl-von-Ossietzky Universität, Oldenburg, Germany

**Sam Jelfs** Smart Sensing and Analysis, Philips Research Europe, Eindhoven, The Netherlands

**Armin Kohlrausch** Human Technology Interaction, Technische Universiteit Eindhoven, Smart Sensing and Analysis Philips Research Europe, Eindhoven, The Netherlands

**Dorothea Kolossa** Institute of Communication Acoustics, Ruhr-Universität Bochum, Bochum, Germany

**Bernardt Laback** Acoustics Research Institute, Austrian Academy of Sciences, Vienna, Austria

**Mathieu Lavandier** Laboratoire Génie Civil et Bâtiment, Ecole Nationale des Travaux Publiques de l'Etat, Université de Lyon, Vaulx-en-Velin, France

**Nicolas Le Goff** Department of Electrical Engineering, Centre for Applied Hearing Research, Technical University of Denmark, Kgs. Lyngby, Denmark

**Christian Luther** Institute of Communication Acoustics, Ruhr-Universität Bochum, Bochum, Germany

**Piotr Majdak** Acoustics Research Institute, Austrian Academy of Sciences, Vienna, Austria

**Tobias May** Department of Electrical Engineering, Centre for Applied Hearing Research, Technical University of Denmark, Kgs. Lyngby, Denmark

**Bernd Meyer** Department of Medical Physics and Acoustics, Carl-von-Ossietzky Universität, Oldenburg, Germany

**John Mourjopoulos** Audio and Acoustic Technology Group, Electrical and Computer Engineering Department, University of Patras, Rio, Patras, Greece

**Michele Nicoletti** Bio-Inspired Information Processing, Institute of Medical Engineering, Technische Universität München, Garching, Germany

**Klaus Obermayer** Neural Information Processing Group, Institut für Softwaretechnik und Theoretische Informatik, Technische Universität Berlin, Berlin, Germany

**Anthony Parks** Graduate Program in Architectural Acoustics, Center of Cognition, Communication and Culture, Rensselaer Polytechnic Institute, Troy, NY, USA

**Torben Pastore** Graduate Program in Architectural Acoustics, Center of Cognition, Communication and Culture, Rensselaer Polytechnic Institute, Troy, NY, USA

**Alban Portello** Laboratory for Analysis and Architecture of Systems, Université Paul Sabatier (Univ. Toulouse), National Centre for Scientific Research (CNRS), Toulouse, France

**Ville Pulkki** Department of Signal Processing and Acoustics, Aalto University, Espoo, Finland

**Alexander Raake** Assessment of IP-based Applications, Telekom Innovation Laboratories (T-Labs), Technische Universität Berlin, Berlin, Germany

**Olli Santala** Department of Signal Processing and Acoustics, Aalto University, Espoo, Finland

**Anton Schlesinger** Institute of Communication Acoustics, Ruhr-Universität Bochum, Bochum, Germany

**Peter Søndergaard** Department of Electrical Engineering, Centre for Applied Hearing Research, Technical University of Denmark, Kgs. Lyngby, Denmark; Acoustics Research Institute, Austrian Academy of Sciences, Vienna, Austria

**Constantin Spille** Department of Medical Physics and Acoustics, Carl-von-Ossietzky Universität, Oldenburg, Germany

**Sascha Spors** Institute of Communications Engineering, Universität Rostock, Rostock, Germany

**Maik Stamm** Chair of Communication Acoustics, Technische Universität Dresden, Dresden, Germany

**Marko Takanen** Department of Signal Processing and Acoustics, Aalto University, Espoo, Finland

**Alexandros Tsilfidis** Audio and Acoustic Technology Group, Electrical and Computer Engineering Department, University of Patras, Rio, Patras, Greece

**Steven van de Par** Acoustics Group, Institute of Physics, Carl-von-Ossietzky Universität, Oldenburg, Germany

**Adam Westermann** National Acoustic Laboratories, Australian Hearing, Department of Linguistics, Macquarie University, Sydney, Australia

**Hagen Wierstorf** Assessment of IP-based Applications, Telekom Innovation Laboratories (T-Labs), Technische Universität Berlin, Berlin, Germany

**Christian Wirtz** MED-EL Deutschland GmbH, Starnberg, Germany

**Ning Xiang** Graduate Program in Architectural Acoustics, Center of Cognition, Communication and Culture, Rensselaer Polytechnic Institute, Troy, NY, USA

# An Introduction to Binaural Processing

**A. Kohlrausch, J. Braasch, D. Kolossa and J. Blauert**

## 1 Introduction

Immanuel Kant has been quoted [63] with the statement that "Blindness separates us from things but deafness from people", which emphasizes that hearing is a prominent social sense of human beings. Further, hearing provides us with relevant information about the state of our environment and activities around us, including those in locations beyond our field of vision. Thus, hearing is indeed of high relevance for our orientation in the world and our situational awareness. Also, in contrast to vision, hearing doesn't ever completely sleep and therefore has an effective warning function.

Engineers like to think of the auditory system as a kind of multi-purpose computer with two input ports. The input ports are the two ears, at equal height on both sides of a solid ellipsoid, the head. The head serves as an effective *antenna holder*, which can move about with six degrees of freedom relative to the main body, whereby the body itself can also move in three-dimensional space and can change its orientation relative to a reference position.

The auditory system receives its input in form of the elastic vibrations and waves of the surrounding fluids and solids, with which it is in mechanical contact. The

A. Kohlrausch
Human-Technology Interaction, TU Eindhoven, Eindhoven, The Netherlands

A. Kohlrausch
Philips Research Europe, Eindhoven, The Netherlands

J. Braasch
Center Cognition, Communication and Culture,
Rensselaer Polytechnic Institute, Troy NY, US

D. Kolossa (✉) · J. Blauert
Institute of Communication Acoustics, Ruhr-Universität Bochum,
Bochum, Germany
e-mail: dorothea.kolossa@rub.de

J. Blauert (ed.), *The Technology of Binaural Listening*, Modern Acoustics and Signal Processing, DOI: 10.1007/978-3-642-37762-4_1,

contact to the receptive organs, the "microphones" of the auditory system, is provided either by air conduction via the ear canals or by bone conduction via the skull. Bone conduction is usually neglected when dealing with listening in air, since it is attenuated by roughly 60 dB with respect to the air-conduction, which represents a power ratio of 1000000.

We can hear with one ear only, for instance, with one ear being impaired or plugged [105], but hearing with two functioning ears, *binaural hearing*, offers a number of important advantages over *monaural hearing*. This is due to the fact that binaural hearing provides additional information, which is encoded in the differences of the input signals to the two ears. In addition, having two ears placed at slightly different positions in the sound field offers the possibility to focus attention to the ear with a better signal-to-noise ratio, indicated as *better-ear listening*.

Under the assumption that these differences are represented by a linear, time-invariant system, the only interaural differences possible are interaural arrival-time differences, ITDs, and interaural level differences, ILDs—both frequency dependent. Yet, note that the assumption of linearity and time invariance is not always sufficiently fulfilled, for instance, when objects in the sound field move quickly, such as sound sources, reflective surfaces and/or the listeners, or when there are turbulences in the air. Nevertheless, even then binaural hearing offers substantial advantages.

## 2 Performance of Binaural Hearing

A selection of more relevant advantages of binaural hearing is listed and discussed in the following—for more details see, for instance, [7]. All these advantages can easily be tested by plugging one ear firmly with a finger and open it again alternatingly.

### *2.1 Sound Localization*

Auditory events, that is, whatever we hear, exist at specific positions and with specific extensions in space. The totality of auditory events defines the aural space. The aural space when listening binaurally is substantially different from the aural space when listening monaurally.

When listening binaurally, the auditory events are less spatially *blurred* than in the monaural case. For instance, two auditory events that are only $1^\circ$ apart in azimuth can be discriminated binaurally for frontal sound incidence, while in monaural listening the respective localization blur is at least 10 times larger. Higher spatial distinction in the binaural case also holds for elevation and distance. Further, in binaural hearing, the spatial extent of auditory events is more clearly defined, that is, there is a clear distinction between spatially compact and spatially diffuse ones. In monaural hearing, the auditory events are much more spatially diffuse in general.

## 2.2 Dealing with Reflections

When listening to a sound source in a room with reflective walls, the auditory system receives the direct sound from the source plus multiple reflections from different directions. Nevertheless, as long as the time delay of the reflections with respect to the direct sound is within certain limits, there is only one auditory event. In other words, the direct sound gains *localization dominance* [73] over the reflected sounds such that only one auditory event appears, called *fusion*. The center of gravity of the auditory event usually lies in the direction of the sound source. This combination of fusion and localization dominance is called the *precedence effect* in psychoacoustics, formerly known as rule of the first wavefront—for details see [8]. The effect is fundamental to proper formation of the aural space in rooms with reflective walls, since it supports the identification of the sound source, among other things. If the time delay between direct and reflected sounds at the ears is less than about 1 ms, fusion occurs as well, but the reflective sounds codetermine the direction of the auditory event. This effect, called *summing localization*, provides the basis for technical applications such as stereophony and surround sound. If the delay is too large, more than one auditory event appears, whereby those originating from the reflections are heard as repetitions, called *echoes*, of the one which originates from the direct sound. The minimum delay at which echoes are heard is called the *echo threshold*. This threshold is signal dependent; it varies from roughly 1 ms for short impulses through 50 ms for ongoing speech to 80 ms for classic/romantic music.

Figure 1 illustrates the three effects for the simple case of a direct sound plus one coherent delayed sound, the reflection. They are, in this case, realized by two loudspeakers. The lead speaker emits the same signal as the lag one, but the latter one is delayed. The signals in this example are broadband sounds of equal level, such as classical music or traffic noise.

At zero delay the auditory event of the listener appears at the midline between the two loudspeakers. Introducing a delay of up to 1 ms causes the auditory event to shift laterally with increasing delay toward the direction of the lead speaker. For delays above 1 ms up to the echo threshold, the auditory event has its center of gravity in the direction of the lead speaker. Thereby, the listener may well sense the presence of reflected sound—the auditory event is louder, its timbre changes, so-called *coloration*, and it becomes more spacious, but its position is dominated by the lead-speaker sound. For delays above the echo threshold, the auditory event splits into a direct part in the direction of the lead speaker and a repetition of it in the direction of the lag speaker.

A further advantage of binaural hearing versus monaural hearing is that the sense of reverberance and coloration is reduced. Reverberance is perceived when the sound field contains manifold reflections of increasing temporal density, usually decaying exponentially when the primary sound ceases. With one ear plugged, these sound fields sound distinctly more reverberant than with two ears open. The effect is called *binaural dereverberation* [99].

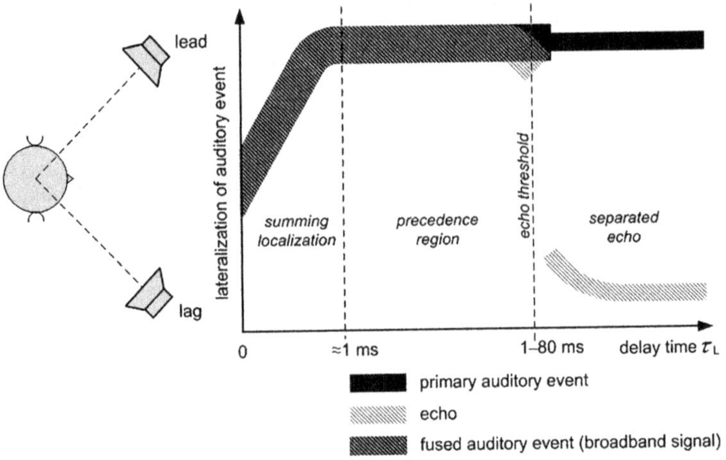


**Fig. 1** Demonstration of summing localization, precedence effect and echo threshold with two coherent sound sources, the lag one emitting with a delay, $\tau_L$, relative to the lead one

Reflections with delays of a few milliseconds give rise to spectral changes, since the addition of the direct sound with these reflections causes comb-filter effects. As a result, the auditory events sound *colored*. This coloration is clearly smaller with two ears open than with one ear plugged, because the exact position of spectral peaks in the two ears will usually differ and the peaks will therefore be somewhat averaged. This effect is known as *binaural decoloration* [28].

## 2.3 Auditory Stream Segregation

Listening in a sound field generated by several different sound sources and their reflections leads to a so-called aural or auditory scene in the aural world. Aural scenes contain multiple concurrent auditory events. The auditory system, in the process of forming these different auditory events, analyzes the acoustic ear-input signals with respect to what belongs to the individual sound sources. This process, called *auditory stream segregation*, is an important area of current research in psychoacoustics and perceptual psychology [26]. With one ear plugged, auditory stream segregation is profoundly impaired, that is, fewer sound sources are identified and fewer distinct auditory events are formed, with the consequence that the aural world becomes *muddy*.

Once auditory stream segregation has been performed successfully and distinct auditory events have been formed, the listeners can focus their attention on specific auditory events while suppressing the information rendered by the other ones. For example, a listener in a concert can concentrate on a specific instrument, such as a clarinet, and attend selectively to the melody that this instrument plays, disregarding the other instruments.

Further, in situations where many concurrent talkers are active simultaneously, listeners are able to concentrate on one voice and, hence, understand it clearly, ignoring the other voices. This effect, known as *cocktail-party effect*—compare, for example, [13]—may provide an enhancement of speech intelligibility relative to monaural listening. This enhancement can be quantified by measuring the *speech-reception threshold gain* or SRT gain and it is often expressed as the BILD—where BILD stands for *binaural intelligibility-level difference* [12]. For this measure the additional amount of noise that can be added to a speech signal to compensate the positive effect of binaural rather than monaural presentation regarding speech recognition is determined. With all localization cues available, that is, ITDs and ILDs, the gain in SRT can amount to 2 to 8 dB, depending on the number and positions of the interfering speech sources [27].

The basic perceptual effect behind the cocktail-party effect is known as *binaural release from masking*. For example, a target signal that would otherwise be masked by other sounds may reach the perception threshold once it is presented 180° out-of-phase to both ears [70]. This effect can be quantified by considering the difference in allowable interferer SNR for the sound to remain perceptible, presenting once monaurally and once binaurally. This change in allowable SNR is termed *binaural masking-level difference*, BMLD [54]. One should, however, be aware that the ability to *detect* the presence of a signal in a binaural scene does not necessarily imply the ability to *identify* its properties, as is needed for speech understanding or source identification—see, for instance, [101]. In general, the BMLD in a given spatial configuration will be larger than the BILD.

The ability to concentrate on relevant auditory events and discount other ones can be exploited by listeners to partially suppress noise and other undesired signals. In these cases, the common term for the effect is *binaural noise suppression*. This may, for instance, become important when dealing with the audibility of warning signals in noisy surroundings.

In order to understand these effects, the ascending auditory pathway is the first place to look—see Sect. 4 of this chapter.

## 3 Application Areas for Binaural Models

The astonishing performance of the binaural auditory system has caught the interest of scientists as well as engineers for a long time. Among other efforts, they have started to build models of the auditory system or parts of it. Although models are not the thing itself, they are able to mimic certain functions of the real auditory system and, thus, help to understand this system better and often trigger further research into it.

Besides this research aspect, models have the potential to be applied, since they mimic specific, technologically interesting functions of the auditory system. As early as 1989,[1] one of the current authors had proposed to apply binaural models for the purposes listed in the following.

> Source-position finders, tools for architectural-acoustics evaluation, (e.g., such as echo detectors, spaciousness meters), tools for sound-quality evaluation (e.g., binaural-loudness meters, binaural sensory-consonance meters), tools for editing binaural recordings (e.g., binaural mixing consoles and control rooms), cocktail-party processor (e.g., for hearing aids and as front ends for speech recognizers), adaptive pick-up devices for hands-free telephones, intelligent microphones for acoustically adverse conditions (e.g., to reduce noise, reverberance and coloration), tele-surveillance systems (automatic source identification and assessment) and, last but not least, experimental tools for psychoacoustic research.

In retrospective, it can be stated that all these application ideas have been tackled by now and a number of them have lead to successful products. However, microelectronics and signal processing have made substantial progress in the meantime and it may be time to reconsider the application possibilities of models of binaural hearing.

To this end, the AABBA grouping, formed in 2009, discussed the following potential applications to define their activities [10], arriving at the following application categories.

*Audio technology* Binaural-cue selector, quality assessment of audio channels, quality assessment of loudspeakers, automatic surveillance of transmission quality.

*Audiology* Assessment of disorders of binaural hearing, assessment of binaural dereverberation and binaural decoloration, assessment of speech-understanding capabilities in acoustically adverse surroundings, binaural-loudness meter.

*Aural virtual environments* Auditory-scene mapping, identification of virtual sources, assessment of the perceived room size.

*Hearing aids*: Fitting of binaural hearing aids, diagnosis of dysfunctions of hearing aids.

*Product-sound quality* Assessment of spatial properties of product sounds.

*Room acoustics* Echo detector, spaciousness meter, detectors of image shifts, assessment of the sense of envelopment and immersion, assessment of the precedence effect, assessment of a global "*quality of the acoustics*".

*Speech technology* Speaker-position mapping, binaural speech intelligibility, assessment of speech recognition in adverse acoustical conditions, assessment of the cocktail-party effect.

*Binaural models as a research tool* To be employed for the evaluation, assessment and analysis of human spatial hearing in a multimodal world, for instance, with listeners moving in space and/or receiving additional visual and/or tactile cues.

Analysis of the results of this discussion lead to the identification of four prominent generic applications areas for binaural models as follows [10].

1. *Spatial scanning and mapping of auditory scenes*: Estimation of the position and the spatial extents of auditory events forming an aural scene. This could

[1] among other occasions, at a public lecture at the University of Florida, Gainesville.

be a natural scene as in room acoustics or a virtual scene as in virtual-reality applications or at the play-back end of audio systems—including spatially diffuse auditory events, often perceived as components of reverberance.

2. *Analysis of auditory scenes with the aim of deriving parametric representations at the signal level*: When estimated, these parameters may be intended to be used, for example,
   (a) For coding and/or re-synthesis of auditory scenes.
   (b) For speech-enhancement in complex acoustic environments—incl. hearing aids.
   (c) For systems to enhance the spatial perception in sound fields, such as better localization, a better sense of envelopment and/or decoloration and dereverberation.
   (d) For the identification of perceptual invariances of auditory scenes.
3. *Analysis of auditory scenes with the aim of deriving parametric representations at the symbolic level*, for example,
   (a) Identification of determinants of meaning contained in binaural-activity maps.
   (b) Assignment of meaningful symbols to the output of binaural models.
4. *Evaluation of auditory scenes in terms of quality*: Whereby quality is judged strictly from the users' point of view, for instance,
   (a) Quality of the so-called *acoustics* of spaces for musical performances.
   (b) Quality of systems for holophonic representation of auditory scenes, such as auditory displays and virtual-reality generators.
   (c) Spatial quality of audio-systems for recording, transmission and play-back, including systems that employ perceptual coding, or
   (d) Performance of speech-enhancement systems—including hearing aids.

Since 2009, AABBA has been active with regard to these application areas, particularly areas 1, 2 and 4. The current book presents a selection of relevant results of this endeavor.

## 4 The Physiology of Binaural Hearing

For engineering models of binaural hearing, the functional adequacy of the system elements and processes is more relevant than biological fidelity. Nevertheless, these models are to a large extent inspired by the respective processes in the human auditory system. Consequently, an introduction into the structure of the human auditory system is offered here.

The auditory system can be divided into two parts, firstly, the "mechanical" side of hearing, that is, the conversion of sounds into firing patterns of the auditory nerve, AN, and secondly, the connectivity and functional mechanisms of neural processing—as

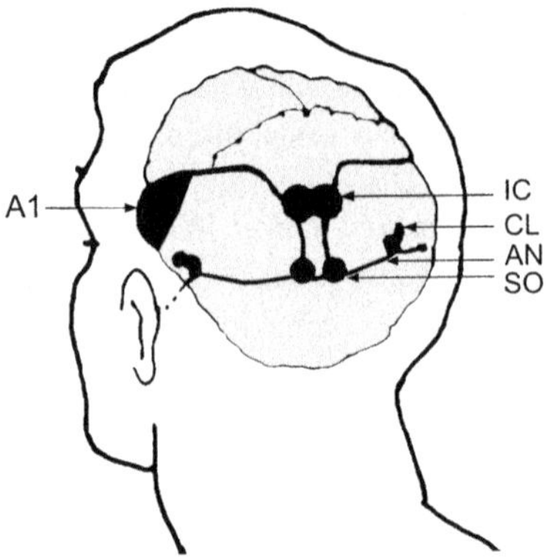


**Fig. 2** Schematic of relevant components of the binaural system.
*CL*...cochlea,
*AN*...auditory nerve,
*SO*...superior olivary complex including LSO and MSO,
*IC*...inferior colliculus,
*A1*...primary auditory cortex

far as a biological description of these can already be given with sufficient confidence. Figure 2 provides an overview of more relevant components of the binaural system.

## *4.1 From Sounds to Firing Patterns*

The human ear can be seen as consisting of three separate parts, the external, the middle, and the inner ear or *cochlea*.

The external ear spectrally shapes the sound according to the direction of incidence and passes it on through the ear canal. The eardrum acoustically connects the external to the middle ear, and protects the sensitive mechanisms of the middle ear, in which some of the smallest bones in the human body provide for the necessary impedance adjustment to transmit sufficient energy from the air-filled external ear to the liquid-filled inner ear, the cochlea. After the sound waves have been transmitted to the oval window of the cochlea via the three bones composing the middle ear—malleus, incus, and stapes—they are propagated onwards in the form of traveling waves, deforming the surface of the *basilar membrane*, which spans the length of the cochlea.

A rather precise mapping of sound frequency to the place of excitation on the basilar membrane is achieved by an elaborate active resonance mechanism, providing feedback resonance enhancement at the boundary between basilar and tectorial membrane. This mapping of frequency to place, the so-called *tonotopy*, has the consequence of a sinusoidal wave leading to neural excitation at only a rather narrow part of the basilar membrane. This neural excitation is the input stimulus to the auditory nerve, which transmits it onwards to the higher stages of neural processing.

## *4.2 From the Auditory Nerve to the Auditory Cortex*

The function of the inner ear can be described quite well by the frequency selectivity observed in the ascending part of auditory nerve fibers connected to it, that is, by their so-called *tuning curves*, together with their refraction times and with some well-studied temporal and spectral masking effects. In contrast, the functional mechanisms of higher-level neural processing have not yet led to a similarly consensual model.

There is, however, some agreement on a suitable intermediate representation at the auditory-nerve level—compare the following section. Using these auditory-nerve firing patterns as input, the first stages of the auditory system where binaural processing takes place, are the superior olivary complex, SO, and the inferior colliculus, IC.

### Intermediate Neural Representations

Throughout the length of the cochlea, about 3500 inner hair cells are arranged along the basilar membrane in the *organ of Corti*. All neural representations of auditory events originate here, where the inner hair cells elicit neural activation patterns in response to the traveling-wave maxima on the basilar membrane. As the cochlea in effect performs a frequency analysis of its input signal, every location on the basilar membrane and thus every inner hair cell has its unique *best frequency* to which it responds most strongly. But since the response to one incoming single frequency is not limited to one point on the basilar membrane, the overall effect of the organ of Corti is often modeled by a set of adjacent bandpass filters—see, for example, [2] for details on the characteristics of these auditory filters.

A good overall prediction of the single-frequency response of one auditory nerve fiber at a specific location on the basilar membrane can be obtained when each bandpass filter is followed by a neural-response model that includes refractoriness and adaptation properties [106]. This model also describes several masking effects [80].

The neural activation is phase-locked to the stimulus maxima at low frequencies, but due to the refraction time of involved neural cells, this phase locking ceases gradually at frequencies between 800 Hz and 2 kHz [92], and response timing becomes more and more determined by refraction times rather than by stimulus maxima. In effect this means that the neural activity represents the carrier signal at low frequencies and the envelope at higher ones.

All in all, the activation of the auditory nerve is therefore usually analyzed in the form of a spectro-temporal response pattern. Examples of these patterns have been measured for many species and can be found in [89] for anesthetized cats. After attempts to separate the temporal and the spectral response characteristic, it was shown that the full spectro-temporal representation encodes significantly more information, and is thus necessary for a full representation of the neural response [44].

In addition to the previously discussed bottom-up processes in the ascending auditory pathway, top-down controlled processes have also been shown to be significant in auditory processing [53]. However, the understanding of such processes is still developing, and it will be of great interest to see how physiological data, such as in [98], can be used to develop a clearer understanding of attention-driven and learned responses in animals and humans. Modeling aspects of top-down processes are addressed in [93], this volume.

In the ascending auditory pathway, the *superior olivary complex*, SO, is the first place where massive binaural interaction takes place, where the *lateral superior olive*, LSO, processes interaural level differences, ILDs, using input from the ipsilateral and the contralateral cochlear nucleus, and the *medial superior olive*, MSO, of non-echolocating mammals predominantly, but not solely, processes interaural time and phase differences [56]. At a higher level of the auditory pathway, as shown in Fig. 2, the *inferior colliculus*, IC, receives input from both cochlear nuclei as well as from the nearby *superior colliculus*, which processes visual inputs—leading to the IC's potential of performing not only aural-object localization but possibly also multi-sensory integration.

There is, therefore, a focus in current research work on attempting to determine and characterize response patterns of IC neurons, in many cases by testing on anesthetized mammals. A typical representation used here to determine the neural response, $r$, as a convolution of the spectro-temporal stimulus, $S$, and a gain function, $g$, is the so-called *spectro-temporal receptive field* [106] according to

$$r(t) = \sum_{f} \sum_{x} g(f, x) S(f, t - x). \tag{1}$$

While this model is monaural in nature, it needs not be formulated as such, but can rather consider both ear signals as inputs. For example, [90] have shown that the ITD sensitivity of single neurons in the IC is compatible with the results of just-noticeable-ITD measurements on human listeners, making accurate IC-neuron models an attractive and simple candidate for explaining localization performance.

## 5 Binaural Modeling

Binaural modeling tries to replicate specific behavioral aspects of binaural hearing by means of computer algorithms. The different models resulting from these efforts can be classified in various ways. One can order such models according to the complexity of the situations they handle. A relatively simple model could be one that localizes a single sound source in an anechoic environment—see Sect. 5.1. A more complex counterpart could be a model that indicates the direction of a sound source in a context of several other competing sound sources and in the presence of reverberation—see [75], this volume.

Another way is to categorize models according to the aspects of spatial scenes they are extracting, such as, spatial direction, distance, spatial sound source and room properties, apparent source width, listener envelopment, the detection of sound source or listener movement, and/or statements about the spatial fidelity of reproduction systems. Finally, a distinction can be made whether models focus on replicating human performance, independent of the technical means, or whether models are strongly inspired by replicating components of the human auditory system. The former models are often strongly based in the discipline of digital signal processing, for instance, blind source separation [103], while models from the latter type are much more linked to knowledge of physiology and modular models of the peripheral hearing system, for example, [60]. The descriptions in this chapter will mostly address models of the latter type and will be organized in terms of their main emphasis, namely, firstly binaural localization models and secondly binaural detection models.

## *5.1 Localization Models*

Acoustic localization models typically mimic the human auditory system to some degree to estimate the positions of sound sources. In robotic applications, knowledge about the human hearing system is only applied as far as it can support the model performance. In other cases, where the goal is to better understand the human auditory system, it is important to accurately simulate the functionality or even the physiological structure of the auditory pathway. Localization models utilize binaural cues, which are divided into interaural cues—those cues that require both ears to be analyzed—and monaural cues that can be extracted using only one ear. Interaural cues are often more robust than monaural cues. They play an important role in judging the lateral position, but provide less salient cues for the discrimination of the front/back direction and elevation.

### Interaural Cues

Interaural cues consist of *interaural time differences*, ITDs, and *interaural level differences*, ILDs. For historic reasons, the models for ITDs are introduced here first, because the ITD-based Jeffress model [58] was the first localization model. The core idea of the Jeffress model is the combination of delay lines and coincidence cells. In this model, two separate delay lines exist for each ear that run parallel. The signals propagate on each line in opposite direction as shown in Fig. 3. At one point the signals traveling along both delay lines meet at a coincidence cell, which then sends a signal to the next stage. If a sound signal is impinging for a sideways direction, the signal arrives first at the ipsilateral side due to path-length differences from the sound source to each of the two ears. The signal at the ipsilateral side enters the delay line first and has more time to travel before it meets the signal from the contralateral ear. Consequently, both signals will evoke a laterally displaced coincidence cell, the location of which is tuned to the lateral angle of the sound source. The Jeffress model is able to predict the lateral position of a sound source, because each cell is tuned to a specific angle of incidence.

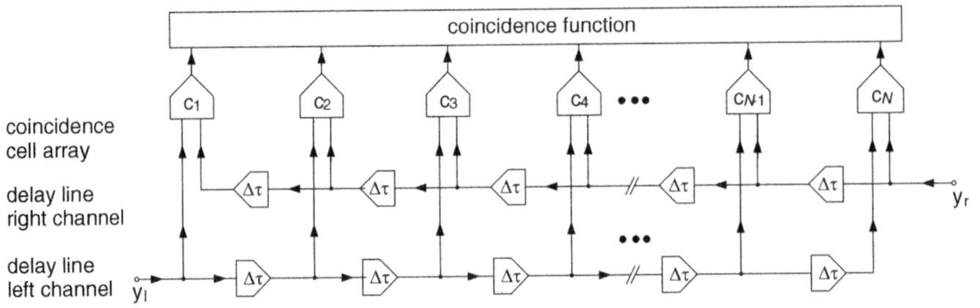


**Fig. 3** Coincidence mechanism as first proposed by Jeffress [58]

Most current localization models, see, for example, [11, 15, 81], use the *interaural cross-correlation*, IACC, method that was introduced by Cherry and Sayers [31] to estimate ITDs. The normalized IACC is defined as:

$$\Psi_{y_{l,r}}(\tau) = \frac{\int\limits_{t=-\infty}^{+\infty} y_l(t) \cdot y_r(t+\tau)\, dt}{\sqrt{\int\limits_{t=-\infty}^{+\infty} y_l^2(t)\, dt \cdot \int\limits_{t=-\infty}^{+\infty} y_r^2(t)\, dt}}, \tag{2}$$

with the internal delay, $\tau$, and the left and right sound pressure signals, $y_l(t)$ and $y_r(t)$.

Stern and Colburn [94] have shown that this method is a good representation of Jeffress' concept, if delay lines and coincidence cells consist of larger cell populations and stochastic processes. Blauert and Cobben [11] and Stern and Colburn [94] started to compare the left and right ear signals within frequency bands of approximately one-third-octave width to simulate the auditory system in greater detail. The segregation of the signals into narrow frequency bands, the so-called auditory bands, simulates the mechanics of the basilar membrane of the auditory pathway. The processing of the hair cells is typically simulated as well, often by applying a half-wave rectifier and a subsequent low-pass filter—usually of first order with a limiting frequency of about 1 kHz.

Figure 4 shows the general structure of a binaural model. The model receives input from both ears that are first processed through middle-ear modules, A. Modules $B_{1..n}$ represent the bandpass filters and hair-cell processing. In the modules C, the IACC functions are computed separately for each frequency band before the ILDs are calculated in modules D. Finally, the model predicts the position of the auditory event(s) based on available frequency-wide cues contained in the binaural-activity map, E.

The left panel of Fig. 5 shows an example of an IACC function of a broadband noise signal for three different positions for a frequency band centered at 434 Hz. The peak of the solid cross-correlation function is located at 0 μs which corresponds

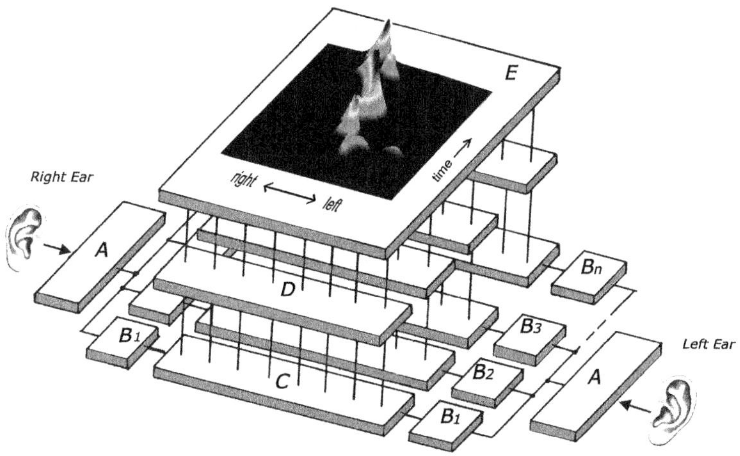


**Fig. 4** Architecture of the bottom-up part of a model of binaural signal processing—from [9]. *A*... middle-ear modules, $B_{1..n}$ ... cochlea modules—working in spectral bands, *C*... modules to identify and assess interaural arrival-time differences, ITDs, in bands, *D*... modules to assess interaural level differences, ILDs, in bands, *E*... running binaural-activity map with the three dimensions intensity, sideward deviation and time

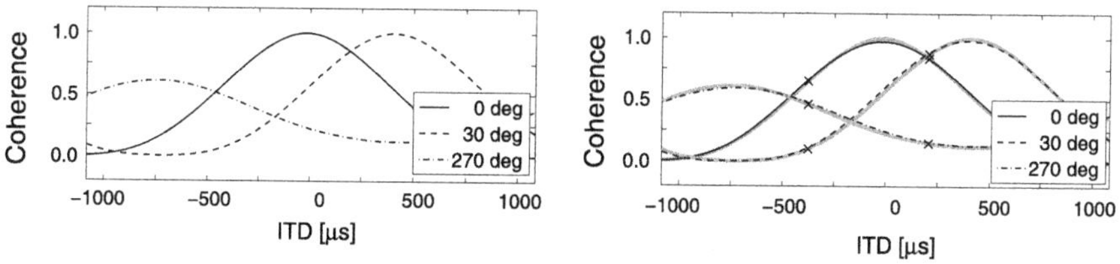


**Fig. 5** *Left* Interaural cross correlation functions for a sound source at three different positions in the horizontal plane. The sound sources at 0° and 30° azimuth are fully correlated, the sound source at 270° is partly decorrelated. *Right* Interaural cross correlation functions for a sound source at three different positions in the horizontal plane. The same stimuli as in the *left panel* are used, but this time a two-channel model was applied with delay lines for ±45°. The actually measured values, $x_-$ for the −45°-phase condition and $x_+$ for the +45°-phase condition, are shown by the '×' symbols. The simulated IACC curves were compensated for half-wave rectification. The *gray curves* show the actual IACC curves from the *left panel*. The normalized cross-correlation curves were estimated using standard trigonometric sine–cosine relationships for magnitude $A = \sqrt{x_-^2 + x_+^2}$ and phase $\phi = \arctan(x_-/x_+)$

to the position at 0°. The peak of the dashed IACC function is located at 400 μs, which indicates an azimuth of 30°. The height of the peak depicts the coherence, that is, the degree to which both signals are similar when shifted by the corresponding internal delay, $\tau$. In both cases, the signal is fully correlated. In the third example, depicted by a dash-dotted line, the signal is partly decorrelated as indicated by the lower peak height of 0.6. The peak location at −750 μs belongs to an azimuth angle of 270°.

A few years ago the Jeffress model and with it the cross-correlation approach was challenged by physiological studies on gerbils and guinea pigs. McAlpine and Grothe [77] and others [49, 76, 78, 83] have shown that the ITD cells for these species are not tuned evenly across the whole physiologically relevant range, but heavily concentrate on two phases of $\pm 45°$. Consequently, their absolute best-ITD values vary with the center frequency that the cells are tuned to. Dietz et al. [37, 38] and Pulkki and Hirvonen [84] developed lateralization models that draw from McAlpine and Grothe's [77] findings. It is still under dispute whether the Jeffress delay-line model or the two-channel model correctly represents the human auditory system, since the human ITD mechanism cannot be studied directly on a neural basis. For other species, such as owls, a mechanism similar to the one proposed by Jeffress has been confirmed by Carr and Konishi [30]. Opponents of the two-channel theory point out that the cross-correlation model has been tested much more rigorously than other ones and is able to predict human performance in great detail—for instance [5]. From a practical standpoint the result for both approaches are not as different as one might think. For the lower frequency bands, the cross-correlation functions always have a sinusoidal shape, due to the narrow width of the auditory bands—see the right panel of Fig. 5. Consequently, the whole cross-correlation function is more or less defined by two phase values 90° apart.

Interaural level differences are the second major localization cue. They occur because of shadowing effects of the head, especially when a sound arrives sideways. Typically, ILDs reach values of up to $\pm 30$ dB at frequencies around 5 kHz and azimuth angles of $\pm 60°$. At low frequencies the shadowing effect of the head is not very effective and ILDs hardly occur, unless the sound sources comes very close to the ear-canal entrance [8, 29]. This led Lord Rayleigh [74] to postulate his *duplex theory*, which states that ILDs are the primary localization cue for high frequencies and ITDs for low frequencies. In the latter case, Lord Rayleigh assumed that unequivocal solutions for the ITDs can no longer exist for high frequencies. Then, the wave length of the incoming sound is much shorter than the width of the head, which determines the physiological range for ITDs of approximately $\pm 800\,\mu$s.

Mills [79] later supported the duplex theory by demonstrating that the auditory system can no longer detect ITDs from the fine structure of signals above 1500 Hz. This effect results from the inability of the human auditory system to phase lock the firing patterns of auditory cells with the waveform of the signal at these frequencies. Meanwhile, however, it has been shown that the auditory system can extract ITDs at high frequencies from the signals' envelopes [61, 62], and the original duplex theory had to be revised accordingly.

ILDs, $\alpha$, can be computed directly from the left and right ear signals—which is typically done for individual frequency bands,

$$\alpha = 10 \log_{10}(P_{\mathrm{l}}) - 10 \log_{10}(P_{\mathrm{r}}), \qquad (3)$$

with $P_{\mathrm{l}}$ the power of the left and $P_{\mathrm{r}}$ the power of the right signal. Reed and Blum [86] introduced a physiologically-motivated algorithm to compute ILDs based on the activity, $E(\alpha)$, of an array of *excitation/inhibition*, EI, cells:

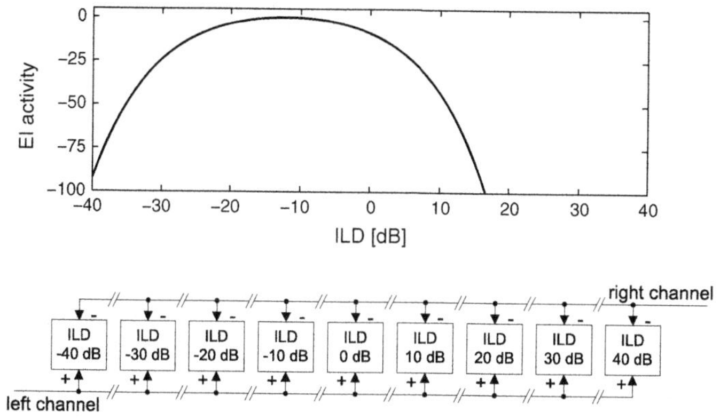


**Fig. 6** *Bottom* EI-cell structure. *Top* output of the EI cells for a signal with an ILD of −12 dB

$$E(\alpha) = \exp\left[\left(10^{\alpha/\mathrm{ILD}_{\max}}\sqrt{P_\mathrm{l}} - 10^{-\alpha/\mathrm{ILD}_{\max}}\sqrt{P_\mathrm{r}}\right)^2\right], \tag{4}$$

with $P_\mathrm{l}$, $P_\mathrm{r}$, being the power in the left and right channels and $\mathrm{ILD}_{\max}$ the maximal ILD magnitude that the cells are tuned to. Each cell is tuned to a different ILD. Figure 6 shows an example for a sound with an ILD of −12 dB. The curve depicts how the response of each cell is reduced the further the applied ILD is away from the value the cell is tuned to.

## Localization in the Horizontal Plane

Several methods to calculate sound source positions from the extracted binaural cues exist. One method of achieving this is to create a database to convert measured binaural cues, namely, ITDs and ILDs, into spherical coordinates. Such a database or map can be derived from a measured catalog of *head-related transfer functions*, HRTFs, of a large number of sound source directions. Here, the binaural cues are calculated frequency-wise from the left- and right-ear HRTFs of each position. Using this database, the measured binaural cues of a sound source with unknown positions can be mapped to spherical angles. The application of the remapping method to localize a signal in the horizontal plane is discussed in detail in [19], this volume. Figure 7 shows the results of remapped cross-correlation functions and ILD-based EI cell-array functions for different frequency bands.

An ongoing challenge has been to figure out how the auditory system combines the individual cues to determine the location of auditory events—in particular to answer how the auditory system performs the following tasks.

- Combining different cue types, such as ILDs and ITDs.
- Integrating information over time.
- Integrating information over frequency.

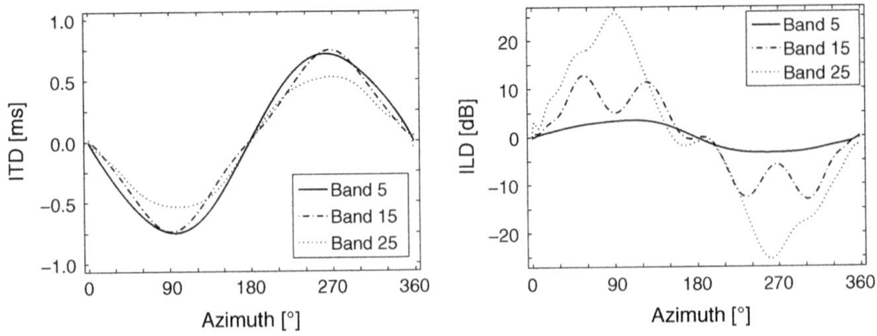


**Fig. 7** *Left* Interaural time differences. *Right* Interaural level differences. Plotted for different frequency bands: Band 5, $f_c = 234$ Hz (*solid line*); Band 15, $f_c = 1359$ Hz (*dashed line*); Band 25, $f_c = 5238$ Hz (*dotted line*)

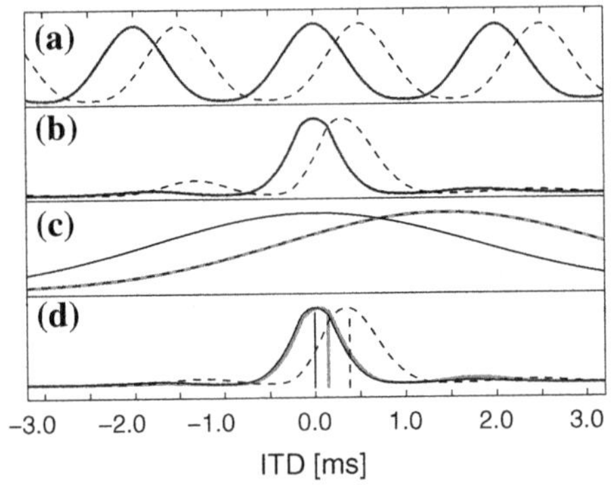


**Fig. 8** Results for the position-variable model [94] for a 500 Hz sinusoidal signal with different combinations of ITD and ILD, namely, 0 ms/0 dB (*black solid line*), 0 ms/15 dB (*grey solid line*), and 0.5 ms/15 dB (*black dashed line*): **a** interaural cross-correlation functions, **b** delay-line weighted function of (**a**) to emphasize small ITD values, **c** ILD functions, **d** combined ITD and ILD analysis by multiplying (**b**) with (**c**) and calculating the centroid as represented by the *vertical lines*

- Discriminating between concurrent sources.
- Dealing with room reflections.

Much of how the auditory system combines cues has been learned from so-called trading experiments [36, 51], where test participants were confronted with conflicting localization cues—for example, the ILD would point to one side, the ITD to the opposite side. It turns out that the auditory system often places the auditory event in between the positions the separate ILD and ITD cues would point to by trading off both cues. Stern and Colburn's [94] position variable model, shown in Fig. 8, is able to simulate cue trading by remapping the ILD cue to a function of internal delay and centering a broad bell-shaped curve at the remapped ILD position. The position of the combined auditory event is then determined by multiplying the ITD-based cross-correlation pattern with the ILD curve and calculating the centroid of this curve to determine the source position. Gaik [47] later demonstrated that test participants can learn to discriminate between ITD and ILD cues to indicate the source position based on either cue. He developed a model based on the Lindemann algorithm [71] that can demonstrate how auditory events fall apart into two separate events if the localization-cue mismatch between ILDs and ITDs exceeds a threshold.

A number of detailed overviews [17, 95, 96] have been written on cue weighting and only a brief introduction will be given here. One big question is how the auditory system weights cues temporally. The two opposing views are that the auditory system primarily focuses on the onset part of the signal versus the belief that the auditory system integrates information over a longer signal duration. Researchers have worked with conflicting cues—such as trade off between early onset and later ongoing cues—and it is generally agreed upon that the early cues carry a heavier weight [46, 50, 109]. This phenomenon can be simulated with a temporal weighting function. More recently it was suggested that the auditory system does not simply blindly combine these cues, but also evaluates the robustness of these cues and discounts unreliable cues. A good example for this approach is a model by Faller and Merimaa [45]. In their model not only the positions of the cross-correlation peaks are calculated to determine the ITDs but also the coherence—as, for example, determined by the maximum value of the interaural cross-correlation function. Coherent time-frequency segments are considered to be more salient and weighted higher assuming that concurrent sound sources and wall reflections that can produce unreliable cues decorrelate the signal and thus show low coherence.

Frequency weighting also applies and, in fact, the duplex theory can be seen as an early model where ITD cues are weighted high at low frequencies, and ILD cues dominate at higher frequencies. Newer models have provided a more detailed view of how ITDs are weighted over frequency. Different curves have been obtained for different sound stimuli [1, 85, 97].

Matters become more complicated if it is not clear how many sound sources currently exist. Then the cues do not only have to be weighted properly but also assigned to the corresponding source. Here one can either take a target+background approach [81], where only the target sound parameters are quantified and everything else is treated as noise, or one can attempt to determine the positions of all sound sources involved [15, 16, 88]. Often in models that segregate the individual sounds from a mixture, the positions of the sources are known a priori, such as in [13, 87].

Dealing with room reflections remains to be one of the biggest challenges in communication acoustics across a large variety of tasks including sound localization, sound source separation as well as speech and other sound feature recognition. Typically, models use a simplified room impulse response, often only consisting of a direct sound and a single discrete reflection to simulate the precedence effect. Lindemann [71, 72] took the following approach to the inhibition of location cues coming from reverberant information. Whenever his contralateral inhibition algorithm detects a signal at a specific interaural time difference, the mechanism starts to suppress information at all other internal delays or ITDs and thus solely focuses on the direct source signal component. The Lindemann model relies on onset cues to be able to inhibit reflections, but fairly recently Dizon and Colburn [39] have shown that the onset of a mixture of an ongoing direct sound and its reflection can be truncated without affecting the precedence effect.

Based on their observation that human test participants can localize the on- and offset-truncated direct sound correctly in the presence of a reflection, Braasch and Blauert recently proposed an autocorrelation-based approach [18]. The model reduces the influence of the early specular reflections by autocorrelating the left and right ear signals. Separate autocorrelation functions for the left and right channels determine the delay times between the direct sound source and the reflection in addition to their amplitude ratios. These parameters are then used to steer adaptive deconvolution filters to eliminate each reflection separately. It is known from research on the *apparent source width* of auditory objects that our central nervous system is able to extract information about early reflections [3], which supports this approach. The model is able to simulate the experiments from Dizon and Colburn's [39] study.

## Localization Using Monaural Cues

A model proposed in 1969/1970 [6] analyzes monaural cues in the median plane as follows. The powers in different frequency bands, the directional bands, are analyzed and compared to each other. Based on the signal's angle of incidence (front, above, or back), the pinnae enhance or de-emphasize the power in certain frequency regions, which are the primary localization cues for sound sources within the median plane. Building on this knowledge, Blauert's model uses a comparator to correctly predict the direction of the auditory event for narrowband signals.

Zakarauskas and Cynader [107] developed an extended model for monaural localization, which is based on the assumption that the slope of a typical sound source's own frequency spectrum only changes gradually with frequency, while the pinnae-induced spectral changes vary more with respect to frequency. The model primarily uses the second-order derivative of the spectrum in frequency to determine the elevation of the sound source—assuming that the sound source itself has a locally constant frequency slope. In this case, an internal, memorized representation of a sound source's characteristic spectrum becomes obsolete.

Baumgartner et al. [4], this volume, created a probabilistic localization model that analyzes inter-spectral differences, ISDs, between the internal representations of a perceived sound and templates calculated for various angles. The model also includes listener-specific calibrations to 17 individual listeners. It had been shown earlier that, for some cases, ISDs can be a better predictor for human localization performance than the second-order derivative of the spectrum [67]. By finding the best ISD match between the analyzed sound and the templates, Baumgartner et al.'s model is able to demonstrate similar localization performance as human listeners.

In contrast to models which do not require a reference spectrum of a sound source before it is altered on the pathway from the source to the ear, it is sometimes assumed that listeners use internal representations of a variety of everyday sounds to which the ear signals are compared to in order to estimate the monaural cues. A database with the internal representations of a high number of common sounds has not been implemented in monaural model algorithms so far. Some models exist, however, that use an internal representation of a single reference sound, that is, for broad-band noise [52] and for click trains [57].

## 5.2 Detection Models

A second major capacity of binaural hearing, besides localizing sound sources, lies in its ability to improve the detectability of signals in spatial scenes. Where localization of individual sound sources is enabled by *similarity* between right and left ear signals [45], improvements in detectability of sound sources in spatial scenes are made possible by *dissimilarities* in these signals. In a spatial aural scene, such reductions in coherence indicate the presence of several simultaneous sound sources at different directions from the position of the listener, or the effect of strong room reflections and reverberation. One of the application-relevant aspects here is the question of how similarity and dissimilarity need to be defined in order to model and mimic human behavior. This question is closely related to the method of how the sound is actually presented to the listeners. Binaural signal detectability has for a long time mainly been studied in headphone experiments. Here, dissimilarity between the signals reaching the right and the left ear coincides with inter-channel dissimilarity. This is in contrast to two-channel loudspeaker reproduction, where due to crosstalk, each channel reaches both ears. This aspect becomes relevant if configurations from headphone experiments are replicated with loudspeakers. A recent application in which this distinction between interaural and inter-channel similarity has become relevant, was the development of efficient stereo and multichannel coding algorithms—see [21].

In headphone experiments, a binaural signal detection paradigm can be realized by combining two stimulus components, a masker and a signal and by presenting these two components with different interaural parameters. For this type of configuration, a particular notation has been introduced, namely, $N_{\alpha_N,\phi_N,\tau_N} S_{\alpha_S,\phi_S,\tau_S}$. Here, $N$ and $S$ indicate the noise and signal components, respectively. The first index indicates the interaural level difference between the noise and signal components at the ears, where $\alpha$ stands for the interaural amplitude ratio, $\phi$ indicates the (broadband) interaural phase difference and $\tau$ indicates an interaural delay. An additional parameter, used to modify noise maskers, is the interaural correlation, indicated by the index $\rho$. The extreme cases of correlation values of $\pm 1$ agree with the extreme values of the interaural phase difference, being 0 or $\pi$.

In many basic signal-detection experiments, interaural differences have been restricted to interaural phase differences, that is, $\alpha = 1$ and $\tau = 0$, and choices for the interaural phase difference focused on the values 0 and $\pi$. A specific case of interaural differences is the presentation of masker or signal only monaurally, a condition, for which the index $m$ is being used. These early experiments focused on the *improvement* in signal detectability and, much less so, on the resulting detection thresholds in terms of signal level or signal-to-noise ratio. Therefore, results of binaural detection experiments were typically reported in terms of threshold differences, using thresholds in a typical monaural condition as reference. *Monaural* in this context refers to any condition in which masker and signal components are presented with the same interaural difference, like $N_0S_0$ $N_\pi S_\pi$, or $N_mS_m$—but see [68] for exceptions of the widely assumed performance identity in the conditions $N_0S_0$ and $N_mS_m$. The most widely studied conditions resulting in a binaural-detection advantage were $N_0S_\pi$, $N_\pi S_0$, $N_0S_m$ and $N_\pi S_m$—see for example [100]. For the usually positive difference in detection thresholds in a true *binaural* condition, the term *binaural masking-level difference*, BMLD, is used.

## Distinguishing Interaural Differences and Similarities

As mentioned in Sects. 4 and 5.1, the binaural system has two different ways of interaural interaction at the level of individual neurons. These are indicated by letter combinations EE and EI to describe purely excitatory interaction or the combination of excitatory and inhibitory interaction. In a functional way, these two modes of interaction can be associated with the mathematical relation of establishing *similarity* via a correlation, or coincidence process, that is, EE-type interaction, and the relation of establishing *dissimilarity*, that is, EI-type interaction, which is strongly associated with the equalization and cancellation (EC) modeling approach—see next section.

These two ways of physiological interaction have channeled two independent families of binaural models. They have been characterized in the extensive overview by [32] as cross-correlation models, EE, on the one hand and as noise suppression models, EI, on the other hand. The resulting binaural mechanisms differ in their phenomenological properties and physiological basis, yet, they can actually be considered as closely related, with an EI-based representation being the activity-inverted version of an EE-based representation. In fact, Colburn and Durlach [32] and Green [48] stated that the decision variables based on a correlation and on an EC mechanism are linear functions of one another, and thus lead to identical predictions of signal detectability. Consequently, as written in [32], the effect of interaural parameters of both the masker and signal can be accounted for independently of whether the decision variable is derived from the interaural correlation or from interaural differences.

This equivalence of the two modeling approaches on the level of a mathematical analysis does, however, not necessarily hold if it comes to a concrete implementation in terms of signal-processing models—for a discussion, see [23]. First of all, several authors showing equivalence between different detection variables make the explicit or implicit assumption that the masker is a Gaussian noise—see, for instance, [40]. Therefore, binaural detection experiments using maskers with non-Gaussian statistics have been instrumental in challenging these conclusions, because they allow for stimulus conditions where, for instance, the interaural correlation and the width of the distribution of ILDs and ITDs can be changed independently; this is in contrast to standard binaural masking conditions using Gaussian noise maskers and sinusoidal signals. In one such experiments using multiplied noise maskers,

[22] investigated masking conditions with both static and dynamically varying interaural differences. The conclusion was that their data could not be explained by a simple cross-correlation model, and also not on the basis of the standard deviations or the rms values of the interaural differences. The authors proposed a model based on the intensity difference between the internal representations in the right and left ear, after equalizing the masker in terms of mean interaural time and level differences. This approach is related to the EC theory [41], see next section, but was an attempt to go beyond the initial theory by not only predicting *differences* in masked thresholds, but to predict *masked thresholds* directly. Although being far removed from a satisfying way to model all their data, this interaural-difference-based approach was the most promising of all tested and strongly supported the authors in choosing an EC-based central interaction component in their later development of a signal-processing model for binaural unmasking [23–25].

For the purpose of this introductory chapter, the description concentrates on binaural detection models based on the EC type of interaction. The primary reason for this is that in the past 15 years much more effort has been put into developing binaural detection models based on this basic function, compared to efforts focused on cross-correlation models of binaural signal detection. This concentration of research efforts is also reflected in the level of detail with which the different model types have been described and evaluated, and in their applicability to real-world tasks—compare, for example, the following Ph.D. theses from the past decade [14, 20, 59, 69, 82, 102, 108]. Last but not least, this development is also reflected in the collection of algorithms included in the AABBA toolbox, as introduced in [91], this volume.

## EC Theory

One of the oldest analytical approaches to model changes in signal detectability in *binaural* conditions is the EC theory, introduced by Durlach [41, 42]. The EC theory in its original form was conceptually a black-box model, as it was primarily defined with the goal to predict BMLD values. Its internal components and elements were not based on known structures of the auditory pathway. It allowed to predict the difference in detection thresholds, that is, the BMLD, relative to a monaural reference condition. This change in detectability was reached by two basic operations. In a first equalization step, indicated by the letter "E", the stimuli presented to the right and the left ear are equalized. Given that, for binaural masking conditions, the signal-to-noise ratio at detection threshold is clearly negative, this equalization step basically results in equalized masker components in the two ears. The equalization transformations that Durlach considered for the conditions investigated in [42] comprise adjustments of the interaural amplitude ratio and of the interaural time difference. In the second step, the cancellation, the signals in the left and right hearing pathways are subtracted from each other, ideally eliminating the masker completely. In consequence, the signal-to-noise ratio after the cancellation step is improved relative to the input, which, in turns, leads to lower detection thresholds.

These transformations in the E step are performed with some random errors to reflect that the whole process of equalization and cancellation is realized in a biological system. In consequence, the two masker components will never be perfectly equal and, after subtraction, a certain amount of *internal noise* remains present in the model. The errors caused in the internal transformation were defined by Durlach as errors in amplitude and in time, and, in fitting the model to a large set of experimental data, these errors were quantified in terms of their standard deviation. The best fit was reached for an amplitude error with $\sigma_\epsilon = 0.25$ and a time error with $\sigma_\delta = 105\,\mu\text{s}$.

This concept has mostly been interpreted as a way for predicting *binaural unmasking*, but Durlach pointed out from the beginning that it had a close link to *localization* [42]. After all, as part of the E step, the interaural differences in time and level of the masker need to be determined first in order to be able to equalize right- and left-ear maskers. Thus, both localization models and a binaural detection algorithm based on the EC concept are dependent on components or processes that determine the values of these two interaural parameters. Given the black-box nature of the concept, this process of determining the optimal transformations in the "E" step was not addressed in the early descriptions of the EC theory.

The outcome of the EC process is a value for the predicted BMLD that can be analytically derived by computing the EC factor for the reference condition and the condition under test, and transforming it into a dB value. For the four main conditions mentioned before, the EC factor is predicted as follows [43].

$$f(N_0 S_\pi) = \frac{K+1}{K-1}, \tag{5}$$

$$\begin{aligned} f(N_0 S_m) &= \frac{K}{2(K-1)} && \text{for } K \leq 2, \\ &= 1 && \text{for } K > 2, \end{aligned} \tag{6}$$

$$f(N_\pi S_0) = \frac{K+1}{K-\gamma(\pi/\omega_0)} \qquad \text{with } \frac{\omega_0}{2\pi} = f_0 = \text{signal frequency}, \tag{7}$$

$$\begin{aligned} f(N_\pi S_m) &= \frac{K}{2(k-\gamma(\pi/\omega_0))} && \text{for } K \leq 2\gamma(\pi/\omega_0), \\ &= 1 && \text{for } K > 2\gamma(\pi/\omega_0). \end{aligned} \tag{8}$$

In these equations, the term $K$ comprises all internal processing errors

$$K = (1+\sigma_\epsilon^2) exp((\omega_0 \sigma_\delta)^2). \tag{9}$$

It is this term that leads to a frequency dependence of the BMLD. This is due to the fact that the timing error component has a frequency-independent value in terms of *time*, which means that this value increases with frequency in terms of signal *phase*.

## Signal-Processing Models Incorporating the EC Concept

As pointed out above, the EC theory was a first, albeit black-box approach, to quantify the binaural advantage in signal detectability. The EC theory was, however, far removed from a true model for binaural processing to be applicable to a wide variety of experimental conditions without restrictions regarding signal properties. Already in 1974, the year in which their seminal review chapter on *Models of Binaural Interaction* was written, Colburn and Durlach had formulated five requirements for binaural detection models, and directly added that, at that point in time, none of the published models fulfilled all these requirements. They stated—[32], p. 514—"Aside from the general fact that none of the existing models is capable of predicting more than a small portion of all the existing data on binaural interaction, they are all deficient in at least one, and often all, of the following areas.

1. Providing a complete quantitative description of how the stimulus waveforms are processed and how this processing is corrupted by internal noise.
2. Deriving all the predictions that follow from the assumptions of the model and comparing these predictions to all the relevant data.
3. Having a sufficiently small number of free parameters in the model to prevent the model from becoming merely a transformation of coordinates or an elaborate curve fit.
4. Relating the assumptions and parameters of the model in a serious manner to known physiological results.
5. Taking account of general perceptual principles in modeling the higher-level, more central portions of the system for which there are no adequate systematic physiological results available".

In the following, some of the steps that have happened since then will be outlined to indicate how some of these deficits are being dealt with in present-day binaural detection models.

One of the questions that has been addressed by several authors is how, in the EC process, the system chooses the compensation values in the "E" step. Durlach [43] had assumed that if the binaural system has complete *a priori* information about the stimulus characteristics it can select the appropriate elements before the test stimulus is presented. Otherwise, it needs to evaluate a variety of alternative choices and, in the case of interaural stimulus parameters, the lack of *a priori* knowledge would affect the amount of binaural unmasking.

In a series of experiments, one of the present authors specifically tested this prediction about the influence of *a priori* information, by presenting subjects with unpredictable dichotic signal-masker configurations [64, 65]. The experiments indicated that subjects had little difficulties to achieve maximum binaural unmasking also in such conditions, implying that, in terms of the EC theory, different equalization transformations must be realized in parallel. Thus, there was no need for an actual adjustment of the optimal equalization strategy and there did not exist an *analyzer sluggishness* in monitoring different EC channels at the same time [65]. These conclusions extended earlier arguments by von Hövel that the binaural system must be capable of realizing *different* equalization transformations simultaneously in *different* frequency bands [104]. From the discussion in [65] it becomes obvious that at that point in time, around 1990, correlation-based models comprising a central binaural-activity map with the dimensions frequency and interaural delay allowed for a much more intuitive explanation for binaural unmasking in dynamic conditions than the EC theory.

A first step forward with respect to transforming the EC theory into a signal-processing model was provided by Culling and Summerfield [34]. They studied the role of across-frequency grouping of interaural time differences in the perceptual separation of concurrent vowel sounds. To describe their experimental findings, they developed a model incorporating the following components.

- A Gammatone filterbank to include peripheral frequency separation.
- A nonlinear haircell model including compression and rectification of the filtered waveforms.
- A linear array of interaural delays.
- The computation of the difference function for each interaural delay, as the integral over an exponentially tapering window of the absolute differences between the corresponding filter outputs.

The output of this model thus corresponds to the residue after the cancellation step of the EC theory [43]. Figure 9, replotted from [34], shows such an output for four

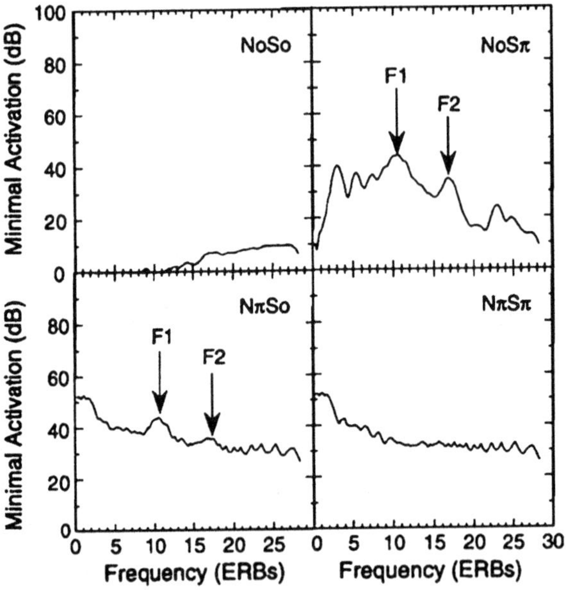


**Fig. 9** Internal spectra generated by the model of Culling and Summerfield [34] for the vowel /a/ presented in pink noise at an S/N of −15 dB for four different binaural conditions. Reprinted with permission from [34], pg. 794. Copyright (1995), Acoustical Society of America

binaural conditions. The stimulus was composed of the vowel /a/ and pink noise at such a low signal-to-noise ratio that insufficient information was available in the two *monaural* conditions. In contrast, for the two *binaural* conditions $N_0S_\pi$ and $N_\pi S_0$, the formant positions are clearly visible along the frequency axis, indicating the possibility to recognize the actual vowel.

An up-to date description of this line of binaural modeling can be found in [33], this volume.

A major step towards a signal-processing binaural model was realized in the work of Breebaart and colleagues [20, 23–25]. They built on the time-domain monaural model by [35] and extended it with a binaural processing unit. This unit was realized as a two-dimensional array of EI elements, where each element was characterized by a specific interaural delay and an interaural amplitude ratio. Thus, this central binaural processing unit integrated

- Parallel processing in auditory-filter subbands.
- The interaural delay axis found in cross-correlation models, including the $p(\tau)$ weighting function—see Fig. 3.
- A second axis of interaural intensity differences inspired by [86]—see (4).
- The computation of a difference intensity, as proposed before by [22] and [34].
- A temporal integration of the intensity differences as motivated by experiments on *binaural sluggishness* [55, 66].
- A compressive nonlinearity working on the temporally-smoothed output, which had a linear transformation for small output values, that is, small deviations from perfect coherence, and converged to a logarithmic transformation for high output values, which indicate low coherence values.
- Finally, to limit the performance of the model, an additive noise component which had the same value for all EI elements.

Figure 10 shows as an example the model output in a typical binaural masking condition. The left panel represents the activity within one auditory filter for a diotic

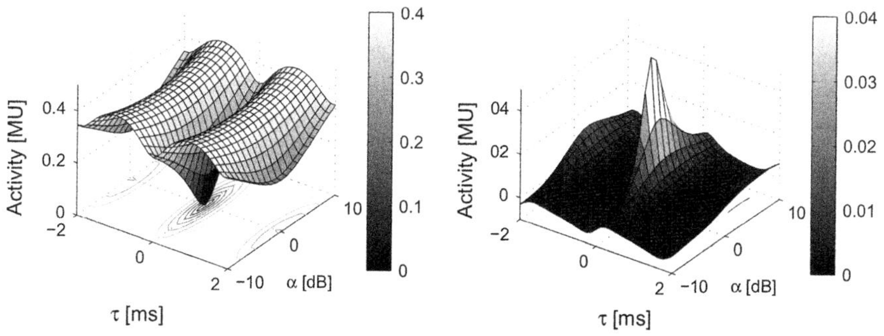


**Fig. 10** *Left* Activity at the output of the binaural processor from [23] within the 500 Hz auditory filter. The stimulus is a wideband diotic noise up to 4000 Hz, with an overall level of 70 dB SPL. *Right Change* in activity when a 500 Hz $S_\pi$ signal is added with a level of 50 dB SPL. Reprinted with permission from [23] pg. 1082. Copyright (2001), Acoustical Society of America

noise, with a minimum activity at internal delay, $\tau$, and internal level difference, $\alpha$, both equal to zero. The right panel shows the increase in activity at the output, when a low-level $S_{\pi}$ signal is added to the masker. It becomes clear that the biggest change is visible at those positions where the masker activity is minimal, reflecting the concept of equalization and cancellation as proposed by Durlach.

This model has been evaluated for a constant set of *all* parameters across a wide range of binaural masking conditions [24, 25] and this analysis revealed that indeed, many aspects of temporal and spectral properties in binaural unmasking could be described with a fixed set of model parameters. By keeping the model parameters constant, the authors could also pinpoint those aspects of the model where the agreement was unsatisfactory and where further improvements were needed.

Instead of a detailed discussion of the strong and the weak points of this model, for which the reader is referred to the original papers from 2001, this section is concluded by considering, to what extent the five requirements for binaural models as formulated by Colburn and Durlach nearly 40 years ago [32] have been realized with this newest generation of binaural signal detection models.

1. *Quantitative description of model stages*: The model was conceived as a signal processing model which allowed computing the multi-dimensional internal representation for any type of binaural stimulus configuration. In fact, including a validated version of this model in the AABBA toolbox is a perfect answer to this requirement.
2. *Applying the model in all relevant conditions*: In the set of three back-to-back papers that described the model and its predictions, the same model with a fixed parameter set was applied to a wide range of experimental conditions, including some where the model failed. Nevertheless, for all practical purposes, this requirement can only be met partially.
3. *Small number of free parameters*: The two evaluation papers indicated how good the model predictions across a wide range of experimental conditions were for a *fixed* value of all model parameters.
4. *Relation to physiological properties*: The peripheral, monaural, part of the model is closely related to properties found in the peripheral neural pathway. An obvious step to further improvements is the inclusion of a nonlinear auditory filterbank. The central, EI-type, interaction is inspired by neural elements found in, for instance, the LSO, although there exists the debate of how prominent EI-type neurons are in the mammalian auditory system.
5. *Modeling principles for central parts without clear physiological models*: For the purpose of signal detectability, the concept of an ideal observer, as realized in the models by Dau and colleagues, was also incorporated in the binaural model. This decision unit needs to be adapted to the tasks of the listeners. Thus, the model as proposed for binaural signal detection cannot directly be applied to signal localization, despite the fact that information about interaural differences of the dominant source are clearly visible in the binaural-activity map rendered by the model.

## 6 Conclusion

Since Jeffress [58] proposed his delay line model in 1948, binaural models have grown in complexity and are now able to address a large set of perceptual phenomena to perform localization, detection, and feature recognition tasks. Current models are able to simulate the auditory periphery in great detail and master the extraction of ITD, ILD as well as monaural cues. A sophisticated set of weighting methods exist for temporal and frequency weighting to simulate auditory event parameters. Specialized algorithms exist to deal with multiple and reverberant sound scenarios. Nevertheless, a holistic model that can simulate a complete set of spatial hearing tasks does not yet exist. Even if such a model would exist, the non-linear behavior of the auditory system would make it extremely difficult to calibrate the model to a very large acoustic data set. Further research is also needed for handling complex room geometries as current models are typically bound to very simple rooms, often consisting of a single reflection, while those that attempt to work with any room geometry are easily outperformed by the human auditory system. Arguably the best current practice is to build on a modular approach, where a task-specific model is generated to meet the given requirements. In [91], this volume, such a modular system, based on the programming environment MATLAB, will be introduced that includes a number of algorithms that have been discussed in this chapter.

**Acknowledgments** The authors would like to thank S. Jelfs, T. Pastore and two anonymous reviewers for their valuable comments on an earlier version of this manuscript.

## References

1. M. A. Akeroyd and Q. Summerfield. A fully temporal account of the perception of dichotic pitches. *Br. J. Audiol.*, 33:106–107, 1999.
2. J. Ashmore. Cochlear outer hair cell motility. *Physiol. Revs.*, 88:173–210, 2008.
3. M. Barron and A. H. Marshall. Spatial impression due to early lateral reflections in concert halls: the derivation of a physical measure. *J. Sound Vib.*, 77:211–232, 1981.
4. R. Baumgartner, P. Majdak, and B. Laback. Assessment of sagittal-plane sound-localization performance in spatial-audio applications. In J. Blauert, editor, *The technology of binaural listening*, chapter 4. Springer, Berlin-Heidelberg-New York NY, 2013.
5. L. Bernstein and C. Trahiotis. Enhancing sensitivity to interaural delays at high frequencies by using "transposed stimuli". *J. Acoust. Soc. Am.*, 112:1026–1036, 2002.
6. J. Blauert. Sound localization in the median plane. *Acustica*, 22:205–213, 1969/70.
7. J. Blauert. An introduction to binaural technology. In R. H. Gilkey and T. R. Anderson, editors, *Binaural and spatial hearing in real and virtual environments*, chapter 28, pages 593–609. Lawrence Erlbaum, Mahwah NJ, 1996.
8. J. Blauert. Spatial hearing: *The psychophysics of human sound localization*. 2nd, revised ed. MIT Press, Berlin-Heidelberg-New York NY, 1997.
9. J. Blauert and J. Braasch. Binaural signal processing. In *Proc. 19th Intl. Conf. Signal Processing*, chapter PID 1877149. IEEExplore, 2011.
10. J. Blauert, J. Braasch, J. Buchholz, H. Colburn, U. Jekosch, A. Kohlrausch, J. Mourjopoulos, V. Pulkki, and A. Raake. Aural assessement by means of binaural algorithms—the AabbA

project. In J. Buchholz, T. Dau, J. Dalsgaard, and T. Poulsen, editors, *Binaural processing and spatial hearing*, pages 113–124. The Danavox Jubilee Foundation, Ballerup, Denmark, 2009.
11. J. Blauert and W. Cobben. Some consideration of binaural cross correlation analysis. *Acustica*, 39:96–104, 1978.
12. M. Bodden. *Binaurale Signalverarbeitung: Modellierung der Richtungserkennung und des Cocktail-Party-Effektes [Binaural signal processing: Modelling the recognition of direction and the cocktail-party effect]*. PhD thesis, Ruhr-Univ. Bochum, Bochum, 1992.
13. M. Bodden. Modeling human sound-source localization and the Cocktail-Party Effect. *Act. Acust./Acustica*, 1:43–55, 1993.
14. J. Braasch. *Auditory localization and detection in multiple-sound-source scenarios*. PhD thesis, Ruhr-Univ. Bochum, Bochum, 2001.
15. J. Braasch. Localization in the presence of a distracter and reverberation in the frontal horizontal plane: II. Model algorithms. *Act. Acust./Acustica*, 88:956–969, 2002.
16. J. Braasch. Localization in the presence of a distracter and reverberation in the frontal horizontal plane: III. The role of interaural level differences. *Act. Acust./Acustica*, 89:674–692, 2003.
17. J. Braasch. Modeling of binaural hearing. In J. Blauert, editor, *Communication acoustics*, pages 75–108. Springer Verlag, 2005.
18. J. Braasch and J. Blauert. Stimulus-dependent adaptation of inhibitory elements in precedence-effect models. In *Proc. Forum Acusticum 2011*, pages 2115–2120, Aalborg Denmark, 2011.
19. J. Braasch, S. Clapp, A. Pars, T. Pastore, and N. Xiang. A binaural model that analyses acoustic spaces and stereophonic reproduction systems by utilizing head rotations. In J. Blauert, editor, *The technology of binaural listening*, chapter 8. Springer, Berlin-Heidelberg-New York NY, 2013.
20. J. Breebaart. *Modeling binaural signal detection*. PhD thesis, Techn. Univ. Eindhoven, 2001.
21. J. Breebaart and C. Faller. *Spatial audio processing: MPEG surround and other applications*. Wiley, Chichester, 2008.
22. J. Breebaart, S. van de Par, and A. Kohlrausch. The contribution of static and dynamically varying ITDs and IIDs to binaural detection. *J. Acoust. Soc. Am.*, 106:979–992, 1999.
23. J. Breebaart, S. van de Par, and A. Kohlrausch. Binaural processing model based on contralateral inhibition. I. Model structure. *J. Acoust. Soc. Am.*, 110:1074–1088, 2001.
24. J. Breebaart, S. van de Par, and A. Kohlrausch. Binaural processing model based on contralateral inhibition. II. Predictions as a function of spectral stimulus parameters. *J. Acoust. Soc. Am.*, 110:1089–1104, 2001.
25. J. Breebaart, S. van de Par, and A. Kohlrausch. Binaural processing model based on contralateral inhibition. III. Predictions as a function of temporal stimulus parameters. *J. Acoust. Soc. Am.*, 110:1105–1117, 2001.
26. A. Bregman. *Auditory scene analysis: The perceptual organization of sound.* MIT Press, 1990.
27. A. Bronkhorst and R. Plomp. Effect of multiple speechlike maskers on binaural speech recognition in normal and impaired hearing. *J. Acoust. Soc. Am.*, 92:3132–3139, 1992.
28. M. Brüggen. *Klangverfärbungen durch Rückwürfe und ihre auditive und instrumentelle Kompensation [Sound coloration due to reflections and its auditory and instrumental compensation]*. dissertation.de-Verlag im Internet, Berlin, 2001.
29. D. Brungart and W. Rabinowtz. Auditory localization of nearby sources. Head-related transfer functions. *J. Acoust. Soc. Am.*, 106:1465–1479, 1999.
30. C. E. Carr and M. Konishi. A circuit for detection of interaural time differences in the brain stem of the barn owl. *J. Neuroscience*, 10(10):3227–3246, 1990.
31. E. C. Cherry and B. M. A. Sayers. "Human 'cross-correlator' "—A technique for measuring certain parameters of speech perception. *J. Acoust. Soc. Am.*, 28(5):889–895, 1956.
32. H. S. Colburn and N. I. Durlach. Models of binaural interaction. In E. Carterette and M. Friedman, editors, *Handb. of perception*, volume IV, pages 467–518. Academic Press, New York, 1978.

33. J. Culling, M. Lavandier, and S. Jelfs. Predicting binaural speech intelligibility in architectural acoustics. In J. Blauert, editor, *The technology of binaural listening*, chapter 16. Springer, Berlin-Heidelberg-New York NY, 2013.
34. J. F. Culling and Q. Summerfield. Perceptual separation of concurrent speech sounds: Absence of across-frequency grouping by common interaural delay. *J. Acoust. Soc. Am.*, 98:785–797, 1995.
35. T. Dau, D. Püschel, and A. Kohlrausch. A quantitative model of the "effective" signal processing in the auditory system: I. Model structure. *J. Acoust. Soc. Am.*, 99:3615–3622, 1996.
36. E. E. David, N. Guttman, and W. A. von Bergeijk. Binaural interaction of high-frequency complex stimuli. *J. Acoust. Soc. Am.*, 31:774–782, 1959.
37. M. Dietz, S. D. Ewert, and V. Hohmann. Auditory model based direction estimation of concurrent speakers from binaural signals. *Speech Communication*, 53(5):592–605, 2011.
38. M. Dietz, S. D. Ewert, V. Hohmann, and B. Kollmeier. Coding of temporally fluctuating interaural timing disparities in a binaural processing model based on phase differences. *Brain Research*, 1220:234–245, 2008.
39. R. M. Dizon and H. S. Colburn. The influence of spectral, temporal, and interaural stimulus variations on the precedence effect. *J. Acoust. Soc. Am.*, 119:2947–2964, 2006.
40. R. H. Domnitz and H. S. Colburn. Analysis of binaural detection models for dependence on interaural target parameters. *J. Acoust. Soc. Am.*, 59:598–601, 1976.
41. N. I. Durlach. Note on the Equalization and Cancellation theory of binaural masking level differences. *J. Acoust. Soc. Am.*, 32:1075–1076, 1960.
42. N. I. Durlach. Equalization and cancellation theory of binaural masking-level differences. *J. Acoust. Soc. Am.*, 35:1206–1218, 1963.
43. N. I. Durlach. Binaural signal detection: Equalization & cancellation theory. In J. Tobias, editor, *Foundations of modern auditory theory*, volume II, pages 369–462. Academic Press, New York, London, 1972.
44. J. Eggermont, A. Aertsen, D. J. Hermes, and P. Johannesma. Spectro-temporal characterization of auditory neurons: Redundant or necessary? *Hearing Research*, 5:109–121, 1981.
45. C. Faller and J. Merimaa. Source localization in complex listening situations: Selection of binaural cues based on interaural coherence. *J. Acoust. Soc. Am.*, 116:3075–3089, 2004.
46. R. L. Freyman, P. M. Zurek, U. Balakrishnan, and Y. C. Chiang. Onset dominance in lateralization. *J. Acoust. Soc. Am.*, 101:1649–1659, 1997.
47. W. Gaik. Combined evaluation of interaural time and intensity differences: Psychoacoustic results and computer modeling. *J. Acoust. Soc. Am.*, 94:98–110, 1993.
48. D. Green. On the similarity of two theories of comodulation masking release. *J. Acoust. Soc. Am.*, 91:1769, 1992.
49. B. Grothe, M. Pecka, and D. McAlpine. Mechanisms of sound localization in mammals. *Physiological Reviews*, 90(3):983–1012, 2010.
50. E. Hafter. Binaural adaptation and the effectiveness of a stimulus beyond its onset. In R. H. Gilkey and T. R. Anderson, editors, *Binaural and spatial hearing in real and virtual environments*, pages 211–232. Lawrence Erlbaum, Mahwah, NJ, 1997.
51. G. G. Harris. Binaural interaction of impulsive stimuli and pure tones. *J. Acoust. Soc. Am.*, 32:685–692, 1960.
52. K. Hartung. *Modellalgorithmen zum Richtungshören, basierend auf Ergebnissen psychoakustischer und neurophysiologischer Experimente mit virtuellen Schallquellen [Model algorithms regarding directional hearing, based on psychoacoustic and neurophysiological experiments with virtual sound sources]*. PhD thesis, Ruhr-Univ. Bochum, Bochum, 1998.
53. J. He and Y. Yu. Role of descending control in the auditory pathway. In A. Rees and A. Palmer, editors, *The Oxford handbook of auditory science: The auditory brain*, pages 247–268. Oxford Univ. Press, 2010.
54. I. J. Hirsh. The influence of interaural phase on interaural summation and inibition. *J. Acoust. Soc. Am.*, 20:536–544, 1948.
55. I. Holube, M. Kinkel, and B. Kollmeier. Binaural and monaural auditory filter bandwidths and time constants in probe tone detection experiments. *J. Acoust. Soc. Am.*, 104:2412–2425, 1998.

56. D. Irvine. Physiology of the auditory brainstem. In A. Popper and R. Fay, editors, *The mammalian auditory pathway: Neurophysiology*, pages 153–232. Springer, 1992.
57. J. Janko, T. Anderson, and R. Gilkey. Using neural networks to evaluate the viability of monaural and inter-aural cues for sound localization. In R. H. Gilkey and T. R. Anderson, editors, *Binaural and spatial hearing in real and virtual environments*, pages 557–570. Lawrence Erlbaum Associates, Mahwah, NJ, 1997.
58. L. A. Jeffress. A place theory of sound localization. *J. Comp. Physiol. Psychol.*, 41:35–39, 1948.
59. S. Jelfs. *Modelling the cocktail party: A binaural model for speech intelligibility in noise.* PhD thesis, Cardiff University, 2011.
60. M. L. Jepsen, S. D. Ewert, and T. Dau. A computational model of human auditory signal processing and perception. *J. Acoust. Soc. Am.*, 124:422–438, 2008.
61. P. Joris. Envelope coding in the lateral superior olive. II. Characteristic delays and comparison with responses in the medial superior olive. *J Neurophysiol*, 76:2137–2156, 1996.
62. P. Joris and T. Yin. Envelope coding in the lateral superior olive. I. Sensitivity to interaural time differences. *J Neurophysiol*, 73:1043–1062, 1995.
63. H. Keller. Letter to Dr. John Kerr Love of March 31. In J. K. Love, editor, *Helen Keller in Scotland, a personal record written by herself.* Methuen, London, 1910.
64. A. Kohlrausch. *Psychoakustische Untersuchungen spektraler Aspekte beim binauralen Hören [Psychoacoustic investigations of spectral effects in binaural hearing].* PhD thesis, Univ. of Göttingen, 1984.
65. A. Kohlrausch. Binaural masking experiments using noise maskers with frequency-dependent interaural phase differences. II: Influence of frequency and interaural-phase uncertainty. *J. Acoust. Soc. Am.*, 88:1749–1756, 1990.
66. B. Kollmeier and R. H. Gilkey. Binaural forward and backward masking: Evidence for sluggishness in binaural detection. *J. Acoust. Soc. Am.*, 87:1709–1719, 1990.
67. E. H. A. Langendijk and A. W. Bronkhorst. Contribution of spectral cues to human sound localization. *J. Acoust. Soc. Am.*, 112(4):1583–1596, 2002.
68. A. Langhans and A. Kohlrausch. Differences in auditory performance between monaural and diotic conditions. I: Masked thresholds in frozen noise. *J. Acoust. Soc. Am.*, 91:3456–3470, 1992.
69. N. Le Goff. *Processing interaural differences in lateralization and binaural signal detection.* PhD thesis, Techn. Univ. Eindhoven, The Netherland, 2010.
70. J. Licklider. The influence of interaural phase relations upon the masking of speech by white noise. *J. Acoust. Soc. Am.*, 20:150–159, 1948.
71. W. Lindemann. Extension of a binaural cross-correlation model by contralateral inhibition. I. Simulation of lateralization of stationary signals. *J. Acoust. Soc. Am.*, 80:1608–1622, 1986.
72. W. Lindemann. Extension of a binaural cross-correlation model by contralateral inhibition. II. The law of the first wave front. *J. Acoust. Soc. Am.*, 80:1623–1630, 1986.
73. R. Litovsky, H. Colburn, W. Yost, and S. Guzman. The precedence effect. *J. Acoust. Soc. Am.*, 106:2219–2236, 1999.
74. Lord Rayleigh. On our perception of sound direction. *Phil. Mag.*, 13:214–232, 1907.
75. T. May, S. van de Par, and A. Kohlrausch. Binaural localization and detection of speakers in complex acoustic scenes. In J. Blauert, editor, *The technology of binaural listening*, chapter 15. Springer, Berlin-Heidelberg-New York NY, 2013.
76. D. McAlpine. Creating a sense of auditory space. *J. Physiol.*, 566(1):21–28, 2005.
77. D. McAlpine and B. Grothe. Sound localisation and delay lines - do mammals fit the model? *Trends in Neuroscience*, 26:347–350, 2003.
78. D. McAlpine, D. Jiang, and A. R. Palmer. A neural code for low-frequency sound localization in mammals. *Nature Neuroscience*, 4(4):396–401, 2001.
79. A. W. Mills. On the minimum audible angle. *J. Acoust. Soc. Am.*, 30:237–246, 1958.
80. B. C. J. Moore. *An introduction to the psychology of hearing.* Emerald Group, 5th edition, 2003.

81. J. Nix and V. Hohmann. Sound source localization in real sound fields based on empirical statistics of interaural parameters. *J. Acoust. Soc. Am.,* 119:463–479, 2006.
82. M. H. Park. *Models of binaural hearing for sound lateralisation and localisation*. PhD thesis, Univ. Southampton, 2007.
83. M. Pecka, A. Brand, O. Behrend, and B. Grothe. Interaural time difference processing in the mammalian medial superior olive: The role of glycinergic inhibition. *J. Neuroscience,* 28(27):6914–6925, 2008.
84. V. Pulkki and T. Hirvonen. Functional count-comparison model for binaural decoding. *Act. Acust./Acustica,* 95(5):883–900, 2009.
85. J. Raatgever. *On the binaural processing of stimuli with different interaural phase relations.* PhD thesis, Techn. Univ. Delft, 1980.
86. M. Reed and J. Blum. A model for the computation and encoding of azimuthal information by the lateral superior olive. *J Acoust. Soc. Am.,* 88:1442–1453, 1990.
87. N. Roman, S. Srinivasan, and D. Wang. Binaural segregation in multisource reverberant environments. *J. Acoust. Soc. Am.,* 120:4040–4051, 2006.
88. N. Roman and D. Wang. Binaural tracking of multiple moving sources. *IEEE Transactions on Audio, Speech, and Language Processing,* 16:728–739, 2008.
89. S. Shamma. Speech processing in the auditory system II: Lateral inhibition and the central processing of speech evoked activity in the auditory nerve. *J. Acoust. Soc. Am.,* 78:1622–1632, 1985.
90. B. C. Skottun, T. M. Shackleton, R. H. Arnott, and A. R. Palmer. The ability of inferior colliculus neurons to signal differences in interaural delay. *Proc. National Acad. Sciences,* 98:14050–14054, 2001.
91. P. Søndergaard and P. Majdak. The auditory modeling toolbox. In J. Blauert, editor, *The technology of binaural listening,* chapter 2. Springer, Berlin-Heidelberg-New York NY, 2013.
92. L. Squire, F. Bloom, and S. McConnell. *Fundamental neuroscience.* Academic Press, 2002.
93. M. Stamm and M. Altinsoy. Employing binaural-proprioceptive interaction in human-machine interfaces. In J. Blauert, editor, *The technology of binaural listening,* chapter 17. Springer, Berlin-Heidelberg-New York NY, 2013.
94. R. Stern and H. Colburn. Theory of binaural interaction based on auditory-nerve data. IV. A model for subjective lateral position. *J. Acoust. Soc. Am.,* 64:127–140, 1978.
95. R. M. Stern and C. Trahiotis. Models of binaural interaction. In B. C. J. Moore, editor, *Hearing,* pages 347–386. Academic Press, New York, 1995.
96. R. M. Stern, D. L. Wang, and G. J. Brown. Binaural sound localization. In D. Wang and G. Brown, editors, *Computational auditory scene analysis: Principles, algorithms, and applications,* pages 147–185. IEEE Press, 2006.
97. R. M. Stern, A. S. Zeiberg, and C. Trahiotis. Lateralization of complex binaural stimuli: A weighted-image model. *J. Acoust. Soc. Am.,* 84:156–165, 1988.
98. N. Suga, Y. Zhang, J. Olsen, and J. Yan. Modulation of frequency tuning of thalamic and midbrain neurons and cochlear hair cells by the descending auditory system in the mustached bat. In M. V. C. Moss and J. Thomas, editors, *Echolocation in bats and dolphins,* pages 214–221. Univ. of Chicago Press, 2002.
99. A. Tsilfidis, A. Westermann, J. Buchholz, E. Georganti, and J. Mourjopoulos. Binaural dereverberation. In J. Blauert, editor, *The technology of binaural listening,* chapter 14. Springer, Berlin-Heidelberg-New York NY, 2013.
100. S. van de Par and A. Kohlrausch. Dependence of binaural masking level differences on center frequency, masker bandwidth and interaural parameters. *J. Acoust. Soc. Am.,* 106:1940–1947, 1999.
101. S. van de Par, A. Kohlrausch, J. Breebaart, and M. McKinney. Discrimination of different temporal envelope structures of diotic and dichotic target signals within diotic wide-band noise. In D. Pressnitzer, A. de Cheveigné, S. McAdams, and L. Collet, editors, *Auditory signal processing: physiology, psychoacoustics, and models,* pages 398–404. Springer, New York, 2005.

102. J. van Dorp Schuitman. *Auditory modelling for assessing room acoustics*. PhD thesis, Techn. Univ. Delft, 2011.
103. E. Vincent, R. Gribonval, and C. Févotte. Performance measurement in blind audio source separation. *IEEE Transaction on Audio, Speech, and Language Processing*, 14:1462–1469, 2006.
104. H. von Hövel. *Zur Bedeutung der Übertragungseigenschaften des Außenohres sowie des binauralen Hörsystems bei gestörter Sprachübertragung [On the relevance of the transfer properties of the external ear and the binaural auditory system for corrupted speech transmission]*. PhD thesis, RWTH Aachen, 1984.
105. F. Wightman and D. Kistler. Monaural sound localization revisited. *J. Acoust. Soc. Am.*, 101:1050–1063, 1997.
106. E. Young. Level and spectrum. In A. Rees and A. Palmer, editors, *The Oxford handb. of auditory science: The auditory brain*, pages 93–124. Oxford University Press, 2010.
107. P. Zakarauskas and M. Cynader. A computational theory of spectral cue localization. *J. Acoust. Soc. Am.*, 94:1323–1331, 1993.
108. C. Zerbs. *Modeling the effective binaural signal processing in the auditory system*. PhD thesis, Carl-von-Ossietzky Univ. Oldenburg, 2000.
109. P. M. Zurek. A note on onset effects in binaural hearing. *J. Acoust. Soc. Am.*, 93:1200–1201, 1993.

# The Auditory Modeling Toolbox

**P. L. Søndergaard and P. Majdak**

## 1 Introduction

An auditory model is a mathematical algorithm that mimics part of the human auditory system. There are at least two main motivations for developing auditory processing models: First, to represent the results from a variety of experiments within one framework and to explain the functioning of the auditory system. In such cases, the models help to generate hypotheses that can be explicitly stated and quantitatively tested for complex systems. Second, models can help to evaluate how a deficit in one or more components affects the overall operation of the system. In those cases, some of the models can be useful for technical and clinical applications, such as the improvement of human-machine communication by employing auditory modeling based processing techniques, or the development of new processing strategies in hearing-assist devices. The *auditory modeling toolbox*, AMToolbox, is a freely available collection of such auditory models.[1]

Often a new auditory model aims at improving an already existing one. Thus, auditory modeling begins with the process of comprehending and reproducing previously published models. Imagine a thesis adviser who wants to integrate a new feature $Y$ into an existing model $X$. The student might spend months on the implementation of $X$, trying to reproduce the published results for $X$, before even being able to integrate the feature $Y$. While already the re-implementation of old models sounds like re-inventing the wheel, sometimes, it is even not possible to validate the new implementation of the old model because of lack of the original

[1] Much of the cooperation on the AMToolbox takes place within the framework of the AABBA group, an open group of scientist dealing with *aural assessment by means of binaural algorithms*.

P. Majdak (✉) · P. L. Søndergaard
Acoustics Research Institute, Austrian Academy of Sciences, Wien, Austria
e-mail: piotr@majdak.com

J. Blauert (ed.), *The Technology of Binaural Listening*, Modern Acoustics and Signal Processing, DOI: 10.1007/978-3-642-37762-4_2,

data as used in the original publication. This problem is not new, it has already been described in [9] as follows.

> An article about computational science in a scientific publication is *not* the scholarship itself, it is merely *advertising* of the scholarship. The actual scholarship is the complete software development environment and the complete set of instructions which generated the figures.

In order to address this problem, the manuscript publication must go with the software publication, allowing to reproduce the published research, a strategy called reproducible research [10]. Reproducible research is becoming more and more popular—see for instance [76]—and the AMToolbox is an attempt to promote the reproducible research strategy within the hearing science by pursuing the following three virtues.

- *Reproducibility* in terms of
  - Valid reproduction of the published outcome like figures and tables from selected publications
  - Trust in the published models with no need for a repetition of the verification
  - Modular model implementation and documentation of each model stage with a clear description of the input and output data format
- *Accessibility*, namely, free and open source software, available to download, use, and contribute by anyone
- *Consistency*, achieved by all functions written in the same style, using the same names for key concepts and conventions for conversion of physical units to numbers

In the past, other toolboxes concerning auditory models have been published [60, 63, 71]. The *auditory toolbox* [71] was an early collection of implementations focused on auditory processing. It contains basic models of the human peripheral processing, but the development of that toolbox seems to have stopped. The *auditory-image-model toolbox*, AIM, [63] comprises a more up-to-date model of the neural responses at the level of the auditory nerve. It seems to be still actively developed. The *development system for auditory modeling*, DSAM, [60], includes various auditory nerve models including the AIM. Written in C, it provides a great basis for the development of computationally efficient applications. Note that while the source code of the DSAM is free, the documentation is only commercially available. In contrast to those toolboxes, the *auditory modeling toolbox*, AMToolbox, comprises a larger body of recent models, provides a rating system for the objective evaluation of the implementations, is freely available—both code and documentation—and offers high proficiency gain when it comes to understanding and further developing existing model implementations.

## 2 Structure and Implementation Conventions

The AMToolbox is published under the GNU general public license version 3, a free and open source license[2] that guarantees the freedom to share and modify it for all its users and all its future versions. The AMToolbox, including its source code, is available from SourceForge.[3] AMToolbox works not only in Matlab,[4] versions 2009b and higher, but is in particular developed for Octave,[5] version 3.6.0 and higher, in order to avoid the need for any commercial software. The development is open and transparent by keeping the source files in the software repository *Git*[6] allowing for independent contributions and developments by many people. AMToolbox has been tested on 64–bit Windows 7, on Mac OSX 10.7 *Lion*, and on several distributions of Linux. Note that for some models, a compiler for *C* or Fortran is required. While Octave is usually provided with a compiler, for Matlab the compiler must be installed separately. Therefore, binaries for major platforms are provided for Matlab.

AMToolbox is build on top of the *large time-frequency-analysis toolbox*, LTFAT, [72]. LTFAT is a Matlab/Octave toolbox for time-frequency analysis and multichannel digital signal processing. LTFAT is free and open source. It provides a stable implementation of the signal processing stages used in the AMToolbox. LTFAT is intended to be used both as a computational tool and for teaching and didactic purposes. Its features are basic Fourier analysis and signal processing, stationary and non-stationary Gabor transforms, time-frequency bases like the modified discrete cosine transform, and filterbanks and systems with variable resolution over time and frequency. For all those transforms, inverse transforms are provided for a perfect reconstruction.

Further, LTFAT provides general, not model-related auditory functions for the AMToolbox. Several phenomena of the human auditory system show a linear frequency dependence at low frequencies, and an approximately logarithmic dependence at higher frequencies. These include the just-noticeable difference in frequency, giving rise to the *mel scale* [74] and its variants [26]. The concept of critical bands giving rise to the *Bark scale* [82], and the equivalent rectangular bandwidth, ERB, of the auditory filters giving rise to the *ERB scale* [58]—later revised in [31]. All these scales, including their revisions, are available in the LTFAT toolbox as frequency-mapping functions.

[2] http://www.gnu.org/licenses/gpl.html, last viewed on 9.1.2013.

[3] http://sourceforge.net/projects/amtoolbox, last viewed on 9.1.2013.

[4] http://www.mathworks.de/products/matlab/ last viewed on 9.1.2013.

[5] http://www.gnu.org/software/octave/, last viewed on 9.1.2013.

[6] http://git-scm.com/, last viewed on 11.1.2013.

## 2.1 Structure

AMToolbox consists of monaural and binaural auditory models, as described in the latter sections of this chapter, complemented with additional resources. The additional resources are

- *Data* from psychoacoustic experiments and acoustic measurements, used in and retrieved from selected publications
- *Experiments*, that is, applications of the models with the goal of simulating experimental runs from the corresponding publications
- *Demonstrations* of a simple kind, for getting started with a model or data

By providing both the data and the experiments, two types of verifications can be applied, namely,

- Verifications where the human data serve to reproduce figures from a given paper showing recorded human data
- Verification where *experiment functions* simulate experimental runs from a given paper and display the requested plots. Data collected from experiments with human can then be compared by visual inspection

Demonstrations are functions beginning with `demo_`. The aim of the demonstrations is to provide examples for the processing and output of a model in order to get quickly into the purpose and functionality of the model. Demonstrations do not require input parameters and provide a visual representation of a model output. Figure 1 shows an example for a demonstration, `demo_drnl`, which plots the spectrograms of the dual-resonance nonlinear, DRNL, filterbank and inner-hair-cell, IHC, envelope extraction of the speech signal [*greasy*].

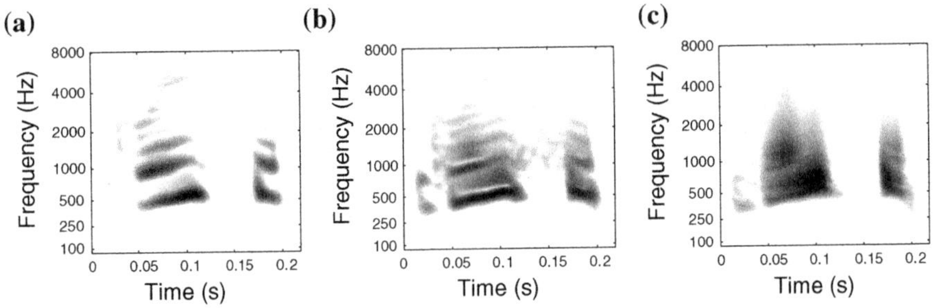


**Fig. 1** Example for a demonstration provided in the AMToolbox. The *three panels* show spectrograms of the DRNL filterbank and IHC envelope extraction of the speech signal [*greasy*] presented at different levels. The figure can be plotted by evaluating the code `demo_drnl`. **a** SPL of 50 dB. **b** SPL of 70 dB. **c** SPL of 90 dB

## Data

The data provide a quick access to already existing data and a target for an easy evaluation of models against a large set of existing data. The data are provided either by a collection of various measurement results in a single function, for example, `absolutethreshold`, where the absolute hearing thresholds as measured with various methods are provided—see Fig. 2—or by refering to the corresponding publication, for example, `data_lindemann1986a`. The latter method provides a very intuitive access of the data to the user, as the documentation for the data is provided in the referenced publication. The corresponding functions begin with `data_`.

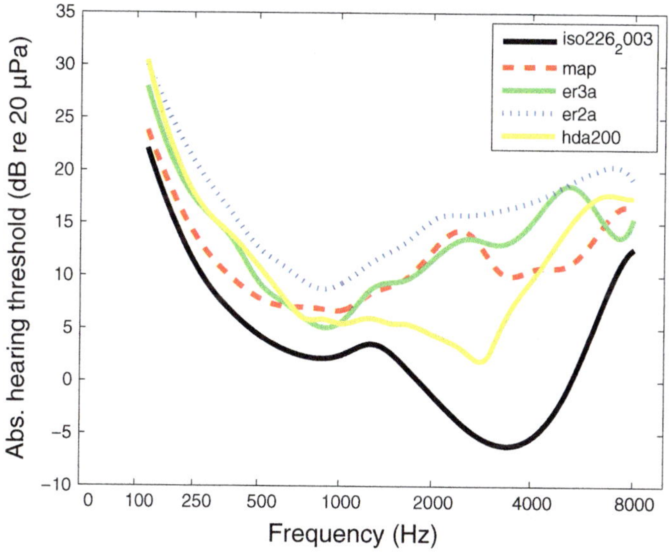


**Fig. 2** Example for data provided in the AMToolbox. The figure shows the absolute hearing thresholds measured under various conditions and returned by the function `absolutethreshold` by evaluating the following code:

```
types = {'iso226_2003','map','er3a','er2a','hda200'};
symbols = {'k' ,'r-{}-' ,'g' ,'b:','y'};
fc=125:125:8000; hold on; box on;
for ii=1:numel(types),
    opt={symbols{ii}, 'LineWidth', 3};
    semiaudplot(fc,absolutethreshold(fc,types{ii}),'opts',opt);
end;
legend(types); xlabel('Frequency (Hz)','FontSize',16);
ylabel('Abs. hearing threshold (dB re 20 Pa)','FontSize',16);
```

Currently, the AMToolbox provides the following publication-specific data.

- `data_zwicker1961`: Specification of critical bands [81]
- `data_lindemann1986a`: Perceived lateral position under various conditions [47]
- `data_neely1988`: Auditory brainstem responses (ABR) wave V latency as function of center frequency [59]
- `data_glasberg1990`: Notched-noise masking thresholds [31]
- `data_goode1994`: Stapes footplate displacement [32]
- `data_pralong1996`: Amplitudes of the headphone and outer ear frequency responses [64]
- `data_lopezpoveda2001`: Amplitudes of the outer and middle ear frequency responses [48]
- `data_langendijk2002`: Sound-localization performance in the median plane [45]
- `data_elberling2010`: ABR wave V data as function of level and sweeping rate [22]
- `data_roenne2012`: Unitary response reflecting the contributions from different cell populations within the auditory brainstem [68]
- `data_majdak2013`: Directional responses from a sound-localization experiment involving binaural listening with listener-specific HRTFs and matched and mismatched crosstalk-cancellation filters [52]
- `data_baumgartner2013`: Calibration and performance data for the sagittal-plane sound localization model [3]

The general data provided by the AMToolbox include data like the *speech intelligibility index* as a function of frequency [12]—`siiweightings`. Further, data for the *absolute threshold of hearing* in a free field [37]—`absolutethreshold`. The absolute thresholds are further provided as the *minimal audible pressures*, MAPs, at the eardrum [4] by using the flag `'map'`. The MAPs are provided for the insert earphones ER-3A (Etymotic) [38] and ER-2A (Etymotic) [33] as well as the circum-aural headphone HDA-200 (Sennheiser) [40]. Absolute thresholds for the ER-2A and HDA-200 are provided for the frequency range up to 16 kHz [39].

## Experiments

AMToolbox provides applications of the models that simulate experimental runs from the corresponding publications, and display the outcome in the form of numbers, figures or tables—see for instance Fig. 3. A model application is called `experiment`, the corresponding functions begin with `exp_`. Currently, the following experiments are provided.

- `exp_lindemann1986a`: Plots figures from [47] and can be used to visualize differences between the current implementation and the published results of the binaural cross-correlation model

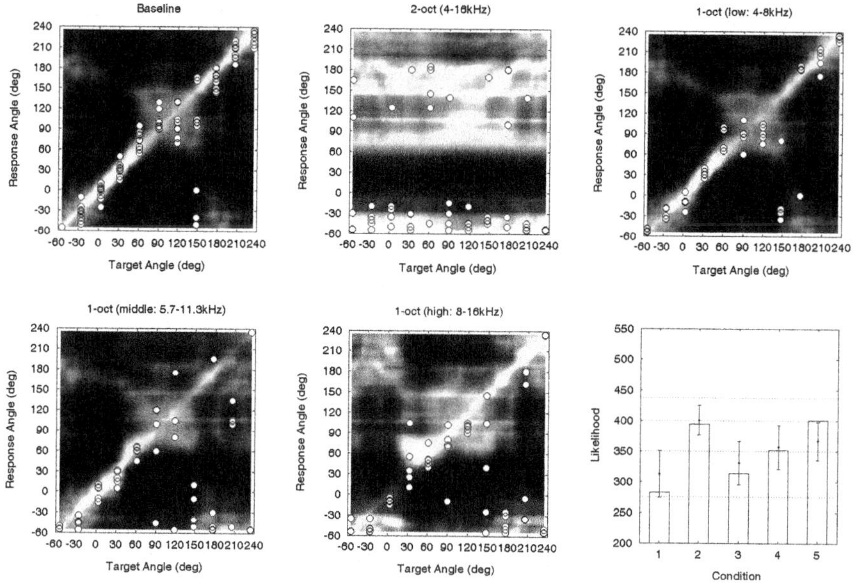


**Fig. 3** Example for an experiment in the AMToolbox. The figure is plotted by evaluating the code `exp_langendijk2002("fig7")` and aims at reproducing Fig. 7 of the article describing the `langendijk2008` model [45]

- `exp_lopezpoveda2001`: Plots figures from [48] and can be used to verify the implementation of the DRNL filterbank
- `exp_langendijk2002`: Plots figures from [45] and can be used to verify the implementation of the median-plane localization model, see Fig. 3
- `exp_jelfs2011`: Plots figures from [42] and can be used to verify the implementation of the binaural model for speech intelligibility in noise
- `exp_roenne2012`: Plots figures from [68] and can be used to verify the implementation of the model of auditory evoked brainstem responses to transient stimuli
- `exp_baumgartner2013`: Plots figures from [3] and can be used to verify the implementation of the model of sagittal-plane sound localization performance

## *2.2 Documentation and Coding Conventions*

In order to ensure traceability of each model and data, each implementation must be backed up by a publication in indexed articles, standards, or books. In the AMToolbox, the models are named after the first author and the year of the publication. This convention might appear unfair to the remaining contributing authors, yet, it establishes a straight-forward naming convention. Similarly, other files necessary for a model are prefixed by the name of the model, that is, first author plus the year, to make it clear to which model they belong.

All function names are lowercase. This avoids a lot of confusion because some computer architectures respect upper/lower casing and others do not. Furthermore, in Matlab/Octave documentation, function names are traditionally converted to uppercase. It is also not allowed to use underscores in variable or function names because they are reserved for structural purposes, for example, as in `demo_gammatone` or `exp_lindemann1986a`. As much as possible, function names indicate what they do, rather than the algorithm they use, or the person who invented it. We do not allow to use global variables since they would make the code harder to debug and to parallelize. Variable names are allowed to be both lower and upper case.

Further details on the coding conventions used in the AMToolbox can be found at the website.[7]

## *2.3 Level Conventions*

Some auditory models are nonlinear and the numeric representation of physical quantities like pressure must be well-defined. The auditory models included in the AMToolbox have been developed with a variety of level conventions. Thus, the interpretation of the numeric unity, that is, the of value 1, varies. For of historical reasons, per default, the unity represents the *sound-pressure level*, SPL, as the *root-mean-square* value, RMS, of 100 dB. The function `dbspl`, however, allows to globally change this representation in the AMToolbox. Currently, the following values for the interpretation of the unity are used by the models in the AMToolbox.

- SPL of 100 dB (default), used in the adaptation loops [13]. In this representation, the signals correspond to pressure in 0.5 Pa
- SPL of 93.98 dB, corresponding to the usual definition of the SPL in dB re 20 μPa. This representation corresponds to the international system of units, SI, namely, the signals are the direct representation of the pressure in Pa
- SPL of 30 dB, used in the inner-hair cell model [56]
- SPL of 0 dB, used in the DRNL filterbank [48] and in the model for binaural signal detection [6]

Note that when using linear models like the linear all-pole Gammatone filterbank, the level convention can be ignored.

# 3 Status of the Models

Ther description of a model implementation in the AMToolbox context can only be a snapshot of the development since the implementations in the toolbox are continuously developed, evaluated, and improved. In order to provide an overview of the

---

[7] http://amtoolbox.sourceforge.net/notes/amtnote003.pdf, last viewed on 9.1.2013.

**Table 1** Model status. Ch: chapter number with the model

| Model Name | Function | Ch | Rating D | C | V |
|---|---|---|---|---|---|
| *Peripheral models* | | | | | |
| Continuous-azimuth HRTFs | `enzner2008` | 3 | ★★☆ | ★★☆ | ★★☆ |
| Directional time-of-arrival | `ziegelwanger2013` | – | ★★★ | ★★★ | ★★★ |
| Gammatone filterbank | `gammatone` | – | ★★★ | ★★★ | ☆☆☆ |
| Invertible Gammatone filterbank | `hohmann2002` | – | ★☆☆ | ★☆☆ | ★★☆ |
| Dual-resonance nonlin. filterbank | `drnl` | – | ★★★ | ★★★ | ★★☆ |
| Cochlear transmission-line model | `verhulst2012` | 13 | ☆☆☆ | ★☆☆ | ☆☆☆ |
| Auditory-nerve filterbank | `zilany2007humanized` | – | ☆☆☆ | ★☆☆ | ☆☆☆ |
| Inner hair cell | `ihcenvelope` | – | ★★★ | ★★★ | ☆☆☆ |
| Adaptation loops | `adaptloop` | – | ★★★ | ★★★ | ☆☆☆ |
| Modulation filterbank | `modfilterbank` | – | ★★★ | ★★★ | ☆☆☆ |
| Auditory brainstem responses | `roenne2012` | – | ★★☆ | ★★☆ | ★★★ |
| *Signal detection models* | | | | | |
| Monaural masking | `dau1997preproc` | – | ★★★ | ★★★ | ☆☆☆ |
| Binaural signal detection | `breebaart2001preproc` | 5 | ★★★ | ★★★ | ☆☆☆ |
| *Spatial models* | | | | | |
| Lateralization, cross-correlation | `lindemann1986` | 10 | ★★★ | ★★★ | ★★☆ |
| Concurrent-speakers lateral dir. | `dietz2011` | 6 | ★★★ | ★★★ | ★★★ |
| Lateralization, supervised training | `may2013` | 15 | ★☆☆ | ★☆☆ | ☆☆☆ |
| Binaural activity map | `takanen2013` | 13 | ★★★ | ★☆☆ | ★☆☆ |
| Median-plane localization | `langendijk2002` | 4 | ★★☆ | ★★★ | ★★★ |
| Sagittal-plane localization | `baumgartner2013` | 4 | ★★★ | ★★★ | ★★★ |
| Distance perception | `georganti2013` | 7 | ★☆☆ | ★☆☆ | ☆☆☆ |
| *Speech perception models* | | | | | |
| Speech intelligibility in noise | `joergensen2011` | – | ★☆☆ | ★☆☆ | ★★★ |
| Spatial unmasking for speech | `jelfs2011` | 16 | ★★☆ | ★★★ | ★★☆ |

The hyphen indicates a general model with no particular assignment to a specific chapter. D: Rating for the model documentation. C: Rating for the model source code. V: Rating for the model verification with experiments

development stage, a rating system is used in the AMToolbox. The rating status for the AMToolbox version 1.0 is provided in Table 1.[8]

First, we rate the implementation of the model by considering its source code and documentation.

☆☆☆ *Submitted* The model has been submitted to the AMToolbox, there is, however, no working code/documentation in the AMToolbox, or there are compilation errors, or some libraries are missing. The model neither appears on the website nor is available for download

[8] The current up-to-date status of the AMToolbox can be found under http://amtoolbox.sourceforge.net/notes/amtnote006.pdf, last viewed on 14.2.2013.

★☆☆ *OK* The code fits the AMToolbox conventions just enough for being available for download. The model and its documentation appear on the website, but major work is still required

★★☆ *Good* The code/documentation follows our conventions, but there are open issues

★★★ *Perfect* The code/documentation is fully up to our conventions, no open issues

Second, the implementation versus the corresponding publication is verified in experiments. In the best case, the experiments produce the same results as in the publication—up to some minor layout issues in the graphical representations. Verifications are rated at the following levels.

☆☆☆ *Unknown* The AMToolbox can not run experiments for this model and can not produce results for the verification. This might be the case when the verification code has not been provided yet

★☆☆ *Untrusted* The verification code is available but the experiments do not reproduce the relevant parts of the publication (yet). The current implementation can not be trusted as a basis for further developments

★★☆ *Qualified* The experiments produce similar results as in the publication in terms of showing trends and explaining the effects, but not necessarily matching the numerical results. Explanation for the differences can be provided, for example, not all original data available, or publication affected by a known and documented bug

★★★ *Verified* The experiments produce the same results as in the publication. Minor differences are allowed if randomness is involved in the model, for instance, noise as input signal, probabilistic modeling approaches, and a plausible explanation is provided

## 3.1 Peripheral Models

This section describes models of auditory processes involved in the periphery of the human auditory system like outer ear, middle ear, inner ear, and the auditory nerve.

### Continuous-Azimuth Head-Related Transfer Functions—`enzner2008`

Head-related transfer functions, HRTFs, describe the directional filtering of the incoming sound due to torso, head, and pinna. HRTFs are usually measured for discrete directions in a system-identification procedure aiming at fast acquisition and high spatial resolution of the HRTFs—compare [51]. The requirement of high-spatial-resolution HRTFs can also be addressed with a *continuous-azimuth* model of HRTFs [23, 24]. Based on this model, white noise is used as an excitation signal and normalized least-mean-square adaptive filters are employed to extract HRTFs from binaural recordings that are obtained during a continuous horizontal rotation of the listener. Recently, periodic perfect sweeps have been used as the excitation signal in order to increase the robustness against nonlinear distortions at the price of a potential time aliasing [2].

Within the AMToolbox, the excitation signal for the playback is generated and the binaurally recorded signal is processed. The excitation signal can be generated either by the means of Matlab/Octave internal functions [23] or with the function `perfectsweep` [2]. For both excitation signals, the processing of the binaural recordings [23] is implemented in `enzner2008` that outputs HRTFs with arbitrary azimuthal resolution.

### Directional Time-of-Arrival—`ziegelwanger2013`

The broadband delay between the incoming sound and the ear-canal entrance depends on the direction of the sound source. The delay, also called time-of-arrival, TOA, can be estimated from an HRTF. A *continuous-direction* TOA model, based on a geometric representation of the HRTF measurement setup has been proposed [78]. In the function `ziegelwanger2013`, TOAs, estimated from HRTFs separately for each direction, are used to fit the model parameters. Two model options are available, the on-axis model where the listener is assumed to be placed in the center of the measurement, and the off-axis model where a translation of the listener is considered. The corresponding functions, `ziegelwanger2013onaxis` and `ziegelwanger2013offaxis`, output the monaural directional delay of the incoming sound as a continuous function of the sound direction. It can be used to further analyze broadband-timing aspects of HRTFs, such as broadband interaural-time differences, ITDs, in the process of sound localization.

### Gammatone Filterbank—`gammatone`

A classical model of the human basilar membrane, BM, processing is the Gammatone filterbank, of which there exist many variations [50]. In the AMToolbox, the original IIR approximation [62] and the all-pole approximation [49] have been implemented for both real- and complex-valued filters in the function `gammatone`. To build a complete filterbank covering the audible frequency range, the center frequencies of the gammatone filters are typically chosen to be equidistantly spaced on an auditory frequency scale like the ERB scale [31], provided in the LTFAT.

### Invertible Gammatone Filterbank—`hohmann2002`

The classic version of the Gammatone filterbank does not provide for a method to reconstruct a signal from the output of the filters. A solution to this problem has been proposed [35] where the original signal can be reconstructed using a sampled all-pass filter and a delay line. The reconstruction is not perfect, but stays within 1 dB of error in magnitude between 1 and 7 kHz and, according to [35], the errors are barely audible. The filterbank has a total delay of 4 ms and uses $4^{\text{th}}$-order complex-valued all-pole Gammatone filters [49]—equidistantly scaled on the ERB scale.

### Dual-Resonance Nonlinear Filterbank—`drnl`

The DRNL filterbank introduces the modeling of the nonlinearities in peripheral processing [48, 57]. The most striking feature is a compressive input-output function, and, consequently, level-dependent tuning. The DRNL function `drnl` supports the parameter set for a human version of the nonlinear filterbank [48].

### Auditory-Nerve Filterbank—`zilany2007humanized`

The auditory-nerve, AN, model implements the auditory periphery to predict the temporal response of AN fibers [79]. The implementation provides a "humanized" parameter set, which can be used to model the responses in human AN fibers [68]. In the AMToolbox, the function is called `zilany2007humanized` and outputs the temporal excitation of 500 AN fibers equally spaced on the BM.

### Cochlear Transmission-Line Model—`verhulst2012`

The model computes the BM velocity at a specified characteristic frequency by modeling the human cochlea as a nonlinear transmission-line and solving the corresponding ordinary differential equations in the time-domain. The model provides the user direct control over the poles of the BM admittance, and thus over the tuning and gain properties of the model along the cochlear partition. The passive structure of the model was designed [80] and a functional, rather than a micro-mechanical, approach for the nonlinearity design was followed with the purpose of realistically representing level-dependent BM impulse response behavior [67, 70]. The model simulates both forward and reverse traveling waves, which can be measured as the otoacoustic emissions, OAEs.

In the AMToolbox, the model is provided by the function `verhulst2012`. The model can be used to investigate time-dependent properties of cochlear mechanics and the generator mechanisms of OAEs. Furthermore, the model is a suitable preprocessor for human auditory perception models where realistic cochlear excitation patterns are required.

### Inner Hair Cells—`ihcenvelope`

The functionality of the IHC is typically described as an envelope extractor. While the envelope extraction is usually modelled by a half-wave rectification followed by a low-pass filtering, many variations to this scheme exist. For example, binaural models typically use a lower cutoff frequency for the low-pass filtering than monaural models.

In the AMToolbox, the IHC models are provided by the function `ihcenvelope` and selected by the corresponding flag. Models based on the low-pass filter with the following cutoff frequency are provided, namely, 425 Hz [5]—flag `'ihc_bernstein'`, 770 Hz [6]—flag `'ihc_breebaart'`, 800 Hz [47]—flag `'ihc_lindemann'`, and 1000 Hz [15]—flag `'ihc_dau'`. Further, the classical envelope extraction by the Hilbert transform is provided [28]—`'hilbert'`. Finally, a probabilistic approach for the synaptic mechanisms of the human inner hair cells [56] is provided—`'ihc_meddis'`.

### Adaption Loops—adaptloop

*Adaptation loops* is a simple method to model the temporal nonlinear properties of the human auditory periphery by using a chain of typically five feedback loops in series. Each loop has a different time constant. The AMToolbox implements the adaptation loops in the function `adaptloop` with the original, linearly spaced constants [66]—flag `'adt_puschel'`. In [6], the original definition was modified to include a minimum level to avoid the transition from complete silence and an overshoot limitation, `'adt_breebaart'`, because it behaved erratically if the input changes from complete silence [13]. Also, the constants from [15], `'adt_dau'`, are provided, which better approximate the forward masking data.

### Modulation Filterbank—modfilterbank

The modulation filterbank is a processing stage that accounts for amplitude modulation, AM, detection and AM masking in humans [13, 27]. In the AMToolbox, the modulation filterbank is provided in the function `modfilterbank`. The input to the modulation filterbank is low-pass filtered using a first-order Butterworth filter with a cutoff frequency at 150 Hz. This filter simulates a decreasing sensitivity to sinusoidal modulation as a function of modulation frequency. By default, the modulation filters have center frequencies of 0, 5, 10, 16.6, 27.77 ... Hz, where each next center frequency is 5/3 times the previous one. For modulation center frequencies below and including 10 Hz, the real values of the filters are returned and, for higher modulation frequencies, the absolute value, that is, the envelope, is returned.

### Auditory Brainstem Responses—roenne2012

A quantitative model describing the formation of human auditory brainstem responses, ABRs, to tone pulses, clicks, and rising chirps as a function of stimulation level is provided in the function `roenne2012`. The model computes the convolution of the instantaneous discharge rates using the "humanized" nonlinear

AN model [79] with an empirically derived unitary response function that is assumed to reflect contributions from different cell populations within the auditory brainstem, recorded at a given pair of electrodes on the scalp. The key stages in the model are (i) the nonlinear processing in the cochlea, including key properties such as compressive BM filtering, IHC transduction and IHC-AN synapse adaptation, and (ii) the linear transformation between the neural representation at the output of the AN and the recorded potential at the scalp.

# 4 Signal-Detection Models

Signal detection models predict the ability to detect a signal or a signal change by human listeners. These models usually rely on a peripheral model and use a framework to simulate the decisions made by listeners. Note that functions with the prefix `preproc` are modeling the preprocessing part of the model only, while excluding the decision framework.

### Preprocessing for Modeling Simultaneous and Nonsimultaneous Masking—`dau1997preproc`

A model of human auditory masking of a target stimulus by a noise stimulus has been proposed [15]. The model includes stages of linear BM filtering, IHC-transduction, adaptation loops, a modulation low-pass filter, and an optimal detector as the decision device. The model was shown to quantitatively account for a variety of psychoacoustical data associated with simultaneous and non-simultaneous masking [16]. In subsequent studies [13, 14], the cochlear processing was replaced by the GM filterbank and the modulation low-pass filter was replaced by a modulation filterbank, which enabled the model to account for AM detection and AM masking. The preprocessing part of this model consisting of the GM filterbank, the IHC stage, the adaptation loops, and the modulation filterbank is provided by the function `dau1997preproc`.

### Preprocessing for Modeling Binaural Signal Detection Based on Contralateral Inhibition—`breebaart2001preproc`

A model of human auditory perception in terms of the binaural signal detection has been proposed [6–8]. The model is essentially an extension of the monaural model [13, 14], from which it uses the peripheral stages, that is, linear BM filtering, IHC-transduction, and adaptation loops, and the optimal detector decision device. The peripheral internal representations for both ears are then fed to an equalization-cancellation binaural processor consisting of excitation-inhibition, EI, elements, resulting in a binaural internal representations that is finally fed into the decision device. Implemented in the function `breebaart2001preproc`, the preprocessing part of the model outputs the EI-matrix, which can be used to predict a large range of binaural detection tasks [8] or to evaluate sound localization performance for stereophonic systems [61].

## 5 Spatial Models

Spatial models consider the spatial position of a sound event in the modeling process. The model output can be the internal representation of the spatial event on a neural level. The output can also be a perceived quality of the event like the sound position, apparent source width, or the spatial distance, also in cases of multiple sources.

### Modeling Sound Lateralization with Cross-Correlation—`lindemann1986`

A binaural model for predicting the lateralization of a sound has been proposed [47]. This model extends the delay line principle [41] by contralateral inhibition and monaural processors. It relies on a running interaural cross-correlation process to calculate the dynamic ITD which are combined with the interaural level differences, ILDs. The peak of the cross-correlation is sharpened by contralateral inhibition and shifted by the ILD.

In the AMToolbox, the model is implemented in the function `lindemann1986` and consists of linear BM filtering, IHC-transduction, cross-correlation, and the inhibition step. The output of the model is the interaural cross-correlation in each characteristic frequency, see Fig. 4. The model can handle stimuli with a combination of ITD and ILD and predict split images for unnatural combinations of the two. An example is given in Fig. 4.

### Modeling Lateral-Deviation Estimation of Concurrent Speakers—`dietz2011`

Most binaural models are based on the concept of place coding, namely, on coincidence neurons along counterdirected *delay lines* [41]. However, recent physiologic evidence from mammals, for example, [55], supports the concept of rate coding [77] and argues against axonal delay lines. In [19], this idea is extended and the derivation of the interaural phase differences, IPDs, from both the temporal fine structure and the temporal envelope without employing mechanisms of delay compensation is proposed. This concept was further developed as a hemispheric rate comparison model in order to account for psychoacoustic data [17] and for auditory model based multi-talker lateralization [18]. The latter is the basis for further applications, such as multi-talker tracking and automatic speech recognition in multi-talker conditions [73]. The model, available in the AMToolbox in the function `dietz2011`, is functionally equivalent to both [18] and [73].

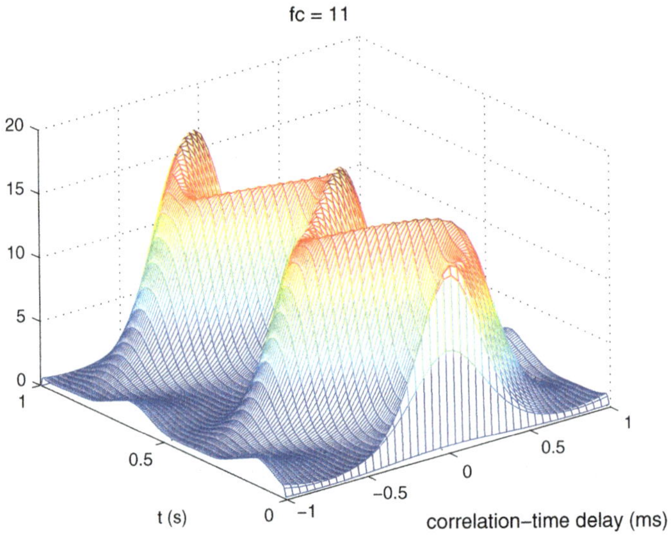


**Fig. 4** Example for a model output in the AMToolbox. The figure shows the modeled binaural activity of the 500-Hz frequency channel in response to a 500-Hz sinusoid with a 2-Hz binaural modulation and a sampling frequency of 44.1 kHz as modeled by the cross-correlation model [47]. The figure can be plotted by evaluating the code below.

```
[cc,t] = lindemann1986(bmsin(500,2,44100),44100,'T_int',6);
plotlindemann(cc,t,'fc',500);
```

## Modeling Sound-Source Lateralization by Supervised Learning—**may2013**

A probabilistic model for sound-source lateralization based on the supervised learning of azimuth-dependent binaural cues in individual frequency channels is implemented in the function `may2013` [54]. The model jointly analyzes both ITDs and ILDs by approximating the two-dimensional feature distribution with a Gaussian-mixture model, GMM, classifier. In order to improve the robustness of the model, a multi-conditional training stage is employed to account for the uncertainty in ITDs and ILDs resulting from complex acoustic scenarios. The model is able to robustly estimate the position of multiple sound sources in the presence of reverberation [54]. The model can be used as a pre-processor for applications in computational auditory scene analysis such as missing data classification.

## Modeling Binaural Activity—**takanen2013**

The decoding of the lateral direction of a sound event by the auditory system is modeled [75] and implemented in `takanen2013`. A binaural signal, processed by a peripheral model, is fed into functional count-comparison-based models of the medial and lateral superior olive [65] which decode the directional information from the binaural signal. In each frequency channel, both model outputs are combined and further processed to create the binaural activity map representing neural activity as a temporal function of lateral arrangement of the auditory scene.

### Modeling Median-Plane Localization—`langendijk2002`

Sound localization in the median planes relies on the analysis of the incoming monaural sound spectrum. The monaural directional spectral features arise due to the filtering of the incoming sound by the HRTFs. A model for the probability of listener's directional response to a sound in the median plane has been proposed [45]. The model uses a peripherally-processed set of HRTFs to mimic the representation of the localization cues in the auditory system. The decision process is simulated by minimizing the spectral distance between the peripherally-processed incoming sound spectrum and HRTFs from the set. Further, a probabilistic mapping is incorporated into the decision process.

In the AMToolbox, the model is provided by the function `langendijk2002`. The model considers the monaural spectral information only and outputs the prediction for the probability of responding at a vertical direction for stationary wideband sounds within the median-sagittal plane.

### Modeling Sagittal-Plane Localization—`baumgartner2013`

The median-plane localization model [45] has been further improved with the focus to provide a good prediction of the localization *performance* for individual listeners in sagittal planes [3]. The model considers adaption to the actual bandwidth of the incoming sound and calibration to the listener-specific sensitivity. It considers a binaural signal and implements binaural weighting [34]. Thus, it allows for predicting target position in arbitrary sagittal planes, namely, parallel shifts of the median plane. The model further includes a stage to retrieve psychophysical performance parameters such as quadrant error rate, local polar RMS error, or polar bias from the probabilistic predictions, allowing to directly predict the localization performance for human listeners. For stationary, spectrally unmodulated sounds, this model incorporates the linear Gammatone filterbank [62], and has been evaluated under various conditions [3]. Optionally, the model can be extended by incorporating a variety of more physiology-related processing stages, for example, the DRNL filterbank [48]—flag `'drnl'` or the humanized AN model [68, 79] in order to model, for example, the level dependence of localization performance [53]—flag `'zilany2007humanized'`.

In the AMToolbox, the model is provided by the function `baumgartner2013`, which is the same implementation as that used in [3]. Further, a pool of listener-specific calibrations is provided, `data_baumgartner2013`, which can be used to assess the impact of arbitrary HRTF-based cues on the localization performance.

### Modeling Distance Perception—**georganti2013**

A method for distance estimation in rooms based on binaural signals has been proposed [29, 30]. The method requires neither *a priori* knowledge of the room impulse response, nor the reverberation time, nor any other acoustical parameter. However, it requires training within the rooms under examination and relies on a set of features extracted from the reverberant binaural signals. The features are incorporated into a classification framework based on GMM classifier. For this method, a distance estimation feature has been introduced exploiting the standard deviation of the interaural spectral level differences in the binaural signals. This feature has been shown to be related to the statistics of the corresponding room transfer function [43, 69] and to be highly correlated with the distance between source and receiver. In the AMToolbox, the model is provided by the function `georganti2013`, which is the same implementation as that used in [30].

## 6 Speech-Perception Models

Speech perception models incorporate the speech information into the modeling process and thus, they usually test the speech intelligibility under various conditions.

### Modeling Monaural Speech Intelligibility in Noise—**joergensen2011**

A model for quantitative prediction of speech intelligibility based on the signal-to-noise envelope-power ratio, SNRenv, after modulation frequency selective processing has been proposed [44]. While the SNRenv-metric is inspired by the concept of the signal-to-noise ratio in the modulation domain [20], the model framework is an extension of the envelope-power-spectrum model for modulation detection and masking [25], and is denoted as the speech-based envelope-power-spectrum model, sEPSM. Instead of comparing the modulation power of clean target speech with a noisy-speech mixture such as the modulation transfer function [36], the sEPSM compares an estimate of the modulation power of the clean speech within the mixture and the modulation power of the noise alone. This means that the sEPSM is sensitive to effects of nonlinear processing, such as spectral subtraction, which may increase the noise modulation power, where the classical speech models fail. In the AMToolbox, the model is provided by the function `joergensen2011`.

### Modeling Spatial Unmasking for Speech in Noise and Reverberation—jelfs2011

A model of spatial unmasking for speech in noise and reverberation has been proposed [46]. It has been validated it against human speech reception thresholds, SRTs. The underlying structure of the model has been further improved [42] and operates directly upon binaural room impulse responses, BRIRs. It has two components, better-ear listening and binaural unmasking, which are assumed to be additive. The BRIRs are filtered into different frequency channels using an auditory filterbank [62]. The better-ear listening component assumes that the listener can select sound from either ear at each frequency according to which one has the better signal-to-noise ratio, SNR. The better-ear SNRs are then weighted and summed across frequency according to Table I of the speech-intelligibility index [1], see `siiweightings`. The binaural unmasking component calculates the binaural masking level difference within each frequency channel based on equalization-cancellation theory [11, 21]. These values are similarly weighted and summed across frequency. The summed output is the effective binaural SNR, which can be used to predict differences in SRT across different listening situations. Implemented in the AMToolbox in the function `jelfs2011`, the model has been validated against a number of different sets of SRTs both from the literature and from [42]. The output of the model can be used to predict the effects of noise and reverberation on speech communication for both normal-hearing listeners and users of auditory prostheses and to predict the benefit of optimal head orientation.

## 7 Working with the AMToolbox

Assuming a working Matlab/Octave environment, the following steps are required for getting started.

1. Download the LTFAT from http://ltfat.sourceforge.net
2. Download the AMToolbox from http://amtoolbox.sourceforge.net
3. Start the LTFAT at the Matlab/Octave prompt: `ltfatstart`
4. Start the AMToolbox: `amtstart`

Further instructions on the setup for the AMToolbox can be found in the file INSTALL in the main directory of the AMToolbox. The further steps depend on the particular tasks and application. As a general rule, a demonstration is a good starting point, thus, a `demo_` function can be used to obtain a general impression of the corresponding model, see for example Fig. 1. Then, an experiment, namely, an `exp_` function can be used to see how the model output compares to the corresponding publication, see for example Fig. 3. Editing the corresponding `exp_` function will help to understand the particular experiment implementation, the call to the model functions, and the parameters used in that experiment and generally available for the model. By modifying the `exp_` function and saving as an own experiment, a new application of the model can be easily created.

Note that some of the models require additional data not provided with the AMToolbox because of size limitations. These data can be separately downloaded with the corresponding link usually being provided by the particular model function.

## 8 Conclusion

AMToolbox is a continuously growing and developing collection of auditory models, human data, and experiments. It is free as in *"free beer"*, that is, *freeware*, and it is free as in *"free speech"*, in other words *liberty*[9] It is available for download,[10] and auditory researchers are welcome to contribute their models to the AMToolbox in order to increase the pool of easily accessible and verified models and, thus, to promote their models in the community.

Much effort has been put to the documentation. The documentation in the software is directly linked with the documentation appearing at the web page,[11] providing a consistent documentation of the models. Finally, a rating system is provided that which clearly shows the current stage of verification of each model implementation. The ratings are continuously updated,[12] and we hope that all the implementations of the models from AMToolbox will reach the state of *Verified* soon.

**Acknowledgments** The authors thank J. Blauert for organizing the AABBA project and all the developers of the models for providing information on the models. They are also indepted B. Laback and two anonymous reviewers for their useful comments on earlier versions of this article.

## References

1. American National Standards Institute, New York. *Methods for calculation of the speech intelligibility index*, ANSI S3.5-1997 edition, 1997.
2. C. Antweiler, A. Telle, P. Vary, and G. Enzner. Perfect-Sweep NLMS for Time-Variant Acoustic System Identification. In *Proc. Intl. Conf. Acoustics, Speech, and Signal Processing, ICASSP*, pages 517–529, Kyoto, Japan, 2012.
3. R. Baumgartner, P. Majdak, and B. Laback. Assessment of sagittal-plane sound-localization performance in spatial-audio applications. In J. Blauert, editor, *The technology of binaural listening*, chapter 4. Springer, Berlin-Heidelberg-New York NY, 2013.
4. R. A. Bentler and C. V. Pavlovic. Transfer Functions and Correction Factors used in Hearing Aid Evaluation and Research. *Ear Hear*, 10:58–63, 1989.

---

[9] http://www.gnu.org/philosophy/free-sw.html, last viewed on 9.1.2013.

[10] from http://amtoolbox.sourceforge.net, last viewed on 9.1.2013.

[11] see http://amtoolbox.sourceforge.net/doc/ last viewed on 9.1.2013.

[12] see http://amtoolbox.sourceforge.net/notes/amtnote006.pdf, last viewed on 9.1.2013.

5. L. Bernstein, S. van de Par, and C. Trahiotis. The normalized interaural correlation: Accounting for NoS$\pi$ thresholds obtained with Gaussian and "low-noise" masking noise. *J Acoust Soc Am*, 106:870–876, 1999.
6. J. Breebaart, S. van de Par, and A. Kohlrausch. Binaural processing model based on contralateral inhibition. I. Model structure. *J Acoust Soc Am*, 110:1074–1088, 2001.
7. J. Breebaart, S. van de Par, and A. Kohlrausch. Binaural processing model based on contralateral inhibition. II. Dependence on spectral parameters. *J Acoust Soc Am*, 110:1089–1104, 2001.
8. J. Breebaart, S. van de Par, and A. Kohlrausch. Binaural processing model based on contralateral inhibition. III. Dependence on temporal parameters. *J Acoust Soc Am*, 110:1105–1117, 2001.
9. J. Buckheit and D. Donoho. *Wavelab and Reproducible Research*, pages 55–81. Springer, New York NY, 1995.
10. J. Claerbout. Electronic documents give reproducible research a new meaning. *Expanded Abstracts, Soc Expl Geophys*, 92:601–604, 1992.
11. J. Culling. Evidence specifically favoring the equalization-cancellation theory of binaural unmasking. *J Acoust Soc Am*, 122:2803–2813, 2007.
12. J. Culling, S. Jelfs, and M. Lavandier. Mapping Speech Intelligibility in Noisy Rooms. In *Proc. 128th Conv. Audio Enginr. Soc. (AES)*, page Convention paper 8050, 2010.
13. T. Dau, B. Kollmeier, and A. Kohlrausch. Modeling auditory processing of amplitude modulation. I. Detection and masking with narrow-band carriers. *J Acoust Soc Am*, 102:2892–2905, 1997.
14. T. Dau, B. Kollmeier, and A. Kohlrausch. Modeling auditory processing of amplitude modulation. II. Spectral and temporal integration. *J Acoust Soc Am*, 102:2906–2919, 1997.
15. T. Dau, D. Püschel, and A. Kohlrausch. A quantitative model of the effective signal processing in the auditory system. I. Model structure. *J Acoust Soc Am*, 99:3615–3622, 1996.
16. T. Dau, D. Püschel, and A. Kohlrausch. A quantitative model of the "effective" signal processing in the auditory system. II. Simulations and measurements. *J Acoust Soc Am*, 99:3623–3631, 1996.
17. M. Dietz, S. D. Ewert, and V. Hohmann. Lateralization of stimuli with independent fine-structure and envelope-based temporal disparities. *J Acoust Soc Am*, 125:1622–1635, 2009.
18. M. Dietz, S. D. Ewert, and V. Hohmann. Auditory model based direction estimation of concurrent speakers from binaural signals. *Speech Comm*, 53:592–605, 2011.
19. M. Dietz, S. D. Ewert, V. Hohmann, and B. Kollmeier. Coding of temporally fluctuating interaural timing disparities in a binaural processing model based on phase differences. *Brain Res*, 1220:234–245, 2008.
20. F. Dubbelboer and T. Houtgast. The concept of signal-to-noise ratio in the modulation domain and speech intelligibility. *J Acoust Soc Am*, 124:3937–3946, 2008.
21. N. I. Durlach. Binaural signal detection: equalization and cancellation theory. In J. V. Tobias, editor, *Foundations of Modern Auditory Theory. Vol. II*, pages 369–462. Academic, New York, 1972.
22. C. Elberling, J. Callø, and M. Don. Evaluating auditory brainstem responses to different chirp stimuli at three levels of stimulation. *J Acoust Soc Am*, 128:215–223, 2010.
23. G. Enzner. Analysis and optimal control of LMS-type adaptive filtering for continuous-azimuth acquisition of head related impulse responses. In *Proc. Intl. Conf. Acoustics, Speech, and Signal Processing, ICASSP*, pages 393–396, Las Vegas NV, 2008.
24. G. Enzner. 3D-continuous-azimuth acquisition of head-related impulse responses using multi-channel adaptive filtering. In *Proc. IEEE Worksh. Appl. of Signal Process. to Audio and Acoustics, WASPAA*, pages 325–328, New Paltz NY, 2009.
25. S. Ewert and T. Dau. Characterizing frequency selectivity for envelope fluctuations. *J Acoust Soc Am*, 108:1181–1196, 2000.
26. G. Fant. Analysis and synthesis of speech processes. In B. Malmberg, editor, *Manual of phonetics*. North-Holland, Amsterdam, 1968.
27. R. Fassel and D. Püschel. *Modulation detection and masking using deterministic and random maskers*, pages 419–429. Universitätsgesellschaft, Oldenburg, 1993.
28. D. Gabor. Theory of communication. *J IEE*, 93:429–457, 1946.

29. E. Georganti, T. May, S. van de Par, and J. Mourjopoulos. Sound source distance estimation in rooms based on statistical properties of binaural signals. IEEE *Trans Audio Speech Lang Proc*, submitted.
30. E. Georganti, T. May, S. van de Par, and J. Mourjopoulos. Extracting sound-source-distance information from binaural signals. In J. Blauert, editor, *The technology of binaural listening*, chapter 7. Springer, Berlin-Heidelberg-New York NY, 2013.
31. B. R. Glasberg and B. Moore. Derivation of auditory filter shapes from notched-noise data. *Hear Res*, 47:103–138, 1990.
32. R. Goode, M. Killion, K. Nakamura, and S. Nishihara. New knowledge about the function of the human middle ear: development of an improved analog model. *Am J Otol*, 15:145–154, 1994.
33. L. Han and T. Poulsen. Equivalent threshold sound pressure levels for Sennheiser HDA 200 earphone and Etymotic Research ER-2 insert earphone in the frequency range 125 Hz to 16 kHz. *Scandinavian Audiology*, 27:105–112, 1998.
34. M. Hofman and J. Van Opstal. Binaural weighting of pinna cues in human sound localization. *Exp Brain Res*, 148:458–70, 2003.
35. V. Hohmann. Frequency analysis and synthesis using a gammatone filterbank. *Acta Acust./ Acustica*, 88:433–442, 2002.
36. T. Houtgast, H. Steeneken, and R. Plomp. Predicting speech intelligibility in rooms from the modulation transfer function. i. general room acoustics. *Acustica*, 46:60–72, 1980.
37. ISO 226:2003. *Acoustics - Normal equal-loudness-level contours*. International Organization for Standardization, Geneva, Switzerland, 2003.
38. ISO 389–2:1994(E). *Acoustics - Reference zero for the calibration of audiometric equipment - Part 2: Reference equivalent threshold sound pressure levels for pure tones and insert earphones*. International Organization for Standardization, Geneva, Switzerland, 1994.
39. ISO 389–5:2006. *Acoustics - Reference zero for the calibration of audiometric equipment - Part 5: Reference equivalent threshold sound pressure levels for pure tones in the frequency range 8 kHz to 16 kHz*. International Organization for Standardization, Geneva, Switzerland, 2006.
40. ISO 389–8:2004. *Acoustics - Reference zero for the calibration of audiometric equipment - Part 8: Reference equivalent threshold sound pressure levels for pure tones and circumaural earphones*. International Organization for Standardization, Geneva, Switzerland, 2004.
41. L. Jeffress. A place theory of sound localization. *J Comp Physiol Psych*, 41:35–39, 1948.
42. S. Jelfs, J. Culling, and M. Lavandier. Revision and validation of a binaural model for speech intelligibility in noise. *Hear Res*, 2011.
43. J. Jetzt. Critical distance measurement of rooms from the sound energy spectral response. *J Acoust Soc Am*, 65:1204–1211, 1979.
44. S. Jørgensen and T. Dau. Predicting speech intelligibility based on the signal-to-noise envelope power ratio after modulation-frequency selective processing. *J Acoust Soc Am*, 130:1475–1487, 2011.
45. E. Langendijk and A. Bronkhorst. Contribution of spectral cues to human sound localization. *J Acoust Soc Am*, 112:1583–1596, 2002.
46. M. Lavandier and J. Culling. Prediction of binaural speech intelligibility against noise in rooms. *J Acoust Soc Am*, 127:387–399, 2010.
47. W. Lindemann. Extension of a binaural cross-correlation model by contralateral inhibition. I. Simulation of lateralization for stationary signals. *J Acoust Soc Am*, 80:1608–1622, 1986.
48. E. Lopez-Poveda and R. Meddis. A human nonlinear cochlear filterbank. *J Acoust Soc Am*, 110:3107–3118, 2001.
49. R. Lyon. All pole models of auditory filtering. In E. Lewis, G. Long, R. Lyon, P. Narins, C. Steele, and E. Hecht-Poinar, editors, *Diversity in Auditory Mechanics: Proc. Intl. Symp., University of California, Berkeley*. World Scientific Publishing, 1996.
50. R. Lyon, A. Katsiamis, and E. Drakakis. History and future of auditory filter models. In *Proc. 2010 IEEE Intl. Symp. Circuits and Systems, ISCAS*, pages 3809–3812, 2010.

51. P. Majdak, P. Balazs, and B. Laback. Multiple exponential sweep method for fast measurement of head-related transfer functions. *J Audio Eng Soc*, 55:623–637, 2007.
52. P. Majdak, B. Masiero, and J. Fels. Sound localization in individualized and non-individualized crosstalk cancellation systems. *J Acoust Soc Am*, 133:2055–2068, 2013.
53. P. Majdak, T. Necciari, B. Baumgartner, and B. Laback. Modeling sound-localization performance in vertical planes: level dependence. In *Poster at the 16th International Symposium on Hearing (ISH), Cambridge*, UK, 2012.
54. T. May, S. van de Par, and A. Kohlrausch. Binaural localization and detection of speakers in complex acoustic scenes. In J. Blauert, editor, *The technology of binaural listening*, chapter 15. Springer, Berlin-Heidelberg-New York NY, 2013.
55. D. McAlpine and B. Grothe. Sound localization and delay lines-do mammals fit the model? *Trends in Neurosciences*, 26:347–350, 2003.
56. R. Meddis, M. J. Hewitt, and T. M. Shackleton. Implementation details of a computation model of the inner hair-cell auditory-nerve synapse. *J Acoust Soc Am*, 87:1813–1816, 1990.
57. R. Meddis, L. O'Mard, and E. Lopez-Poveda. A computational algorithm for computing nonlinear auditory frequency selectivity. *J Acoust Soc Am*, 109:2852–2861, 2001.
58. B. Moore and B. Glasberg. Suggested formulae for calculating auditory-filter bandwidths and excitation patterns. *J Acoust Soc Am*, 74:750–753, 1983.
59. S. Neely, S. Norton, M. Gorga, and J. W. Latency of auditory brain-stem responses and otoacoustic emissions using tone-burst stimuli. *J Acoust Soc Am*, 83:652–656, 1988.
60. P. O'Mard. Development system for auditory modelling. Technical report, Centre for the Neural Basis of Hearing, University of Essex, UK, 2004.
61. M. Park, P. A. Nelson, and K. Kang. A model of sound localisation applied to the evaluation of systems for stereophony. *Acta Acustica/Acust.*, 94:825–839, 2008.
62. R. Patterson, I. Nimmo-Smith, J. Holdsworth, and P. Rice. An efficient auditory filterbank based on the gammatone function. *APU report*, 2341, 1988.
63. R. D. Patterson, M. H. Allerhand, and C. Giguère. Time-domain modeling of peripheral auditory processing: A modular architecture and a software platform. *J Acoust Soc Am*, 98:1890–1894, 1995.
64. D. Pralong and S. Carlile. The role of individualized headphone calibration for the generation of high fidelity virtual auditory space. *J Acoust Soc Am*, 100:3785–3793, 1996.
65. V. Pulkki and T. Hirvonen. Functional count-comparison model for binaural decoding. *Acta Acustica/Acust.*, 95:883–900, 2009.
66. D. *Püschel. Prinzipien der zeitlichen Analyse beim Hören*. PhD thesis, Universität Göttingen, 1988.
67. A. Recio and W. Rhode. Basilar membrane responses to broadband stimuli. *J Acoust Soc Am*, 108:2281–2298, 2000.
68. F. Rønne, J. Harte, C. Elberling, and T. Dau. Modeling auditory evoked brainstem responses to transient stimuli. *J Acoust Soc Am*, 131:3903–3913, 2012.
69. M. Schroeder. Die statistischen Parameter der Frequenzkurven von grossen Räumen. *Acustica*, 4:594–600, 1954.
70. C. Shera. Intensity-invariance of fine time structure in basilar-membrane click responses: Implications for cochlear mechanics. *J Acoust Soc Am*, 110:332–348, 2001.
71. M. Slaney. Auditory toolbox, 1994.
72. P. L. Søndergaard, B. Torrésani, and P. Balazs. The Linear Time Frequency Analysis Toolbox. *Int J Wavelets Multi*, 10:1250032 [27 pages], 2012.
73. C. Spille, B. Meyer, M. Dietz, and V. Hohmann. Binaural scene analysis with multi-dimensional statistical filters. In J. Blauert, editor, *The technology of binaural listening*, chapter 6. Springer, Berlin-Heidelberg-New York NY, 2013.
74. S. Stevens, J. Volkmann, and E. Newman. A scale for the measurement of the psychological magnitude pitch. *J Acoust Soc Am*, 8:185–190, 1937.
75. M. Takanen, O. Santala, and V. Pulkki. Binaural assessment of parametrically coded spatial audio signals. In J. Blauert, editor, *The technology of binaural listening*, chapter 13. Springer, Berlin-Heidelberg-New York NY, 2013.

76. P. Vandewalle, J. Kovacevic, and M. Vetterli. Reproducible research in signal processing - what, why, and how. *IEEE Signal Proc Mag*, 26:37–47, 2009.
77. G. von Békésy. Zur theorie des hörens; Über das Richtungshören bei einer Zeitdefferenz oder Lautstärkenungleichheit der beiderseitigen Schalleinwirkungen. *Phys Z*, 31:824–835, 1930.
78. P. Ziegelwanger, H Majdak. Continuous-direction model of the time-of-arrival in the head-related transfer functions.*J Acoust Soc Am*, submitted.
79. M. S. A. Zilany and I. C. Bruce. Representation of the vowel /$\epsilon$/ in normal and impaired auditory nerve fibers: Model predictions of responses in cats. *J Acoust Soc Am*, 122:402–248, 2007.
80. G. Zweig. Finding the impedance of the organ of corti. *J Acoust Soc Am*, 89:1229–1254, 1991.
81. E. Zwicker. Subdivision of the audible frequency range into critical bands (frequenzgruppen). *J Acoust Soc Am*, 33:248–248, 1961.
82. E. Zwicker and H. Fastl. *Psychoacoustics: Facts and models*. Springer Berlin, 1999.

# Trends in Acquisition of Individual Head-Related Transfer Functions

**G. Enzner, Chr. Antweiler and S. Spors**

## 1 Introduction

*Head-related transfer functions*, HRTFs, describe the filtering effect of head, pinna and torso when sound from an acoustic point source is received at a defined position in the ear-canal of a listener under free-field acoustic conditions. The time-domain equivalent of the HRTF is termed *head-related impulse response*, HRIR. The definition of HRTF/HRIR in particular relies on the free-space, that is, anechoic acoustic conditions to separate the human receiver characteristic from other room acoustic characteristics.[1]

The HRIR intuitively represents the *interaural time differences*, ITDs, between both ears as a predominant feature of spatial hearing [113]. Due to the ambivalence in time-difference estimation at high frequencies, human hearing further exploits *interaural level differences*, ILDs, as a second important criterion for sound-source localization. On a locus where neither time or level differences are of significance, for instance, in the median plane, or, more generally, on the cones-of-confusion around the interaural axis, the auditory localization finally relies on *spectral cues*, which are represented by the HRTF as well [20].

Primary application of HRTF has been headphone-based or loudspeaker-based binaural rendering of virtual auditory spaces—see, for example, [5, 23, 53, 75, 105].

[1] If an acoustic enclosure is meant to be involved in the HRIR, this is indicated by using the terms *binaural room impulse response*, BRIR, or *binaural room transfer function*, BRTF.

G. Enzner (✉)
Institute of Communication Acoustics, Ruhr-Universität Bochum, Bochum, Germany
e-mail: gerald.enzner@rub.de

Chr. Antweiler
Institute of Communication Systems and Data Processing, RWTH Aachen University, Aachen, Germany

S. Spors
Institute of Communications Engineering, Universität Rostock, Rostock, Germany

J. Blauert (ed.), *The Technology of Binaural Listening*, Modern Acoustics and Signal Processing, DOI: 10.1007/978-3-642-37762-4_3,

In this context, the perceptual relevance of different HRTF characteristics was initially studied a lot by researchers—see Sect. 1.1 for more details. According to their results, several HRTF databases were then created via extensive measurement projects in order to provide the HRTF characteristics most effectively and comprehensively to hearing researchers and virtual-auditory-display designers—see Sect. 1.2. The consideration of HRTFs eventually seems to be an *evergreen* in the audio and acoustics domain and, as a result, continued research activity in hi-fidelity acquisition and representation of HRTFs is observed, as shown by the main parts of this chapter.

## 1.1 HRTF Characteristics and Their Perceptual Relevance

A first important factor of the human binaural receiver concerns the individual characteristics of HRTFs and the related best-practice definition regarding the actual point of measurement at the ears. In [52], the entire acoustic transfer from a sound source in free space to the eardrum of a listener is therefore divided into three parts as follows,

- transmission from the free-field to the blocked entrance of the ear-canal,
- impedance conversion related to ear-canal blocking, and
- transmission along the ear-canal.

All three parts of the transmission were found to be highly individual. However, a comparison of measurements at the open entrance, blocked entrance, and eardrum revealed the smallest inter-individual deviations for the case of a blocked-entrance HRTF. The blocked-ear condition was therefore identified to provide a suitable measurement point in the sense that complete spatial information is supposed to be included in this least-individual HRTF variant. An extensive measurement on 40 individuals was presented by the same authors in [76].

The relevance of individual HRTFs is often judged with respect to the corresponding localization accuracy in virtual auditory spaces. A well known study [111, 112], using broadband-noise stimuli and no head-tracking support, reports accurate localization for both real free-field sources and virtual sources generated from the listeners own HRTF. However, with non-individual HRTFs the same authors observed high rates of front-back and up-down confusions, so-called reversals, as compared to real free-field stimuli [110]. While the interaural cues for horizontal location seemed to be robust, their data in particular suggests that the spectral cues have to be considered as important for resolving location along the cones-of-confusion around the interaural axis. These spectral cues are distorted in non-individual binaural synthesis. With the confusions being resolved, there is again a close correspondence between real free-field and virtual-source conditions. In a different work [77], an increased confusion of localization in the median plane was confirmed when non-individual rather then individual binaural recordings were presented to listeners. Across different studies and the authors' own experience, a particular confusion is often given by the perception of increased elevation for stimulation in the horizontal plane [16, 17].

A bit later, the comprehensive study in [18] evaluated the effect of HRTF individualization in direct comparison with the effects of head movements and room characteristics using speech stimuli. In this context, and in line with [26], HRTF individualization played the least significant role regarding accurate localization of azimuth and elevation and also regarding the reduction of reversals. Head-tracking and the related supply of head-motion cues to the listener was figured out as the most effective means for reducing reversals. This result confirms the earlier observed importance of real and virtual head movements for binaural localization of sound sources [109, 114]. Regarding the desired *externalization* [55] of virtual sources, in other words *out-of-the-head-localization* [83]—related to distance perception—it was found that the simulation of room characteristics, such as reverberation, had the most significant effect [18]. This finding supports other work [35, 104]. It is finally intuitive to argue that multi-modal, such as audio-visual or audio-tactile stimulation, will further help externalization of sound.

After the observation of localization and externalization ability using individual HRTF, it was investigated how accurately the HRTFs must be reproduced in spectral terms to achieve true three-dimensional perception of auditory signals. It was found that magnitude HRTFs can be smoothed significantly via truncation of their Fourier-series representation without affecting the perceived location of sound. It turned out that the first 16 Fourier components of the magnitude HRTF still provided enough information for all listeners and all source directions. Yet stronger spectral smoothing was reported to cause a perception of increased elevation, what is consistent with the fact that magnitude-HRTF spectra from high elevations are relatively smooth as compared to lower elevations. According to the same study, an HRTF phase-response representation in terms of a broadband interaural time delay is, in most of the cases, indistinguishable from natural sounds. Similar conclusions were drawn in [24], where magnitude- and phase-response smoothing was implemented via gammatone filterbanks, that is, filters that mimic the spectral selectivity of the human cochlea. Refinement of these results and comprehensive extension towards the applicability, for instance, in MPEG-Surround audio coding, is presented in [25, 117]. For specific speech stimuli, [19] found that lowpass filtering at 8 kHz dramatically degraded localization regarding the polar angle and increased cone-of-confusion errors, while the preservation of information above 8 kHz was explicitly demonstrated to be essential for accurate localization of the source direction.

From the engineering point of view, particularly the HRIR filter length used for HRIR representation in binaural rendering is an important parameter. It is generally accepted that impulse-response durations of 5–10 ms seem to be sufficient. Theoretically, a much larger choice could be considered to capture the exact and full spectral detail of the HRTF. However, psychophysical tests did not reveal a perceptual relevance of the HRIR length beyond 256 coefficients at 44.1 kHz sampling frequency. In conjunction with binaural room-simulation systems, it has been reported that even 128 coefficients or less can be perceptually sufficient [97]. Another important engineering parameter, the spatial resolution of the HRTF, is treated in the following section together with the properties of various HRTF databases. More comprehensive and somewhat recent overviews of the various aspects of HRTFs, such as the measurement point, the measurement signal, the duration of the HRIR, the required spatial density, the need for individualization, headphone auralization, and some results of psychophysical validations, are found in [79, 115].

## 1.2 HRTF Databases

On the one hand, the HRTF is an important element of the technology of binaural hearing and listening. On the other hand, the acquisition of actual HRTFs has been recognized as a tedious and delicate lab task of its own. It typically requires an anechoic chamber to obtain free-space HRTFs and, moreover, a large time budget of both the operator and the subject of interest to complete a measurement with sufficient spatial resolution. The implementation of the measurement procedure further requires one or more loudspeakers, in-ear microphones, and audio reproduction and recording software. In most of the cases, the subject of interest is steered mechanically into different directions w.r.t. the loudspeaker or, alternatively, the loudspeaker position is adjusted in discrete steps, that is, typically a *stop-&-go approach* is applied, which uses a lot of time on measurement pause and recalibration. Various systems of this kind exist, but they are usually not transportable to other locations. As a result, many organizations have chosen to provide their measurements as publicly available databases to the community. In the following, only a few are listed in roughly chronological order. If not noted otherwise, the impulse responses are provided with 44.1 kHz temporal sampling.

*KEMAR—the MIT-Media-Lab HRTF Database*: This early and still very popular database [45] for the *Knowles-Electronics Mannequin for Acoustic Research*, KEMAR, represents an extensive, but non-individual recording. The database and good documentation are available online.[2] In total, 710 different positions were sampled at elevations ranging from $-40°$ to $+90°$ in $10°$ increments with regard to the horizontal plane and roughly $5°$ azimuth spacing per elevation at 1.4 m distance between loudspeaker and KEMAR. All HRTF pairs are provided as individual *.wav files.

*AUDIS—the AUDIS Catalog of Human HRTFs*: In the context of the European-Union, EU, funded project *Auditory Displays*, AUDIS, [22], which heavily relied on binaural technology and reliable human HRTF data, a special program for collecting HRTF sets was undertaken. Here, 2.4 m distance to the loudspeaker, $10°$-spaced elevations from $-10°$ to $+90°$, and an azimuth spacing of $15°$ was used. The total measurement then comprised 122 directions for each of about 20 individuals. Moreover, round-robin tests have been performed with four contributing partners to analyze differences in the data across different laboratories.

[2] http://sound.media.mit.edu/resources/KEMAR.html

This resulted in a compact set of recommendations, termed *Golden Rules* for HRTF measurements.[3]

*CIPIC—the CIPIC Lab HRTF Database*: This database contains measured HRTFs at high spatial resolution for more than 90 individuals with 45 publicly available, including KEMAR with large and small pinna [7]. The spatial sampling is mostly uniform with 5° spacing in both elevation and azimuth, resulting in 1250 sampling points on the 1 m radius auditory sphere. The database[4] additionally includes a set of individual anthropometric measurements for each subject. The latter can be useful to perform scaling studies on HRTFs. Additional documentation and Matlab utility programs are provided with the database.

*LISTEN—the IRCAM HRTF Database*: Again developed in an EU project, this database contains blocked-meatus HRTFs at elevations from −45° to +90° in 5° increments with roughly 15° azimuth spacing, resulting in 187 positions in total. It provides raw HRTF measurements, optional diffuse-field compensation and morphological data. The database for about 50 individuals and documentation are available online.[5] The impulse responses come as individual *.wav files or in more compact Matlab format.

*ARI—the Acoustics-Research Institute HRTF Database*: It comprises high-resolution HRTFs of more than 70 individuals. Most of them were measured using in-ear microphones, but for a few further ones behind-the-ear microphones placed in hearing-aid devices were employed. 1550 positions were then measured for each listener, including the full azimuth-circle—with 2.5° spacing in the horizontal plane—and elevations from −30° to +80°. The multiple exponential-sweep method [69] was applied to reduce the measurement time to the minimum deemed to be acceptable for the listeners. Database, tools and documentation are available.[6]

*FIU—the Florida-International-Univ. DSP-Lab HRTF Database*: This recent database contains HRTF data from 15 individuals at twelve different azimuths and six different elevations [51]. It further includes 3-D images of the persons' pinnae and related anthropometric measures of the various parts of the pinnae. The database is unique in the sense that it uses a higher audio-sampling frequency as compared to other databases, namely, 96 kHz. Download of the database is provided.[7]

A general consequence arising from those spatially sampled HRTF databases is the need for spatial interpolation of HRTF data in 3-D-sound systems [16], particularly, when dynamic auditory virtual environments are desired. The issue of interpolation will be treated in more detail in Sect. 2 of this chapter.

---

[3] http://dx.doi.org/10.5278/VBN/MISC/AUDIS

[4] http://interface.cipic.ucdavis.edu

[5] http://recherche.ircam.fr/equipes/salles/listen/index.html

[6] http://www.kfs.oeaw.ac.at/content/view/608/606

[7] http://dsp.eng.fiu.edu/HRTFDB

## 1.3 Alternative Trends in HRTF Acquisition

All the previous databases represented single-range HRIR data, that is, such with a constant measurement radius from the head-center of the subject. However, besides the directional information, the distance of the virtual source to the listener is an important cue in binaural sound rendering. Since simple HRTF-level adjustments are insufficient to achieve spatial realism in the near-field of the listener, in particular, the exact wavefront curvature, the concept of a range-dependent HRTF database was evaluated [65]. The reported database, which is, however, not publicly available, was sampled at distances of 0.2, 0.3, 0.4, 0.5, 0.75, 1.0, 1.5, and 2.0 m, and the directional resolution was chosen to 1° azimuth spacing and 5° elevation spacing over the full sphere. A similar configuration was applied for the more recent and downloadable 3-D-HRTF recordings in [87].

The databases as mentioned in the previous section were obtained from measurement systems differing regarding their particular hardware, software, and lab room architecture. In order to achieve at least some harmonization of HRTF measurement, the *Club Fritz* has launched a round-robin study in which the HRTF of the particular dummy head Neumann KU-100 is being measured at different laboratories [63]. The involved labs partially coincide with the aforementioned database providers. While previous works have focused on inter-positional and inter-individual HRTF variations a lot, see also [88], the study by *Club Fritz* is devoted to the analysis of variations due to the measurement system. According to the *Club-Fritz* authors, their preliminary analysis of spectral characteristics and interaural time-differences revealed some differences the significance of which still need to be clarified.

Contrasting the usual way of setting up an HRTF measurement system—see, for instance, [20, 76, 86]—a reciprocal configuration for HRTF measurements is proposed in [121]. By applying the acoustic principle of reciprocity, one can swap the loudspeaker and microphone positions as compared to the conventional direct lab arrangement, namely, by inserting a miniature speaker into the test person's ear and placing several microphones around this person. The reciprocity principle implies that swapped loudspeaker and microphone positions, unless other changes were made to the setup, yield identical impulse responses as compared to the direct arrangement [21]. The advantages of this method include a reduction of inter-equipment reflections due to the placement of small-size microphones instead of the larger loudspeakers around the head. More importantly, HRTF acquisition for many positions can be done in parallel—and thus quickly—by playing the probe sound via a single in-ear loudspeaker and recording the received sound simultaneously at all microphones. One frequent concern regarding the reciprocal method is the possibly weak or distorted low-frequency output of the miniature loudspeaker in the ear canal. The proposed remedy to this issue is offered by low-frequency augmentation of the reciprocal method with an analytical HRTF solution based on a simple head and torso model—see, for instance, [4, 6]. A related drawback of reciprocal HRTF acquisition is the poor signal-to-noise ratio which results for reasons of maintaining comfort and physiological safety during in-ear reproduction. Longer probe-signal reproduction and more averaging is thus required to achieve the desired quality.

Apart from purely-acoustic measurement, an individual HRTF dataset can be generated from knowledge of the anthropometric data of the individual of interest and on the basis of several acoustically measured non-individual HRTF training sets. Here, a regression model first needs to be constructed between the anthropometric data and the features of the respective person's measured HRTFs. The anthropometry of a new person is then mapped to individual HRTFs by the principle of best selection or by means of least-squares regression [91, 95]. However, to determine the anthropometry of a person with suitable precision is a challenging task and often requires exhausting procedures or special equipment [92]. With the availability of bio-morphological data in the form of scanned 3-D-images, the numerical boundary-element method could also be applied for the calculation of individual HRTFs [42, 62, 82]. Here, high accuracy and computational efficiency of the calculation are the primary aspects of the ongoing optimization [48, 50]. In [43], a systematic analysis of the dependency of HRTFs and binaural cues on the anthropometric data of children and adults of all ages is found.

## 1.4 HRTF Representation

In order to achieve straightforward HRTF exchange, for example, for mutual support and HRTF comparison between different laboratories, harmonization regarding the HRTF database file format is urgently required. To this end, a very useful option has recently been suggested, namely, the *open directional-audio file format*, OpenDAFF [106]. OpenDAFF is an open-source software package for storage and realtime usage of directional audio data such as loudspeaker directivities, microphone directivities and, last but not least, HRTFs. Each particular radius of the HRTF sphere is represented by a single *.DAFF file with uniform angular sampling. The package provides free Matlab routines for writing and reading OpenDAFF, a free C++ reader and a graphical-viewer tool for DAFF data. It is supposed to facilitate exchange of directional audio data. It has been developed in the context of real-time auralization of auditory virtual realities. For instance, the advanced *3-D-continuous-azimuth* HRTF data to be described in detail in Sect. 3 of this chapter are available in OpenDAFF [107]. While OpenDAFF is immediately applicable for the direct storage of measured HRIRs/HRTFs, more sophisticated HRTF representations, for instance, such as based on the pole-zero model [54], would require more tailored file formats.

Starting from seminal work in [40], representation of HRTFs via *spherical-harmonics transformation* has become a very popular strategy. Especially the inherent interpolation ability and, thus, continuous representation of 3-D-HRTFs via the inverse transformation has attracted researchers a lot—see, for instance, [34, 85, 119]. However, at the same time, it was noticed that this transformation is very sensitive to spatially sampled or missing data, such as discrete HRTF with polar gaps. Therefore, a regularization of the transformation was suggested—see, for example, [122]. Still another approach utilizes the *continuous-azimuth HRTF measurements* as described in Sect. 3.7 in order to avoid both a regularization of the transformation and spatial aliasing in the representation of 3-D-HRTF fields [39]. Very recently, a systematic approach to 3-D-HRTF sampling was presented. It answers the two intercoupled questions regarding the computation of the spherical harmonics transformation [120], namely,

- What is the required angular resolution?
- What is the most suitable sampling scheme?

In this study, a sampling scheme termed IGLOO, known from astrophysics, was identified as the most suitable one after comparison of different candidates, see [123], and it was stated that 2209 HRTF measurements across the sphere are required in this case. In contrast to the direct storage of HRTFs, such as with OpenDAFF, a file format dedicated to high-order-spherical-harmonics representation of HRTFs is not yet available.

# 2 Discrete Measurement, Interpolation, Extrapolation

## 2.1 *Sequential Capturing of HRTFs*

A set of left- and right-ear HRTFs depends on source position, listener position, head- and torso-orientation and various other parameters. Typical measurement setups allow only to vary one or two spatial degrees of freedom. In most cases, the incidence angle of the source is varied with respect to a fixed head-and-torso orientation for a constant source distance. For a fixed orientation of the head with respect to the torso and under acoustic free-field conditions it is equivalent if either the source is mechanically moved on a spherical surface, for instance, a loudspeaker, or if the listener or head-and-torso simulator is rotated. In most of the currently available datasets the head orientation has not been varied with respect to the torso orientation or incidence angle of the source. The importance of these degrees of freedom have, for example, been investigated in [49].

For a particular angle of incidence, the HRTFs can be measured by emitting a specific measurement signal by the source positioned at the particular position and capturing the response at the left and right ear, respectively. In order to capture an entire dataset of HRTFs, traditional techniques sequentially repeat this procedure for each source position. Hence, the considered spatial degrees of freedom are spatially sampled in typical measurement setups and databases. In contrast, the novel techniques discussed in Sect. 3 allow for a spatially-continuous measurement of HRTF datasets.

In order to capture HRIRs or BRIRs efficiently, head-and-torso simulators have been developed that allow to mechanically rotate and even tilt the head by means of software controlled servo motors [33, 66, 74]. In Fig. 1a the modified KEMAR [30] mannequin of the Quality-and-Usability Lab at Technische Universität Berlin is shown. It enables a horizontal rotation of the head, while the torso remains fixed in the room-coordinate system. The industrial servomotor allows for a rotation of the

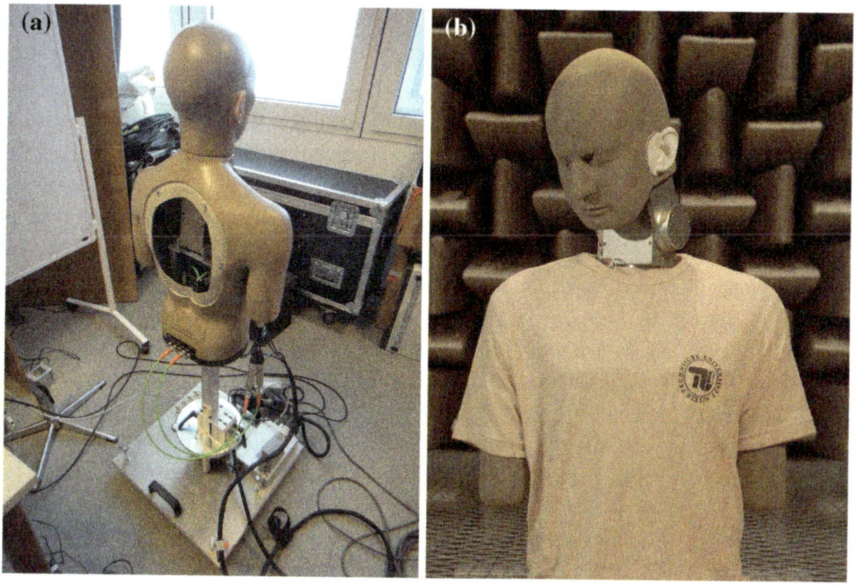


**Fig. 1** Head-and-torso simulators—right photo courtesy of A. Lindau, TU Berlin: **a** Modified KEMAR, **b** FABIAN

head in the range of $\pm 180°$ with a precision of $0.01°$. Figure 1b shows the *instrument for fast and automatic binaural impulse-response acquisition*, FABIAN [66], built by the Audio-Communication Group at Technische Universität Berlin. This system allows for horizontal rotation of the head in the range of $\pm 90°$ and a vertical tilting in the range of $-45°$ to $+90°$, both with a mechanical precision of $0.02°$. The drive mechanism can be modified to cover rotation and lateral tilting alternatively. Both head-and-torso simulators can be placed on a software-controlled turntable in order to rotate the torso independently from the head.

## 2.2 Signal Acquisition and HRTF Calculation

For static scenarios, an HRTF represents the transfer function of the *linear and time-invariant*, LTI, acoustic system between the source and the ear. The choice of the excitation signal interestingly has decisive influence on the results. Different or partly even contradictory constraints have to be met. The *optimal* excitation depends primarily on the algorithm for HRIR extraction, but also on the measurement scenario—for example, static or dynamic—the acoustic environment, and the hardware setup. To achieve a high signal-to-noise ratio, SNR, over the frequency range of interest, the excitation signal must have high energy, as compared to the system noise, everywhere in this frequency range. As the maximum amplitude is limited, a higher energy level of the signal is possible if the energy is additionally spread over time rather than using impulse-like excitation.

Besides the question of how to feed as much energy as possible into the measurement system, also distortions due to non-ideal converters, loudspeaker non-linearities or amplifiers have to be considered in practice. In the digital-to-analog converter, for example, a steep low-pass filter for suppressing frequency components above the Nyquist frequency might introduce severe distortions due to overshooting effects [78]. In order to keep this influence low, the digital amplitude has to be reduced accordingly, which in turn results in a loss of SNR.

A wide variety of methods have been developed to estimate the transfer path or impulse response of static LTI systems. A good overview on various methods applied in the field of acoustics can be found in [78]. In this context, *maximum-length sequences*, MLS, [89] and nonperiodic sweeps are used most commonly. In the sense of maximum energy efficiency the MLS seems to be ideal. However, its rectangular waveform is difficult to reproduce by D/A converters and loudspeakers. As a consequence, its level has to be reduced by 5–8 dB below full scale to avoid clipping effects [78, 100]. In contrast, sweep signals provide a lower energy efficiency, but due to their specific waveform they better avoid nonlinear distortions such that higher energy, roughly 3 dB more, can be brought into the system with sweeps.

Sweeps can be created efficiently in the time or frequency domain, both resulting in specific properties. Linear and logarithmic sweeps are the most common types used in acoustic measurements. The former having a white spectrum and the latter a pink spectrum decaying with 3 dB/Octave. Additional preemphasis may be sensible in actual setups, for instance, to tailor the signal to the maximum power that a speaker can handle at higher frequencies or to fit it to the empirical noise floor, in turn retrieving measurements with frequency independent SNR. In the time domain, sweep signals have a constant amplitude, allowing for easy adjustment of the level to avoid clipping. The practical crest factor is around 3 dB. One remarkable property of the sweep technique is that the contributions resulting from the nonlinear behavior of the measurement chain can be partly identified and removed by time-domain windowing of the resulting impulse response [102]. When using logarithmic sweeps, one can identify the so-called *harmonic impulse responses* [41, 81], which represent the frequency responses corresponding to the nonlinear components. Overall, in static scenarios, sweep signals as excitation signals show a clear advantage over other ones when considering practical aspects like noise, time variance and nonlinear distortions [78].

Once the sweep signal is created, it is emitted through the measurement loudspeaker. The response at the ear-canal has to be captured synchronously. The frequency response is calculated by linear deconvolution, preferably in the frequency domain—that is, by spectral division. Note that, without countermeasures being applied, the frequency response of the electro-acoustic equipment will be part of the measured HRTF. In practice, it is thus advisable to equalize at least the loudspeaker response. Equalization can be incorporated into the preemphasis during sweep generation. Some authors define the HRTF as the ratio of the transfer functions from the source to the ear and the transfer function form the source to a position at the center of the head without the head being present—see, for example, [20]. This definition already incorporates the compensation of the electro-acoustic equipment used.

The achieved signal-to-noise ratio, SNR, of the calculated impulse response depends, amongst other things, on the length of the sweep signal, repetitions of the measurement, and the background noise. However, the required SNR range is still discussed amongst researchers. The target range is somewhere between 60–90 dB.

In recent years, the importance of *individual* HRIR data is growing rapidly, for example, for studies concerning acoustic source localization. For the acquisition of *individual* HRIRs it is of special interest to keep the measurement procedure as short as possible. One approach to speed up the procedure is to measure HRTFs from multiple directions of sound incidence simultaneously. The basic sweep technique has been extended for the simultaneous measurement of multiple transfer functions using overlapping or interleaved exponential sweeps [47, 69, 71, 84, 108]. An alternative approach considering dynamic measurement of HRTFs is discussed in Sect. 3.

## 2.3 Angular Interpolation and Range Extrapolation

It is obvious that the measurement of HRIR datasets becomes a complex task for a densely-sampled space of source positions. Therefore, most of the currently available datasets consider only a single measurement distance with a limited, sometimes not constant, angular resolution. Typical distances are in the range of 1–3 m with an angular resolution of 5° to 10°—see Sect. 1.2. It is generally assumed that for source distances above approximately 1 m the characteristics of HRTFs do not change significantly [20, 29]. In contrast, it is however known that HRIRs vary substantially with distance for nearby sources [28, 29, 58, 64, 96]. Such HRIRs are typically termed as *near-field* or *proximal* HRIRs.

A number of techniques have been proposed to compute HRIRs for arbitrary source positions from HRIR datasets measured at fixed distance and discrete angles. Two different classes can be distinguished, (i) angular interpolation and (ii) range/distance extrapolation techniques.

The first class considers computing HRIRs at angles in between two measured angles using the HRIRs from these two or even more angles. A wide range of interpolation techniques has been proposed in the past decades. The simplest method is to compute a weighted average of two or more neighboring HRIRs [73]. A number of alternatives to this basic approach can be found in the literature, of which only a few shall be mentioned here, namely, *frequency-domain techniques* [32] relying on minimum-phase representations of HRIRs, techniques using methods from *data-analysis and representation* [93], techniques using *functional representations* [118], or the approaches based on *virtual sound field synthesis* [80]. A comparisons of various interpolation methods can, for instance, be found in [31, 56].

In practical systems with moving virtual sources or dynamic auralization based on head-tracking, *cross-fading* of HRTFs is often applied [16]. Here the signals resulting from the convolution of a source signal with two different HRTFs are cross-faded in order to avoid audible artifacts. *Linear* or *raised-cosine windows* are often applied in this context. It has been shown that this method provides good results for horizontal-plane HRIRs with an angular spacing of up to 2° or more [67, 73]. Due to its simplicity, cross-fading is used in many practical implementations of auditory virtual environments based on HRIRs [46, 105].

Regarding the second class, range extrapolation techniques, approaches have been presented that modify the spectrum of measured HRTFs, for instance, by using simplified head models in order to cope for the spectral changes that occur for nearby sources [27, 61, 90]. However, these techniques neglect the structural changes for different source distances.

A number of advanced approaches have been published that are based upon extrapolation using principles from wave physics. In [34, 118] two techniques are presented which are based upon the expansion of HRIRs into *surface spherical harmonics*. These spherical harmonics form an orthogonal basis that can be used for joint angular interpolation and range extrapolation under free-field acoustic conditions. Therefore, extrapolating the HRIRs is only possible if no scattering objects are located within the extrapolation region. The application of these techniques to the computation of near-field HRIRs is, hence, limited to distances not including reflections from the upper torso. Methods based on spherical harmonics expansions are also challenging with respect to numerical implementation. The method reported in [34] was recently verified in [85] by comparing extrapolation results with the radial HRTF measurements known from [65] and with analytic HRTF solutions .

As an alternative to the approaches above, the distribution of measurement positions of a given HRTF dataset can be interpreted as a virtual loudspeaker array. When this array is considered as a unit, where each virtual loudspeaker is driven by an individual signal, the HRTFs from a desired virtual source to both ears can be synthesized. Various techniques have been published which are based upon the theory of *higher-order Ambisonics*, HOA, for the computation of the driving signals [72, 80]. The synthesis of virtual sources at closer distances than the virtual loudspeaker array—so-called focused sources—is, however, subject to numerical instabilities [1] in HOA. Most of these issues can be overcome by using the driving functions derived from *wave-field synthesis*, WFS, for the virtual secondary sources [98, 99].

## 3 Continuous-Azimuth HRTF Measurement

Depending on the desired resolution of an HRTF table, the discrete *stop-&-go* measurement procedure can be very time-consuming. Durations in the order of hours were reported. In individual HRTF acquisition, this would be an unacceptable burden on the person to be measured. The long duration is at least partly caused by the fruitless measurement pause during mechanical recalibration of the apparatus to each new location, that is, to each new solid angle in space. In this section, recent developments are thus outlined to overcome this fundamental limitation of discrete measurement technology. The development essentially consists of two concepts, while the goal is to achieve fast and comprehensive and, at the same time, consistent measurement in the azimuth and the elevation coordinate of the auditory sphere. The first concept in Sect. 3.1, a concept of mechanical nature, translates the desired high-resolution spatial measurement into a time-varying identification problem via dynamical apparatus. The corresponding time-varying-system model is developed in Sect. 3.2 and the related adaptive signal processing for HRIR extraction in Sect. 3.3. The second concept, in Sect. 3.4, uses multichannel adaptive identification for simultaneous measurement from several directions. Sects. 3.5–3.7 outline the results of an experimental validation, a set of guidelines for practical usage, and an HRTF-field representation of comprehensive HRTF data in a perspective plot.

### 3.1 Dynamical Apparatus

Several publications have expressed that discrete measurement in space is a necessary characteristic in the configuration of HRTF measurement systems—for example, [115, 119]. However, at least three other systems were proposed recently to overcome this *de-facto* standard in the context of HRTF acquisition [3, 36, 44]. All of them rely on the same and probably sole way to overcome the stop-&-go nature of discrete measurements, that is, via dynamical lab apparatus—generally using moving microphones or, more specifically to HRTF, continuously rotating persons or dummy heads equipped with binaural in-ear microphones.

The system to be discussed here is illustrated in Fig. 2 and represents a large-scale extension of the one proposed in [37]. It uses a circular loudspeaker array the individual channels of which are assigned to a set of discrete elevations. The subject of interest, an artificial head or real person, is placed in the center of the array at 1–2 m distance from the loudspeakers. The recording microphones are placed at the entrances of the blocked ear canals [52].

By rotating the person of interest or artificial head with uniform angular speed, while the loudspeakers continuously reproduce probe noise, the HRTFs of the whole azimuth-circle can be observed in finite time, that is, after completion of one revolution. The observation time per angle $\phi$ will be infinitesimally short in this configuration, and this naturally raises the question of how to extract the HRTF under these circumstances? In what follows, a novel time-varying-system model is described—developed by one of the current authors [36]—that is intended to pave the way for comprehensively extracting the HRTF or, more precisely, the *plenacoustic* HRTF on the circle [2].

### 3.2 Formal System Model: State-Space Model

At first, single-loudspeaker activity in the system is considered, that is, via activity caused by the signal $x_{\theta_\nu}(k) = x_{0^\circ}(k)$ in the horizontal plane, $\theta_\nu = 0^\circ$, at discrete

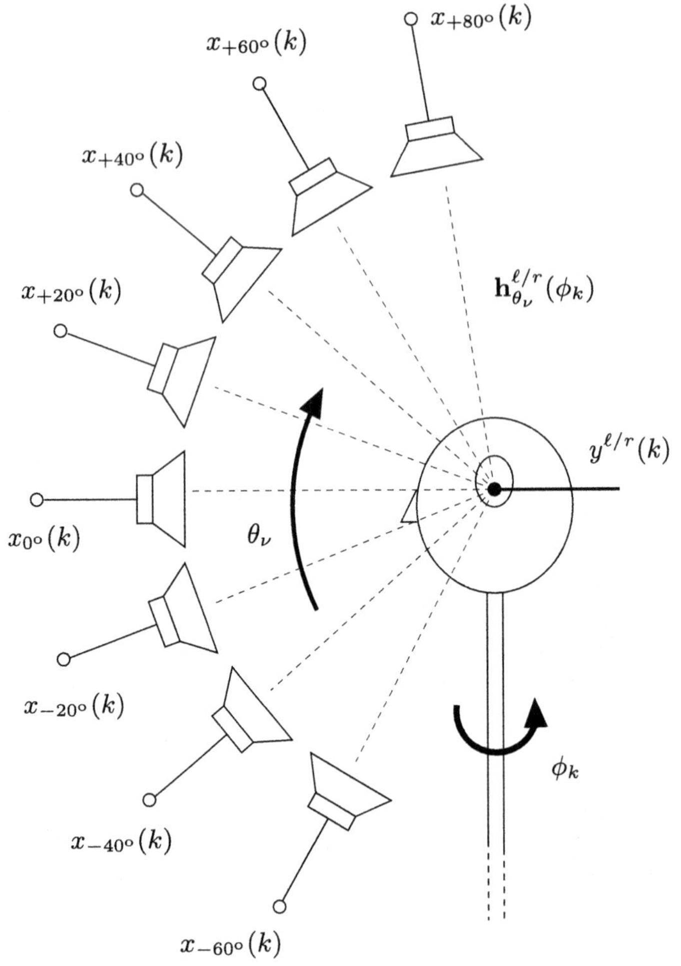


**Fig. 2** System with continuously rotating azimuth at Ruhr-Universität Bochum. *Left-side* view on the median plane of the auditory sphere. Loudspeakers are fixed

time $k$. The chosen loudspeaker emits a broadband probe signal, for instance, *white noise* or *perfect sequences*—see Sect. 4—and the signals $y^{\ell}_{\theta_\nu}(k)$ and $y^{r}_{\theta_\nu}(k)$ at the left ($\ell$) and right ($r$) ears are recorded. After a continuous and full 360° rotation of the test person, this binaural recording represents an observation of all possible HRIRs on the azimuth-circle under consideration. Generalization to multiple-loudspeaker activity follows in Sect. 3.4.

Assuming sufficiently linear and broadband transducers, the acoustic transmission of the loudspeaker signal to the rotating binaural receiver can be described as a slowly time-varying linear system with impulse responses, $h^{\ell/r}_{\kappa,\theta_\nu}(\phi_k)$, at the impulse-response index $\kappa$. These HRIRs then formally relate input and output of the corresponding system according to a linear convolution model using the convolution index $\kappa$, that is,

$$y_{\theta_\nu}^{\ell/r}(k) = \sum_{\kappa=0}^{N} x_{\theta_\nu}(k-\kappa) h_{\kappa,\theta_\nu}^{\ell/r}(\phi_k) + n^{\ell/r}(k), \tag{1}$$

where $N$ denotes the length or, in other words, the memory of the HRIRs. The symbol $n^{\ell/r}(k)$ models independent observation noises, for instance, microphone noise, microphone-amplifier noise, or ambient noise at the recording positions. The quasi-continuous azimuth, $\phi_k = \omega_o k T_s$, at time $k$, evolves with constant angular speed, $\omega_o = 2\pi / T_{360}$, where $T_{360}$ is the duration of a 360° revolution and $T_s = 1/f_s$ is the temporal sampling interval. In this time-varying-system model, every time index $k$ thus corresponds to an angle $\phi_k$, or, vice versa, every angle $\phi_k$ corresponds to a time index $k$. Assuming a revolution time on the order of 1 min and using a sampling frequency of $f_s = 44.1$ kHz, it can be easily found that the effective azimuth-spacing, $\partial\phi = \phi_{k+1} - \phi_k = 2\pi / T_{360} / f_s$, then lies on the order of $\partial\phi \approx 10^{-4}$ degrees, which is extremely small.

Then, let

$$\mathbf{x}_{\theta_\nu}(k) = \left(x_{\theta_\nu}(k), x_{\theta_\nu}(k-1), \ldots, x_{\theta_\nu}(k-N+1)\right)^T \tag{2}$$

denote a vector of $N$ most recent HRIR input samples, that is, the loudspeaker-input samples, and let

$$\mathbf{h}_{\theta_\nu}^{\ell/r}(\phi_k) = \left(h_{0,\theta_\nu}^{\ell/r}(\phi_k), h_{1,\theta_\nu}^{\ell/r}(\phi_k), \ldots h_{N-1,\theta_\nu}^{\ell/r}(\phi_k)\right)^T \tag{3}$$

represent the corresponding vector of the HRIR coefficients as defined by (1). Using this notation, the binaural-signal model in (1) can be formally written via the inner vector product, that is, with dropping the convolution index,

$$y_{\theta_\nu}^{\ell/r}(k) = \mathbf{x}_{\theta_\nu}^T(k) \mathbf{h}_{\theta_\nu}^{\ell/r}(\phi_k) + n^{\ell/r}(k). \tag{4}$$

Looking at the time-varying HRIR, $\mathbf{h}_{\theta_\nu}^{\ell/r}(\phi_k)$, as the unknown state of the apparatus and at $y_{\theta_\nu}^{\ell/r}(k)$ as the state observation, (4) can be readily termed a *state-observation equation*.

In order to express the time-varying nature of the unknown system in formal terms, the concept of *first-order recursive Markov modeling* is made use of, namely,

$$\mathbf{h}_{\theta_\nu}^{\ell/r}(\phi_{k+1}) = a \cdot \mathbf{h}_{\theta_\nu}^{\ell/r}(\phi_k) + \Delta\mathbf{h}_{\theta_\nu}^{\ell/r}(\phi_k), \tag{5}$$

where two consecutive states at times $k$ and $k+1$ are related to each other by the transition coefficient, $0 \leq a \leq 1$, and the independent process-noise quantity, $\Delta\mathbf{h}_{\theta_\nu}^{\ell/r}(\phi_k)$, with covariance $\boldsymbol{\sigma}_\Delta^2 = \mathrm{E}\{\Delta\mathbf{h}_{\theta_\nu}^{\ell/r}(\phi_k)\Delta\mathbf{h}_{\theta_\nu}^{\ell/r}(\phi_k)^T\}$. This Markov model ideally represents a quantity $\mathbf{h}_{\theta_\nu}^{\ell/r}(\phi_k)$ which gradually changes between time instants $k$ and $k+1$ in an unpredictable way. This behavior may not be exactly in agreement with the nature of continuously rotating HRIRs. Lacking a more specific and deterministic model, this simple stochastic framework will in the following be considered as a good basis with presumably sufficient modeling ability.

Equations (4) and (5) together form a linear dynamical model, that is a state-space model, of the unknown state $\mathbf{h}_{\theta_\nu}^{\ell/r}(\phi_k)$. Further assuming normally distributed observation and process noises, $n^{\ell/r}(k)$ and $\Delta\mathbf{h}_{\theta_\nu}^{\ell/r}(\phi_k)$, respectively, (4) and (5) represent a *Gauss-Markov* dynamical model. The linear minimum mean-square error estimate, $\widehat{\mathbf{h}}_{\theta_\nu}^{\ell/r}(\phi_k)$, of the unknown state, $\mathbf{h}_{\theta_\nu}^{\ell/r}(\phi_k)$, at time $k$, given the observations, $y_{\theta_\nu}^{\ell/r}(k)$, up to and including time $k$, that is, the conditional mean, subject to the Gauss-Markov model, $\widehat{\mathbf{h}}_{\theta_\nu}^{\ell/r}(\phi_k) = \mathrm{E}\{\mathbf{h}_{\theta_\nu}^{\ell/r}(\phi_k) \mid y_{\theta_\nu}^{\ell/r}(k),\ y^{\ell/r}(k-1),\ \ldots\}$, can then be computed on the basis of conventional *Kalman filtering* [57, 60, 94, 103].

## 3.3 *Single-Channel Adaptive Algorithm for Spatially-Continuous HRIR Extraction*

While the established Kalman filter is known as the best linear state-tracking device, both the underlying model as well as the Kalman-filter algorithm and the corresponding resource requirement turn out to be a bit oversized for the problem at hand. Thus, we revert here to a special case of the Kalman filter, the *broadband Kalman filter*, BKF, as introduced in [38]. This filter type is ideally suited for the current application with broadband input, namely, white-noise, $x_{\theta_\nu}(k)$, into the unknown acoustic system. It consists of the following set of recursive and iteratively coupled equations, in which $\widehat{\mathbf{h}}_{\theta_\nu}^{\ell/r}(\phi_k)$ denotes the predicted, that is, *a-priori*, and $\widehat{\mathbf{h}}_{\theta_\nu}^{\ell/r,+}(\phi_k)$ the corrected, that is, *a-posteriori*, state estimate,

$$\widehat{\mathbf{h}}_{\theta_\nu}^{\ell/r}(\phi_{k+1}) = a \cdot \widehat{\mathbf{h}}_{\theta_\nu}^{\ell/r,+}(\phi_k), \tag{6}$$

$$p(k+1) = a^2 \cdot p^+(k) + \sigma_\Delta^2, \tag{7}$$

$$e^{\ell/r}(k) = y_{\theta_\nu}^{\ell/r}(k) - \mathbf{x}_{\theta_\nu}^T(k)\widehat{\mathbf{h}}_{\theta_\nu}^{\ell/r}(\phi_k), \tag{8}$$

$$\widehat{\mathbf{h}}_{\theta_\nu}^{\ell/r,+}(\phi_k) = \widehat{\mathbf{h}}_{\theta_\nu}^{\ell/r}(\phi_k) + \mathbf{k}(k)e^{\ell/r}(k), \tag{9}$$

$$p^+(k) = \Big(1 - \mathbf{x}_{\theta_\nu}^T(k)\mathbf{k}(k)/N\Big)p(k), \tag{10}$$

$$\mathbf{k}(k) = p(k)\mathbf{x}_{\theta_\nu}(k)\Big(p(k)\mathbf{x}_{\theta_\nu}^T(k)\mathbf{x}_{\theta_\nu}(k) + \sigma_{n,\ell/r}^2(k)\Big)^{-1}. \tag{11}$$

In the BKF, the quantity $\sigma_\Delta^2$ represents a diagonal process-noise covariance, $\boldsymbol{\sigma}_\Delta^2 = \sigma_\Delta^2\mathbf{I}$, in place of a generally fully-populated process-noise-covariance matrix in the original Kalman filter. The symbol $\sigma_{n,\ell/r}^2(k)$ denotes the observation-noise power at the left and right ears. As a result of the aforementioned assumptions in the derivation of the BKF, a scalar $p(k)$ is sufficient instead of the state-error covariance matrix in a full-fledged Kalman filter.

These properties of the broadband Kalman filter render the desired state estimation highly efficient, both computationally and in terms of memory requirements. The depicted algorithm simply consists of inner vector products and scalar multiplications/additions instead of full-fledged matrix algebra. The expected ability to handle time-varying unknown systems in the continuous presence of possibly time-varying observation-noise levels is, however, fully preserved in this simplification of the original Kalman filter.

Before actually running this algorithm, a further Kalman filter simplification shall be applied, which is justified by the fact that HRIR/HRTF acquisition usually takes place in a controlled acoustic environment, such as anechoic chambers and, thus, very low acoustic-observation noise can be assumed. Formally, this allows to neglect the observation-noise power, $\sigma^2_{n,\ell/r}(k)$, in the computation of the Kalman gain, $\mathbf{k}(k)$, according to (11). This, in turn, will entirely cancel the state-error covariance, $p(k)$, from the Kalman gain and, thus, avoid the need for the state-error recursion in (7) and (10). In this way, even the relevance of the process-noise-covariance parameter, $\sigma^2_\Delta$, in (7) disappears. By finally considering a state-transition factor, $a$, close to unity in the Markov model (5), as said in [57], even more degeneration of the algorithm can be easily achieved. After formal rearrangements, that is, by substituting (8) and (11) into (6), the Kalman filter eventually collapses into the celebrated *normalized least-mean square*, NLMS, adaptive algorithm as has been used intuitively for iterative extraction of rotating HRIRs [36]. In order to compensate for the neglected observation-noise power, $\sigma^2_{n,\ell/r}(k)$, a fixed stepsize factor, $0 < \mu_0 < 1$, is added to restore at least some smoothing and, thus, noise-rejection ability of the original Kalman filter as follows,

$$\widehat{\mathbf{h}}^{\ell/r}_{\theta_\nu}(\phi_{k+1}) = \widehat{\mathbf{h}}^{\ell/r}_{\theta_\nu}(\phi_k) + \mu_0 \frac{e^{\ell/r}(k)\mathbf{x}_{\theta_\nu}(k)}{||\mathbf{x}_{\theta_\nu}(k)||_2^2}, \tag{12}$$

$$e^{\ell/r}(k) = y^{\ell/r}_{\theta_\nu}(k) - \mathbf{x}^T_{\theta_\nu}(k)\widehat{\mathbf{h}}^{\ell/r}_{\theta_\nu}(\phi_k). \tag{13}$$

Results of this *1-channel* HRIR-extraction algorithm are reported after generalization of the current *horizontal* HRIR-setting to the full auditory sphere.

## 3.4 Multichannel 3-D Continuous-Azimuth Algorithm

Generalization is achieved by activating all elevation speakers in Fig. 2 to simultaneously reproduce uncorrelated broadband probe noises. Let

$$\mathbf{x}(k) = \left(\mathbf{x}^T_{+80^\circ}(k), \ldots, \mathbf{x}^T_{-60^\circ}(k)\right)^T \tag{14}$$

denote a length $8N$ stacked version of all HRIR input signals and let

$$\mathbf{h}^{\ell/r}(\phi_k) = \left(\mathbf{h}_{+80^\circ}^{\ell/r,T}(\phi_k), \ldots, \mathbf{h}_{-60^\circ}^{\ell/r,T}(\phi_k)\right)^T \tag{15}$$

be the corresponding stacked version of all HRIR channels. These definitions then support a multichannel linear-convolution model of the observation, namely,

$$y^{\ell/r}(k) = \sum_{\theta_\nu} \mathbf{x}_{\theta_\nu}^T(k)\mathbf{h}_{\theta_\nu}^{\ell/r}(\phi_k) + n^{\ell/r}(k) \tag{16}$$

$$= \mathbf{x}^T(k)\mathbf{h}^{\ell/r}(\phi_k) + n^{\ell/r}(k), \tag{17}$$

in line with the previous single-channel model in (4). The stacked multichannel quantities can, therefore, be substituted in place of the respective single-channel quantities in (12) and (13) in order to obtain a multichannel NLMS algorithm for simultaneous HRIR identification. After undoing the stacking one arrives at

$$\widehat{\mathbf{h}}_{\theta_\nu}^{\ell/r}(\phi_{k+1}) = \widehat{\mathbf{h}}_{\theta_\nu}^{\ell/r}(\phi_k) + \mu_0 \frac{e_c^{\ell/r}(k)\mathbf{x}_{\theta_\nu}(k)}{\sum_{\theta_\nu} ||\mathbf{x}_{\theta_\nu}(k)||_2^2} \quad \forall\ \theta_\nu, \tag{18}$$

$$e_c^{\ell/r}(k) = y^{\ell/r}(k) - \sum_{\theta_\nu} \mathbf{x}_{\theta_\nu}^T(k)\widehat{\mathbf{h}}_{\theta_\nu}^{\ell/r}(\phi_k). \tag{19}$$

The lesson from this derivation is that the multi-channel algorithm can be executed almost independently for each elevation/channel, $\theta_\nu$, except for utilization of the summing-normalization, $\sum_{\theta_\nu} ||\mathbf{x}_{\theta_\nu}(k)||_2^2$, and the compound error signal, $e_c^{\ell/r}(k)$, which are in common with the individual update equations (18). It is further instructive to notice that simultaneous operation of multiple loudspeakers will not cause mutual noise or disturbance in the cannel-wise HRIR identification. The multiple adaptive filters, $\mathbf{h}_{\theta_\nu}^{\ell/r}$, $\forall\theta_\nu$, rather cooperate in minimizing the error signal according to (19).

## *3.5 Numerical Results for Validation*

As a first validation of the dynamical HRIR measurements, Fig. 3 depicts the achievable error-signal attenuation, ESA $= 10\log_{10}\sigma_e^2/\sigma_y^2/N$ dB, as a function of the revolution time for different numbers of active elevation channels. Here, $\sigma_e^2$ denotes the variance of signal $e(k)$. The ESA basically describes the success of the algorithm in reducing the recorded signal, $y^{\ell/r}(k)$, or, in other words the usability of the HRIR estimate, $\widehat{\mathbf{h}}_{\theta_\nu}^{\ell/r}$, to regenerate the recorded signal and, for this reason, the accuracy of the HRIR estimate. Lower ESA is desirable as it indicates more successful HRIR

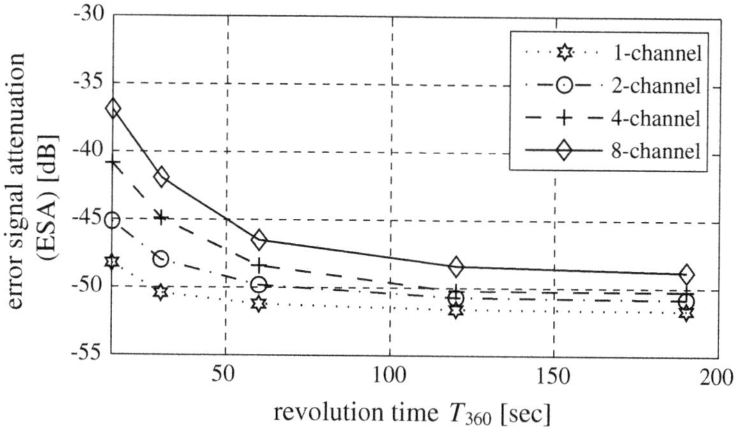


**Fig. 3** Error-signal attenuation, ESA, for dynamical HRIR measurement. Results for the left ear. Measurement loudspeaker located in the left-front quadrant at 2 m distance from a rotating artificial head. 1-channel result in the horizontal plane. $\mu_0 = 0.5$, $N = 308$, $f_s = 44.1\,\text{kHz}$

identification. The excitation signal used for achieving the experimental results in Fig. 3 is white noise.

The results demonstrate, with slight dependency on the number of simultaneously measured elevation channels, that a revolution time in the order of 1 min represents an interesting working point. Faster rotation significantly degrades ESA, because the adaptive algorithm looses track of the time-varying system when it varies too heavily. Slower rotation than 1 min does not yield significant advantages either. Note that the particular saturation of the results at $-52\,\text{dB}$ is due to undermodeled low-frequency acoustic-wave propagation in the anechoic chamber when using finite adaptive-filter length [107]. For the results shown here, $N = 308$ at 44.1 kHz sampling was employed. While considering the ESA, it might be helpful to argue that $\sigma_e^2$ not only evaluates the HRIR and its accuracy but also its spectral counterpart, the HRTF—according to Parseval's theorem. Finally, it should be noted that the angular HRIR resolution is quasi-infinite due to continuous-azimuth identification.

## *3.6 Usability of Continuous-Azimuth HRIRs*

Due to the different nature of continuous HRIR measurements, as compared to discrete HRIR acquisition, some practical aspects have to be addressed in order to make use of the continuous HRIR data. Consider the following typical example. Assuming a sampling frequency of $f_s = 44.1\,\text{kHz}$ and a measurement duration of $T_{360} = 20\,\text{s}$ for a 360° rotation, a set of $T_{360} \cdot f_s = 882000$ HRIRs becomes available just for the horizontal plane. This corresponds to an extremely high resolution of 2450 HRIRs/degree. Complete storage of the HRIRs for two channels, assuming 16 bit resolution and an HRIR length of $N = 308$ coefficients, would result in a memory requirement of $f_s \cdot T_{360} \cdot N \cdot 2 \cdot 2\,\text{Byte} = 1.1\,\text{GB}$. This huge amount of data is not feasible in most of the practical applications. Direct storage of continuous-azimuth HRIRs is thus not recommended.

An option to handle the tremendous amount of HRIR data more efficiently is to calculate HRIRs off-line from the ear signals and to sample out and store, for instance, every $N = 308$th HRIR only. As a result, still 8 HRIRs/degree are available. Due to this still very high HRIR resolution, a simple linear interpolation of the impulse responses itself is absolutely sufficient when intermediate HRIRs are needed for auralization or other purpose. This strategy basically resembles the one known from spatially discrete HRIR measurement and usage of HRIR tables. Hence, from here, all the other tool-chain known to make use of discrete HRIRs in binaural sound technology can be employed. The discrete and continuous HRIR/HRTF technologies therefore converge in the sense that the benefits of both worlds merge towards a fast and individual measurement with efficient storage and ease-of-use.

As an example of successful usage of continuous HRTF measurements, we refer to the implementation of a binaural rendering engine for multiple moving sound sources. Figure 4 depicts the graphical user-interface of a realtime demonstrator with six rotating audio objects, such as a singer, a guitar, etc., and a stereophonic virtual-loudspeaker arrangement. This system relies on a frame-based HRIR-filtering concept with signal-based cross-fading as described in Sect. 2.3. Alternatively, fast convolution in the block-frequency domain can be applied when further reduction

**Fig. 4** Real-time binaural rendering demonstrator based on continuously measured HRIRs . Presented at Intl. Workshop on Acoustic Signal Enhancement, Aachen, Sept. 2012

of the computational complexity is desired. In any case, the sampled version of the continuous-HRIR data in conjunction with *simple and uncritical* cross-fading was highly appreciated in informal listening demonstrations.

## 3.7 HRTF-Field Representation

With typically $\partial\phi \approx 10^{-4}$ degrees azimuth spacing, as calculated in Sect. 3.2 for about 1 min revolution time, the adaptive algorithm performs a quasi-continuous HRIR extraction. This finally offers an interesting opportunity of representing the HRIR data as a continuous HRIR-field in space.

By theoretically considering a sound impulse emitted from the left or right ear canal, and by using the principle of acoustic reciprocity [21], the *HRIR-field*, $h^{\ell/r}_{\kappa,\theta}(\phi, r)$, would be observed as the response in the $(\theta, \phi, r)$-space, where $r$ denotes distance from the head-center. The field exists independently for the left and right ear and is illustrated in Fig. 5 by the spherical wave, centered at the left-ear position. From now on, just the horizontal plane is considered for simplicity.

On this basis, the corresponding soundfield, $p_{\theta=0^\circ}(\omega, \phi, r)$, in the frequency-domain, that is,

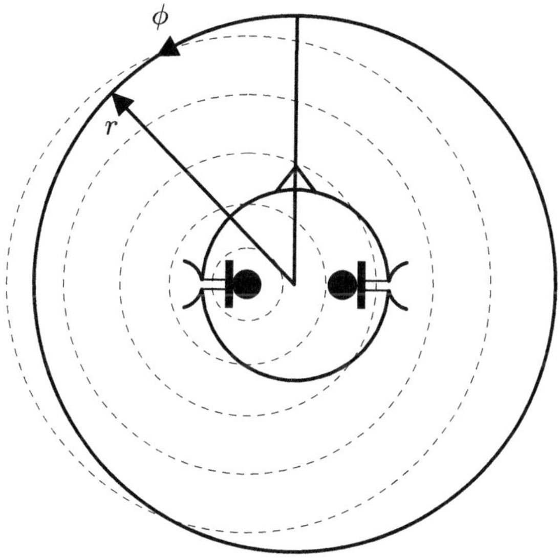


**Fig. 5** The left-ear reciprocal HRTF-field, *dashed circles*, and the head-centered cylindrical $(\phi, r)$-coordinate system, *solid*—adapted from [39]

$$p_{0^\circ}(\omega, \phi, r) = \mathcal{F}\left\{h_{\kappa,0^\circ}(\phi, r)\right\} = \sum_{\kappa=0}^{N-1} h_{\kappa,0^\circ}(\phi, r)e^{-j\omega T_s \kappa}, \tag{20}$$

is obtained via the temporal Fourier transform and then defines our *reciprocal HRTF-field* in the horizontal plane. Here, left- and right-ear indices, $\ell/r$, were dropped for brevity. The actual HRTF-field, $p_{0^\circ}(\omega, \phi, r_m)$, on the HRIR-measurement radius $r_m$ from the head center can be calculated by substituting the measured HRIR, $\widehat{h}_{\kappa,0^\circ}(\phi, r_m)$, from Sect. 3.3 in place of $h_{\kappa,0^\circ}(\phi, r)$.

Relying on this concept of the HRIR- or HRTF-field, the acoustic wave-equation can be invoked for this exterior problem, and its general solution can be written as a modal series in polar coordinates [116], namely,

$$p_{0^\circ}(\omega, \phi, r) = \sum_{n=-\infty}^{\infty} A_n(\omega) H_n^{(2)}(kr) e^{jn\phi}, \tag{21}$$

where $H_n^{(2)}(x)$ is the complex Hankel function of the 2nd kind, $k = \omega/c$ the acoustic wavenumber, $c$ the speed of sound, $n$ the modal index, and $A_n$ the independent set of modal coefficients. This HRTF-field representation is tailored to just the horizontal plane, as it assumes independence from the height-coordinate of the $(\phi, r)$-cylinder.

For a given radius, $r = r_m$, and a frequency, $\omega$, the complex Fourier series in (21), with $2\pi$-periodicity in $\phi$, then corresponds to a Fourier analysis of the HRTF-field, $p_{0^\circ}(\omega, \phi, r_m)$, along the azimuth coordinate, that is,

$$A_n(\omega) H_n^{(2)}(kr_m) = \frac{1}{2\pi} \int_0^{2\pi} p_{0^\circ}(\omega, \phi, r_m) e^{-jn\phi}\, d\phi, \tag{22}$$

where $P_n(\omega, r) = A_n(\omega) H_n^{(2)}(kr)$ is termed the *angular wave spectrum*. The modal coefficients, $A_n$, can be finally determined via normalization to the known Hankel function, $H_n^{(2)}(kr_m)$.

Using the quasi-continuous HRIR data, a quasi-continuous and direct evaluation of the Fourier integral in (22) can be achieved. This has been very uncommon in the past, however, it circumvents the earlier issues of spatial aliasing known from spatial Fourier analysis of discrete HRIRs. Figure 6 depicts an example, $P_n(\omega = 2\pi f\,,\ r = r_m = 2\,\text{m})$, obtained from a single-channel continuous-azimuth HRIR measurement in the horizontal plane. A band-limited angular wave spectrum, $P_n(\omega, r)$, in other words, a frequency-dependent finite number of modal coefficients, seems to be sufficient to represent continuous HRIR data very well. Conversely, the availability of this band-limited angular spectrum, for instance, taken from a computer file, would allow complete reconstruction of the continuous HRTF-field via (21), see for example, [39]. However, it should at least be noticed that the spherical or angular wave spectrum theoretically exhibit infinite spatial bandwidth, especially in the near-field due to evanescent components [116].

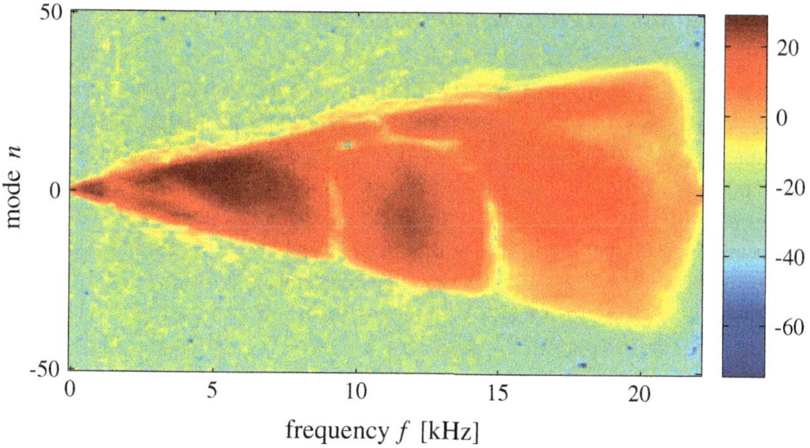


**Fig. 6** Left-ear HRTF-field coefficients, $P_n(\omega, r = 2\,m)$, in dB. Horizontal plane. See [39]

# 4 Excitation Signals for Dynamic HRTF Measurements

In Sect. 3, the continuous-azimuth-HRTF acquisition was treated as a time-varying system-identification problem based on dynamical measurements. Thus, compared to traditional static measurement scenarios of HRTFs, the excitation signals discussed below represent new trends as they are optimized in the dynamic context. Besides the constraints regarding SNR and nonlinear distortion, the excitation signals now have to be designed such that they meet the requirements of the NLMS adaptation process, too. In particular, a rapid-tracking ability of the NLMS process is essential in order to cope with the *time-variant*, that is, dynamic, measurement scenario.

The tracking ability of the NLMS algorithm is determined by the stepsize, $\mu_0$, the filter length, $N$, and the correlation properties of the excitation signal. The choice of a smaller stepsize smoothes the adaptation process. This leads to more robustness against observation noise, yet, at the expense of slower convergence. Another conflicting requirement arises in dimensioning the filter length. On the one-hand, the filter length $N$ should be large enough to avoid undermodeling of acoustic late reflections. In case of an unknown system with memory longer than $N$, the adaptive filter, $\widehat{\mathbf{h}}_{\theta_\nu}^{\ell/r}$, would otherwise suffer from effective noise, in case that white-noise excitation is used, or converge to a modulo-$N$ time-aliased version of $\mathbf{h}_{\theta_\nu}^{\ell/r}$, when $N$-periodic excitation is employed [10]. On the other-hand, a larger filter length generally slows down the identification process [57]. Finally, the correlation properties of the excitation signal have significant effect on the performance of NLMS.

## 4.1 Orthogonal Periodic Excitation Signals

It is well understood that colored input signals reduce the convergence speed [57]. In the extreme, if frequency components are not excited, they cannot be identified. Thus, a broadband excitation signal is essential. Besides the classical white-noise stimulus, an alternative class of excitation signals, the so-called *perfect sequences*, PSEQs, has been proposed [9, 10]. PSEQs are periodic, pseudo-noise sequences with an ideal impulse-like periodic autocorrelation function. Let $M$ be the length of the period and set $M = N$. Then, $M$ consecutive excitation vectors, $\mathbf{x}_{\theta_\nu}(k), \mathbf{x}_{\theta_\nu}(k-1), \ldots, \mathbf{x}_{\theta_\nu}(k-M+1)$, of length $M$ are ideally orthogonal in the $M$-dimensional vector space. Thus, the DFT of one period, $x_{\theta_\nu}(k-M+1)\ldots x_{\theta_\nu}(k)$, of the PSEQ results in an exactly constant frequency spectrum. With the specific correlation properties of the PSEQ excitation, the NLMS algorithm exhibits the same rapid tracking ability as the *recursive least-squares*, RLS, algorithm. In other words, PSEQs represent the *optimal* excitation signal of the NLMS algorithm in the sense of fast convergence and rapid tracking.

In the context of time-varying acoustic-system identification, PSEQs have already been applied in different kinds of applications. Two examples of previous works are briefly mentioned, namely:

- Room impulse responses, RIRs, with time-varying fluctuations were measured via PSEQs and then reused to reproduce real room conditions, such as for benchmarking of acoustic echo-cancellation prototypes [11].
- In a medical application, the dynamic behavior of transfer functions between nose and ears were measured during yawning or swallowing. This allows for real-time monitoring of the Eustachian-tube activity under physiological conditions [14, 15, 100].

So far, two different types of PSEQs have been accomplished, the *odd-perfect ternary sequences* [8, 9] and, more recently, the *perfect sweeps* [13, 100].

### Odd-Perfect Ternary Sequences

*Odd-perfect ternary sequences*, Odd-PSEQs, [68], while actually belonging to the more general class of ternary sequences, are characterized by being symmetric with only a single zero per period. As a result, Odd-PSEQs are almost binary and exhibit a high energy efficiency $\eta$. Generally

$$\eta = \left( \sum_{\kappa=0}^{M-1} x_{\theta_k}^2(\kappa) \right) \Big/ \left( M \max_{\forall \kappa} x_{\theta_k}^2(\kappa) \right), \tag{23}$$

which amounts to $\eta = (M-1)/M$ for Odd-PSEQs, that is, $\eta = 0.9968$ for $M = 308$. This equivalently means a low crest factor, $1/\eta$. Odd-PSEQs can be constructed for every length, $M = q^w + 1$, where $q$ denotes a prime number, $q > 2$, and $w \in \mathbb{N}$.

As the length $M$ has to match the length of the adaptive filter, cf. [9], it is important that they exist for a wide range of lengths.

In contrast to classical ternary PSEQs, such as *Ipatov* sequences [59], Odd-PSEQs have to be applied periodically in an *odd-cyclic* manner, which means that the sign is alternated in each period and it possesses a periodic *odd* autocorrelation as follows,

$$\varphi_{x_{\theta_\nu} x_{\theta_\nu}}(\lambda) = \sum_{i=0}^{M-1} x_{\theta_\nu}(i)\, x_{\theta_\nu}(\lambda + i) = \begin{cases} \|\mathbf{x}_{\theta_\nu}(\lambda)\|^2 & \lambda \bmod 2M = 0 \\ -\|\mathbf{x}_{\theta_\nu}(\lambda)\|^2 & \lambda \bmod 2M = M \\ 0 & \text{otherwise}, \end{cases} \tag{24}$$

which depicts their qualification as an optimal orthogonal excitation signal.

## Perfect Sweeps

Sweep signals, too, can be constructed such that they meet the orthogonality requirements for an optimal excitation. For construction in the frequency domain, a constant magnitude is first assumed. Furthermore, the group delay, that is, the instantaneous frequency, of a linear sweep has to increase linearly. Its phase therefore rises quadratically with frequency according to

$$X_n = \begin{cases} e^{-j\frac{2\pi}{M}n^2}, & 0 \le n \le \frac{M}{2} \\ X^*_{M-n}, & \frac{M}{2} < n < M, \end{cases} \tag{25}$$

where $X_n$ denotes the DFT coefficient at frequency bin $n$, and $X^*$ being the complex conjugate. The so-called *perfect sweep* is eventually obtained via IFFT [100, 101]. In contrast to Odd-PSEQs, perfect sweeps can be designed for any length $M$.

Based on this explicit design procedure, each frequency bin of the unknown system will be uniformly excited and an impulse-like periodic autocorrelation function is obtained. Thus, in the sense of *optimal* excitation for the NLMS algorithm, all requirements are fulfilled and, as already mentioned and following our previous terminology, these special sweep signals are named perfect sweeps—representing a special class of PSEQs.

In contrast to Odd-PSEQs, the energy efficiency of perfect sweeps is much lower, for instance, $\eta = 0.4888$ for $M = 308$. Thus, using the same digital amplitude for both excitation signals, a loss of 3 dB in SNR is encountered for the perfect sweep. However, for sound reproduction, the perfect sweep is superior in terms of robustness against nonlinear acoustic distortions or, in other words, perfect sweeps can be passed with less distortion through the measurement system than Odd-PSEQs. In [78, 100], it is shown that for pseudo-random sequences, such as MLS or Odd-PSEQs, a headroom of at least 5–8 dB below full scale has to be maintained to avoid large distortion. The use of perfect sweeps will allow to excite the unknown acoustic system, in our case the continuous HRIR measurement setup, with considerably more power while keeping the influence of non-linear distortions reasonably small. This is demonstrated by the following, application independent, basic experiment.

### Experiment

As an example, the effect of distortions due to level-dependent loudspeaker nonlinearities is investigated. To this end, static measurements of RIRs are performed in a professional studio box with different types of loudspeakers at 44.1 kHz sampling rate. The digital amplitude of the excitation signal is chosen such that the D/A converter of the measurement setup can be assumed to be sufficiently linear. White noise, Odd-PSEQs, and perfect sweeps are considered as excitation signals. They are normalized to the same signal power, thus showing different maximum amplitudes, but providing for the same SNR at the recording microphone.

Figure 7 depicts the achievable *error signal attenuation*, ESA, of an NLMS adaptive RIR filter as a function of the *sound pressure level*, SPL. Note that increasing the SPL, for instance, by 5 dB, goes along with a 5 dB SNR increase at the microphone. Consequently, a 5 dB lower ESA value would be expected in case of linear systems [57]. It can be seen that the results for Odd-PSEQ excitation exhibit such a linear ESA behavior only for the Genelec high-end loudspeaker. However, the ESA curves obtained for perfect-sweep excitation always show almost linear behavior, despite the nonlinearity at large SPL. White-noise excitation, as a reference, limits the ESA with both speakers.

Since nonlinear distortion is generally less pronounced at lower SPL, the curves for the different excitation signals naturally approach each other. However, this obviously causes higher ESA values due to lower SNR at the microphone. By using perfect-sweep excitation, considerably more power can be fed into the loudspeakers in order to maintain high SNR and, thus, low ESA, but still without suffering too much from nonlinear distortions in the identification process. These results demonstrate a significant influence of the excitation signal, which is relevant for HRIR measurements likewise.

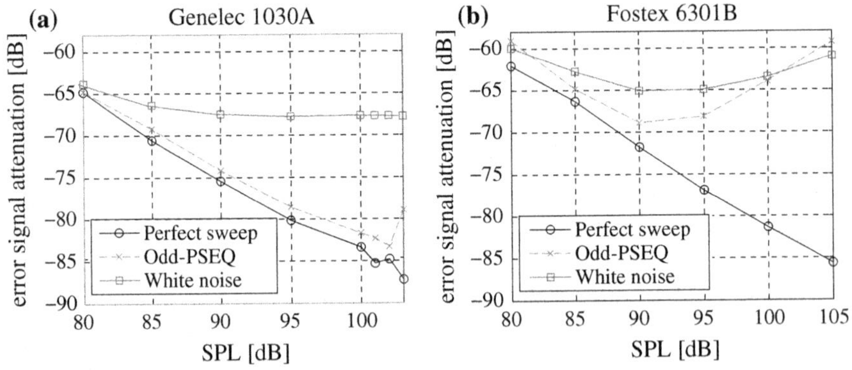


**Fig. 7** Error-signal attenuation, ESA, for static RIR measurements with different excitation signals for two different loudspeakers with $N = 3000$, $\mu_0 = 1$

## 4.2 Performance of PSEQs in Dynamic HRIR Measurements

In this section the influence of different excitation signals in dynamic HRIR measurements is investigated [10, 13]. The measurements are performed with the rotating system according to Fig. 2, however, using only the loudspeaker in the horizontal plane, that is, the excitation signal $x_{\theta_\nu} = x_{0^\circ}$. The loudspeaker is again stimulated with our three different excitation signals, namely, white noise, Odd-PSEQ, and perfect sweep, using the same signal power and 44.1 kHz sampling frequency. The revolution time of the measurement apparatus is set to $T_{360} = 20$ s. In order to focus on dynamical characteristics of HRTF acquisition, the digital amplitude and the sound-pressure level are carefully chosen to keep the effect of nonlinear distortion as small as possible for all excitation signals—based on the previous experiment. The reaction of the system is obtained in terms of the two microphone signals recorded at the left and right ear-canal entrances of an artificial head.

Figure 8 shows local ESA results for a complete continuous revolution of a dummy head, including some rotation in advance for initialization and some overrun at the end. Over the 360°, characteristic fluctuation of the curves, with symmetry between the ears, is observed. Lowest ESA values are achieved at 270° for the left ear and 90° for the right ear, since, in these cases, the respective microphone picks up the most direct sound from the loudspeaker. Highest SNR in these directions in fact explains the lowest ESA values.

In the comparison of different excitation signals, the perfect sweep outperforms the other excitation signals by far. This is due to the fact that a perfect sweep unites the orthogonality requirements for *optimal* NLMS excitation with the favorable characteristics of sweep signals in general. As a result, perfect sweeps enable rapid system tracking as well as high robustness against nonlinear distortions. This confirms the theoretical considerations from Sect. 2.2 and the experimental results from Sect. 4.1.

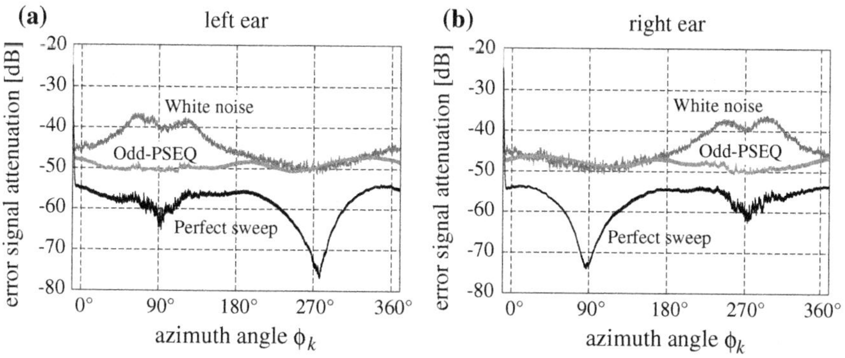


**Fig. 8** Error-signal attenuation, ESA, for dynamic HRIR measurements with $N = M = 308$, $\mu_0 = 1$, $T_{360} = 20$ s

Using the same HRIR measurement setup as before, static measurements were additionally performed for different angles. At $\phi = 270°$, left ear, for instance, ESA values of −79.6 dB, −51.3 dB, and −49.1 dB were obtained with perfect sweep, Odd-PSEQ, and white noise, respectively. A comparison of these numbers with those in Fig. 8a verifies that fast dynamic measurements do line up with traditional static measurements in terms of precision.

## 4.3 3-D Continuous-Azimuth HRIR Acquisition

In order to generalize the previous single-channel considerations, the multichannel case will now finally be discussed, namely, 3-D continuous-azimuth HRIR acquisition according to Fig. 2, with all elevations activated.

It is of special interest how the class of perfect excitation signals, namely, the perfect sweeps, tackles the problem of multichannel-system identification, which requires low cross-correlation between the parallel acoustic inputs. Starting from the one-channel approach, a strategy has been developed in [8, 12] to construct a set of *simultaneous and optimal* excitation signals for all channels, based on one prototype PSEQ. In the case of $P$ loudspeakers, a prototype PSEQ, $p(k)$, of period $PM$ is designed independently and supplied to the first channel. The other channels are excited with phase-shifted versions

$$x_{\theta_1}(k) = p(k) \tag{26}$$

$$x_{\theta_2}(k) = p(k - M) \tag{27}$$

$$\vdots$$

$$x_{\theta_P}(k) = p(k - (P-1)M). \tag{28}$$

In conjunction with the multichannel NLMS algorithm in (18) and (19), this creates the possibility of *uniquely* identifying the *true* HRIRs of all $P$ acoustic channels simultaneously, without cross-talk, and with a single comprehensive measurement, that is, only one rotation—for more details see [8, 12, 37].

Figure 9 exemplarily compares the results of single- and multichannel measurements. With a revolution time of $T_{360} = 15$ s in the single-channel case, the loudspeaker in the horizontal plane is excited with white-noise and perfect-sweep signals of the period $M = 308$, respectively. In the multichannel case, however, all $P = 8$ loudspeakers are used simultaneously. In case of white-noise excitation, all loudspeakers are fed with independent white noises, while for the perfect sweep the excitation scheme according to (26)–(28) is chosen, namely, with a prototype sequence, $p(k)$, of period $8 \cdot M = 2464$. In order to suitably balance the time-constant of the NLMS algorithm with the HRIR variability in both the single- and multichannel cases, the revolution time for the multichannel measurement has to be increased to $T_{360} = 8 \cdot 15\,\text{s} = 120\,\text{s}$. The single- and multichannel results for white-noise

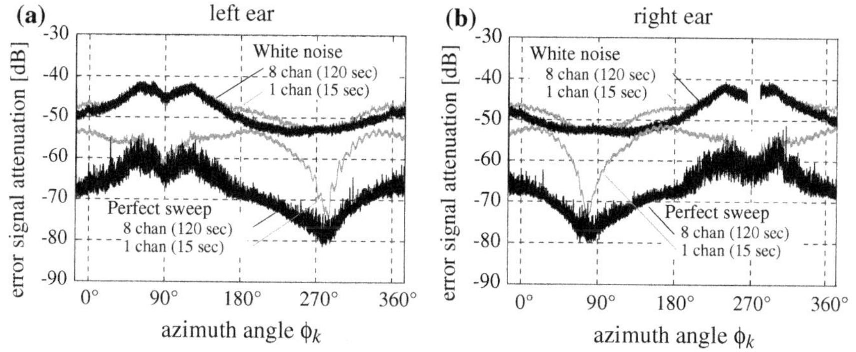


**Fig. 9** Error-signal attenuation, ESA, for dynamic 3-D HRIR measurements with $N = M = 308$, $\mu_0 = 1$, $T_{360} = 15\,\text{s}$ (single-channel case) or $T_{360} = 120\,\text{s}$ (multichannel case)

excitation then coincide in Fig. 9 up to minor deviations. Although the measurement duration is increased according to the number of acoustic channels to be identified, it should be noted that simultaneous multichannel HRIR acquisition still pays off in terms of *synchronous and consistent* measurement results for all elevations, since they are excited in parallel with the person to be measured naturally staying in one and the same position.

Overall, multichannel excitation with phase-shifted perfect sweeps shows the best results. As compared to white-noise stimuli, improvements in a range of 20–30 dB are observed. Due to perfectly-zero cross-correlation between the channel inputs, no degradation due to a multichannel issue is encountered. The ideal autocorrelation of perfect sweeps as well as the possibility to design orthogonal sets of perfect sweeps for multichannel excitation make this class of signals *optimal* for NLMS-driven acoustic system identification. In case of time-variant systems, the tracking ability related to perfect sequences is especially appreciated. Parameters such as the sampling frequency, the stepsize, or the length of the adaptive filter, can be freely adjusted. Beyond these features, perfect sweeps exhibit robustness against nonlinear distortions and provide in all experiments the best results in terms of error-signal attenuation.

## 5 Conclusions

After the manifold of previous work in HRTF acquisition and representation, one of the driving forces behind ongoing research is still the simple and compelling idea of creating accurate virtual reality using precise HRTFs. The extensiveness of plenacoustic HRTFs, however, requires deep understanding of the perceptual relevance of specific HRTF properties in order to achieve efficient acquisition, smoothing, sampling, representation, and usage of HRTFs. It turns out that by far not all issues are fully resolved. Regarding the perceptual relevance, some of the previous studies, for instance, used individual HRTFs with white-noise stimulus, while others combined *good-localizer* HRTFs with speech stimulus, to draw their respective conclusions.

As a major theme of this chapter, conceptual comparison of spatially discrete and continuous HRTF acquisition was addressed. The discrete version is more established and mechanically easy to handle. High measurement accuracy is reported, but the procedure used to consume a lot of time for sufficient spatial resolution. In recent years a footrace is taking place to overcome the long measurement duration with accelerated systems for individual HRTF acquisition. The spatially-continuous version, as was described in much detail in this chapter, exhibits somewhat opposite properties. It uses a dynamical apparatus and adaptive filters for HRTF extraction. It is, thus, by concept extremely fast, namely, on the order of 1 min/person, but HRTF storage may require huge memory or special care. One option for convergence of both variants is the down-sampling of the continuous data and then using the same HRTF file formats. Advantages of the continuous method, however, still include its flexibility regarding the down-sampling and its inherent property of overcoming spatial aliasing and interpolation issues.

While many options exist for HRTF acquisition and representation, many issues regarding the actual requirements on HRTF, for example, impulse-response SNR, spatial resolution, spatial interpolation, and optimal representation, are still under discussion. This also comprises the more general question of what is the most appropriate metric of HRTF fidelity. Based on such metrics, in a next step, a thorough experimental comparison of discrete and continuous HRTF measurements and its illustration is required. A first step into this direction has been made by a direct comparison of linear-interpolation error versus adaptive-system-identification misalignment [70]. In a future context, it is, however, expected that accurate HRTFs alone will not be a sufficient technology and requirements on HRTFs will have to be determined in conjunction with a multitude of technologies being applied in the context of binaural hearing and listening—compare Chaps. 4–17, this volume.

**Acknowledgments** The authors appreciate the contributions of M. Weinert to the development of the continuous-azimuth technique via his student-research projects from 2008 to 2011. They further thank their two internal and five external reviewers for many useful suggestions during the writing phase of this chapter.

## References

1. J. Ahrens and S. Spors. Focusing of virtual sound sources in higher order ambisonics. In *124th Conv. AES*, Amsterdam, The Netherlands, May 2008.
2. T. Ajdler, L. Sbaiz, and M. Vetterli. Plenacoustic function on the circle with application to HRTF interpolation. In Proc. IEEE *Intl. Conf. Acoust., Speech, Signal Proc. (ICASSP)*, Philadelphia, PA, March 2005.
3. T. Ajdler, L. Sbaiz, and M. Vetterli. Dynamic measurement of room impulse responses using a moving microphone. *J. Acoust. Soc. Am.*, 122:1636–1645, 2007.
4. V. Algazi, C. Avendano, and R. Duda. Elevation localization and head-related transfer function analysis at low frequencies. *J. Acoust. Soc. Am.*, 109:1110–1122, 2001.

5. V. Algazi and R. Duda. Headphone-based spatial sound. *IEEE Signal Process. Mag.*, 28:33–42, 2011.
6. V. Algazi, R. Duda, R. Duraiswami, N. Gumerov, and Z. Tang. Approximating the head-related transfer function using simple geometric models of the head and torso. *J. Acoust. Soc. Am.*, 112:2053–2064, 2002.
7. V. Algazi, R. Duda, D. Thompson, and C. Avendano. The CIPIC HRTF Database. In *Proc. IEEE Worksh. Applications of Signal Process. to Audio and Acoust., WASPAA*, pages 99–102, New Paltz, NY, Oct. 2001.
8. C. Antweiler. Multi-channel system identification with perfect sequences. In *Advances in Digital Speech Transmission (Martin, R., Heute, U. und Antweiler, C., eds.)*, pages 171–198, Chichester, UK: John Wiley & Sons, Ltd., Jan. 2008.
9. C. Antweiler and M. Antweiler. System identification with perfect sequences based on the NLMS algorithm. *AEÜ. Intl. J. Electron. Commun.*, 3:129–134, 1995.
10. C. Antweiler and G. Enzner. Perfect-sequence LMS for rapid acquisition of continuous-azimuth head related impulse responses. In *Proc. IEEE Worksh. Applications of Signal Process. to Audio and Acoust., WASPAA*, pages 281–284, New Paltz, NY, USA, Oct. 2009.
11. C. Antweiler and H.-G. Symanzik. Simulation of time variant room impulse responses. In *Proc. IEEE Intl. Conf. Acoust., Speech, Signal Proc. (ICASSP)*, pages 3031–3034, Detroit, USA, May 1995.
12. C. Antweiler, A. Telle, and P. Vary. NLMS-type system identification of MISO systems with shifted perfect sequences. In *Proc. Intl. Worksh. Acoustic Echo and Noise Control. IWAENC*, Seattle, USA, Sept. 2008.
13. C. Antweiler, A. Telle, P. Vary, and G. Enzner. Perfect-sweep NLMS for time-variant acoustic system identification. In *Proc. IEEE Intl. Conf. Acoust., Speech, Signal Proc. (ICASSP)*, pages 517–529, Kyoto, Japan, March 2012.
14. C. Antweiler, A. Telle, P. Vary, and E. D. Martino. A new otological diagnostic system providing a virtual tube model. In *IEEE Biomedical Circuits Syst. Conf. BIOCAS*, London, 2006.
15. C. Antweiler, A. Telle, P. Vary, and E. D. Martino. Virtual time-variant model of the Eustachian tube. In *Proc. Intl. Symp. Circuits Syst. (ISCAS)*, pages 5559–5562, Island of Kos, Greece, May 2006.
16. D. Begault. *3D Sound for Virtual Reality and Multimedia*. Academic Press, Cambridge, MA, 1994.
17. D. Begault and E. Wenzel. Headphone localization of speech. *Human Factors*, 35:361–376, 1993.
18. D. Begault, E. Wenzel, and M. Anderson. Direct comparison of the impact of head tracking, reverberation, and individualized head-related transfer functions on the spatial perception of a virtual speech source. *J. Audio Engr. Soc.*, 49:904–916, 2001.
19. V. Best, S. Carlile, C. Jin, and A. van Schaik. The role of high frequencies in speech localization. *J. Acoust. Soc. Am.*, 118:353–363, 2005.
20. J. Blauert. *Spatial Hearing*. MIT Press, Cambridge, 1997.
21. J. Blauert. *Acoustics for Engineers*. Springer, Berlin, 2008.
22. J. Blauert, M. Brüggen, K. Hartung, A. Bronkhorst, R. Drullmann, G. Reynaud, L. Pellieux, W. Krebber, and R. Sottek. The AUDIS Catalog of Human HRTFs. In *16th Intl. Congr. Acoust. and 135th Meet. Acoust. Soc. Am.*, pages 2901–2902, Seattle, June 1998.
23. J. Blauert, H. Lehnert, J. Sahrhage, and H. Strauss. An interactive virtual-environment generator for psychoacoustic research i: Architecture and implementation. *Acta Acust./Acustica*, 86:94–102, 2000.
24. J. Breebaart and A. Kohlrausch. The perceptual (ir)relevance of HRTF magnitude and phase spectra. In *110th Conv. AES, Paper No. 5406*, Amsterdam, NL, May 2001.
25. J. Breebaart, F. Nater, and A. Kohlrausch. Spectral and spatial parameter resolution requirements for parametric, filter-bank-based HRTF processing. *J. Audio Engr. Soc.*, 58:126–140, 2010.

26. A. Bronkhorst. Localization of real and virtual sound sources. *J. Acoust. Soc. Am.*, 98:2542–2553, 1995.
27. D. Brungart. Near-field virtual audio displays. In *IMAGE 2000*, Scottsdale, USA, July 2000.
28. D. Brungart, N. Durlach, and W. Rabinowitz. Auditory localization of nearby sources. II. Localization of a broadband source. *J. Acoust. Soc. Am.*, 106:1956–1968, 1999.
29. D. Brungart and W. Rabinowitz. Auditory localization of nearby sources. Head-related transfer functions. *J. Acoust. Soc. Am.*, 106:1465–1479, 1999.
30. M. Burkhard and R. Sachs. Anthropometric manikin for acoustic research. *J. Acoust. Soc. Am.*, 58:214–222, 1975.
31. B. Carty. *Movements in Binaural Space: Issues in HRTF Interpolation and Reverberation, with applications to Computer Music*. PhD thesis, NUI Maynooth, 2010.
32. B. Carty and V. Lazzarini. Frequency-domain interpolation of empirical HRTF data. In 126th *Conv. AES*, Munich, Germany, May 2009.
33. F. Christensen, G. Martin, P. Minnaar, W. Song, B. Pedersen, and M. Lydolf. A listening test system for automotive audio - part 1: System design. In *118th Conv. AES*, Barcelona, Spain, May 2005.
34. R. Duraiswami, D. Zotkin, and N. Gumerov. Interpolation and range extrapolation of HRTFs. In *Proc. IEEE Intl. Conf. Acoust., Speech, Signal Proc. (ICASSP)*, Montreal, Canada, May 2004.
35. N. Durlach, A. Rigopulos, X. Pang, W. Woods, A. Kulkkarni, H. Colburn, and E. Wenzel. On the externalization of auditory images. *Presence: Teleoperators and Virtual Environments*, 1:251–257, Spring 1992.
36. G. Enzner. Analysis and optimal control of LMS-type adaptive filtering for continuous-azimuth acquisition of head related impulse responses. In *Proc. IEEE Intl. Conf. Acoust., Speech, Signal Proc. (ICASSP)*, Las Vegas, NV, April 2008.
37. G. Enzner. 3D-continuous-azimuth acquisition of head-related impulse responses using multi-channel adaptive filtering. In *Proc. IEEE Worksh. Applications of Signal Process. to Audio and Acoust., WASPAA, New Paltz*, NY, Oct. 2009.
38. G. Enzner. Bayesian inference model for applications of time-varying acoustic system identification. In *Proc. Europ. Signal Process. Conf. EUSIPCO*, Aalborg, DN, August 2010.
39. G. Enzner, M. Krawczyk, F.-M. Hoffmann, and M. Weinert. 3D reconstruction of HRTF-fields from 1D continuous measurements. In *Proc. IEEE Worksh. Applications of Signal Process. to Audio and Acoust., WASPAA*, New Paltz, NY, Oct. 2011.
40. M. Evans. Analysing head-related transfer function measurements using surface spherical harmonics. *J. Acoust. Soc. Am.*, 104:2400–2411, 1998.
41. A. Farina. Simultaneous measurement of impulse response and distortion with a swept-sine technique. In *108th Conv. AES*, Feb. 2000.
42. J. Fels, P. Buthmann, and M. Vorländer. Head-related transfer functions of children. *Acta Acoust./Acustica*, 90:918–927, 2004.
43. J. Fels and M. Vorländer. Anthropometric parameters influencing head-related transfer functions. *Acta Acust./Acustica*, 95:331–342, 2009.
44. K. Fukudome, T. Suetsugu, T. Ueshin, R. Idegami, and K. Takeya. The fast measurement of head related impulse responses for all azimuthal directions using the continuous measurement method with a servo-swiveled chair. *Applied Acoustics*, 68:864–884, 2007.
45. W. Gardner and K. Martin. HRTF measurements of a KEMAR. *J. Acoust. Soc. Am.*, 97:3907–3908, 1995.
46. M. Geier, J. Ahrens, A. Möhl, S. Spors, J. Loh, and K. Bredies. The soundscape renderer: A versatile framework for spatial audio reproduction. In *Proceedings of the DEGA WFS Symposium*, Ilmenau, Germany, Sept. 2007.
47. A. González, P. Zuccarello, G. P. nero, and M. de Diego. Simultaneous measurement of multichannel acoustic systems. *J. Audio Engr. Soc.*, 52:26–42, 2004.
48. R. Greff and B. Katz. Round robin comparison of HRTF simulation systems: Preliminary results. In *123rd Conv. AES*, Oct. 2007.

49. M. Guldenschuh, A. Sontacchi, and F. Zotter. HRTF modelling in due consideration variable torso reflections. In *Proc. EAA Acoust.*, Paris, France, July 2008.
50. N. Gumerov, A. O'Donovan, R. Duraiswami, and D. Zotkin. Computation of the head-related transfer function via the fast multipole accelerated boundary element method and its spherical harmonic representation. *J. Acoust. Soc. Am.*, 127:370–386, 2010.
51. N. Gupta, A. Barreto, M. Joshi, and J. Agudelo. HRTF Database at FIU DSP Lab. In *Proc. IEEE Intl. Conf. Acoust., Speech, Signal Proc. (ICASSP)*, pages 169–172, Dallas, TX, March 2010.
52. D. Hammershøi and H. Møller. Sound transmission to and within the human ear canal. *J. Acoust. Soc. Am.*, 100:408–427, 1996.
53. D. Hammershøi and H. Møller. Binaural Technique - Basic Methods for Recording, Synthesis, and Reproduction. In J. Blauert, editor, *Communication Acoustics*. Springer-Verlag, 2005.
54. Y. Haneda, S. Makino, Y. Kaneda, and N. Kitawaki. Common-acoustical-pole and zero modeling of head-related transfer functions. *IEEE Trans. Speech Audio Proc.*, 7:188–196, 1999.
55. W. Hartmann and A. Wittenberg. On the externalization of sound images. *J. Acoust. Soc. Am.*, 99:3678–3688, 1996.
56. K. Hartung, J. Braasch, and S. Sterbing. Comparison of different methods for the interpolation of head-related transfer functions. In *16th Intl. Conf. AES*, Rovaniemi, Finnland, April 1999.
57. S. Haykin. *Adaptive Filter Theory*. Prentice-Hall, Upper Saddle River, NJ, 4th edition, 2002.
58. J. Huopaniemi and K. Riederer. Measuring and modeling the effect of source distance in head-related transfer functions. In *Intl. Congr. Acoust.*, Seattle, USA, June 1998.
59. V. Ipatov. Ternary sequences with ideal periodic autocorrelation properties. *Radio Engineering and Electronic Physics*, 24:75–79, 1979.
60. R. Kalman. A new approach to linear filtering and prediction problems. *Trans. ASME, J. Basic Enggr.,* 82:35–45, 1960.
61. A. Kan, C. Jin, and A. van Schaikc. A psychophysical evaluation of near-field head-related transfer functions synthesized using a distance variation function. *J. Acoust. Soc. Am.*, 125:2233–2242, 2009.
62. B. Katz. Boundary element method calculation of individual head-related transfer function. *J. Acoust. Soc. Am.,* 110:2440–2455, 2001.
63. B. Katz and D. Begault. Round robin comparison of HRTF measurement systems: Preliminary results. In *19th Intl. Congr. Acoust.*, Madrid, Sept. 2007.
64. T. Lentz. *Binaural Technology for Virtual Reality*. PhD thesis, RWTH Aachen, 2007.
65. T. Lentz. Near-field HRTFs. In *Proc. of Deutsche Jahrestagung für Akustik (DAGA)*, Stuttgart, March 2007.
66. A. Lindau, T. Hohn, and S. Weinzierl. Binaural resynthesis for comparative studies of acoustical environments. In *122th Conv. AES*, Vienna, Austria, 2007.
67. A. Lindau and S. Weinzierl. On the spatial resolution of virtual acoustic environments for head movements in horizontal vertical and lateral direction. In *Proc. EAA Symp. Auralization,* Espoo, Finland, 2009.
68. H.-D. Lüke and H. Schotten. Odd-perfect, almost binary correlation sequences. *IEEE Trans. Aerosp. Electron. Syst.,* 31:495–498, 1995.
69. P. Majdak, P. Balazs, and B. Laback. Multiple exponential sweep method for fast measurement of head-related transfer functions. *J. Audio Engr. Soc.,* 55:623–637, 2007.
70. S. Malik, J. Fligge, and G. Enzner. Continuous HRTF acquisition vs. HRTF interpolation for binaural rendering of dynamical auditory virtual environments. In *Proc. ITG Conf. Speech Comm.,* Bochum, Germany, Oct. 2010.
71. B. Masiero, M. Pollow, and J. Fels. Design of a fast broadband individual head-related transfer function measurement system. In *Proc. Forum Acusticum,* Aalborg, Denmark, 2011. Acta Acoust./Acustica, vol. 97, supp. 1, p. 136, Hirzel, 2011.
72. D. Menzies and M. Al-Akaidi. Nearfield binaural synthesis and Ambisonics. *J. Acoust. Soc. Am.,* 121:1559–1563, 2007.
73. P. Minnaar, J. Plogsties, and F. Christensen. Directional resolution of head-related transfer functions required in binaural synthesis. *J. Audio Engr. Soc.,* 53:919–929, 2005.

74. C. Moldrzyk, W. Ahnert, S. Feistel, T. Lentz, and S. Weinzierl. Head-tracked auralization of acoustical simulation. In *117th Conv. AES*, San Fransisco, USA, Oct. 2004.
75. H. Møller. Fundamentals of binaural technology. *Applied Acoustics*, 36:171–218, 1992.
76. H. Møller, M. Sorensen, D. Hammershøi, and C. Jensen. Head-related transfer functions of human subjects. *J. Audio Engr. Soc.*, 43:300–321, 1995.
77. H. Møller, M. Sorensen, C. Jensen, and D. Hammershøi. Binaural technique: Do we need individual recordings? *J. Audio Engr. Soc.*, 44:451–469, 1996.
78. S. Müller and P. Massarani. Transfer-function measurement using sweeps. *J. Audio Engr. Soc.*, 49:443–471, 2001.
79. R. Nicol. *Binaural Technology*. AES Monograph, New York, 2010.
80. M. Noisternig, A. Sontacchi, T. Musil, and R. Höldrich. A 3D Ambisonic based binaural sound reproduction system. In *AES 24th Intl. Conf. Multichannel Audio*, Banff, Canada, June 2003.
81. A. Novák, L. Simon, F. Kadlec, and P. Lotton. Nonlinear system identification using exponential swept-sine signal. *IEEE Trans. Instrum. Meas.*, 59:2220–2229, 2010.
82. M. Otani and S. Ise. Fast calculation system specialized for head-related transfer function based on boundary element method. *J. Acoust. Soc. Am.*, 119:2589–2598, 2006.
83. G. Plenge. On the differences between localization and lateralization. *J. Acoust. Soc. Am.*, 56:944–951, 1974.
84. M. Pollow, B. Masiero, P. Dietrich, J. Fels, and M. Vorländer. Fast measurement system for spatially continuous individual HRTFs. In *4th Int. Symposium on Ambisonics and Spherical Acoustics, 25th AES UK Conference*, University of York, UK, March 2012.
85. M. Pollow, K. Nguyen, O. Warusfel, T. Carpentier, M. Müller-Trapet, M. Vorländer, and M. Noisternig. Calculation of head-related transfer functions for arbitrary field points using spherical harmonics decomposition. *Acta Acoust./Acustica*, 98:72–82, 2012.
86. D. Pralong and S. Carlile. Measuring the human head-related transfer functions: A novel method for the construction and calibration of a miniature in-ear recording system. *J. Acoust. Soc. Am.*, 95:3435–3444, 1994.
87. T. Qu, Z. Xiao, M. Gong, Y. Huang, X. Li, and X. Wu. Distance-dependent head-related transfer functions measured with high spatial resolution using a spark gap. *IEEE Trans. Audio, Speech, and Lang. Process.*, 17:1124–1132, 2009.
88. K. Riederer. *Objective and Subjective Evaluation of Measured Head-Related Transfer Functions*. PhD thesis, Report 76, Helsinki University of Technology, 2005.
89. D. D. Rife and J. Vanderkooy. Transfer function measurement with maximum-length sequences. *J. Acoust. Soc. Am.*, 37:419–444, 1989.
90. D. Romblom and B. Cook. Near-field compensation for HRTF processing. In *125th Conv. AES*, San Francisco, USA, Oct. 2008.
91. M. Rothbucher, M. Durkovic, H. Shen, and K. Diepold. HRTF customization using multiway array analysis. In *Proc. Europ. Signal Process. Conf. EUSIPCO*, 2010.
92. M. Rothbucher, T. Habigt, J. Habigt, T. Riedmaier, and K. Diepold. Measuring anthropometric data for HRTF personalization. In *Proc. 6th Intl. Conf. on Signal-Image Technology and Internet-Based Systems (SITIS)*, 2010.
93. J. Scarpaci and S. Colburn. Principal components analysis interpolation of head related transfer functions using locally-chosen basis functions. *J. Acoust. Soc. Am.*, 117:2561–2562, 2005.
94. L. L. Scharf. *Statistical Signal Processing*. Addison-Wesley Publishing Company, 1991.
95. D. Schönstein and B. Katz. HRTF selection for binaural synthesis from a database using morphological parameters. In *Proc. 20th Intl. Congr. Acoust. (ICA)*, Sidney, Australia, August 2010.
96. B. Shinn-Cunningham, N. Kopco, and T. Martin. Localizing nearby sound sources in a classroom: Binaural room impulse responses. *J. Acoust. Soc. Am.*, 117:3100–3115, 2005.
97. A. Silzle. Länge von Außenohr-Impulsantworten und Auswahl von Reflexionen für interaktive auditive virtuelle Umgebungen. In *Proc. of Deutsche Jahrestagung für Akustik (DAGA)*, Braunschweig, Germany, March 2006.

98. S. Spors and J. Ahrens. Efficient range extrapolation of head-related impulse responses by wave field synthesis techniques. In *Proc. IEEE Intl. Conf. Acoust., Speech, Signal Proc. (ICASSP)*, May 2011.
99. S. Spors and J. Ahrens. Interpolation and range extrapolation of head-related transfer functions using virtual local sound field synthesis. In *130th Conv. AES*, May 2011.
100. A. Telle. *Sonotubometrie mit Methoden der digitalen Signalverarbeitung*. PhD thesis, RWTH Aachen, 2012.
101. A. Telle, C. Antweiler, and P. Vary. Der perfekte Sweep - ein neues Anregungssignal zur adaptiven Systemidentifikation zeitvarianter akustischer Systeme. In *Proc. of Deutsche Jahrestagung für Akustik (DAGA)*, pages 341–342, Berlin, Germany, March 2010.
102. A. Torras-Rosell and F. Jacobsen. A new interpretation of distortion artifacts in sweep measurements. *J. Audio Engr. Soc.*, 59:283–289, 2011.
103. R. Unbehauen. *Systemtheorie 1*. R. Oldenburg Verlag, München Wien, 7th edition, 1997.
104. F. Völk, F. Heinemann, and H. Fastl. Externalization in binaural synthesis: Effects of recording environment and measurement procedure. In *Proc. EAA Acoust.*, Paris, France, July 2008.
105. M. Vorländer. *Auralization: Fundamentals of Acoustics, Modelling, Simulation, Algorithms and Acoustic Virtual Reality*. Springer, 2007.
106. F. Wefers. OpenDAFF - A free, open-source software package for directional audio data. In *Proc. of Deutsche Jahrestagung für Akustik (DAGA)*, March 2010.
107. M. Weinert, G. Enzner, J.-M. Batke, P. Jax, and C. Antweiler. Komfortable Messung und Bereitstellung individueller kopfbezogener Impulsantworten als OpenDAFF. In *Proc. of Deutsche Jahrestagung für Akustik (DAGA)*, Darmstadt, Germany, 2012.
108. S. Weinzierl, A. Giese, and A. Lindau. Generalized multiple sweep measurement. In *126th Conv. AES*, May 2009.
109. E. Wenzel. The relative contribution of interaural time and magnitude cues to dynamics sound localization. In *Proc. IEEE Worksh. Applications of Signal Process. to Audio and Acoust., WASPAA*, New Paltz, NY, Oct. 1995.
110. E. Wenzel, M. Arruda, D. Kistler, and F. Wightman. Localization using non-individualized head-related transfer functions. *J. Acoust. Soc. Am.*, 94:111–123, 1993.
111. F. Wightman and D. Kistler. Headphone simulation of free-field listening I: Stimulus synthesis. *J. Acoust. Soc. Am.*, 85:858–867, 1989.
112. F. Wightman and D. Kistler. Headphone simulation of free-field listening II: Psychophysical validation. *J. Acoust. Soc. Am.*, 85:868–878, 1989.
113. F. Wightman and D. Kistler. The dominant role of low-frequency interaural time differences in sound localization. *J. Acoust. Soc. Am.*, 91:1648–1661, 1992.
114. F. Wightman and D. Kistler. Resolution of front-back ambiguity in spatial hearing by listener and source movements. *J. Acoust. Soc. Am.*, 105:2841–2853, 1999.
115. F. Wightman and D. Kistler. Measurement and validation of human HRTFs for use in research. *Acta Acust./Acustica*, 91:429–439, 2005.
116. E. G. Williams. *Fourier Acoustics: Sound Radiation and Nearfield Acoustical Holography*. Academic Press, San Diego, London, 1999.
117. B. Xie and T. Zhang. The audibility of spectral detail of head-related transfer functions at high frequency. *Acta Acust./Acustica*, 96:328–339, 2010.
118. W. Zhang, T. Abhayapala, and R. Kennedy. Modal expansion of HRTFs: continuous representation in frequency-range-angle. In *Proc. IEEE Intl. Conf. Acoust., Speech, Signal Proc. (ICASSP)*, Taipei, Japan, 2009.
119. W. Zhang, R. Kennedy, and T. Abhayapala. Efficient continuous HRTF model using data independent basis functions: Experimentally guided approach. *IEEE Trans. Audio, Speech, and Lang. Process.*, 17:819–829, 2009.
120. W. Zhang, M. Zhang, R. Kennedy, and T. Abhayapala. On high-resolution head-related transfer function measurements: An efficient sampling scheme. *IEEE Trans. Audio, Speech, and Lang. Process.*, 20:575–584, 2012.
121. D. Zotkin, R. Duraiswami, E. Grassi, and N. Gumerov. Fast head-related transfer function measurement via reciprocity. *J. Acoust. Soc. Am.*, 120:2202–2215, 2006.

122. D. Zotkin, R. Duraiswami, and N. Gumerov. Regularized HRTF fitting using spherical harmonics. In *Proc. IEEE Worksh. Applications of Signal Process. to Audio and Acoust., WASPAA,* New Paltz, NY, Oct. 2009.
123. F. Zotter. *Analysis and Synthesis of Sound-Radiation with Spherical Arrays.* PhD thesis, University of Music and Performing Arts, Graz, Austria, 2009.

# Assessment of Sagittal-Plane Sound Localization Performance in Spatial-Audio Applications

**R. Baumgartner, P. Majdak and B. Laback**

## 1 Sound Localization in Sagittal Planes

### *1.1 Salient Cues*

Human normal-hearing, NH, listeners are able to localize sounds in space in terms of assigning direction and distance to the perceived auditory image [26]. Multiple mechanisms are used to estimate sound-source direction in the three-dimensional space. While interaural differences in time and intensity are important for sound localization in the lateral dimension, left/right, [53], monaural spectral cues are assumed to be the most salient cues for sound localization in the sagittal planes, SPs, [27, 54]. Sagittal planes are vertical planes parallel to the median plane and include points of similar interaural time differences for a given distance. The monaural spectral cues are essential for the perception of the source elevation within a hemifield [2, 22, 24] and for front-back discrimination of the perceived auditory event [46, 56]. Note that also the binaural pinna disparities [43], namely, interaural spectral differences, might contribute to SP localization [27].

The mechanisms underlying the perception of lateral displacement are the main topic of other chapters. This chapter focuses on the remaining directional dimension, namely, the one along SPs. Because interaural cues and monaural spectral cues are thought to be processed largely independently of each other [27], the interaural-polar coordinate system is often used to describe their respective contributions in the two dimensions. In the interaural-polar coordinate system the direction of a sound source is described with the lateral angle, $\phi \in [-90°, 90°]$, and the polar angle, $\theta \in [-90°, 270°)$—see Fig. 1, left panel. Sagittal-plane localization refers to the listener's assignment of the polar angle for a given lateral angle and distance of the sound source.

R. Baumgartner · P. Majdak (✉) · B. Laback
Acoustics Research Institute, Austrian Academy of Sciences, Vienna, Austria
e-mail: piotr@majdak.com

J. Blauert (ed.), *The Technology of Binaural Listening*, Modern Acoustics and Signal Processing, DOI: 10.1007/978-3-642-37762-4_4,

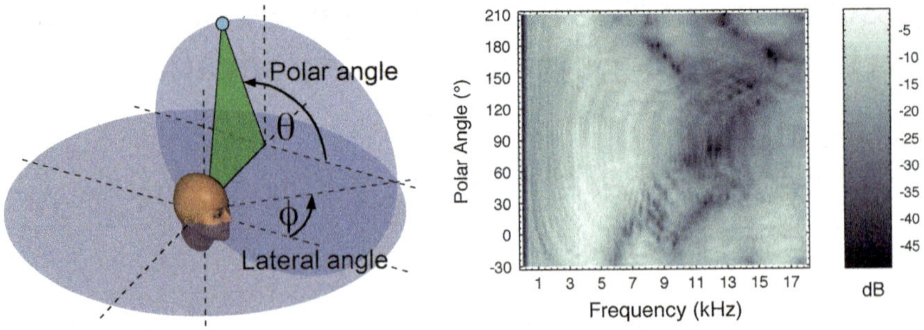


**Fig. 1** *Left* Interaural-polar coordinate system. *Right* HRTF magnitude spectra of a listener as a function of the polar angle in the median SP—left ear of NH58

Although spectral cues are processed monaurally, the information from both ears affects the perceived location in most cases [39]. The ipsilateral ear, namely, the one closer to the source, dominates and its relative contribution increases monotonically with increasing lateral angle [12]. If the lateral angle exceeds about 60°, the contribution of the contralateral ear becomes negligible. Thus, even for localization in the SPs, the lateral source position, mostly depending on the broadband binaural cues [27], must be known in order to determine the binaural weighting of the monaural spectral cues.

The nature of the spectral features important for sound localization is still subject of investigations. Due to the physical dimensions, the pinna plays a larger role for higher frequencies [36] and the torso for lower frequencies [1]. Some psychoacoustic studies postulated that macroscopic patterns of the spectral features are important rather than fine spectral details [2, 10, 16, 22–24, 28, 44]. On the other hand, other studies postulated that SP sound localization is possibly mediated by means of only a few local spectral features [17, 37, 52, 56]. Despite a common agreement, according to which the amount of the spectral features can be reduced without substantial reduction of the localization performance, the perceptual relevance of particular features has not been fully clarified yet.

## 1.2 Head-Related Transfer Functions

The effect of the acoustic filtering of torso, head and pinna can be described in terms of a linear time-invariant system by the so-called head-related transfer functions, HRTFs, [4, 38, 45]. The right panel of Fig. 1 shows the magnitude spectra of the left-ear HRTFs of an exemplary listener, NH58,[1] along the median SP.

HRTFs depend on the individual geometry of the listener and thus listener-specific HRTFs are required to achieve accurate localization performance for binaural

[1] These and all other HRTFs are from http://www.kfs.oeaw.ac.at/hrtf.

synthesis [6, 35]. Usually, HRTFs are measured in an anechoic chamber by determining the acoustic response characteristics between loudspeakers at various directions and microphones inserted into the ear canals. Currently, much effort is put also into the development of non-contact measurement methods for capturing HRTFs like numerical calculation of HRTFs from optically scanned geometry [20, 21] and on customization of HRTFs basing on psychoacoustic tests [16, 34, 46].

Measured HRTFs contain both direction-dependent and direction-independent features and can be thought of as a series of two acoustic filters. The direction-independent filter, represented by the common transfer function, CTF, can be calculated from an HRTF set comprising many directions [34] by averaging the log-amplitude spectra of all available HRTFs of a listener's ear. The phase spectrum of the CTF is the minimum phase corresponding to the amplitude spectrum of the CTF.

In the current study, the topic of interest is the directional aspect. Thus, the directional features are considered, as represented by the directional transfer functions, DTFs. The DTF for a particular direction is calculated by filtering the corresponding HRTF with the inverse CTF. The CTF usually exhibits a low-pass filter characteristic because the higher frequencies are attenuated for many directions due to the head and pinna shadow—see Fig. 2, left panel. Compared to HRTFs, DTFs usually pronounce frequencies and thus spectral features above 4 kHz—see Fig. 2, right panel. DTFs are commonly used to investigate the nature of spectral cues in SP localization experiments with virtual sources [10, 30, 34].

In the following, the proposed model is described in Sect. 2 and the results of its evaluation are presented in Sect. 3, based on recent virtual-acoustics studies that used listener-specific HRTFs. In Sect. 4, the proposed model is applied to predict localization performance for different aspects of spatial-audio applications that involve spectral localization cues. In particular, a focus is put on the evaluation of non-individualized binaural recordings, the assessment of the quality of spatial cues for the design of hearing-assist devices, namely, in-the-ear versus behind-the-ear microphones and the estimation and improvement of the perceived direction of phantom

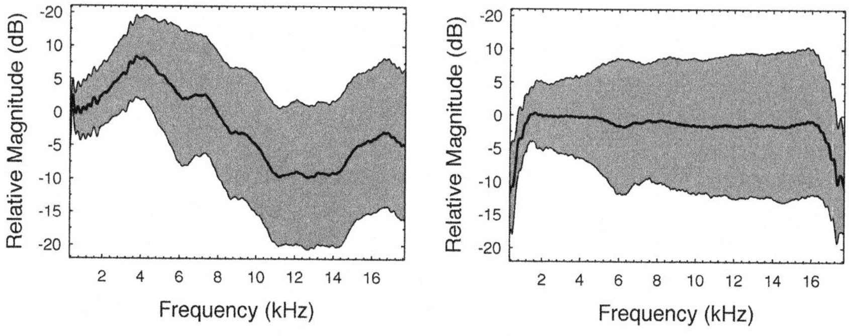


**Fig. 2** *Left* Spatial variation of HRTFs around CTF for listener NH58, left ear. *Right* Corresponding DTFs, i.e. HRTFs with CTF removed. *Solid line* Spatial average of transfer function. *Grey area* ± 1 standard deviation

sources in surround-sound systems, namely, 5.1 versus 9.1 versus 10.2 surround. Finally, Sect. 5 concludes with a discussion of the potential of the model for both evaluating audio applications and improving the understanding of human sound-localization mechanisms.

## 2 Models of Sagittal-Plane Localization

This section considers existing models aiming at predicting listener's polar response angle to the incoming sound. These models can help to explain psychoacoustic phenomena or to assess the spatial quality of audio systems while avoiding the running of costly and time-consuming localization experiments.

In general, machine-learning approaches can be used to predict localization performance. Artificial neural networks, ANNs, have been shown to achieve rather accurate predictions when trained with large datasets of a single listener [19]. However, predictions for a larger subpopulation of human listeners would have required much more effort. Also, the interpretation of the ANN parameters is not straight forward. It is difficult to generalize the findings obtained with an ANN-based model to other signals, persons and conditions and thus to better understand the mechanisms underlying spatial hearing.

Hence, the focus is laid on a *functional* model where model parameters should correspond to physiological and/or psychophysical localization parameters. Until now, a functional model considering both spectral and temporal modulations exists only as a general concept [50]. Note that in order to address a particular research question, models dealing with specific types of modulations have been designed. For example, models for narrow-band sounds [37] were provided in order to explain the well-known effect of directional bands [4]. In order to achieve a sufficiently good prediction as an effect of the modification of the spectral cues, it is assumed that the incoming sound is a *stationary broadband* signal, explicitly disregarding spectral and temporal modulations.

Note that localization models driven by various signal-processing approaches have also been developed [3, 32, 33]. These models are based on general principles of biological auditory systems, they do not, however, attempt to predict human-listener performance—their outcome shows rather the potential of the signal-processing algorithms involved.

In the following, previous developments on modeling SP localization performance are reviewed and a functional model predicting sound localization performance in arbitrary SPs for broadband sounds is proposed. The model is designed to retrieve psychophysical localization performance parameters and can be directly used as a tool to assess localization performance in various applications. An implementation of the model is provided in the AMToolbox, as the `baumgartner2013` model [47].

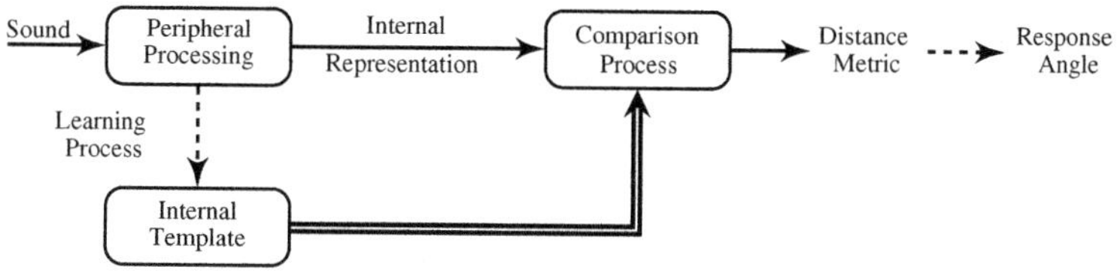


**Fig. 3** General structure of a template-based comparison model for predicting localization in SPs

## *2.1 Template-Based Comparison*

A common property of existing sound localization models based on spectral cues is that they compare an internal representation of the incoming sound with a template [13, 24, 55]—see Fig. 3. The internal template is assumed to be created by means of learning the correspondence between the spectral features and the direction of an acoustic event [14, 49], based on feedback from other modalities. The localization performance is predicted by assuming that in the sound localization task, the comparison yields a distance metric that corresponds to the polar response angle of the listener. Thus, template-based models include a stage modeling the peripheral processing of the auditory system applied to both the template and incoming sound and a stage modeling the comparison process in the brain.

### Peripheral Processing

The peripheral processing stage aims at modeling the effect of human physiology while focusing on directional cues. The effect of the torso, head and outer ear are considered by filtering the incoming sound by an HRTF or a DTF. The effect of ear canal, middle ear and cochlear filtering can be considered by various model approximations. In the early HRTF-based localization models, a parabolic-shaped filter bank was applied [55]. Later, a filter bank averaging magnitude bins of the discrete Fourier transform of the incoming sound was used [24]. Both filter banks, while being computationally efficient, were drastically simplifying the auditory peripheral processing. The Gammatone, GT, filter bank [40] is a more physiology-related linear model of auditory filters and has been used in localization models [13]. A model accounting for the nonlinear effect of the cochlear compression is the dual-resonance nonlinear, DRNL, filter bank [25]. A DRNL filter consists of both a linear and a non-linear processing chain and is implemented by cascading GT filters and Butterworth low-pass filters, respectively. Another non-linear model uses a single main processing chain and accounts for the time-varying effects of the medial-oliviocochlear reflex [57]. All those models account for the contribution of outer hair cells to a different degree and can be used to model the movements of the basilar membrane at a particular frequency. They are implemented in the AMToolbox [47]. In the localization model described in this chapter, the GT filter bank is applied focusing on applications where the absolute sound level plays a minor role.

The filter bank produces a signal for each center frequency and only the relevant frequency bands are considered in the model. Existing models used frequency bands with constant relative bandwidth on a logarithmic frequency scale [24, 55]. In the model described in this chapter, the frequency spacing of the bands corresponds to one equivalent rectangular bandwidth, ERB, [9]. The lowest frequency is 0.7 kHz, corresponding to the minimum frequency thought to be affected by torso reflections [1]. The highest frequency considered in the model depends on the bandwidth of the incoming sound and is maximally 18 kHz, approximating the upper frequency limit of human hearing.

Further in the auditory system, the movements of the basilar membrane at each frequency band are translated into neural spikes by the inner hair cells, IHCs. An accurate IHC model has not been considered yet and does not seem to be vital for SP localization. Thus, different studies used different approximations. In this model, the IHC is modeled as half-wave rectification followed by a second-order Butterworth low-pass with a cut-off frequency of 1 kHz [8]. Since the temporal effects of SP localization are not considered yet, the output of each band is simply temporally averaged in terms of RMS amplitude, resulting in the internal representation of the sound. The same internal representation and therefore peripheral processing is assumed for the template.

### Comparison Stage

In the comparison stage, the internal representation of the incoming sound is compared with the internal template. Each entry of the template is selected by a polar angle denoted as template angle. A distance metric is calculated as a function of the template angle and can be interpreted as a potential descriptor for the response of the listener.

An early modeling approach proposed to compare the spectral derivatives of various orders in terms of a band-wise subtraction of the derivatives and then averaging over the bands [55]. The comparison of the first-order derivative corresponds to the assumption that the overall sound intensity does not contribute to the localization process. In the comparison of the second-order derivatives, the differences in spectral tilt between the sound and the template do not contribute. Note that the plausibility of these comparison methods had not been investigated at that time. As another approach, the cross-correlation coefficient has been proposed to evaluate the similarity between the sound and the template [13, 37]. Later, the inter-spectral differences, ISDs, namely, the differences between the internal representations of the incoming sound and the template, calculated for each template angle and frequency band, were used [34] to show a correspondence between the template angle yielding smallest spectral variance and the actual response of human listeners. All these comparison approaches were tested in [24] who, distinguishing zeroth-, first- and second-order derivatives of the internal representations, found that the standard deviation of ISDs best described their results. This configuration corresponds to an average of the first-order derivative from [55], which is robust against changes in the overall level in the comparison process.

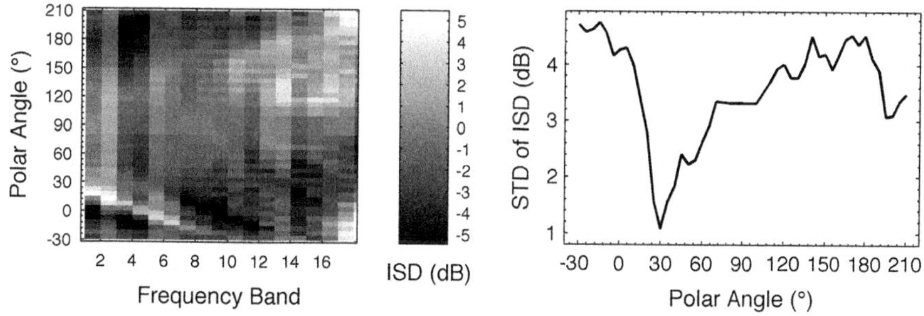


**Fig. 4** Example of the comparison process for a target polar angle of 30°. *Left* Inter-spectral differences, ISDs, as a function of the template angle. *Right* Spectral standard deviation, STD, of ISDs as a function of the template angle

The model proposed in this study also relies on ISDs calculated for a template angle and for each frequency band—see Fig. 4, left panel. Then, the spectral standard deviations of ISDs are calculated for all available template angles—see Fig. 4, right panel. For band-limited sounds, the internal representation results in an abrupt change at the cut-off frequency of the sound. This change affects the standard deviation of the ISDs. Thus, in this model, the ISDs are calculated only within the bandwidth of the incoming sound.

The result of the comparison stage is a distance metric corresponding to the prediction of the polar response angle. Early modeling approaches used the minimum distance to determine the predicted response angle [55], which would nicely fit the minimum of the distance metric used in the example reported here—see Fig. 4, right panel. Also, the cross-correlation coefficient has been used as a distance metric and its maximum has been interpreted as the prediction of the response angle [37]. Both approaches represent a deterministic interpretation of the distance metric, resulting in exactly the same predictions for the same sounds. This is rather unrealistic. Listeners, repeatedly listening to the same sound, often do not respond to exactly the same direction [7]. The actual responses are known to be scattered and can be even multimodal. The scatter of one mode can be described by the Kent distribution [7], which is an elliptical probability distribution on the two-dimensional unit sphere.

## 2.2 Response Probability

In order to model the probabilistic response pattern of listeners, a mapping of the distance metric to polar-response probabilities via similarity indices, SIs, has been proposed [24]. For a particular target angle and ear, they obtained a monaural SI by using the distance metric as the argument of a Gaussian function with a mean of zero and a standard deviation of two—see Fig. 5, $U = 2$. While this choice appears to be somewhat arbitrary, it is an attempt to model the probabilistic relation between the

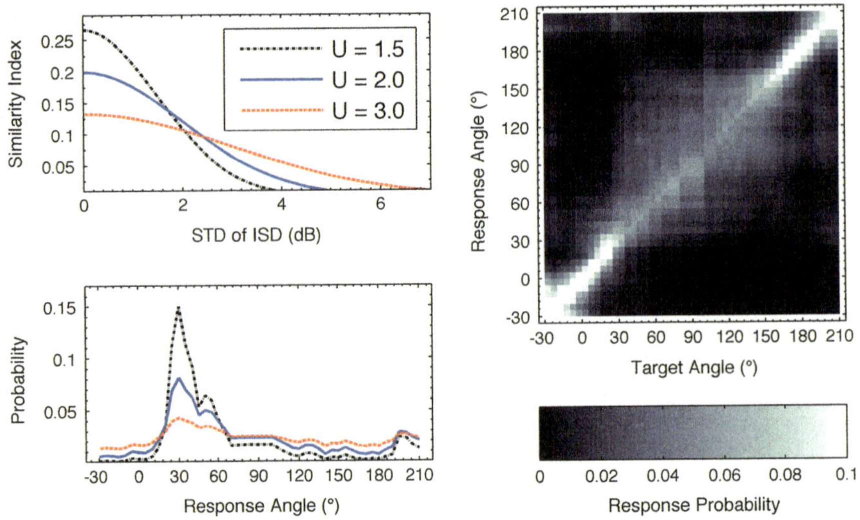


**Fig. 5** *Left* Mapping function of similarity index, *top*, for various uncertainties, $U$, and the resulting PMVs, *bottom*—corresponding to the example shown in Fig. 4. *Right* Predicted response PMV of the localization model as a function of the target angle, i.e. prediction matrix, for the baseline condition in the median SP for listener NH58. Response probabilities are encoded by brightness

distance metric and the probability of responding to a given direction. Note that the resulting SI is bounded by zero and one and valid for the analysis of the incoming sound at one ear only.

The width of the mapping function, $U$ in Fig. 5, actually reflects a property of an individual listener. A listener being more precise in the response to the same sound would need a more narrow mapping than a less precise listener. Thus, in contrast to the previous approach [24], in the model described in this chapter, the width of the mapping function as a listener-specific uncertainty, $U$, is considered. It accounts for listener-specific localization precision [34, 42, 56] due to factors like training and attention [14, 51]. Note that for simplicity, direction-dependent response precision is neglected. The lower the uncertainty, $U$, the higher the assumed sensitivity of the listener to distinguish spectral features. In the next section, this parameter will be used to calibrate the model to listener-specific performance.

The model stages described so far are monaural. Thus, they do not consider binaural cues and have been designed for the median SP where the interaural differences are zero and thus binaural cues do not contribute. In order to take into account the contribution of both ears, the monaural model results for both ears are combined. Previous approaches averaged the monaural SIs for both ears [24] and thus were able to consider the contribution of both ears for targets placed in the median SP. In the model described in this chapter, the lateral target range is extended to arbitrary SPs by applying a binaural weighting function [12, 29], which reduces the contribution of the contralateral ear, depending on the lateral direction of the target sound. Thus, the binaural weighting function is applied to each monaural SI, and the sum of the weighted monaural SIs yields the binaural SI.

For an incoming sound, the binaural SIs are calculated for all template entries selected by the template angle. Such a binaural SI as a function of the template angle is related to the listener's response probability as a function of the response angle. It can be interpreted as a discrete version of a probability density function, namely, a probability mass vector, PMV, showing the probability of responding at an angle to a particular target. In order to obtain a PMV, the binaural SI is normalized to have a sum of one. Note that this normalization assumes that the template angles regularly sample an SP. If this is not the case, regularization by spline interpolation is applied before the normalization.

The PMVs, calculated separately for each target under consideration, are represented in a prediction matrix. This matrix describes the probability of responding at a polar angle given a target placed at a specific angle. The right panel of Fig. 5 shows the prediction matrix resulting for the exemplary listener, NH58, in a baseline condition where the listener uses his/her own DTFs, and all available listener-specific DTFs are used as targets. The abscissa shows the target angle, the ordinate shows the response angle and the brightness represents the response probability. This representation is used throughout the following sections. It also allows for a visual comparison between the model predictions and the responses obtained from actual localization experiments.

## 2.3 Interpretation of the Probabilistic Model Predictions

In order to compare the probabilistic results from the model with the experimental results, likelihood statistics, calculated for actual responses from sound localization experiments and for responses resulting from virtual experiments driven by the model prediction, can be used—see Eq. (1) in [24]. The comparison between the two likelihoods allows one to evaluate the validity of the model, because only for similar likelihoods the model is assumed to yield valid predictions. The likelihood has, however, a weak correspondence with localization performance parameters commonly used in psychophysics.

Localization performance in the polar dimension usually considers local errors and hemifield confusions [35]. Although these errors derived by geometrical aspects cannot sufficiently represent the current understanding of human hearing, they are frequently used and thus enable comparison of results between studies. Quadrant errors, QEs, that is the percentage of polar errors larger or equal to 90°, represent the confusions between hemifields—for instance, front/back or up/down—without considering the local response pattern. Unimodal local responses can be represented as a Kent distribution [7], which, considering the polar dimension only, can be approximated by the polar bias and polar variance. Thus, the local errors are calculated only for local responses within the correct hemifield, namely, without the responses

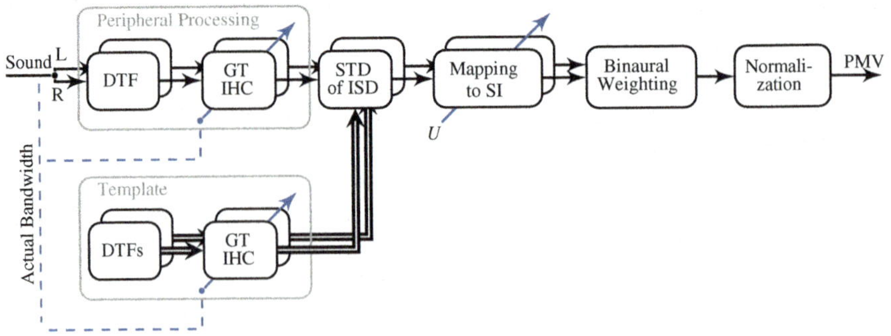


**Fig. 6** Structure of the proposed SP localization model—see text for the description of the stages

yielding the QEs. A single representation of the local errors is the local polar RMS error, PE, which combines localization bias and variance in a single metric.

In the proposed model, QEs and PEs are calculated from the PMVs. The QE is the sum of the PMV entries outside the local polar range defined by the response-target difference greater or equal to 90°. The PE is the discrete expectancy value within the local polar range. In the visualization of prediction matrices—see for example right column of Fig. 5—bright areas in the upper left and bottom right corners would indicate large QEs, a strong concentration of the brightness at the diagonal would indicate small PEs. Both errors can be calculated either for a specific target angle or as the arithmetic average across all target angles considered in the prediction matrix.

Figure 6 summarizes the final structure of the model. It requires the incoming signal from a sound source as the input and results in the response probability as a function of response angle, namely PMV, for given template DTFs. Then, from PMVs calculated for the available target angles, QEs and PEs are calculated for a direct comparison with the outcome of a sound-localization experiment.

## 3 Listener-Specific Calibration and Evaluation

Listeners show an individual localization performance even when localizing broadband sounds in free field [31]. While the listener-specific differences in the HRTFs may play a role, also other factors like experience, attention, or utilization of auditory cues might be responsible for differences in the localization performance. Thus, this section is concerned with the calibration of the model for each particular listener. By creating calibrations for 17 listeners, a pool of listener-specific models is provided. In order to estimate the use of this pool in future applications, the performance of this pool is evaluated in two experiments. In Sect. 4, the pool is applied to various applications.

## 3.1 Calibration: Pool of Listener-Specific Models

The SP localization model is calibrated to the baseline performance of a listener in terms of finding an optimal uncertainty, $U$. Recall that the lower the uncertainty, $U$, the higher the assumed efficiency of the listener in evaluating spectral features. An optimal $U$ minimizes the difference between the predicted and the listener's actual baseline performance in terms of a joint metric of QE and PE, namely, the $\mathcal{L}^2$-norm.

The actual baseline performance was obtained in localization experiments where a listener was localizing sounds using his/her own DTFs presented via headphones. Gaussian white noise bursts with a duration of 500 ms and a fade-in/out of 10 ms were used as stimuli. The acoustic targets were available for elevations from $-30°$ to $80°$ in the lateral range of at least $\pm 30°$ around the median SP. Listeners responded by manually pointing to the perceived direction of a target. For more details on the experimental methods see [10, 30, 51].

The model predictions were calculated considering SPs within the lateral range of $\pm 30°$. The targets were clustered to SPs with a width of $20°$ each. For the peripheral processing, the lower and upper corner frequency was 0.7 and 18 kHz, respectively, resulting in 18 frequency bands with a spacing of one ERB.

Table 1 shows the values of the uncertainty, $U$, for the pool of 17 listeners. The impact of the calibration becomes striking by comparing the predictions based on the listener-specific, calibrated pool with the predictions basing on the pool using $U = 2$ for all listeners as in [24]. Figure 7 shows the actual and predicted performance as a comparison with a pool calibrated to $U = 2$ for all listeners and a listener-specific calibrated pool. Note the substantially higher correlation between the prediction with the actual results in the case of the listener-specific calibration. The correlation coefficients in the order of $r = 0.85$ provide evidence for sufficient power in the predictions for the pool.

**Table 1** Values of the uncertainty $U$ for the pool of listener-specific models identified by NH$n$

| NH$n$ | 12 | 15 | 21 | 22 | 33 | 39 | 41 | 42 | 43 | 46 | 55 | 58 | 62 | 64 | 69 | 71 | 72 |
|---|---|---|---|---|---|---|---|---|---|---|---|---|---|---|---|---|---|
| $U$ | 1.6 | 2.0 | 1.8 | 2.0 | 2.3 | 2.3 | 3.0 | 1.8 | 1.9 | 1.8 | 2.0 | 1.4 | 2.2 | 2.1 | 2.1 | 2.1 | 2.2 |

## 3.2 Evaluation

In order to evaluate the SP localization model, the experimental data from two studies investigating stationary broadband sounds are modeled and compared to the experimental results. Only two studies were available because both the listener-specific HRTFs and the corresponding responses are necessary for the evaluation. For each of these studies, two predictions are calculated, namely, one for the listeners who actually participated in that experiment and one for the whole pool of listener-specific,

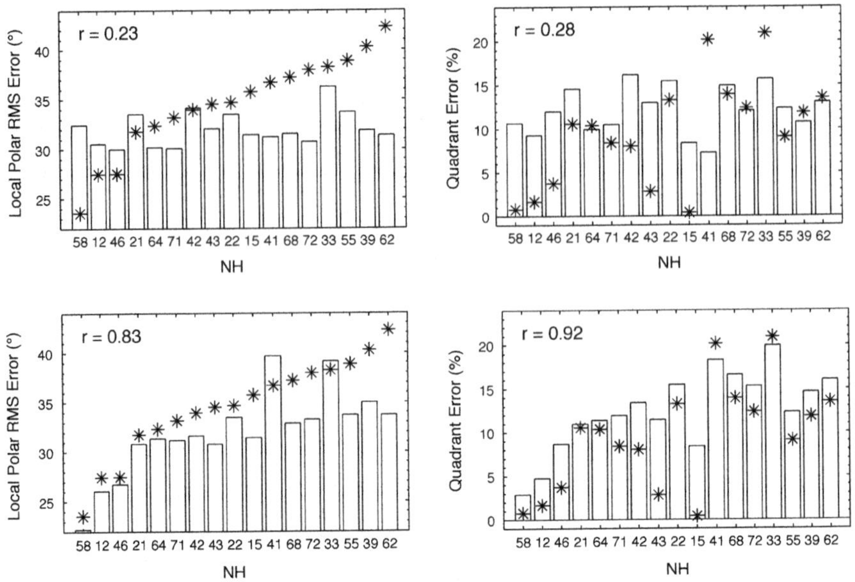


**Fig. 7** Localization performance in baseline condition. *Bars* Model predictions. *Asterisks* Actual performance obtained in sound localization experiments. *Top* Model predictions for $U = 2$ as in [24]. *Bottom* Model predictions for listener-specific calibration. r…Pearson's correlation coefficient with respect to actual and predicted performance

calibrated models. For the participants, the predictions are done on the basis of the actual targets, whereas for the pool, all targets are considered by randomly drawing from the available DTFs.

## Effect of the Number of Spectral Channels

A previous study tested the effect of the number of spectral channels on the localization performance in the median SP [10]. While that study was focused on cochlear-implant processing, the localization experiments were done on listeners with normal hearing using a Gaussian-envelope tone vocoder—for more details see [10]. The frequency range of 0.3–16 kHz was divided into 3, 6, 9, 12, 18, or 24 channels, equally spaced on the logarithmic frequency scale. The top row of Fig. 8 shows three channelized DTFs from an exemplary listener.

The bottom row of Fig. 8 shows the corresponding prediction matrices including the actual responses for this particular listener. Note the correspondence of the localization performance for that particular listener between the actual responses, A, and the model predictions, P. Good correspondence between the actual responses and prediction matrices was found for most of the tested listeners, which is supported by

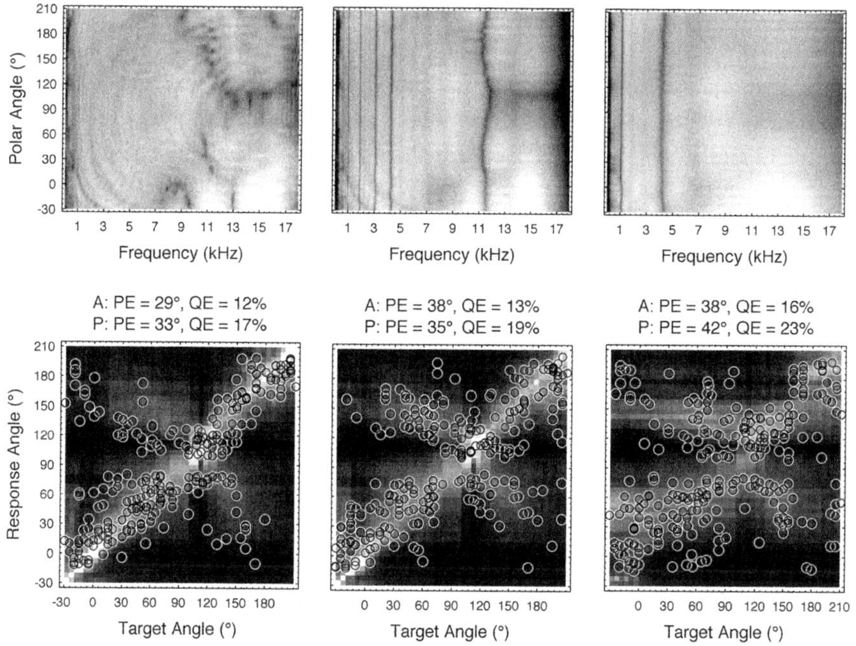


**Fig. 8** Effect of the number of spectral channels for listener, NH42. *Top* Channelized left-ear DTFs of median SP with brightness-encoded magnitude as in Fig. 1, *right panel*. *Bottom* Prediction matrices with brightness-encoded probability as in Fig. 5, *right panel*, and actual responses, *open circles*. *Left* Unlimited number of channels. *Center* 24 spectral channels. *Right* 9 spectral channels. A…actual performance from [10], P…predicted performance

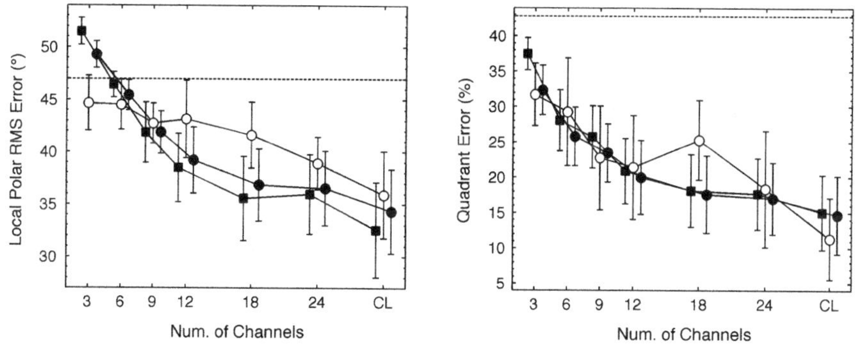


**Fig. 9** Localization performance for listener groups as functions of the number of spectral channels. *Open circles* Actual performance of the listeners replotted from [10]. *Filled circles* Performance predicted for the listeners tested in [10] using the targets from [10]. *Filled squares* Performance predicted for the listener pool, using randomly chosen targets. *Error bars* ±1 standard deviations of the average over the listeners. *Dashed line* Chance performance corresponding to guessing the direction of the sound. CL…unlimited number of channels, broadband clicks

the overall response-prediction-correlation coefficients of 0.62 and 0.74 for PE and QE, respectively.

Figure 9 shows the predicted and the actual performance as averages over the listeners. In comparison to the actual performance, the models underestimated the PEs for 12 and 18 channels and overestimated them for 3 channels. The predictions for the pool seem to follow the predictions for the actually tested listeners showing generally similar QEs but slightly smaller PEs. While the analysis of the nature of these errors is outside of the focus of this chapter, both predictions, those for the actual listeners and those for the pool, seem to well represent the actual performance in this localization experiment.

### Effect of Band Limitation and Spectral Warping

In another previous study, localization performance was tested in listeners using their original DTFs, band-limited DTFs and spectrally warped DTFs [51]. The band limitation was done at 8.5 kHz. The spectral warping compressed the spectral features in each DTF from the range 2.8–16 kHz to the range 2.8–8.5 kHz. While the focus of that study was to estimate the potential of re-learning sound localization with drastically modified spectral cues in a training paradigm, the experimental *ad-hoc* results from the pre-experiment are used to evaluate the proposed model. Note that, for this purpose, the upper frequency of the peripheral processing stage was configured to 8.5 kHz for the band-limited and warped conditions.

The top row of Fig. 10 shows the DTFs and the bottom row the prediction matrices for the original, band-limited and warped conditions for the exemplary listener, NH12. The actual responses show a good correspondence to the prediction matrices. Figure 11 shows group averages of the experimental results and the corresponding predictions. The group averages show a good correspondence between the actual and predicted performance. The correlation coefficient between the actual responses and predictions was 0.81 and 0.85 for PE and QE, respectively. The predictions of the pool well reflect the group averages of the actual responses.

## 4 Applications

The evaluation from the previous section shows response-prediction correlation coefficients in the order of 0.75. This indicates that the proposed model is reliable in predicting localization performance when applied with the listener-specific calibrations. Thus, in this section, the calibrated models are applied to predict localization performance in order to address issues potentially interesting in spatial-audio applications.

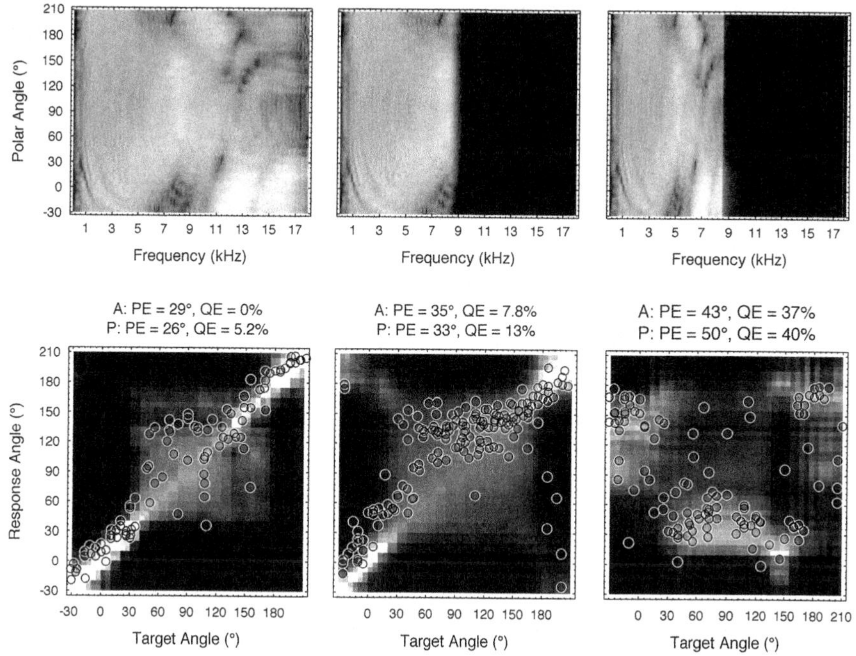


**Fig. 10** Listener, NH12, localizing with different DTFs, namely, original, *left column*, band-limited, *center column*, and spectrally warped, *right column*. *Top* Left-ear DTFs in the median SP. *Bottom* Prediction matrices with actual responses from [51], /open circles/. All other conventions are as in Fig. 8

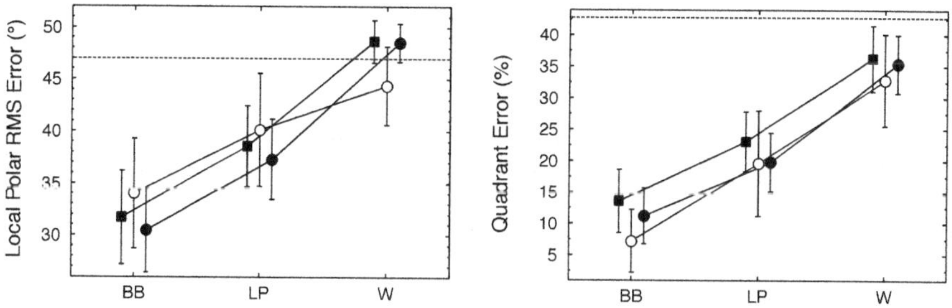


**Fig. 11** Localization performance for listener groups in conditions broadband, BB, band-limited, LP, and spectrally warped, W. *Open circles* Actual performance of the tested listeners from [51]. All other conventions are as in Fig. 9

## 4.1 Non-Individualized Binaural Recordings

Binaural recordings aim at creating a spatial impression when listening via headphones. They are usually created using either an artificial head or mounting microphones into the ear canal of a listener and, thus, implicitly use HRTFs. When listening

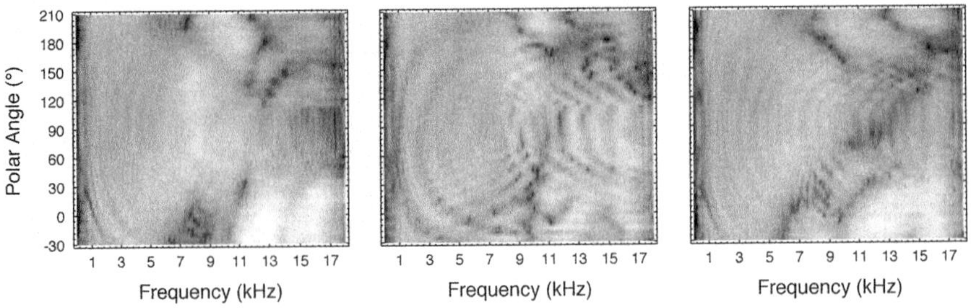


**Fig. 12** Left-ear DTFs of different listeners in the median SP. *Left* NH12. *Center* NH58. *Right* NH33. *Brightness* Spectral magnitude—for code see Fig. 1, *right panel*

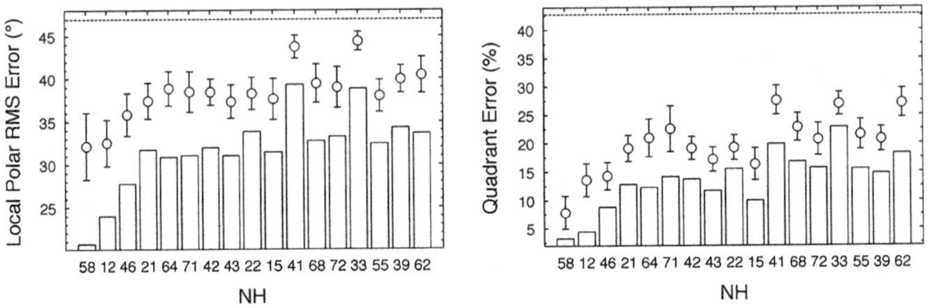


**Fig. 13** Listeners' localization performance for non-individualized versus individualized DTFs. *Bars* Individualized DTFs. *Circles* Non-individualized DTFs averaged over 16 DTF sets. *Error bars* $\pm 1$ standard deviation of the average. *Dashed line* Chance performance corresponding to guessing the direction of the sound

to binaural recordings, the HRTFs of the listener do not necessarily correspond to those used in the recordings. HRTFs are, however, generally highly listener-specific and the relevant spectral features differ across listeners—see Fig. 12. Usually, SP localization performance degrades when listening to binaural signals created with non-individualized HRTFs [34]. The degree of the performance deterioration can be expected to depend on the similarity of the listener's DTFs with those actually applied. Here, the proposed model is used to estimate the localization performance for non-individualized binaural recordings. Figure 13 compares the performance when listening to individualized recordings with the average performance when listening to non-individualized recordings created from all other 16 listeners. It is evident that, on average, listening with other ears results in an increase of predicted localization errors.

Thus, the question arises of how a pool of listeners would localize a binaural recording from a particular listener, for instance, NH58. Figure 14 shows the listener-specific *increase* in the predicted localization errors when listening to a binaural recording spatially encoded using the DTFs from NH58 with respect to the errors predicted for using individualized DTFs. Some of the listeners like NH22 show only little increase in errors, while others like NH12 show large increase.

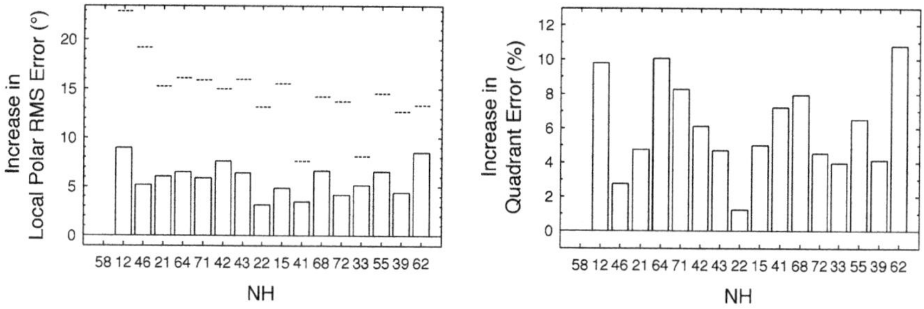


**Fig. 14** *Bars* Listener-specific increase in predicted localization errors when listening to the DTFs from NH58 with respect to the errors predicted when listening to individualized DTFs. *Dashed lines* Chance performance, not shown if too large

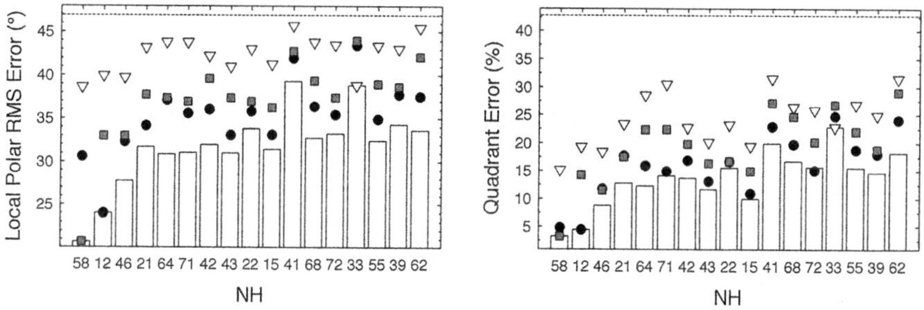


**Fig. 15** Localization performance of the pool listening to different DTFs. *Bars* Individualized DTFs. *Circles* DTFs from NH12. *Squares* DTFs from NH58. *Triangles* DTFs from NH33. *Dashed line* Chance performance

Generally, one might assume that the different anatomical shapes of ears produce more or less distinct directional features. Thus, the quality of the HRTFs might vary, having effect on the ability to localize sounds in the SPs. Figure 15 shows the performance of the pool, using the DTFs from NH12, NH58 and NH33. The DTFs from these three listeners provided best, moderate and worst performance, respectively, predicted for the pool listening to binaural signals created with one of those DTF sets.

This analysis demonstrates how to evaluate across-listener compatibility of binaural recordings. Such an analysis can also be applied for other purposes like the evaluation of HRTFs of artificial heads for providing sufficient spatial cues for binaural recordings.

## 4.2 Assessing the Quality of Spatial Cues in Hearing-Assist Devices

In the development of hearing-assist devices, the casing, its placement on the head, and the placement of the microphone in the casing play an important role for the

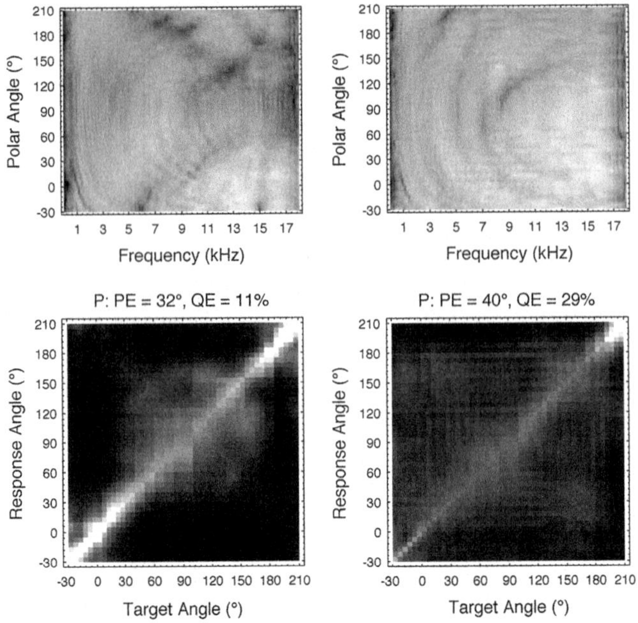


**Fig. 16** Impact of the microphone placement. *Top* Left-ear DTFs of median SP from NH10. *Bottom* Prediction matrices. *Left* ITE microphone. *Right* BTE microphone. All other conventions are as in Fig. 8

effective directional cues. The proposed SP localization model can be used to assess the quality of the directional cues picked up by the microphone in a given device. Figure 16 shows DTFs resulting from behind-the-ear, BTE, compared to in-the-ear, ITE, placement of the microphone for the same listener. The BTE microphone was placed above the pinna, pointing to the front, a position commonly used by the BTE processors in cochlear-implant systems. The bottom row of Fig. 16 shows the corresponding prediction matrices and the predicted localization performance, namely, PE and QE. For this particular listener, the model predicts that if NH10 were listening with the BTE DTFs, his/her QE and PE would increase from 12 to 30% and from 32 to 40°, respectively. This can be clearly related to the impact of degraded spatial cues. Note that in this analysis it was assumed that NH10 fully adapted to the particular HRTFs. This was realized by using the same set of DTFs for the targets and the template in the model.

The impact of using BTE DTFs was also modeled for the pool of listeners using the calibrated models. Two cases are considered, namely, *ad-hoc* listening where the listeners are confronted with the DTF set without any experience in using it, and trained listening where the listeners are fully adapted to the respective DTF set. Figure 17 shows the predictions for the pool. The BTE DTFs result in performances close to guessing and the ITE DTFs from the same listener substantially improve the performance. In trained listening, the performance for the ITE DTFs is at the level of

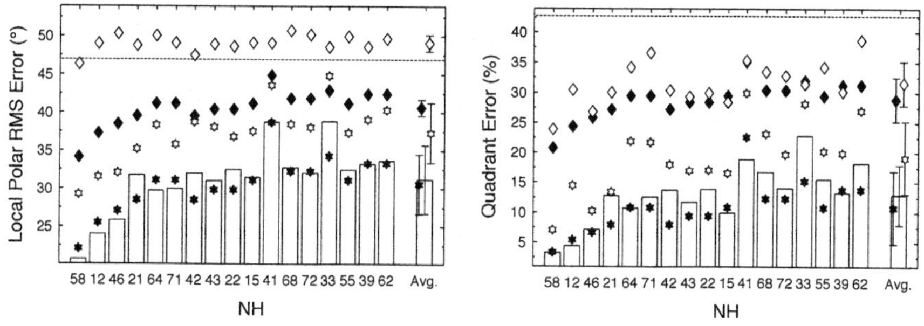


**Fig. 17** Localization performance of the pool listening to different DTFs. *Bars* Individualized DTFs. *Open symbols Ad-hoc* listening. *Filled symbols* Trained listening. *Hexagrams* ITE DTFs from NH10. *Diamonds* BTE DTFs from NH10. Avg... average performance over all listeners. *Error bars* ±1 standard deviation. *Dashed line* Chance performance

the individualized DTFs, consistent with the potential of the plasticity of the spectral-to-spatial mapping [13]. The BTE DTFs, however, do not allow performance at the same level as the ITE DTFs, even when full adaptation is considered.

This analysis shows a model-based method to optimize the microphone placement with respect to the salience of directional cues. Such an analysis might be advantageous in the development of future hearing-assist devices.

## *4.3 Phantom Sources in Surround-Sound Systems*

Sound synthesis systems for spatial audio have to deal with a limited number of loudspeakers surrounding the listener. In a system with a small number of loudspeakers, vector-based amplitude panning, VBAP [41], is commonly applied in order to create phantom sources perceived between the loudspeakers. In a surround setup, this method is also commonly used to position the phantom source along SPs, namely, to pan the source from the front to the back [11] or from the eye level to an elevated level [41]. In this section, the proposed model is applied to investigate the use of VBAP within SPs.

### Amplitude Panning Along a Sagittal Plane

Now a VBAP setup with two loudspeakers is assumed, which are placed at the same distance, in the horizontal plane at the eye level, and in the same SP. Thus, the loudspeakers are in the front and in the back of the listener, corresponding to polar angles of 0° and 180°, respectively. While driving the loudspeakers with the same signal, the amplitude panning ratio can be varied from 0, front speaker only, to 1, rear speaker only, with the goal of panning the phantom source between the two loudspeakers.

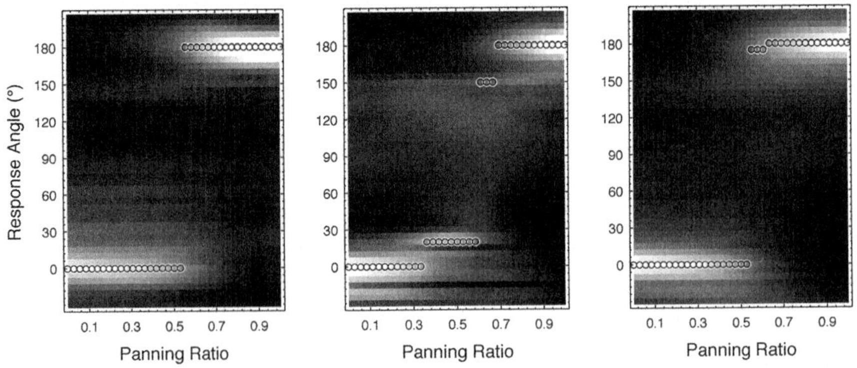


**Fig. 18** Predicted response probabilities, PMVs, as a function of the amplitude panning ratio. *Left* Results for NH22. *Center* Results for NH64. *Right* Results for the pool of listeners. *Circle* Maximum of a PMV. Panning ratio of 0: Only front loudspeaker active. Panning ratio of 1: Only rear loudspeaker active. All other conventions are as in Fig. 5, *right panel*

Figure 18 shows the predicted listener-specific response probabilities in terms of the PMV as a function of the panning ratio for two loudspeakers placed at the lateral angle of 30°. The PMVs are shown for two individual listeners and also for the pool of listeners. The directional stability of phantom sources varies across listeners. For NH22, the prediction of perceived location abruptly changes from front to back, being bimodal only around the ratio of 0.6. For NH64, the transition seems to be generally smoother, with a blur in the perceived sound direction. Note that for NH64 and a ratio of 0.5, the predicted direction is elevated even though the loudspeakers were placed in the horizontal plane. The results for the pool predict an abrupt change in the perceived direction from front to back, with a blur indicating a listener-specific unstable representation of the phantom source for ratios between 0.5 and 0.7.

### Effect of Loudspeaker Span

The unstable synthesis of phantom sources might be reduced by using a more adequate distance in the SP between the loudspeakers. Thus, it is shown how to investigate the polar span between two loudspeakers required to create a stable phantom source in the synthesis. To this end, a VBAP setup of two loudspeakers placed in the median SP, separated by a polar angle and driven with the panning ratio of 0.5, is used. Note that a span of 0° corresponds to a synthesis with a single loudspeaker and thus to the baseline condition. In the proposed SP localization model, the target angle describes the average of the polar angles of both loudspeakers, which, in VBAP, is thought to correspond to the direction of the phantom source. The model was run for all available target angles resulting in the prediction of the localization performance.

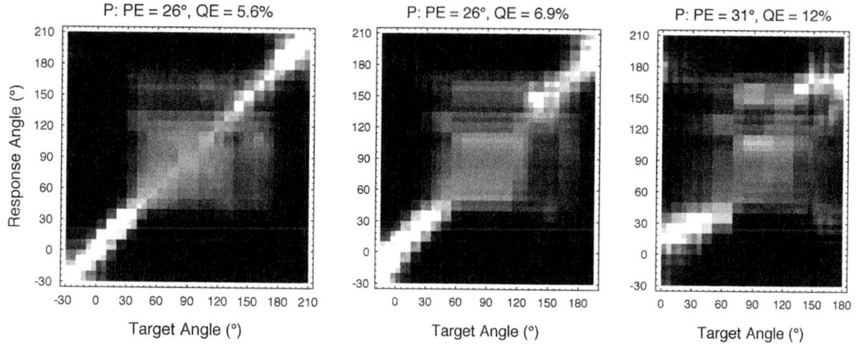


**Fig. 19** Predictions for different loudspeaker spans and NH12. *Left* Span of 0°, single-loudspeaker synthesis, baseline condition. *Center* Span of 30°. *Right* Span of 60°. All other conventions are as in Fig. 8

Figure 19 shows prediction matrices and predicted localization performance for NH12 and three different loudspeaker spans. Note the large increase of errors from 30 to 60° of span, consistent with the results from [5]. Figure 20 shows the average increase in localization error predicted for the pool of listeners as a function of the span. The increase is shown relative to the listener-specific localization performance in the baseline condition. Note that not only the localization errors but also the variances across the listeners increase with increasing span.

This analysis shows how the model may help in choosing the adequate loudspeaker span when amplitude panning is applied to create phantom sources. Such an analysis can also be applied when more sophisticated sound-field reproduction approaches like Ambisonics or wave-field synthesis are involved.

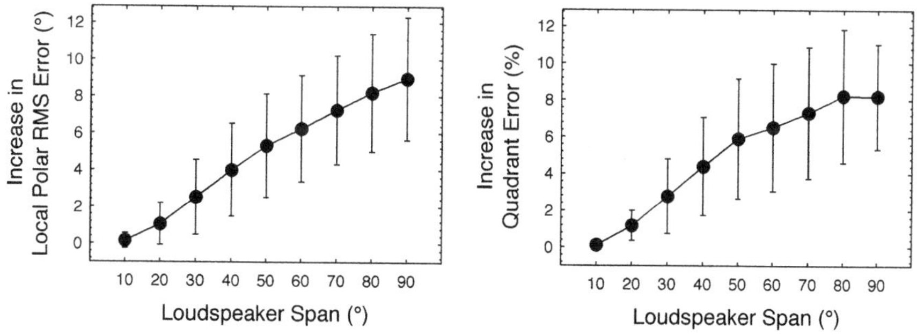


**Fig. 20** Increase in localization errors as a function of the loudspeaker span. *Circles* Averages over all listeners from the pool. *Error bars* ±1 standard deviation

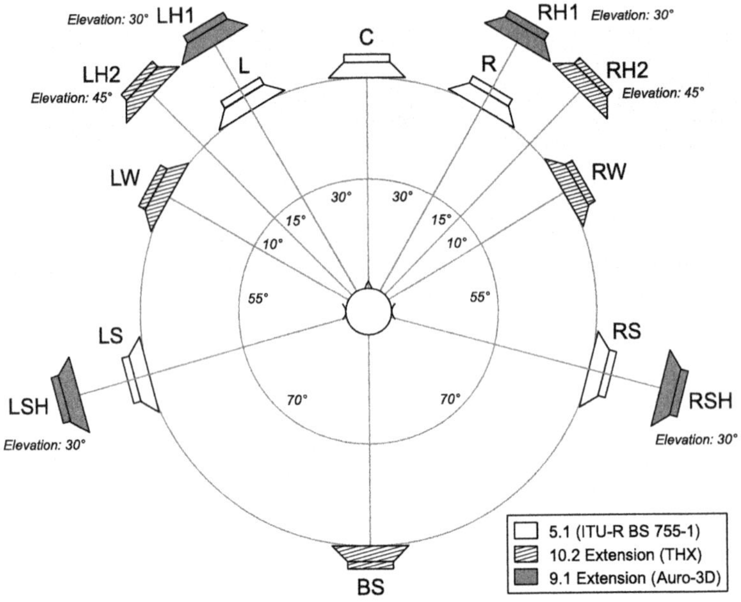


**Fig. 21** Loudspeaker positions of three typical surround-sound systems. Drivers for the low-frequency effect, LFE, channels not shown

## Results for Typical Surround-Sound Setups

The most common standardized surround-sound setup is known as the 5.1 setup [18]. In this setup, all loudspeakers are placed in the horizontal plane at a constant distance around the listener. Recently, other schemes have been proposed to include elevated speakers in the synthesis systems. The 10.2 setup, known as *Audyssey DSX* [15] and the 9.1 setup, known as *Auro-3D* [48], consider two and four elevated loudspeakers, respectively. Figure 21 shows the positions of the loudspeakers in those three surround-sound setups. The model was applied to evaluate the localization performance when VBAP is used to pan a phantom source at the left hand side from front, L, to back, LS. While in the 5.1 setup only loudspeakers L and LS are available, in 10.2 and 9.1 the loudspeakers LH2 and LH1 & LSH, respectively, may also contribute even to create an elevated phantom source.

VBAP was applied between the closest two loudspeakers by using the law of tangents [41]. For a desired polar angle of the phantom source, the panning ratio was $R = \frac{1}{2} - \frac{\tan(\delta)}{2\tan(0.5\beta)}$ with $\beta$ denoting the loudspeaker span in polar dimension and $\delta$ denoting the difference between the desired polar angle and the polar center angle of the span. The contributing loudspeakers were not always in the same SP, thus, the lateral angle of the phantom source was considered for the choice of the SP in the modeling by applying the law of tangents on the lateral angles of the loudspeakers for the particular panning ratio, $R$.

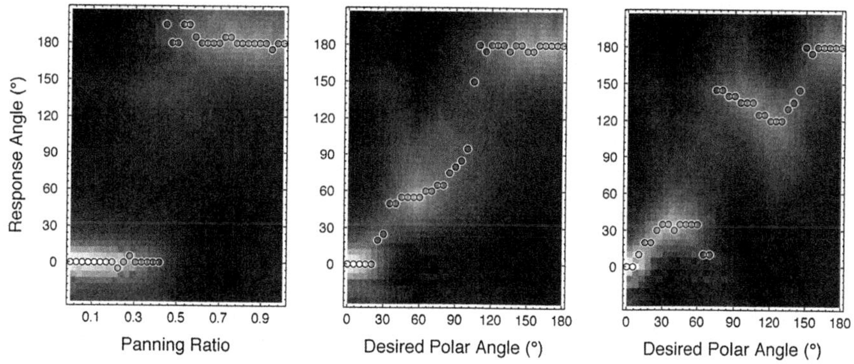


**Fig. 22** Predictions for VBAP applied to various surround-sound systems. *Left* 5.1 setup, panning between the loudspeakers L and LS. *Center* 10.2 DSX setup panning from L, polar angle of 0°, via LH2, 55°, to LS, 180°. *Right* 9.1 Auro-3D setup panning from L, 0°, via LH1, 34°, and LSH, 121°, to LS, 180°. *Desired polar angle* Continuous scale representing VBAP across pair-wise contributing loudspeakers. All other conventions are as in Fig. 18

Figure 22 shows the predicted pool averages of the PMVs as a function of the desired polar angle of the phantom source. The improvements due to the additional elevated loudspeakers in the 10.2 and 9.1 setups are evident. Nevertheless, the predicted phantom sources are far from perfectly following the desired angle. Especially for the 9.1 setup, in the rear hemifield, the increase in the desired polar angle, namely, *decrease* in the elevation, resulted in a decrease in the predicted polar angle, namely, *increase* in the elevation.

The proposed model seems to be well-suited for addressing such a problem. It is easy to show how modifications of the loudspeaker setup would affect the perceived angle of the phantom source. As an example, the positions of the elevated loudspeakers in the 9.1 setup were modified in two ways. First, the lateral distance between the loudspeakers, LH1 and LSH, was decreased by modifying the azimuth of LSH from 110 to 140°. Second, both loudspeakers, LSH and LS, were placed to the azimuth of 140°. Figure 23 shows the predictions for the modified setups. Compared to the original setup, the first modification clearly resolves the problem described above. The second modification, while only slightly limiting the lateral range, provides an even better representation of the phantom source along the SP.

## 5 Conclusions

Sound localization in SPs refers to the ability to estimate the sound-source elevation and to distinguish between front and back. The SP localization performance is usually measured in time-consuming experiments. In order to address this disadvantage, a model predicting SP localization performance of individual listeners has been

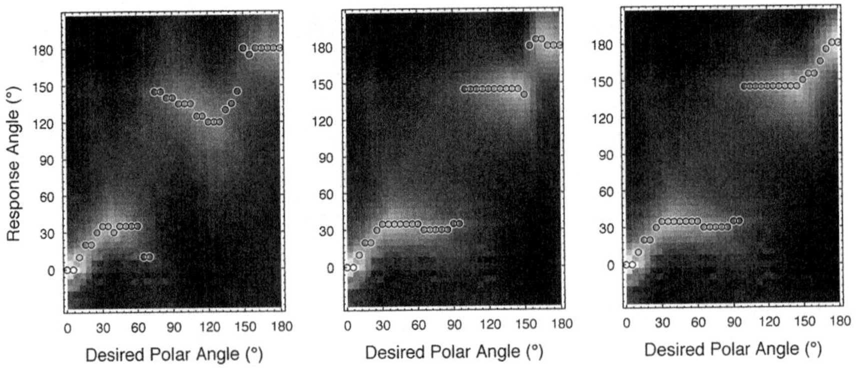


**Fig. 23** Predictions for two modifications to the 9.1 Auro 3D setup. *Left* Original setup, loudspeakers LS and LSH at azimuth of 110°. *Center* LSH at azimuth of 140°. *Right* LS and LSH at azimuth of 140°. All other conventions are as in Fig. 22

proposed. Listener-specific calibration was performed for a pool of 17 listeners, and the calibrated models were evaluated using results from psychoacoustic localization experiments. The potential of the calibrated models was demonstrated for three applications, namely,

1. The evaluation of the spatial quality of binaural recordings
2. The assessment of the spatial quality of directional cues provided by the microphone placement in hearing-assist devices
3. The evaluation and improvement of the loudspeaker position in surround-sound systems

These applications are examples of situations where SP localization cues, namely, spectral cues, likely play a role. The model is, however, not limited to those applications and it hopefully will help in assessing spatial quality in other applications as well.

**Acknowledgments** The authors would like to thank H. S. Colburn and two anonymous reviewers for constructive comments on earlier versions of this manuscript. This research was supported by the Austrian Science Fund, FWF, project P 24124–N13.

## References

1. V. R. Algazi, C. Avendano, and R. O. Duda. Elevation localization and head-related transfer function analysis at low frequencies. *J Acoust Soc Am*, 109:1110–1122, 2001.
2. F. Asano, Y. Suzuki, and T. Sone. Role of spectral cues in median plane localization. *J Acoust Soc Am*, 88:159–168, 1990.
3. E. Blanco-Martin, F. J. Casajus-Quiros, J. J. Gomez-Alfageme, and L. I. Ortiz-Berenguer. Estimation of the direction of auditory events in the median plane. *Appl Acoust*, 71:1211–1216, 2010.

4. J. Blauert. *Räumliches Hören (Spatial Hearing)*. S. Hirzel Verlag Stuttgart, 1974.
5. P. Bremen, M. M. van Wanrooij, and A. J. van Opstal. Pinna cues determine orientation response modes to synchronous sounds in elevation. *J Neurosci*, 30:194–204, 2010.
6. A. W. Bronkhorst. Localization of real and virtual sound sources. *J Acoust Soc Am*, 98:2542–2553, 1995.
7. S. Carlile, P. Leong, and S. Hyams. The nature and distribution of errors in sound localization by human listeners. *Hear Res*, 114:179–196, 1997.
8. T. Dau, D. Püschel, and A. Kohlrausch. A quantitative model of the "effective" signal processing in the auditory system. I. Model structure. *J Acoust Soc Am*, 99:3615–3622, 1996.
9. B. R. Glasberg and B. C. J. Moore. Derivation of auditory filter shapes form notched-noise data. *Hear Res*, 47:103–138, 1990.
10. M. J. Goupell, P. Majdak, and B. Laback. Median-plane sound localization as a function of the number of spectral channels using a channel vocoder. *J Acoust Soc Am*, 127:990–1001, 2010.
11. J. Hilson, D. Gray, and M. DiCosimo. *Dolby Surround Mixing Manual*. Dolby Laboratories, Inc., San Francisco, CA, 2005. chapter 5—Mixing techniques.
12. M. Hofman and J. Van Opstal. Binaural weighting of pinna cues in human sound localization. *Exp Brain Res*, 148:458–470, 2003.
13. P. M. Hofman and A. J. V. Opstal. Spectro-temporal factors in two-dimensional human sound localization. *J Acoust Soc Am*, 103:2634–2648, 1998.
14. P. M. Hofman, J. G. A. van Riswick, and A. J. van Opstal. Relearning sound localization with new ears. *Nature Neurosci*, 1:417–421, 1998.
15. T. Holman. *Surround Sound: Up and Running*. Focal Press, 2008.
16. S. Hwang and Y. Park. Interpretations on pricipal components analysis of head-related impulse responses in the median plane. *J Acoust Soc Am*, 123:EL65-EL71, 2008.
17. K. Iida, M. Itoh, A. Itagaki, and M. Morimoto. Median plane localization using a parametric model of the head-related transfer function based on spectral cues. *Appl Acoust*, 68:835–850, 2007.
18. Int Telecommunication Union, Geneva, Switzerland. *Multichannel stereophonic sound system with and without accompanying picture*, 2012. Recommendation ITU-R BS.775-3.
19. C. Jin, M. Schenkel, and S. Carlile. Neural system identification model of human sound localization. *J Acoust Soc Am*, 108:1215–1235, 2000.
20. B. F. Katz. Boundary element method calculation of individual head-related transfer function. I. Rigid model calculation. *J Acoust Soc Am*, 110:2440–2448, 2001.
21. W. Kreuzer, P. Majdak, and Z. Chen. Fast multipole boundary element method to calculate head-related transfer functions for a wide frequency range. *J Acoust Soc Am*, 126:1280–1290, 2009.
22. A. Kulkarni and H. S. Colburn. Role of spectral detail in sound-source localization. *Nature*, 396:747–749, 1998.
23. A. Kulkarni and H. S. Colburn. Infinite-impulse-response models of the head-related transfer function. *J Acoust Soc Am*, 115:1714–1728, 2004.
24. E. H. A. Langendijk and A. W. Bronkhorst. Contribution of spectral cues to human sound localization. *J Acoust Soc Am*, 112:1583–1596, 2002.
25. E. A. Lopez-Poveda and R. Meddis. A human nonlinear cochlear filterbank. *J Acoust Soc Am*, 110:3107–3118, 2001.
26. F. R. S. Lord Rayleigh. On our perception of sound direction. *Philos Mag*, 13:214–232, 1907.
27. E. A. Macpherson and J. C. Middlebrooks. Listener weighting of cues for lateral angle: The duplex theory of sound localization revisited. *J Acoust Soc Am*, 111:2219–2236, 2002.
28. E. A. Macpherson and J. C. Middlebrooks. Vertical-plane sound localization probed with ripple-spectrum noise. *J Acoust Soc Am*, 114:430–445, 2003.
29. E. A. Macpherson and A. T. Sabin. Binaural weighting of monaural spectral cues for sound localization. *J Acoust Soc Am*, 121:3677–3688, 2007.

30. P. Majdak, M. J. Goupell, and B. Laback. 3-D localization of virtual sound sources: Effects of visual environment, pointing method, and training. *Atten Percept Psycho*, 72:454–469, 2010.
31. J. C. Makous and J. C. Middlebrooks. Two-dimensional sound localization by human listeners. *J Acoust Soc Am*, 87:2188–2200, 1990.
32. M. I. Mandel, R. J. Weiss, and D. P. W. Ellis. Model-based expectation-maximization source separation and localization. *IEEE Trans Audio Speech Proc*, 18:382–394, 2010.
33. T. May, S. van de Par, and A. Kohlrausch. A probabilistic model for robust localization based on a binaural auditory front-end. *IEEE Trans Audio Speech Lang Proc*, 19:1–13, 2011.
34. J. C. Middlebrooks. Individual differences in external-ear transfer functions reduced by scaling in frequency. *J Acoust Soc Am*, 106:1480–1492, 1999.
35. J. C. Middlebrooks. Virtual localization improved by scaling nonindividualized external-ear transfer functions in frequency. *J Acoust Soc Am*, 106:1493–1510, 1999.
36. J. C. Middlebrooks and D. M. Green. Sound localization by human listeners. *Annu Rev Psychol*, 42:135–159, 1991.
37. J. C. Middlebrooks and D. M. Green. Observations on a principal components analysis of head-related transfer functions. *J Acoust Soc Am*, 92:597–599, 1992.
38. H. Møller, M. F. Sørensen, D. Hammershøi, and C. B. Jensen. Head-related transfer functions of human subjects. *J Audio Eng Soc*, 43:300–321, 1995.
39. M. Morimoto. The contribution of two ears to the perception of vertical angle in sagittal planes. *J Acoust Soc Am*, 109:1596–1603, 2001.
40. R. Patterson, I. Nimmo-Smith, J. Holdsworth, and P. Rice. *An efficient auditory filterbank based on the gammatone function*. APU, Cambridge, UK, 1988.
41. V. Pulkki. Virtual sound source positioning using vector base amplitude panning. *J Audio Eng Soc*, 45:456–466, 1997.
42. B. Rakerd, W. M. Hartmann, and T. L. McCaskey. Identification and localization of sound sources in the median sagittal plane. *J Acoust Soc Am*, 106:2812–2820, 1999.
43. C. L. Searle and I. Aleksandrovsky. Binaural pinna disparity: Another auditory localization cue. *J Acoust Soc Am*, 57:448–455, 1975.
44. M. A. Senova, K. I. McAnally, and R. L. Martin. Localization of virtual sound as a function of head-related impulse response duration. *J Audio Eng Soc*, 50:57–66, 2002.
45. E. A. Shaw. Transformation of sound pressure level from the free field to the eardrum in the horizontal plane. *J Acoust Soc Am*, 56:1848–1861, 1974.
46. R. H. Y. So, B. Ngan, A. Horner, J. Braasch, J. Blauert, and K. L. Leung. Toward orthogonal non-individualised head-related transfer functions for forward and backward directional sound: cluster analysis and an experimental study. *Ergonomics*, 53:767–781, 2010.
47. P. Søndergaard and P. Majdak. The auditory modeling toolbox. In J. Blauert, editor, *The technology of binaural listening*, chapter 2. Springer, Berlin-Heidelberg-New York NY, 2013.
48. G. Theile and H. Wittek. Principles in surround recordings with height. *In Proc. 130th AES Conv. Audio Engr. Soc.*, page Convention Paper 8403, London, UK, 2011.
49. M. M. van Wanrooij and A. J. van Opstal. Relearning sound localization with a new ear. *J Neurosci*, 25:5413–5424, 2005.
50. J. Vliegen and A. J. V. Opstal. The influence of duration and level on human sound localization. *J Acoust Soc Am*, 115:1705–1703, 2004.
51. T. Walder. Schallquellenlokalisation mittels Frequenzbereich-Kompression der Außenohrübertragungsfunktionen (sound-source localization through warped head-related transfer functions). Master's thesis, University of Music and Performing Arts, Graz, Austria, 2010.
52. A. J. Watkins. Psychoacoustical aspects of synthesized vertical locale cues. *J Acoust Soc Am*, 63:1152–1165, 1978.
53. F. L. Wightman and D. J. Kistler. The dominant role of low-frequency interaural time differences in sound localization. *J Acoust Soc Am*, 91:1648–1661, 1992.
54. F. L. Wightman and D. J. Kistler. Monaural sound localization revisited. *J Acoust Soc Am*, 101:1050–1063, 1997.

55. P. Zakarauskas and M. S. Cynader. A computational theory of spectral cue localization. *J Acoust Soc Am*, 94:1323–1331, 1993.
56. P. X. Zhang and W. M. Hartmann. On the ability of human listeners to distinguish between front and back. *Hear Res*, 260:30–46, 2010.
57. M. S. A. Zilany and I. C. Bruce. Modeling auditory-nerve responses for high sound pressure levels in the normal and impaired auditory periphery. *J Acoust Soc Am*, 120:1446–1466, 2006.

# Modeling Horizontal Localization of Complex Sounds in the Impaired and Aided Impaired Auditory System

**Nicolas Le Goff, J. M. Buchholz and T. Dau**

## 1 Introduction

Ambient noise, room reflections or the presence of multiple interfering talkers are acoustic factors that can make daily communication challenging. Normal-hearing, NH, people can nevertheless typically communicate almost effortlessly in such adverse conditions [1]. In contrast, hearing-impaired, HI, people often experience major speech-communication difficulties, even when they use hearing aids, HAs. The auditory system uses various acoustic cues, such as common temporal onsets and offsets, spectral content, harmonicity as well as spatial information, to decompose an acoustic scene into its components belonging to the different sound sources [1].

The underlying processes involved in auditory scene analysis are still not very well understood; in particular, the effects of hearing impairment on speech communication in adverse listening conditions have not yet been clarified. The present study focuses on the processing of spatial cues in the auditory system. Whereas the processing of relatively simple spatial sounds is reasonably well-understood in NH listeners, the consequences of hearing impairment for the processing and perception of spatial sounds is not well understood. An important question is, for example, how sensorineural hearing loss, occurring at a peripheral processing level and representing

N. Le Goff (✉)
Center for Applied Hearing Research, Department of Electrical Engineering, Technical University of Denmark, Kongens Lyngby, Denmark
e-mail: nlg@elektro.dtu.dk

J. M. Buchholz
National Acoustic Laboratories, Chatswood, NSW, Australia

J. M. Buchholz
Department of Linguistics—Audiology, Macquarie University, Sydney, Australia

T. Dau
Center for Applied Hearing Research, Department of Electrical Engineering, Technical University of Denmark, Kongens Lyngby, Denmark

J. Blauert (ed.), *The Technology of Binaural Listening*, Modern Acoustics and Signal Processing, DOI: 10.1007/978-3-642-37762-4_5,

the most common type of impairment, affects the processing of spatial cues and the perception of localization. A better understanding of the representation of spatial information in the impaired auditory system could also help understanding the difficulties that the HI listeners experience when communicating in adverse conditions.

For NH listeners, localization of single anechoic sounds has been widely studied and can be well predicted by existing binaural, cross-correlation-based models—compare [2–4]. In contrast, only very few studies have considered auditory localization in noisy conditions—for instance, [5–11]—as well as in reverberant environments—such as [12–15]. Most of these studies found that localization is mainly affected at low signal-to-noise ratios, SNR, and largely limited by audibility, but a more detailed understanding of the limiting factors are still missing. It is generally assumed that the auditory system has developed various strategies for robust localization in these challenging conditions, which includes the (weighted) integration of localization information across time, frequency, and auditory cues, as well as the suppression of wall reflections. The localization of broadband noise with interaural time differences, ITD, that vary across frequency has been described by [16–18], among others. Several studies have shown that localization accuracy is modified by signals preceding and succeeding in time—see [19]. Based on a binaural localization model [20] proposes that, when background noise is present, integration of localization information across time and/or frequency is required. The integration of localization information across ITDs and interaural level differences, ILD, is, for example, reflected in trading experiments—compare [3]. Different auditory mechanisms that help a listener to localize sounds in reverberant environments are summarized by the precedence effect, PE,—see, for instance [21]. In order to model some aspects of these more advanced auditory processes, mechanisms such as contra-lateral inhibition [22] as well as cue selection [23] have been proposed.

Localization in HI listeners can be reduced as a result of different factors [24–27]. ITD discrimination thresholds may be decreased in HI listeners [28, 29] although thresholds vary largely across subjects and may be even as low as for NH listeners [24, 30, 31]. Localization performance in quiet is only reduced in listeners with unilateral hearing loss or substantial low-frequency hearing loss, for instance, [32–34], or with conductive hearing loss [27]. In [35] it is shown that horizontal localization of click-trains in HI listeners is basically as good as in NH listeners when the stimuli are presented at most comfortable levels. However, the performance drops at lower levels due to reduced sensitivity—as measured by an audiogram. Considering horizontal localization, [36] showed that HI listeners show poorer performance than NH listeners when localizing click trains in directional white noise. This decrease in performance is most likely linked to a decrease in audibility, which is caused by a decrease in sensitivity as well as an increase in the amount of temporal and spectral masking—see [25, 27]. Several studies have investigated the PE in HI listeners by measuring the localization dominance of the direct sound over a reflection. In general, the results vary largely across both HI and NH listeners, such that most studies could not reveal any significant reduction in localization dominance in HI listeners. However, a recent study by [37] showed a significant reduction in localization dominance in a large number of HI subjects which was highly correlated with hearing loss. In [38], it is pointed out that, for ongoing sounds, the fine structure of the signal is important for the PE to operate. Given that fine-structure processing is often impaired in HI listeners, it is expected that the PE is less effective. Since cognitive processes may affect localization in natural environments [39], it may be speculated that reduced cognitive capabilities as well as reduced working memory capacity, as often observed in elderly HI listeners will also compromise localization in complex acoustic environments—see [27]. In addition to the above laboratory-based psychoacoustic evidence, questionnaire-based field studies revealed that HI listeners experience localization difficulties in their daily life, which is more severe with increasing hearing loss [40, 41]. Hence, although auditory localization in HI listeners has been widely studied, the detailed underlying processes, particularly in challenging acoustic environments, are still poorly understood.

Sound localization with bilateral HAs, has been widely studied with different device types, listening configurations, source signals, algorithms, and microphone positions. For moderate to severe hearing losses, HA amplification effectively restores audibility up to about 4 kHz and, therefore, enables localization—see, for instance [26]. Subjective studies based on questionnaires generally reported a clear benefit in localization when a second HA is applied [41, 42]. When audibility is not an issue, most psychoacoustic studies agree that, when compared to the unaided case, bilateral amplification slightly deteriorates sound localization performance in the frontal horizontal plane [26, 43–45]. Undisturbed horizontal localization performance has only been reported for HI listeners with rather normal low-frequency hearing and open HA fittings [45, 46], namely, listeners that have access to natural localization cues at low frequencies. According to [47], front-back localization is disturbed by any type of HA due to the obstruction of the pinna by the earmold or the actual HA module. In [47] and [26] it is reported that horizontal localization, in particular front-back confusions, improved for listeners with remaining low- and high-frequency hearing, when no earmold occluded the ear canal and, therefore, natural localization cues were available. In [46] and [48], neither a benefit nor a detrimental effect on front-back confusions has been found for various types of HAs. This contradiction to the findings of [26] may be due to the fact that the later studies applied HAs with an increased bandwidth of 7–8 kHz instead of 5–6 kHz. Both studies also reported that behind-the-ear devices, BTE, produce an increased number of front-back confusions relative to in-the-canal, ITC, in-the-ear, ITE, or in-the-pinna, ITP, devices. In [44] it is reported that the front-back confusions as observed in BTE–HAs can be reduced by applying directional microphones, such as cardiods, instead of omni-directional microphones—in particular at high-frequencies. Moreover, [44] showed that independent bilateral wide-dynamic-range compression, WDRC, can slightly reduce horizontal localization performance due to compressed ILDs. This effect should be removed when a binaural signal link is applied to coordinate the WDRCs at the two HAs. Super-directional beamformers, which combine the signals arriving at all microphones of the left and right HAs, are able to preserve localization in simple acoustical environments—compare [48–50]. However, it is still unclear how well they preserve localization in complex and, in particular, reverberant environments. Hence, HAs have a significant effect on auditory localization, but how a given device, type of fitting, compensation strategy, and/or advanced signal processing, affect an individual HI listener, is not well understood.

In order to better understand the consequences of an individual's hearing loss, including the effect of a HA, an auditory localization model framework would be useful. If successful, such a framework could then also be used to test and optimize the signal processing and compensation strategies applied in a HA for an individual HI listener. As a first step into this direction, a binaural signal processing model framework is proposed in this chapter. The approach is novel as it analyses NH versus individualized HI processing in conditions of signal detection and localization in noisy and/or reverberant situations. The peripheral processing in the model includes the dual-resonance nonlinear, DRNL filterbank proposed in earlier monaural studies [51]. The characteristics of the peripheral processor can be adjusted in both channels—representing the left and the right ear—such that it accounts for sensorineural hearing impairment due to a loss of cochlear compression. Such compression loss is commonly associated with broader auditory filters, i.e., decreased frequency selectivity and reduced sensitivity in terms of audibility. The binaural processor of the model is an equalization-and-cancellation, EC, processor similar to the one proposed by [52]. For the back end, depending on the particular task and outcome measure, the model assumes an optimal detector in conditions of signal detection or discrimination, or a localization unit for the prediction of sound source localization. The localization back end also includes a cue-selection mechanism inspired by the work proposed in [23], selecting reliable spatial information from a complex spatial-excitation pattern.

The overall structure of the model and the main properties of the individual processing steps will be presented first, including the simulation of hearing loss in the peripheral processing stage of the model. The model will be evaluated in classical conditions of binaural signal detection, and then be applied to localization conditions with two concurrent talkers in anechoic and reverberant rooms, assuming either NH or specific types of hearing impairment. Finally, effects of HA signal processing on sound localization will be considered and the benefits and limitations of current compensation strategies for restoring spatial cues will be discussed in the framework of the presented model.

## 2 Binaural Signal Processing Model

### *2.1 Overall Structure*

A block diagram of the processing model is shown in Fig. 1. When appropriate, incoming signals entering the ears can be convolved with binaural room impulse responses, BRIR, and/or applied to individual left- and right-channel HA processing. The structure of the model is similar to previous signal-driven binaural models and can be divided into three parts, namely, peripheral, binaural and central process-

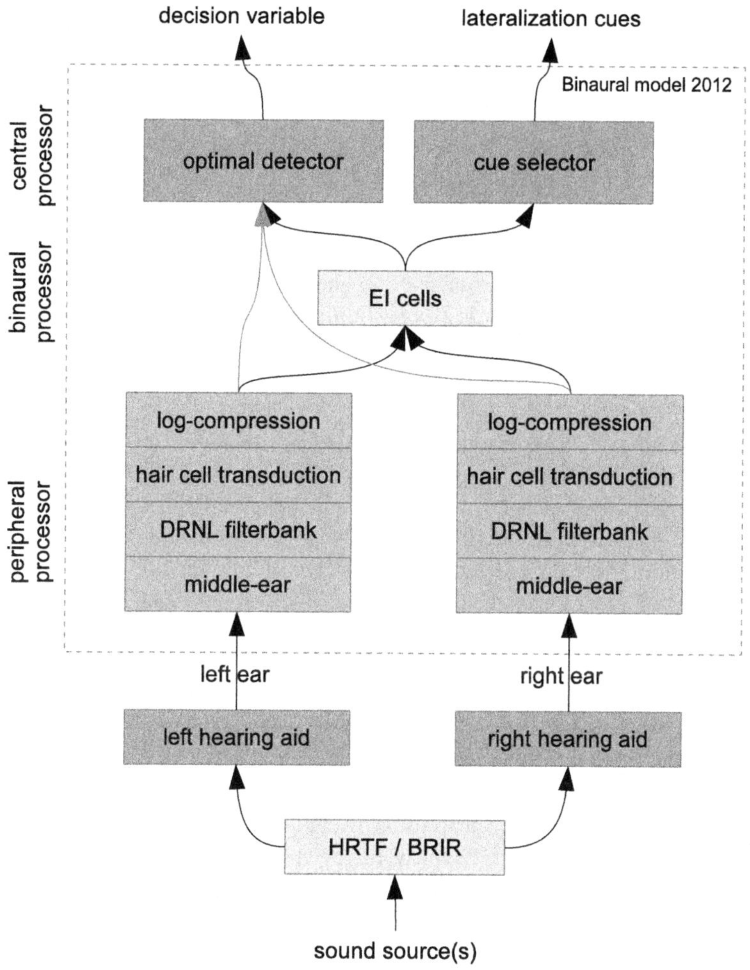


**Fig. 1** Block diagram of the proposed model. At the input to the model, the signals can be processed by a HA algorithm or convolved with BRIRs. The left- and it right-ear input signals are processed in the peripheral processor, that is, middle-ear transformation, basilar-membrane filtering, filterbank, inner hair-cell transformation, and logarithmic compression, and then compared in an EC-type binaural processor. Two different pathways to the central processor are considered. The optimal detector deals with signal-detection conditions. The cue selector extracts the spatial positions of external sound sources

ing. The peripheral processing includes middle-ear transformation, spectral filtering, hair-cell transduction and logarithmic compression. The output of the left and right channels feed an EC type binaural processor that outputs binaural-activity maps along an internal ITD, and ILD, dimension, denoted by $\tau$ and $\alpha$, respectively. The information on the binaural-activity map is then either evaluated by an optimal detector, to predict binaural detection, such as ITD and ILD thresholds or tone-in-noise detection thresholds, or by a cue selection mechanism that outputs localization information. The details of the processing are presented in the following sections.

## 2.2 Peripheral Processing

The peripheral processing of the model consists of four main stages. The middle-ear filter simulates the middle-ear transduction and corresponds to the processing proposed by [51]. It is realized by a 512-point finite-impulse-response filter of which the coefficients have been fitted to empirical stapes-displacement measurements reported by [53]. Essentially, it is a symmetric bandpass filter that peaks at 800 Hz and has 20-dB/decade slopes.

The processing on the basilar membrane, BM, is simulated by a DRNL filterbank, as proposed in [51]. The nonlinearities of the filterbank are reflected as a change in filter bandwidth with level and by the nonlinear input/output, I/O, function in response to tones. The bandwidth of the filter is equal to that of a 4th-order gammatone filter at low stimulation levels and increases with level. Typical I/O functions for on-frequency stimulation, i.e., at the characteristic frequency, consists of three segments a linear segment for input levels below 30–40 dB SPL, a compressive segment for levels between 30–40 and 60–70 dB SPL, and a linear or near-linear segment for higher input levels. The transition levels between the linear and compressive parts depend on the frequency and the individual listener to which the DRNL parameters are fitted to—outlined further below. An illustration of the auditory filter shapes and I/O functions can be found in Fig. 2 of [54].

The hair-cell transduction process is roughly estimated by a half-wave rectification and a second-order lowpass filter with a cutoff frequency of 1 kHz, as used in several previously described auditory models—for instance, [52, 54, 55]. The main effect of this stage is to reduce information on the temporal fine structure, TFS, and extract the envelope of the stimulus representation at high frequencies.

The last stage of the peripheral processor is a static logarithmic compression. This reflects a simplification of the processing assumed in the models of [55, 56] and [52], where a combination of adaptation loops has been considered to result in either a close-to-logarithmic compression for the processing of stationary portions in the stimuli or in linear processing for the fluctuating portions of the stimuli. The adaptation loops enabled the original models to account for intensity discrimination as well as forward masking data. Such adaptive properties were not considered in the model proposed here. The logarithmic compression used in the present model generates a linear internal mapping of the sound pressure levels of the stimuli. Such a mapping, in combination with the level equalization stage in the binaural processor—described below—realizes an internal representation of the ILDs that corresponds to the ILDs of the acoustic input signals.

A key feature of the proposed model is that the peripheral part can be adapted to approximate the processing of individual NH and HI listeners. The increase in

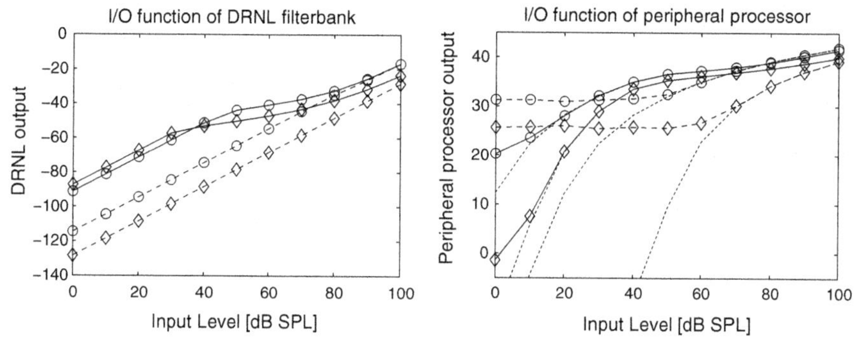


**Fig. 2** *Left* I/O functions of the DRNL module. *Right* I/O functions of the entire peripheral processor Functions are shown for NH (*solid curves*) and HI8 (*dashed curves*) at 500 Hz (*circles*) and 4 kHz (*diamonds*). The *dotted curves* in the right panel represent results obtained without the presence of audibility noise

absolute hearing threshold, a reduction or loss of peripheral, BM, compression and a broadening of the auditory filters are accounted for in the peripheral processor. The absolute hearing threshold is partly determined in the model by the gain in the DRNL. A frequency-dependent noise, referred to here as *audibility noise*, was added to the input signals to simulate increased absolute thresholds. The audibility noise was assumed to be independent across ears and across frequency. The level of the noise was adjusted using the optimal-detector back end of the model framework, such that the detection thresholds for 300-ms pure tones corresponded to the considered individual listener's audiogram. The loss of compression and the associated increase in auditory filter bandwidths were accounted for through a fit of several parameters of the DRNL model to results from temporal masking curves in different conditions, as described in detail in [57].

An average NH listener with an audiogram as reflected in the ISO standard [58] was chosen to represent NH. The DRNL parameters for normal and impaired hearing were taken from [57]. In particular, subject HI8 was chosen for the present study. This listener was considered to have lost compression and had elevated hearing thresholds of 40 and 60 dB HL at 500 and 4 kHz, respectively.

Figure 2 depicts the I/O functions of the DRNL stage for NH and for HI8 at 500 Hz and 4 kHz. The absence of compression for subject HI8 is reflected by the linear functions at both frequencies. The associated reduced frequency selectivity for this listener can be seen in Fig. 9 of [57]. Further, Fig. 2 shows the corresponding I/O functions for the entire peripheral processing block, where the dotted curves represent the I/O functions without the presence of audibility noise, reflecting logarithmic functions of the form $\{output = a*\log(b*input + 1)\}$, with $a$ and $b$ depending on frequency and the DRNL parameters. The presence of the audibility noise produces a constant output value for input levels below the absolute hearing thresholds at the respective frequency for the individual listener.

## 2.3 The Binaural Processor

The binaural stage represents an EC processor—see, for instance [52, 59] or [60]. It is implemented in a similar way as in the binaural processing model of [52]. It differs, however, from the model in [52] in the way that ILDs are coded. The preprocessed left- and right-ear signals are compared in the binaural stage at their respective frequency channels. The binaural processor consists of an array of excitation-and-inhibition, EI, units, each assigned with a characteristic ITD, $\tau$[ms], and a characteristic ILD, $\alpha$ [dB]. As a result, the binaural processor realizes a discrete equalization in time and level, followed by the cancellation operation that calculates the difference between the equalized left- and right-ear signals. The difference is squared to obtain the power of the binaural signal. The output of a single EI unit is thus expressed as follows,

$$E(i, t, \tau, \alpha) = ((L_i, (t + \tau/2) + \alpha/2) - (R_i(t - \tau/2) - \alpha/2))^2 . \quad (1)$$

Here, $L_i$(t) denotes the left-ear preprocessed signal from frequency channel i, and $R_i$(t) denotes the corresponding right-ear preprocessed signal. In contrast to the binaural processor in [52], internal ILDs, $\alpha$, are linearly processed. This processing, in combination with the logarithmic mapping of level realized by the logarithmic compression of in the peripheral processor, provides a mapping of the physical ILDs into $\alpha$-values. Furthermore, a 30-ms long double-sided exponential window limits the temporal resolution of the binaural processor, a phenomenon often referred to as binaural sluggishness [61].

The binaural processor outputs frequency- and time-dependent binaural-activity maps along the $\tau$- and $\alpha$-axes, whereby the activity, *E*, provides an estimate of the interaural coherence. A null activity corresponds to interaural signals that are perfectly correlated, and an increasing activity reflects a reduced interaural coherence. The location of the minimum activity, thus, provides an estimate of the considered source direction. The resolution limit of the model is provided by a constant-variance internal noise added to the binaural-activity map as an independent signal at the output of each EI elements.

## 2.4 Central Processing

Two central processors are integrated in the model and applied to the output of the monaural and binaural preprocessing: an optimal detector, effective in conditions of signal detection, and a localization device including a spatial-cue selection process for the prediction of sound source localization in adverse conditions.

The optimal detector corresponds to the one described in [52]. The binaural detector retrieves the information provided in the binaural-activity map. Following the monaural signal paths, a 10-ms-long double-sided exponential window is applied before the detection stage to limit monaural temporal resolution in both channels. Furthermore, a similar integrator with a time constant of 30 ms is part of the binaural-activity map generation. Both monaural and binaural information are optimally combined across time and frequency channels. Details of the implementation can be found in [52].

The activity represented in the binaural-activity map is minimal at those values of $\tau$ and $\alpha$ that correspond to the ITD and ILD created by the sound sources. The localization of a sound source is therefore achieved by finding the $\tau$ and $\alpha$ values corresponding to points of minimum activity, PMAs, on the binaural-activity map. For convenience, the extracted $\tau$- and $\alpha$-value of the PMAs will be referred to as ITD and ILD throughout this chapter. The amount of activity at the PMA provides a measure of the interaural cross-correlation[1] or coherence whereby a low activity refers to a high coherence. Based on this coherence measure, a cue-selection mechanism that is similar to the one proposed by [23] is implemented here. The cue selector is used to disentangle spatial information in adverse acoustic conditions including room reverberation, multiple talkers or background noise. Consequently, only instantaneous localization predictions that provide a coherence that is above a pre-defined threshold are considered, that is, PMAs on the binaural-activity map that are below a pre-defined threshold. The threshold depends on frequency and the considered acoustic environment—see [23]. The cue-selection operation is thus assumed to increase the reliability of the localization estimation.

# 3 Model Evaluation

## *3.1 ITD and ILD Detection*

Here the model's ability to predict just-noticeable difference for ITDs and ILDs is studied. Model predictions for ITD and ILD detection thresholds are shown in Fig. 3. The left panel of Fig. 3 shows ITD detection thresholds as obtained with 1-ERB-wide noise band stimuli as a function of the center frequency of the noise between 150 and 1.5 kHz. For the measured and simulated data, a 2-down 1-up adaptive procedure was used in combination with a 3-alternative forced-choice paradigm. Each presentation interval was 300 ms long. The predicted frequency dependence of the ITD threshold corresponds well to that observed in the measured data. The threshold function shows the well-known pattern with a minimum at about 1 kHz. The threshold decrease with increasing frequencies up to 1 kHz results from the constant absolute-phase sensitivity of the model. The threshold increases for frequencies above 1 kHz, which is due to the increasing loss of temporal fine structure that leads to loss of information regarding the carrier signal. The effects is modeled by low-pass filtering in the hair-cell transduction stage.

[1] The term interaural *coherence* denotes the amplitude of the normalized interaural cross-correlation function for maximum ITDs of $\pm 1$ ms.

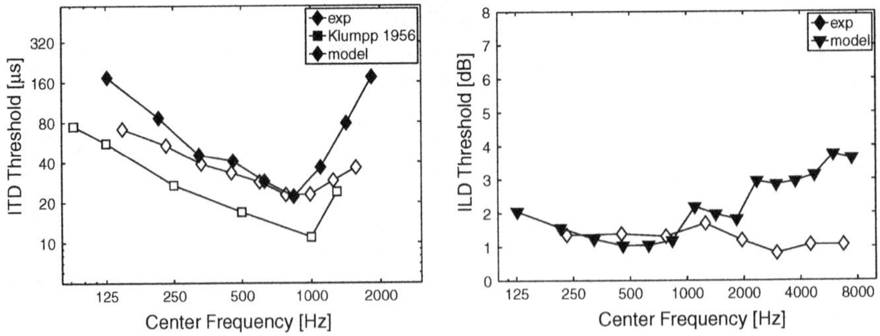


**Fig. 3** *Left* ITD-detection thresholds as a function of the center frequency of one-ERB-wide noise bands. Own data represented by diamonds and data from [62] by squares. *Right* ILD detection thresholds. Experimental data are represented by *open symbols* and model data by *filled symbols*

Figure 3 shows measured and simulated ILD thresholds. The model accounts for the data very well for frequencies up to about 1 kHz but discrepancies can be observed at higher frequencies. Above 2 kHz, the model predicts an increasing ILD threshold up to about 4 dB, with increasing frequency, whereas the measured data stay roughly constant within 1–2 dB. The properties of the assumed middle-ear transfer function, the DRNL filter and the hair-cell transduction stage cause the continuous increase of the simulated ILD threshold. While the discrepancies should be minimized in future model developments, they are considered moderate for the purpose of the present study. However, one of the consequences of the mismatch is that the cue-selection mechanism will slightly underestimate the reliability of ILD information above 1 kHz.

## 3.2 Localization with Normal Hearing

The baseline localization ability of the model is presented for a NH listener in an anechoic environment and in a room. Similar to [23], the acoustic scenario consists of two concurrent talkers located in the horizontal plane. Each talker utters a different phonetically balanced sentence taken from [63] recorded by the same male speaker. The waveforms of the anechoic-speech excerpts of the two talkers are shown in Fig. 4. The speech signals are 2 s long and were sampled at 16 kHz. Due to the intrinsic amplitude modulations in speech signals, the energy fluctuations over time can greatly differ between the two talkers. These specific waveforms were chosen to have instants where only one speaker dominates as well as instants when the speech signals overlap.

Two acoustic environments were considered an anechoic environment and a reverberant room. In the anechoic environment, the location of the speaker was simulated by imposing ITDs corresponding to the difference in arrival time at the two ears [64] and imposing ILDs taken from measurements of [65]. The two talkers were symmetrically located at ±45° azimuth angle in the horizontal plane, which corresponds

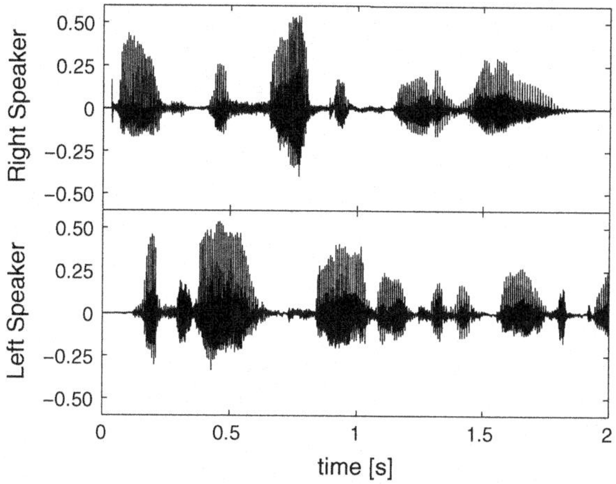


**Fig. 4** Temporal waveforms of the *left* and *right* interfering speaker—taken from [63]

to ITDs of ±0.375 ms and ILDs of ±5 dB assuming a head diameter of 18 cm. All localization simulations were conducted without internal binaural noise.

The information as received, processed, and selected by the cue-selection mechanism as a function of time is shown in Fig. 5 at 500 Hz and 4 kHz. The minimum activity, MA, on the binaural-activity map is shown in the top panels. The middle and bottom panels show the position of the PMAs, map along the $\tau$-axis—corresponding to the estimated ITD of the sound sources—and along the $\alpha$-axis—corresponding to the estimated ILD of the sound sources. As in [23], a cue-selection threshold was arbitrarily chosen for each frequency, represented by the horizontal dashed line in the

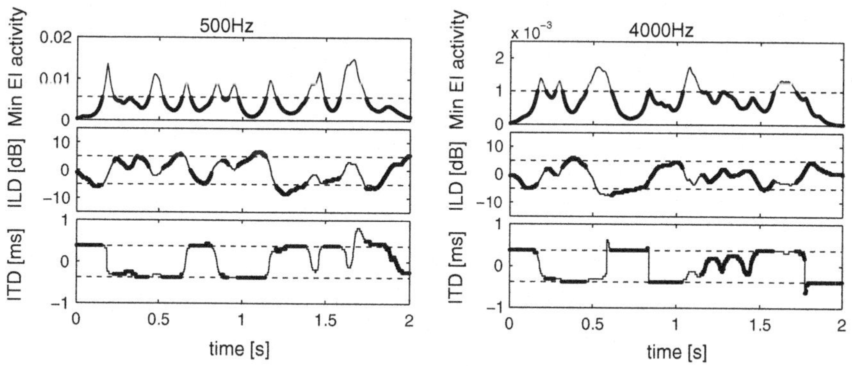


**Fig. 5** *Top* Activity on binaural-activity maps at the PMAs. *Middle* The same but for ILD positions of the PMAs. *Bottom* The same for ITD positions. Data are shown for a NH listener at 500 Hz (*left*) and 4 kHz (*right*). *Bold sections* of the curves represent the information selected by the cue-selection mechanism. The *horizontal dashed line* in the top panels represents the cue selection threshold. *Horizontal dashed lines* in the middle and bottom panels represent the actual locations of the two talkers

top panels. Bold lines indicate the EI activity below the selection threshold and the corresponding time instants were considered as moments at which the localization information was reliable. At these time instants, the estimated ITD and ILD were selected by the cue-selection mechanism—bold sections of the curve in the middle and bottom panels. For this anechoic acoustical scenario, 65–75 % of the localization information was selected by the cue-selection mechanism. Localization information was rejected when the MA was higher than the selection thresholds, which occurred when both speakers were simultaneously speaking, for example at time 0.5, 1.4 or 1.6 s.

The ITDs and ILDs that were estimated as reliable by the cue-selection mechanism were accumulated in time into histograms—Fig. 6. ITD and ILD histograms are shown for 500 Hz and 4 kHz. Both histograms show two peaks matching the actual locations of the two speakers represented by the dashed lines. The two-dimensional histograms, shown in the top-right part of each panel, combine the individual ITD and ILD histograms. *Grey shades* code the frequency of the selected ITDs and ILDs, with *darker areas* representing a large number of occurrences. At both frequencies, two distinct *darker spots* are visible and represent the localization estimates of the two speakers. The size of the darker spots, reflecting the localization blur, strongly depends on the duration of the temporal integrator of the binaural processor for this simulation, which was conducted in an anechoic environment and for a NH listener. A shorter integration time would increase the localization blur. Moreover, the statistic of the ITDs and ILDs could also be used as an input parameter to other processing stages as for instance a speech processor, as discussed in [66], this volume.

The localization of the two speakers was also simulated for a reverberant room. The same two speech signals as considered for the simulations in the anechoic environment were convolved with binaural room impulse responses, BRIR. The speakers were placed at a distance of 1 m and, as in the anechoic environment, at ±45° on the horizontal plane in front of the listener. The BRIRs were derived from a simulated

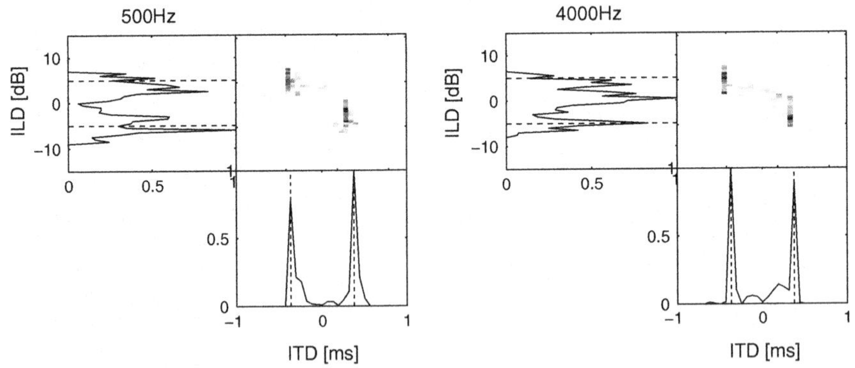


**Fig. 6** Histograms of the selected ITDs and ILDs at 500 and 4 kHz. The stimuli consisted of two talkers symmetrically separated by 45° on the horizontal plane in an anechoic environment with the actual sound locations indicated by the *dashed lines*. The source location estimates are shown for a NH listener in the *top-right window* of each panel by the distinct *darker spots*

cafeteria with a reverberation time of about 0.45 s. The cafeteria was simulated with the room-acoustic simulation software ODEON and translated into binaural signals using a HRTF data set measured with a Bruel and Kjær 4128C head-and-torso simulator, HATS, with purpose built behind-the-ear, BTE, HA dummies from Phonak mounted above the HATS' pinnae [67]. Here, the BRIRs recorded with the HATS in-ear microphones were used—mind that in Sect. 4, which considers the effect of HA, the corresponding BRIRs recorded with the frontal BTE microphones are used.

Figure 7 shows the ITD and ILD histograms as well as the localization estimates for the simulation in the reverberant environment. The layout of the figure corresponds to that of Fig. 6. As in the anechoic environment, the two-dimensional histograms show two distinct darker spots representing the location estimates of the two speakers, although the localization blur is larger here. This increase is due to the reflections on hard surfaces that occur in the reverberant room and reduce the correlation between the left and right ear signals [68]. Due to the room reflections, the activity on the binaural-activity map is elevated and the cue selection thresholds were increased by about 30 % relative to simulations in the anechoic condition. For this room condition, the cue-selection mechanism evaluated that only 40–60 % of the information were reliable, that is, about 10–20 % less than under the anechoic condition.

## 3.3 Localization with Hearing Impairment

The effect of peripheral hearing impairment on localization is discussed in the following. First, the effects of a loss of sensitivity in anechoic environment are analyzed in the framework of the model. Then, the effect of reduced frequency selectivity on

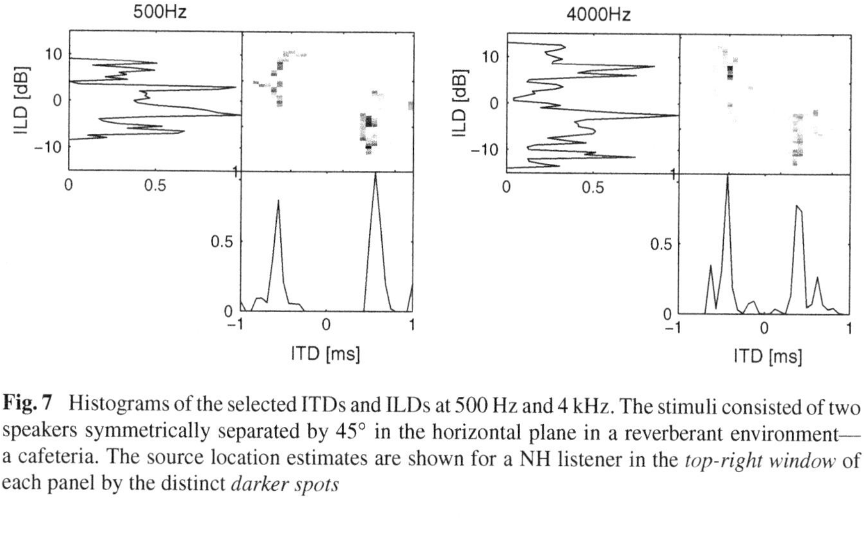


**Fig. 7** Histograms of the selected ITDs and ILDs at 500 Hz and 4 kHz. The stimuli consisted of two speakers symmetrically separated by 45° in the horizontal plane in a reverberant environment—a cafeteria. The source location estimates are shown for a NH listener in the *top-right window* of each panel by the distinct *darker spots*

spectral cues and on the internal representation of the ILDs is studied. The acoustical scenarios are the same as the ones considered above for NH.

## Effects of Reduced Sensitivity

The simulations estimate a HI impaired listener with hearing thresholds of 40 and 60 dB HL at 500 and 4 kHz, respectively, corresponding to HI8 in [57].

Figure 8 indicates the information received, processed and selected by the cue-selection mechanism in the simulations for the listener with reduced sensitivity. The corresponding results for the NH listener were shown in Fig. 5. As visible in the top panels, the loss of sensitivity leads to an overall increase and a compression of the dynamic range of activity in the PMAs. Consequently, the activity fluctuates only slightly around its average value which lies above the cue-selection threshold, effectively de-activating the localization ability of the model. Assuming the same cue-selection threshold for NH and HI listeners, the simulations suggest that it should be difficult to reliably evaluate the localization of the two speakers for the considered HI listener. Furthermore, for the simulation of the HI listener, even if the cue-selection threshold was increased, the cue-selection mechanism would still be greatly impaired due to compression in the activity.

Figure 9 shows the time-accumulated histograms of the ITDs and ILDs. Because the reduced sensitivity de-activated the cue-selection mechanism, all ITDs and ILDs of the PMAs were considered here. At 500 Hz the ITD and ILD histograms are more similar to a single Gaussian distribution rather than a bimodal distribution. The two-dimensional histogram in the top-right corner shows a single *cloud* of spatial information, indicating a low interaural correlation of the input signals of the binaural

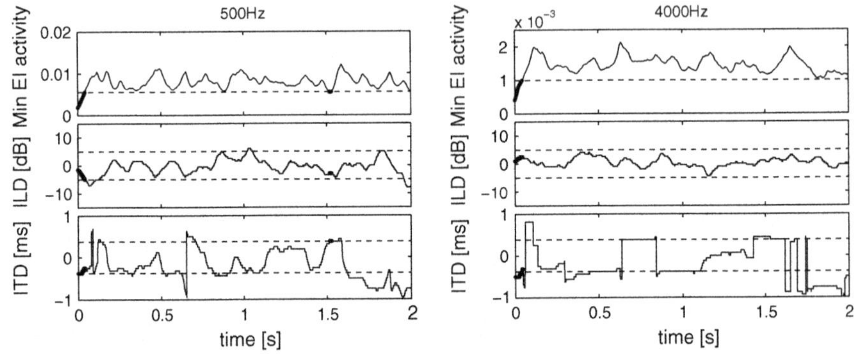


**Fig. 8** *Top* Activity on binaural-activity maps at the PMAs. *Middle* ILD positions of the PMAs. *Bottom* ITD positions of the PMAs. Data are shown for an HI listener with hearing thresholds of 40 and 60 dB HL at 500 Hz (*left*) and 4 kHz (*right*). *Bold lines* represent the information selected by the cue-selection mechanism. The *horizontal dashed line* in the *top panels* represents the cue-selection threshold. *Horizontal dashed lines* in the *middle* and *bottom panels* represent the actual locations of the two talkers

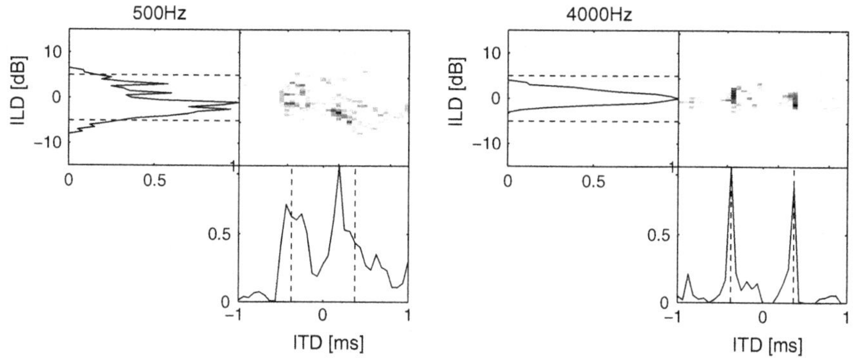


**Fig. 9** Histograms of the ITDs and ILDs present on the binaural-activity maps for a simulated HI listener with hearing thresholds of 40 and 60 dB HL at 500 Hz and 4 kHz, respectively. The stimuli were the same as in Fig. 6. The source location estimates are shown in the *top-right window* of each panel by the *darker spots*

processor. Thus, this estimate of a single diffuse sound source is a result of the reduced sensitivity modeled by the audibility noise, which was assumed independent between the left and right ear. In a condition of a severe sensitivity loss, the audibility noise dominates the internal representations and causes the decorrelated pattern on the binaural-activity maps.

The audibility noise is less dominant at 4 kHz where the ITD histogram shows two peaks at the actual positions of the two talkers. The ILD histogram, however, is very narrow and centered on the median position. It should be mentioned that impaired temporal coding or impaired binaural processing, as for example represented by an increased ITD jitter, has not been considered in the present study. These factors could further deteriorate the representation of ITD cues, as further discussed in Sect. 5. Additional simulations considering various degrees of sensitivity loss—not reported here—showed that the blur on sound source localization estimates gradually increases with increasing sensitivity loss until—as shown here—the audibility noise dominates and individual sources become undistinguishable. Since most listeners with a symmetric sensorineural hearing loss are able to successfully localize anechoic sound sources—when presented at moderate sound levels—only very little localization information seems to be sufficient for accurate localization. As the amount of localization information is significantly reduced in NH listeners when moving from the anechoic to the reverberant condition—see Sect. 3.2—the already reduced localization information available in HI listeners in anechoic conditions suggests that localization will be further reduced in reverberant condition and localization would break down. However, in order to further investigate the effect of HL on localization in reverberant conditions, experimental data are required that allows quantitative comparison with corresponding model predictions.

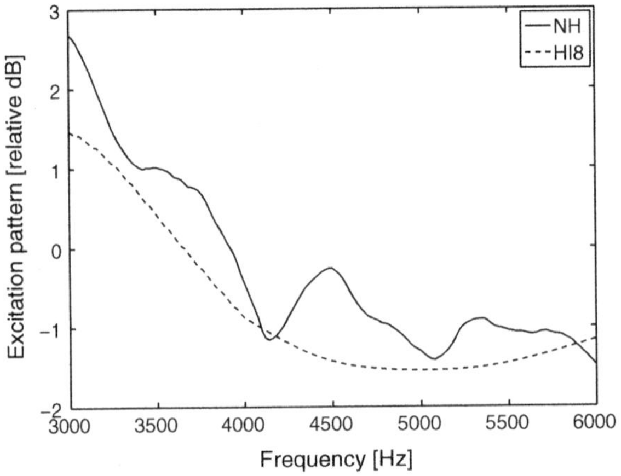


**Fig. 10** Excitation pattern of the BRIR calculated with the DRNL filterbank, considered for an NH listener and for an HI listener, assuming 20 % broader auditory filters

## Effects of Reduced Frequency Selectivity

Beyond the loss of sensitivity, the model can be used to analyze the consequences of broadening of the auditory filters on localization in rooms where the filtering realized by HRTFs or BRIRs provides spectral cues that contribute to the estimation of the sound-source direction—for instance [3].

HRTFs have a very fine spectral resolution. Auditory filters have, however, bandwidths ranging from about 25 Hz at low frequencies to 2200 Hz at high frequencies [69]. Spectral cues are therefore not available to the binaural system with as fine a spectral resolution as represented in HRTFs. This reduction of spectral cue resolution could be even stronger for HI listeners, whose auditory filters are typically broader [25].

In order to investigate the effect of reduced frequency selectivity on spectral cues, the spectral decomposition of the left ear response of one of the BRIR used for localization predictions in Sect. 3.2 was calculated with the DRNL filterbank considered either for a NH listener or for a HI listener with 20 % broader auditory filters—see listener HI8 in [57].

As auditory-filter bandwidth increases with level in the DRNL, the function was amplified to reach a level corresponding to 65 dB SPL. The RMS power at the output of each filter of the filterbank was calculated as a function center frequency of the filter to form an auditory-based excitation pattern. To isolate the effect of reduced frequency selectivity, no loss of sensitivity was included. Thus, the audibility noise was the same in the NH and HI listener models.

The excitation patterns of the BRIR for both the NH and HI listener are shown in Fig. 10. Due to the difference in gain in the DRNL between the two listeners—compare left panel of Fig. 2—the curve for the HI listener has been shifted up by

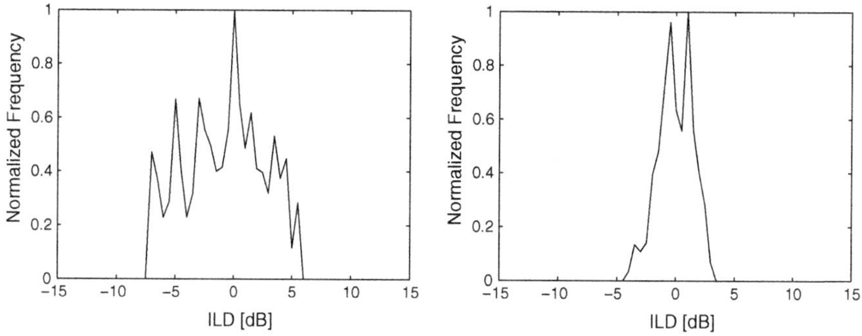


**Fig. 11** Histograms of the internal representation of all ILDs present on the binaural-activity map. *Left* NH listener. *Right* HI listener with 20 % broader auditory filter. Stimuli consisted of two talkers symmetrically separated by 45° on the horizontal plane in a reverberant room

8 dB to simplify the comparison. In line with previous studies simulating the effect of spectral smearing in HI listeners—for instance [70]—the present simulations show that the excitation pattern obtained for the HI listener is much smoother than the one calculated for the NH listener. The difference between the two patterns can amount to 1–2 dB at individual frequencies. Due to the nonlinear nature of the DRNL, this difference will be even larger for lower levels. This spectral smoothing suggests that the broadening of the auditory filters will reduce spatial spectral cues.

The consequence of the auditory-filter broadening is also represented in the binaural-activity maps. Figure 11 shows the ILD histogram at 4 kHz for a NH listener and a HI listener with a normal audiogram and 20 % broader filter. These histograms contain all ILDs of the PMAs present on the binaural-activity map without cue-selection. See the right panel of Fig. 7 for a representation of the selected ILDs and the estimated sound source locations for the NH listener when cue-selection is employed. For the NH listener model, the ILD histogram has a Gaussian-like shape with a range that spans 14 dB from −7 to +7 dB. However, for the HI listener model, the ILD range spans 8 dB, that is, −4.5 to +3.5 dB. The analysis provided in the present study also suggests that ITDs remain largely unaltered by cochlear broadening, which is consistent with the idea that they are more determined by properties of the temporal waveform of the stimuli. Although these results indicate that filter broadening impairs the internal representation of ILDs, further investigations are required to study the effect of frequency selectivity on sound localization in more detail.

# 4 Evaluation of Hearing-Aid Signal Processing Regarding Localization

The proposed model framework may be used to evaluate the effect of HA, processing on the internal representation of spatial cues and on the localization performance in adverse conditions. The evaluation reported in this section uses the same adverse conditions as already used before, namely, two talkers placed at ±45° azimuth angle in the horizontal plane in an anechoic environment.

The basic function of a HA is to provide amplification and to restore audibility. Due to the limited range of input levels for HI listeners, this process is combined with a WDRC in modern HAs. Some potential benefits and drawbacks of the amplification and WDRC processing are evaluated here in the context of localization and in the framework of the presented model.

The same HI listener model with hearing thresholds of 40 and 60 dB HL at 500 Hz and 4 kHz as in Sect. 3.3 was considered. The only change was the addition of HA processing of the input signals of the model. The HA processing was simulated by a model of two binaurally unlinked 30-channel compressive HAs, which were implemented according to [71] and fitted to the audiogram of the modeled HI listener—see HI8 in [57]—using the NAL-NL2 prescription [27]. The simulated HAs were BTE–HAs with omnidirectional directivity and fitted with tight ear molds. The attack time constant was set to 5 ms, the release time constant to 50 ms, and moderate channel coupling was applied, that is, the parameters were set to $c_{\mathrm{D}} = 0.2$ and $c_{\mathrm{U}} = 0.7$. With regard to the model parameters, the cue-selection threshold was the same as in the simulations for the NH and HI listeners, and the audibility noise level was the same as for the simulations with the non-aided HI listener with reduced audibility—Sect. 3.2.

The corresponding internal representations of the stimuli are shown in Fig. 12. A comparison with Figs. 5 and 8 illustrates the effect of HA processing. Restoring the audibility partially recovers the dynamic range in the EI activity at the PMAs and causes an overall decrease of activity, with information passing below the cue-selection thresholds (top panels). This outcome is the result of the amplification of the signals by the HAs, which increases the level of the internal representation of the speech signal as compared to the audibility noise. The ILDs and ITDs of the PMAs, shown in the middle and bottom panels, are also more similar to those observed in Fig. 5 for the NH subject, even though some differences remain.

As a result of the HA processing, the EI activity has decreased and the cue-selection mechanism can successfully operate for the HI listener. The corresponding localization results are shown in Fig. 13. The ITDs of the two speakers can be successfully predicted at both frequencies. However, the ILDs are only partially recovered and generally underestimated. Assuming the same cue-selection threshold as used for simulations with the NH listener, only 20–30 % of the information was selected for this aided HI listener, suggesting a less reliable localization estimation than for the NH listener.

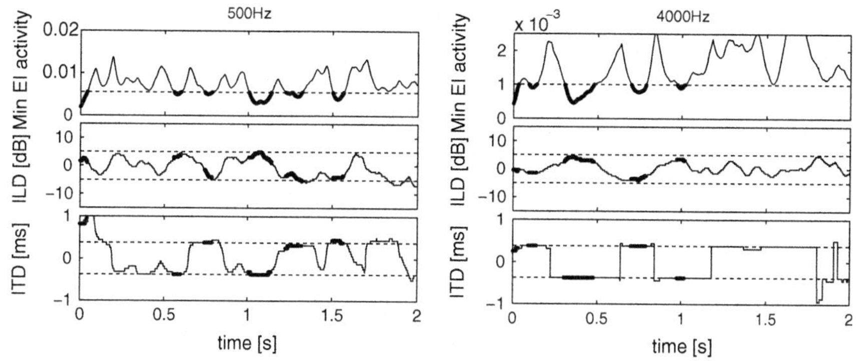


**Fig. 12** *Top* Activity on binaural-activity maps at the PMAs. *Middle*: (*top*), ILD positions of the PMAs. *Bottom* ITD positions of the PMAs. Data are shown for an HI listener with hearing thresholds of 40 and 60 dB HL at 500 Hz and 4 kHz, respectively, aided by compressive binaurally-unlinked HAs. Data are shown for 500 Hz (*left*) and 4 kHz (*right*). *Bold lines* represent the information selected by the cue-selection mechanism. The *horizontal dashed line* in the *top panels* represents the cue selection threshold. *Horizontal dashed lines* in the *middle* and *bottom panels* represent the actual locations of the two talkers

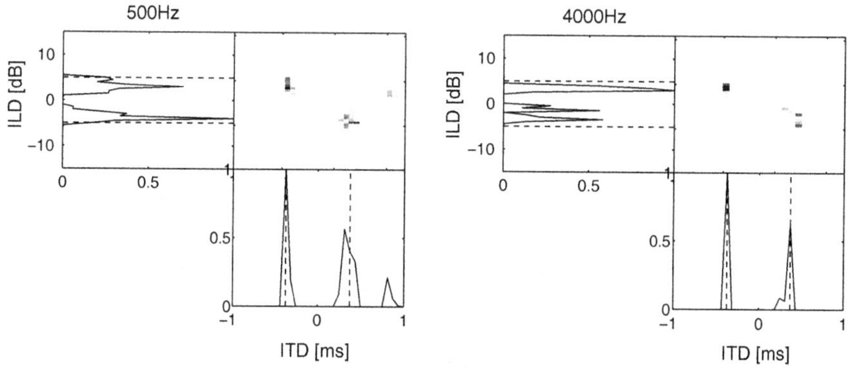


**Fig. 13** Histograms of the ITDs and ILDs present on the binaural-activity maps for a simulated aided HI listener with hearing thresholds of 40 and 60 dB HL at 500 Hz and 4 kHz, respectively. The stimuli were the same as in Figs. 6 and 9. The source-location estimates are shown in the *top-right window* of each panel by the *distinct darker spots*

Although not reported here, in the case that two linear HAs or two binaurally linked HAs are fitted to the HI listener, the estimated ILDs are slightly increased but no systematic improvement in the cue-selection mechanism can be observed. Further research is required to better understand the effect of compression in general as well as binaurally coordinated compression.

## 5 Summary and Conclusion

In the present study, a new binaural model framework was proposed that aims at predicting sound localization in NH, HI and aided HI listeners. The preprocessing of the model consists of a non-linear peripheral processor and an EC binaural processor. EC based binaural processors have traditionally been used to predict binaural tone-in-noise detection—for example, [60, 72]—as well as ITD and ILD detection and discrimination—[72]. Since the proposed model applies a monaural and binaural processor that is very similar to the ones described in [57] and [52], the model framework is able to predict various aspects of monaural and binaural detection. In the present study, the EC binaural processor was applied to predict localization. In the binaural-activity maps generated by the EC processor, a low activity reflects a high interaural correlation and estimates of the positions of sound sources were derived from the positions of the PMAs. In the model, localization in adverse conditions was achieved with the EC-processor followed by a cue-selection mechanism inspired by the cross-correlation- based framework proposed by [23]. The cue-selection mechanism was used to evaluate the reliability of the localization information estimated by the positions of the PMAs.

The proposed framework was used to estimate the effects of peripheral hearing impairment on localization. In the current version of the model, hearing impairment was simulated in terms of either reduced audibility or loss of cochlear compression associated with reduced frequency selectivity. Limited audibility was simulated by adding an internal *audibility noise* to the model input signals. The corresponding predictions showed that the localization estimates, became increasingly blurry with decreasing input-signal to audibility-noise ratio. The simulations also showed that a reduction in audibility impaired the assumed cue-selection mechanism due to an increase and a compression of activity in the PMAs. Future research needs to address a broader range of hearing losses, including asymmetric HLs.

The effects of a simulated loss of compression and the resulting broadening of the auditory filters on localization were tested under a reverberant-room condition. The reduced frequency selectivity led to a spectral smearing of the spectral cues provided by the BRIRs and an alteration of the locations of the PMAs on the binaural-activity map such that the internal representation of the ILDs was mostly located around the median plane. The internal representation of the ITD was, however, hardly affected by the loss of compression.

The study also considered the potential benefit of basic HAs for localization. While the primary function of a HA is to restore audibility, it was shown that including HA processing to the inputs of the model of hearing impairment improves the cues available for localization. In the model, restoring audibility leads to an increase of the signal to-noise ratio at the input and, thus, to an increased reliability of the spatial cues. The analysis also indicated that restoring audibility was beneficial for the assumed cue-selection process as it helped to restore the dynamic range in the PMA activity to that seen in the NH model. Further research is required to better understand the effect of more advanced HAs on localization. In particular, the effect of coordinated compression of bilaterally-fitted HAs should be considered as well as the effect of signal-enhancement strategies such as super-directional beamformers or dereverberation algorithms—compare [27, 73].

The gradual removal of TFS in the hair-cell stage was found to be well suited to predict the increase of ITD detection thresholds above 1 kHz. However, as the ITD of the PMAs is present in both the signal envelope and TFS and, because no binaural noise was considered in the localization simulations, the ITD of the PMAs at high frequency was not limited. This model prediction should be experimentally investigated for NH and HI listeners in future work. Furthermore, the model currently evaluates localization information in individual auditory channels. However, most signals, such as speech, are broadband, and a spectral integration of the spatial information is performed by the hearing system [18]. The nature and modeling of this spectral integration requires further study.

The proposed model framework is a first step towards the development of a dedicated research tool for investigating and understanding the processing of spatial cues in adverse listening conditions, with the long-term goal of contributing to solving of the cocktail-party problem for NH and HI listeners. Future work will be pursued to extend the model and to validate it experimentally.

**Acknowledgments** The authors are indebted to two anonymous external reviewers for constructive suggestions with regard to an earlier version of the manuscript.

## References

1. A. W. Bronkhorst, The Cocktail party phenomenon: A review of research on speech intelligebility in multiple talker conditions, *Acta Acustica united with Acustica*, 86:117–128, 2000.
2. H. S. Colburn, Computational models of binaural processing In: *Auditory Computation*, H. Hawkins and T. McMullen, Eds., Springer, Berlin, 1996, 332–400.
3. J. Blauert, *Spatial Hearing: the psychophysics of human sound localization*, The MIT Press, 1997.
4. M. Dietz, S. D. Ewert and V. Hohmann, Auditory model based direction estimation of concurrent speakers from binaural signals, *Speech Comm.*, 53:592–605, 2011.
5. M. D. Good and R. H. Gilkey, Sound localization in noise: The effect of signal-to-noise ratio, *Am. J. Oto.*, 99:1108–1117, 1996.
6. K. S. Abouchacra, D. C. Emanuel, I. M. Blood and T. R. Letowski, Spatial perception of speech in various signal to noise ratios, *Ear Hear.*, 19:298–309, 1998.
7. C. Lorenzi, S. Gatehouse and C. Lever, Sound localization in noise in normal-hearing listeners, *J. Acoust. Soc. Am.*, 105:1810–1820, 1999.
8. M. L. Hawley, R. Y. Litovsky and H. S. Colburn, Speech intelligibility and localization in a multi-source environment, *J. Acoust. Soc. Am.*, 105:3436–3448, 1999.
9. R. Drullman and A. W. Bronkhorst, Multichannel speech intelligibility and speaker recognition using monaural, binaural and 3D auditory presentation, *J. Acoust. Soc. Am.*, 107:2224–2235, 2000.
10. E. H. A. Langendijk, D. J. Kistler and F. L. Wightman, Sound localization in the presence of one or two distractors, *J. Acoust. Soc. Am.*, 109:2123–2134, 2001.
11. N. Kopco, V. Best and S. Carlile, Speech localization in a multitalker mixture, *J. Acoust. Soc. Am.*, vol. 127:1450–1457, 2010.

12. N. Kopco and B. Shinn-Cunningham, Auditory Localization in Rooms: Acoustic analysis and behavior, in *32nd Intern. Acoust. Conf. - EAA symp.*, Slovakia., 2002.
13. M. Rychtáriková, T. van den Bogaert, G. Vermeir and J. Wouters, Binaural sound source localization in real and virtual rooms, *J. Aud. Eng. Soc.* 57:205–220, 2009.
14. M. Rychtáriková, T. van den Bogaert, G. Vermeir and J. Wouters, Perceptual validation of virtual room acoustics: Sound localisation and speech understanding, *Appl. Acoust.*, 72:196–204, 2011.
15. J. M. Buchholz, V. Best and G. Keidser, Auditory localization in reverberant multi-source environments by normal-hearing and hearing-impaired listeners, in *IHCON Conf., Lake Tahoe, USA*, 2012.
16. J. Raatgever, On the binaural processing of stimuli with different interaural phase relations, PhD Thesis, Techn. Univ. Delft, The Netherlands, 1980.
17. R. M. Stern, A. S. Zeiberg and C. Trahiotis, Lateralization of complex binaural stimuli: A weighted-image model, *J. Acoust. Soc. Am.*, 84:156–165, 1988.
18. N. Le Goff, J. M. Buchholz and T. Dau, Spectral integration of interaural time differences in auditory localization, in *Proc. 21st Intern. Congr. Acoust., ICA 2013*, 2013.
19. N. Le Goff, "Processing interaural differences in lateralization and binaural signal detection", PhD thesis, Techn. Univ. Eindhoven, The Netherland, 2010.
20. J. Nix and V. Hohmann, Sound source localization in real sound fields based on empirical statistics of interaural parameters, *J. Acoust. Soc. Am.*, 119:463–479, 2006.
21. R. Y. Litovsky, H. S. Colburn, W. A. Yost and S. J. Guzman, The precedence effect, *J. Acoust. Soc. Am.*, 106:1633–1654, 1999.
22. W. Lindemann, Extension of a binaural cross-correlation model by means of contralateral inhibition. I. Simulation of lateralization of stationary signals., *J. Acoust. Soc. Am.*, 80:1608–1622, 1986.
23. C. Faller and J. Merimaa, Source localization in complex listening situations: Selection of binaural cues based on interaural coherence, *J. Acoust. Soc. Am.*, 116: 3075–3089, 2004.
24. N. I. Durlach, C. L. Thompson and H. S. Colburn, Binaural interaction of impaired listeners. A review of past research, *Audiology*, 20:181–211, 1981.
25. B. C. J. Moore, *Cochlear Hearing Loss*, Wiley, 2007.
26. D. Byrne and W. Noble, Optimizing sound localization with hearing aids, *Trends Amplif.*, 3:51–73, 1998.
27. H. Dillon, *Hearing Aids*, Boomrang Press, 2012.
28. D. B. Hawkins and F. L. Wightman, Interaural time discrimination ability of listeners with sensorineural hearing loss, *Audiology*, 19:495–507, 1980.
29. L. Smith-Olinde, J. Koehnke and J. Besing, Effects of sensorineural hearing loss on interaural discrimination and virtual localization, *J. Acoust. Soc. Am.*, 103:2084–2099, 1998.
30. H. S. Colburn, Binaural interaction and localization with various hearing impairments, *Scand. Audiol. Suppl.*, 15:27–45, 1982.
31. K. J. Gabriel, J. Koehnke and H. S. Colburn, Frequency dependence of binaural performance in listeners with impaired binaural hearing, *J. Acoust. Soc. Am.*, 91: 336–347, 1992.
32. R. Haeusler, H. S. Colburn and E. Marr, Sound localization in subjects with impaired hearing. Spatial-discrimination and interaural-discrimination tests, *Acta Otolaryngol. Suppl.*, 400:1–62, 1983.
33. U. Rosenhall, The influence of hearing loss on directional hearing, *Scand. Audiol.*, 14:187–189, 1985.
34. W. Noble, D. Byrne and B. Lepage, Effects on sound localization of configuration and type of hearing impairment, *J. Acoust. Soc. Am.*, 95:992–1005, 1994.
35. W. Noble, D. Byrne and K. T. Horst, Auditory localization, detection of spatial separateness, and speech hearing in noise by hearing impaired listeners, *J. Acoust. Soc. Am.*, 102:2343–2352, 1995.
36. C. Lorenzi, S. Gatehouse and C. Lever, Sound localization in noise in hearing-impaired listeners, *J. Acoust. Soc. Am.*, 105:3454–3463, 1999.

37. M. A. Akeroyd and F. H. Guy, The effect of hearing impairment on localization dominance for single-word stimuli, *J. Acoust. Soc. Am.*, 130:312–323, 2011.
38. B. U. Seeber and E. R. Hafter, Failure of the precedence effect with a noise-band vocoder, *J. Acoust. Soc. Am.*, 129:1509–1521, 2011.
39. R. R. Leech, B. Gygi, J. Aydelott and F. Dick, Informational factors in identifying environmental sounds in natural auditory scenes, *J. Acoust. Soc. Am.*, 126:3147–3155, 2009.
40. W. Noble, K. Ter-Horst and D. Byrne, Disabilities and handicaps associated with impaired auditory localization, *J. Am. Acad. Audiol.*, 6:129–140, 1995.
41. W. Noble and S. Gatehouse, Effects of bilateral versus unilateral hearing aid fitting on abilities measured by the speech, spatial, and qualities of hearing scale (SSQ), *Int. J. Audiol.*, 45:172–181, 2006.
42. M. Boymans, S. T. Govers, S. E. Kramer, J. M. Festen and W. A. Dreschler, Candidacy for bilateral hearing aids: a retrospective multicenter study, *J. Speech Language Hear. Res.*, 52:130–140, 2009.
43. T. Van den Bogaert, T. J. Klasen, M. Moonen, L. V. Deun and J. Wouters, Horizontal localization with bilateral hearing aids: Without is better than with, *J. Acoust. Soc. Am.*, 119:515–526, 2006.
44. G. Keidser, K. Rohrseits, H. Dillon, V. Hamacher, L. Carter, U. Rass and E. Convery, The effect of multi-channel wide dynamic range compression, noise reduction, and the directional microphone on horizontal localization performance in hearing aid wearers, *Inter. J. Audiol.*, 45:563–579, 2006.
45. T. Van den Bogaert, E. Carette and J. Wouters, Sound source localization using hearing aids with microphones placed behind-the-ear, in-the-canal, and in-the-pinna, *Inter. J. Audiol.*, 50:164–176, 2011.
46. V. Best, S. Kalluri, S. McLachlan, S. Valentine, B. Edwards and S. Carlile, A comparison of CIC and BTE hearing aids for three-dimensional localization of speech, *Int. J. Audiol.*, 49:723–732, 2010.
47. W. Noble, S. Sinclair and D. Byrn, Improvement in aided sound localization with open earmolds: observations in people with high-frequency hearing loss, *J. Am. Acad. Audiol.*, 9: 25–34, 1998.
48. T. Van den Bogaert, S. Doclo, J. Wouters and M. Moonen, The effect of multi-microphone noise reduction systems on sound source localization by users of binaural hearing aids, *J. Acoust. Soc. Am.*, 124:484–497, 2008.
49. T. J. Klasen, T. V. d. B., M. Moonen and J. Wouters, Binaural noise reduction algorithms for hearing aids that preserve interaural time delay cues, *IEEE Trans. Signal Process*, 55:1579–1585, 2007.
50. J. Mejia, G. Keidser, H. Dillon, CV. Nguyen, and E. Johnson, The effect of a linked bilateral noise reduction processing on speech in noise performance. In *Speech Perception and Auditory Disorders*, ed. by T. Dau, J.C. Dalsgaard, M.L. Jepsen, and T. Poulsen, 2011
51. E. A. Lopez-Poveda and R. Meddis, A human nonlinear cochlear filterbank, *J. Acoust. Soc. Am.*, 110:3107–3118, 2001.
52. J. Breebaart, S. van de Par and A. Kohlrausch, Binaural processing model based on contralateral inhibition. I. Model structure, *J. Acoust. Soc. Am.*, 110:1074–1088, 2001.
53. R. L. Goode, M. L. Killion, K. Nakamura and S. Nishihara, New knowledge about the function of the human middle ear: Development of an improved analogue model, *Am. J. Otol.*, 15:145–154, 1994.
54. M. L. Jepsen, S. D. Ewert and T. Dau, A computational model of human auditory signal processing and perception, *J. Acoust. Soc. Am.*, 124:422–438, 2008.
55. T. Dau, D. Püschel and A. Kohlrausch, A quantitative model of the "effective" signal processing in the auditory system. II. Simulations and measurements, *J. Acoust. Soc. Am.*, 99:3623–3631, 1996.
56. T. Dau, B. Kollmeier and A. Kohlrausch, Modeling auditory processing of amplitude modulation. I. Detection and masking with narrow banc carriers, *J. Acoust. Soc. Am.*, 102:2892–2905, 1997.

57. M. L. Jepsen and T. Dau, Characterizing auditory processing and perception in individual listeners with sensorineural hearing loss, *J. Acoust. Soc. Am.*, 129:262–281, 2011.
58. *ISO 226:2003 Normal equal-loudness-level contours*. International Organization for Standardization, ISO, Geneva.
59. A. Kohlrausch, J. Braasch, D. Kolossa and J. Blauert. An introduction to binaural processing. In J. Blauert, editor, *The technology of binaural listening*, chapter 1. Springer, Berlin-Heidelberg-New York NY, 2013.
60. N. Durlach, Equalization and cancellation theory of binaural masking level-level differences, *J. Acoust. Soc. Am.*, 35:1205–1218, 1963.
61. D. W. Grantham and F. L. Wightman, Detectability of varying interaural temporal differences, *J. Acoust. Soc. Am.*, 63: 511–523, 1978.
62. R. Klumpp and H. Eady, Some Measurements of Interaural Time Difference Thresholds, *J. Acoust. Soc. Am.*, 28:859–860, 1956.
63. *IEEE recommended practice for speech quality measurements*, 1969.
64. B.C. Moore, *An introduction to the psychology of hearing*, 4$^{th}$ Ed. Academic Press, London, 1997.
65. W. E. Feddersen, T. T. Sandel, D. C. Teas and L. A. Jeffress, Localization of High-Frequency Tones, *J. Acoust. Soc. Am.*, 29:988–991, 1957.
66. A. Schlesinger and C. Luther, Optimization of binaural algorithms for maximum predicted speech intelligibility. In J. Blauert, editor, *The technology of binaural listening*, chapter 11. Springer, Berlin-Heidelberg-New York NY, 2013.
67. C. Orinos and J. Buchholz, Measurement of a complete set of HRTFs for in-ear and hearing aid microphones on a Head and Torso Simulator, *J. Acoust. Soc. Am.*, submitted, 2013.
68. H. Kutruff, *Room acoustics*, Elsevier, 1973.
69. B. R. Glasberg and B. C. J. Moore, Derivation of auditory filter shapes from notched noise data, *Hear. Res.*, 47:103–138, 1990.
70. T. Baer and B. C. J. Moore, Effects of spectral smearing on the intelligibility of sentences in noise, *J. Acoust. Soc. Am.*, 94:1229–1241, 1993.
71. J. M. Buchholz, A real-time hearing-aid research platform (HARP): realization, calibration, and evaluation, *Acta Acustica united with Acustica*, under revision, 2013.
72. J. Breebaart, S. van de Par and A. Kohlrausch, Binaural processing model based on contralateral inhibition. II. Dependence on spectral parameters, *J. Acoust. Soc. Am.*, 110:1089–1104, 2001.
73. J. M. Kates, *Digital Hearing Aids*, Plural Publishing, 2008.

# Binaural Scene Analysis with Multidimensional Statistical Filters

C. Spille, B. T. Meyer, M. Dietz and V. Hohmann

## 1 Introduction and Overview

Binaural hearing in humans has been investigated for more than a century. Thompson [63] and Rayleigh [52] identified differences in arrival time and in intensity between the left and the right ear to be the dominating cues for direction estimation. These cues are commonly termed as interaural time differences (ITD) and interaural level differences (ILD). Thompson [62] suggested that binaural sensitivity is not caused by acoustic interference, for instance, via the Eustachian tubes, but rather by neural processing in the brain. It took more than 50 years before the first conceptual model of neural ITD coding was suggested by von Békésy [65]. He suggested a model, where a population of neurons is excited by signals from one ear and inhibited by those from the other. The total population response then codes the interaural differences; this concept is referred to as *rate code*. Another two decades later Jeffress [28] suggested an alternative coding concept stating that the neural signals from each ear are delayed on counterdirected pathways, which act as *delay lines*. Along the pathways the two differently delayed signals are compared by coincidence neurons, which are activated if the signals arrive in coincidence. Due to the increase in relative delay along the delay lines, the position of the active coincidence neuron along the line indicates the ITD, what is known as *place coding*. Coincidence detection along counterdirected delay lines mathematically resembles cross-correlating the left and right signals. Probably because of these conceptual and mathematical simplicities the Jeffress model became the standard model type for developing and evaluating binaural processing models—see, for example, [35, 59, 60]—which became computationally tractable with the advent of digital computer. These models were able to explain a vast range of the binaural phenomena known experimentally from binaural psychoacoustics in humans. Variants of this model concept, for instance a subtraction of left

C. Spille · B. T. Meyer · M. Dietz · V. Hohmann (✉)
Department of Medical Physics and Acoustics, University of Oldenburg,
26111 Oldenburg, Germany
e-mail: volker.hohmann@uni-oldenburg.de

J. Blauert (ed.), *The Technology of Binaural Listening*, Modern Acoustics and Signal Processing, DOI: 10.1007/978-3-642-37762-4_6,

and right input along counterdirected delay lines have also been applied successfully [5] but all of these models are based on delay lines.

Physiological evidence for the existence of axonal delay lines was first found by Carr and Konishi [8] in the brainstem of barn owls. Although the auditory system of mammals was known to be significantly different from that of birds, the Jeffress model remained the standard approach for modeling binaural processing in mammals—see [41] for a review. However, recent evidence from physiology [42], functional magnetic resonance imaging [61], evoked potential measurements [53] and psychoacoustics [50] indicates that it is difficult to explain the mammalian data with the Jeffress model of place coding. Physiological data [4, 42] suggest that ITD might be coded in terms of the rate of firing of binaurally sensitive neurons—so-called *rate coding*—thus rather supporting the original hypothesis of von Békésy [65]. Based on these findings, Dietz et al. [17] developed a computational rate-coding model of binaural processing in humans that did not use any delay lines and that was based on the interaural phase difference (IPD) instead of the ITD. In line with recent models—for example, [20]—and psychoacoustical findings, but in contrast to earlier models that included an explicit temporal integration to model *binaural sluggishness*, this model reflects the high temporal resolution of the binaural system. Furthermore, it processes envelope and carrier IPDs in different channels, explaining psychoacoustic experiments that traded binaural cues of the envelope and carrier, respectively [14]. A further binaural processing model with the same motivation of rate coding has recently been suggested by Pulkki and Hirvonen [51].

All models of binaural processing in humans simulate the frequency selectivity initially provided by the cochlea by processing binaural information in frequency subbands. For this, many models use a linear Gammatone filterbank with auditory frequency and time resolution—see [27]. From a signal processing point of view, this renders the distinction between delay-line models and the IPD model by Dietz et al. [17] difficult: The subband signals are, on short time scales, almost sinusoidal, which means that ITD, as measured by cross-correlation, and IPD, as measured by the IPD model, are almost indistinguishable. A recent study designed to disambiguate these two approaches by studying the ability to lateralize stimuli with an ITD only in the second order envelope [16] does indeed hint towards the non-existence of long internal delays. Van der Heijden and Trahiotis [64] deduced from human psychoacoustic data of tone detection in double delayed noise stimuli that short internal delays up to 750 μs are indeed in operation. However, neither of these psychoacoustic experiments can distinguish between the general concepts of rate and/or place coding.

In the experimental study presented below, the IPD model [17] is used as a front-end, because it provides the required high temporal resolution for detecting short *glimpses* of robust binaural information. Similar results, however, can also be achieved with a cross-correlation model [20, 37, 54] or with the subtraction-based Breebaart model—[34], this volume.

Whereas binaural models mimic many of the distinguished capabilities of binaural hearing and have extensively been applied to investigate and simulate psychoacoustic data of binaural perception of artificial stimuli—see references above—as well as of speech—for example, [2]—their technical application to audio processing in hearing aids, mobile audio devices, robotics, hands-free audio communication and speech-based computer interfaces is still limited. The hypothesis underlying this chapter is that this limitation is due to a *missing link* between basic binaural capabilities, as modeled by the binaural models mentioned above, and audio processing and interpretation of difficult acoustic conditions characterized by superposed speech, noise and reverberation. The latter requires inference about the causes of the observed auditory input in order to be able to *decode* the acoustic scene, that is, to identify and segregate different sources, or to select desired sound sources. This *cognitive* part of the processing is the missing link and constitutes enabling technologies for technical applications of binaural models. In this study, the possibilities of filling the missing gap by extending a binaural model towards interpreting the acoustic scene, that is, *computational binaural scene analysis*, are demonstrated by improving automatic speech recognition (ASR), of superposed spatially-moving speech signals.

*Auditory scene analysis* (ASA), in the sense of low-level cognitive processing of acoustic input in humans, has extensively been investigated in the literature, and many inference principles of the auditory system have been identified [6]. The most important acoustic cues used by the hearing system were identified to be harmonicity, periodicity, common onset—namley, synchronous increase in level across several auditory frequency bands—and the frequency-dependent binaural cues. Each of these cues only provides a small part of the information needed to decode the acoustic scene, namely, regarding sources being present, spatial configuration of the sources, room characteristics etc. Therefore, evidence from many of these different cues has to be integrated for scene interpretation, including integration across auditory frequency subbands and time. Evidence from electrophysiological, EEG data and functional magnetic resonance imaging (fMRI) in humans has led to the interpretation that the cognitive system, including the hearing system, performs cue integration by comparing the current sensory input to hypotheses about the expected observation [47, 70]. The system adapts via neural adaption to the expected input and codes only deviations of the input from the expectation—see [45]. This means that basically only *novelty* is processed by the nervous system, that is, sensory information that deviates from the hypotheses. Deriving hypotheses about the expected observation from earlier observations of the input requires a dynamic predictive model of the auditory scene and is thus part of recent cognitive inference models—see [66] for a recent approach of modeling novelty processing in auditory evoked EEG responses, and [21] for modeling cortical inference circuits.

Recent evidence from fMRI experiments in humans show that the premotor cortex is active during the perception of distorted speech, but not active when music is played. This suggests that premotor activity may facilitate interpreting speech when the input is sparse [46]. In the light of the proposed hypothesis-driven inference model, results are consistent with the notion that the premotor areas responsible for speech production might be used to generate the hypotheses when perceiving speech in difficult conditions—compare the motor theory of speech perception [69]. In other fMRI experiments, Bushara et al. [7] examined the neural binding between hearing and vision. Correlation of auditory and visual stimuli was found to be correlated with reduced brain activity when binding occurred. This suggested that the cognitive system explores several hypotheses about the expected observation across modalities at a time, that is, generates *competing hypotheses*, and that their representations have mutually inhibitory interactions.

From a signal processing point of view, the principle of *novelty processing* agrees well with Bayesian inference methods, which are frequently used in computational-auditory-scene analysis (CASA) approaches—compare [68]. In particular, inference from competing hypotheses can be implemented numerically by sequential Monte-Carlo methods, which will be briefly introduced in the framework of computational binaural scene analysis in the next section. It is argued that CASA based on hypothesis-driven computing using predictive models is a key to successfully applying binaural models to decoding the acoustic scene. Three major limitations requiring this approach will be briefly outlined in the following.

- *Ambiguity of the source separation problem* If the acoustic scene is composed of more than two sources and is received by only two sensors, left and right ear, the separation problem becomes ambiguous. In other words, the source signals cannot be reconstructed from the superposition using linear methods, such as linear microphone-array processing. Disambiguation is possible, however, by predictive models that limit the number of possible explanations of the scene—see, for example, [44].
- *Random fluctuations of signal-derived parameters* Diffuse background noise and reverberation impose statistical fluctuations on all signal-derived features, strongly reducing the statistical evidence provided by a single observation of the respective feature—see, for example, [43]—for a quantification of the fluctuations of binaural features in real acoustic conditions. Predictive statistical models explicitly model the noise and perform a statistical combination of several noise-deteriorated parameters, allowing a noise-robust extraction of information [43].
- *Missing information* Superposed daily-life signals overlap significantly in the time-frequency domain. Thus, a significant part of the information on the different sources is completely masked in the time-frequency domain. In order to separate the sources, predictive models are required to fill the missing information—compare [11].

Every natural or artificial cognitive system has to deal with these limitations and thus requires some structure that collects evidence from noisy, ambiguous and partly missing information. In this study, the principle of competing hypotheses based on a predictive model is applied, in order to achieve this task.

The remainder of this chapter is organized as follows: First, the basic approach to implement principles of computational binaural scene analysis is described in Sect. 2, and existing studies on the subject are reviewed in Sect. 3. Then, the approach to improve ASR using computational binaural scene analysis is introduced in Sect. 4. For this, the binaural model of Sect. 4.1, the statistical properties of its output with respect to sound source localization, see Sect. 4.3, the multidimensional statistical filter that tracks the location of superposed moving speakers, Sect. 4.4, the beamformer that is directed to the desired speaker and its adaptation by the location tracks, Sect. 4.5 and, finally, the ASR system that recognizes speech at the output of the beamformer, Sect. 4.6, are described. Results from an ASR task are detailed in Sect. 4.7. Last, the chapter is summarized and conclusions are given in Sect. 5.

## 2 Computational Binaural Scene Analysis

In this section the focus is on computational binaural scene analysis systems that are based on competing hypothesis. Figure 1 shows the principle processing blocks of the proposed system that builds upon the model proposed by Nix and Hohmann [44]. Key to this approach is the use of *a priori* knowledge about the sound sources S that compose the current acoustic scene to generate a set of hypotheses H. Each hypothesis develops in time and represents a possible state of the sound sources, that is, a set of parameters that describe the exact configuration of all sources, such as, source position, pitch, formant frequencies or vocal tract parameters—depending on the type of source. In each time frame, hypotheses are checked against the observation, which is composed of a number of signal features O computed from the binaural audio input.[1] The likelihood of the observation to occur under the assumption that the hypothesis is true is computed and assigned to each hypothesis. This means that

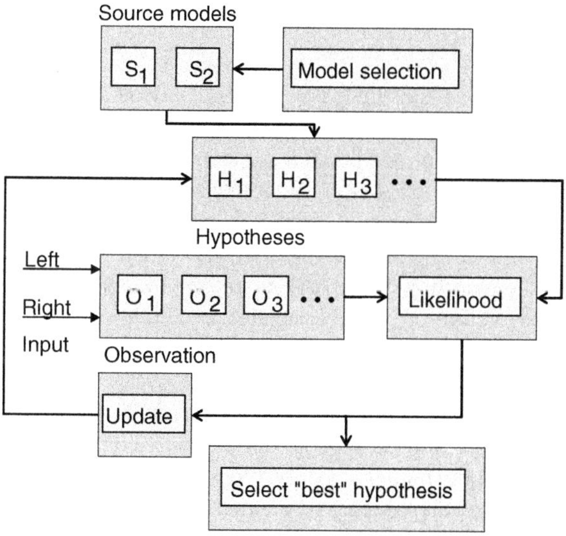


**Fig. 1** Block diagram of a computational binaural scene analysis system based on competing hypothesis. See the text for details

[1] Note that this approach can be extended to more inputs, for example, multiple microphones or audiovisual input, or might be restricted to a single input.The current study covers its application to binaural input signals like recordings from a dummy head.

evidence from many features O is integrated and merged into a single likelihood value, such as feature integration across frequency. The definition of this likelihood function is challenging and its complexity depends on the number of state parameters and the number of observed features. It should reflect the relation between source parameters and observation as good as possible. Note that the likelihoods are stochastically distributed, because the observation is generally fuzzy, even if the state of the sound source is fixed, for instance, binaural features fluctuate even for fixed location of the source, as pointed out above. Note also that assigning likelihoods can be described as extracting the novelty about the sound sources embedded in the observation. If the novelty is high for a specific hypothesis, the observation does not match the expectation set by the hypothesis, and it will be assigned a low likelihood. If the novelty for a specific hypothesis is low, the hypothesis will be assigned a high likelihood. The set of hypotheses and their assigned likelihoods represents the distribution of possible states and thus the current estimate of the auditory scene.

Finally, based on the likelihood and *a priori* known dynamics of the sound sources, which is restricted by physical constraints, such as smooth pitch and location contours or limited rate of change of vocal tract parameters, each hypothesis is updated, that is, the parameter set associated with the hypothesis is changed. By this, the expectation about the state present in the next time step is established. This update function also employs a stochastic factor in most applications, for instance, the location of a sound source might be updated according to a random-walk process. Note that two identical hypotheses develop differently in time due to this random component.

The set of hypotheses and their assigned likelihoods represents the inference about the causes of the sensory input. In many applications, the hypothesis that was assigned the maximum likelihood is taken as the *best* hypothesis. The functioning of the approach very much depends on whether the applied source models S match the sources present in the audio input. A mismatch automatically means a fundamental misinterpretation of the scene. For example, if the system would erroneously select a speech signal to be present in the scene, parameters like formant frequencies, pitch and the temporal evolution of these parameters would be estimated and interpreted as the state of a speech signal—which is actually not present. Therefore, the *model-selection* block in Fig. 1 is most relevant. A biological system has to select or estimate the appropriate models based on information present in the sensory input. Many technical applications assume that the appropriate source models are *a priori* known, that is to say, they omit the model-selection block. In this case, models are fixed and only the hypotheses are being updated dynamically.

Common mathematical approaches to implementing a system according to Fig. 1 are so-called sequential Monte-Carlo methods, in particular particle filtering. Arulampalam et al. [1] provide a tutorial on generic particle filters, which shall be introduced briefly here—for mathematical details the reader is referred to the literature. Figure 2 shows a block diagram of a generic particle filter. The circle of processing blocks is performed for each time instance. A *state* is a mathematical description of the current configuration of each sound source and corresponds to the

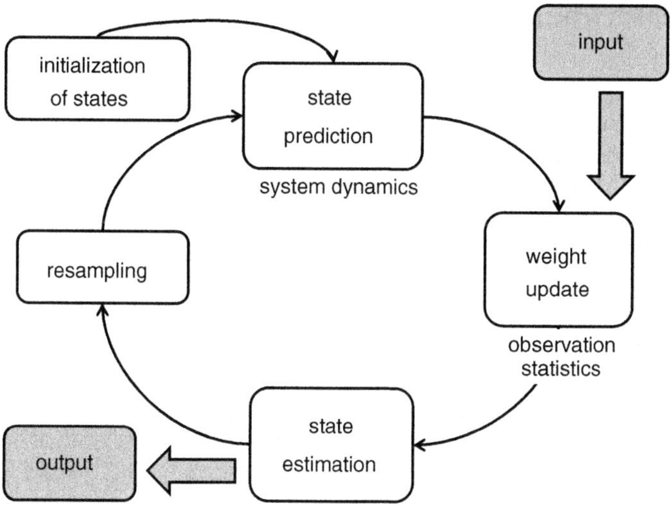


**Fig. 2** Block diagram of a particle filtering algorithm. See the text for details

hypothesis from Fig. 1. It is assigned a *weight* that corresponds to the likelihood from Fig. 1. The combination of a state and its weight is called a *particle*. At system onset, before the first input is taken, a set of particles is initialized, for instance, by sampling the states from an equal distribution across all possible states and assigning random weights. *System dynamics* implements a dynamic model of each sound source by a random mathematical function and represents the update function from Fig. 1. Updating the hypotheses—Fig. 1—is implemented by predicting the future state from the current state and the system dynamics separately for each particle. In the next step, the weight of each particle is updated by an *observation statistics*, which corresponds to the likelihood computation from Fig. 1. The observation statistics links the input, namely, the observation, and the weights by increasing the weight for states that are likely given the input, and vice versa. Even if the observation statistics is not identical to the *true* likelihood function, the set of particles, namely, the states and their normalized assigned weights, represents a sampled version of the *true* likelihood function under very general constraints [19]. *State estimation* means the selection of the particle with the highest weight, which denotes the filter output. Some implementations output the expected value across the set of particles instead. The last step in the processing chain is the resampling step, which does not find a correspondence in Fig. 1. Resampling means that particles are discarded if their weights fall below a certain minimum and are replaced by randomly selected particles that are similar to the particles with the highest weights. The processing chain then begins again with the prediction step, that is, time is taken one step forward.

## 3 Examples from Literature

In the following, some examples of computational binaural scene analysis are briefly reviewed. Note that this review is far from complete; the studies presented here, however, will ease the access to the large body of literature on this topic. A comprehensive overview of CASA techniques—not necessarily binaural—can be found in the book edited by Wang and Brown [68].

A very early approach of binaural scene analysis was introduced by Lyon [36]. Peaks of the subband interaural cross-correlation function identified the location of sounds and time-varying filters steered by these peaks segregated a directional source. The approach pursued here is similar, but relies on a different binaural model and includes particle filtering for modeling source continuity.

Nix and Hohmann [44] presented a binaural-scene-analysis algorithm that tracked the azimuthal direction of arrival (DOA), and the power spectrogram of each of two superposed and moving speech signals. As the observation, short-term FFT-spectra and frequency-specific binaural parameters, ILD and ITD, derived from the spectra were used. As a source model, a first-order Markov process was used that could generate a plausible succession of speech spectra from a random process. For this, typical speech spectra and their transition probabilities were derived using a cluster analysis method from a large speech database that contained many hours of speech. $N = 10{,}000$ typical spectra were used in the Markov process, that is, the transition matrix contained $N \times N$ entries. For tracking the time-course of source location and spectra, a particle filtering approach was used. 100,000–1,000,000 particles were used in the different experiments. Each particle contained the source configuration, namely, azimuth directions and current short-time spectra of both sources. The authors reported that, on the same signal, some runs of the filter succeeded in tracking the sources correctly, whereas other runs failed. This shows the dependence of the system on the random initialization of the particles and shows that the particle filter may fail even if the source model perfectly matches the sources present in the input signal. For the successful runs of the filter, the algorithm was able to track azimuth and magnitude spectra from two superposed speech signals. Evaluations of the signal-to-noise ratio (SNR) showed that the algorithm was able to improve the SNR at input SNRs around zero or below, which is difficult to achieve with algorithms that do not use speech models—compare [44]. The computational effort, however, was very high.

Dietz et al. [15] used the perceptually and physiologically inspired IPD model for estimating the azimuthal DOA of superposed directional sound sources, including free field conditions with up to five concurrent speakers, three concurrent speakers in background noise and one speaker in reverberation. Key to the IPD model is that only those time-frequency segments contribute to the DOA estimate that have a high interaural coherence, similar to Faller and Merimaa [20] and [34] this volume. Those segments usually occur during short instances of time, often in the order of a few tens of milliseconds, when one sound source dominates the binaural input. By processing each of these high-coherence segments as a single event called *glimpse*, a sparse representation of the binaural features is generated, which is in accordance with recent physiological evidence. A glimpsing representation is especially reasonable in strongly modulated signals such a speech [11]. Tracking of the sound sources, that is, estimating its DOA from the sequences of *glimpses* was achieved using a particle filter by Särkkä et al. [55] that handles sparse input. This filter implementation solves the linear parts of the estimation process with a Kalman filter and leaves the nonlinear parts to the particle filter. This approach is called *Rao-Blackwellized particle filter*. The IPD model in combination with the particle filter by [55] is used as the basis for the experimental study presented in this chapter. An elaborate analysis of DOA estimates from the model is given below.

Woodruff and Wang [71] describe a binaural localization framework for multiple sources in noisy and reverberant conditions. Monaural source segregation was used to increase the robustness of azimuth estimates from a binaural input and was shown to improve performance relative to binaural-only methods. This framework also allows model selection or adaption in the sense that an azimuth-dependent model of binaural features allows for adaptation to new environments.

Christensen et al. [10] introduce a speech fragment approach to localizing multiple speakers in reverberant environments. Key to this approach is that binaural and pitch information is sampled from time-frequency regions, so-called *fragments*, that are likely to be dominated by one of the speakers. This method is reported to improve localization performance by up to 24 % compared to a state-of-the-art localizer.

## 3.1 Application to Automatic Speech Recognition

Mel-frequency cepstral coefficients (MFCCs) are one of the standard features for today's ASR systems [13]. They effectively encode the spectral envelope of short-time segments of speech, perform well for acoustically clean conditions and reflect properties of the auditory system only to a limited extent. Auditory-inspired pre-processing of speech signals, however, has also been shown to be a useful approach in automated speech processing tasks. Applications include the identification of speakers [40], this volume and [38, 39] as well as automatic speech recognition, ASR, for which auditory frontends have been shown to increase the robustness in the presence of noise and reverberation [58]. Examples of the large number of studies following this approach range from the integration of signal processing strategies known to be employed in the inner ear [26] to the application of filters resembling pattern observed in the primary auditory cortex of mammals [31].

Across-frequency binaural processing has also been investigated in the framework of binaural speech recognition. Palomäki and Brown [48] compare across-frequency and within-frequency processing in combination with internal noise in a computational model of binaural speech recognition. Palomäki et al. [49] use the statistics of binaural features to identify unreliable spectro-temporal segments. Unreliable segments are treated as *missing data* by the speech recognition system. In other words, no evidence is provided by this segment and the system's speech model re-generates the missing data in the estimation process. These missing data techniques have been elaborated further [23, 32] and have been shown to be very successful in rendering ASR more robust in noisy conditions. In this chapter, these techniques are not employed, but it would be possible to use a missing data recognizer directly on the output of the binaural model, which establishes a means to define missing data due to its sparseness. Instead, following the philosophy of the AABBA project, auditory-inspired processing is employed in form of the binaural model described in the next section. Note that [40], this volume, elaborate further on missing data techniques.

# 4 ASR in Multi-Speaker Conditions Using Binaural Scene Analysis

Figure 3 shows a block diagram of the whole processing chain from the raw speech data to the ASR system. Speech data is used to generate moving speakers by convolving it with recorded 8-channel HRIRs—2 in-ear channels and 3 channels from each of two behind-the-ear (BTE) hearing aids. The in-ear signals are fed into the binaural model that is employed to estimate the direction of arrival of spatially distributed speakers—compare Fig. 1 block *Observation*. A particle filter is then used to keep track of the positions of the moving speakers. This relates to the blocks *Source Model*, *Hypotheses*, *Likelihood*, *Update* and *Select best hypotheses* in Fig. 1. The particle filter's output is used to steer a beamformer, enhancing the 6-channel speech signal that is to be transcribed by an ASR system. In the following sections each of these processing steps is described in more detail.

## 4.1 *Binaural Model Structure*

The main aim of this study was to apply an auditory binaural model, the IPD model [17], to automatic speech recognition. The IPD model has previously been extended and applied to direction of arrival (DOA) estimation [15] which, in turn, has been

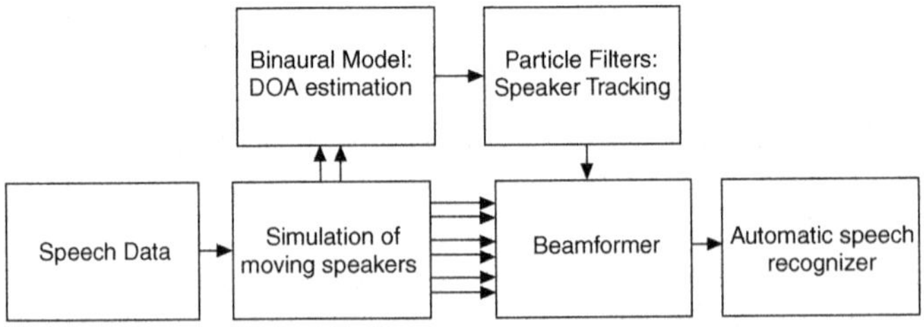


**Fig. 3** Block diagram of the experimental setup. See the text for details

applied for binaural synthesis [57]. The model extracts binaural parameters such as IPD and ILD in a way that mimics the performance of the human auditory system. Four specific aspects of temporal auditory processing were of specially in the focus of the IPD model, particularly,

- High temporal resolution.
- Limited phase-locking range.
- Use of temporal envelope disparities.
- A limited internal ITD range.

For the sake of consistency the implementation was kept unchanged from Dietz et al. [15], even though further improvements such as DC-offset free modulation filters [16] and aspects of pre-binaural adaptation—for example, [18, 30]—have recently been suggested to model psychoacoustic performance of envelope ITD sensitivity and binaural tuning of single cells more realistically.

Figure 4 gives an overview of the processing stages of the IPD model. For the stages up to the extraction of the interaural transfer function, ITF, the IPD and the ILD were adopted from [17]. Later stages of Fig. 4, from interaural vector strength (IVS) to DOA glimpse extraction, belong to the binaural cue selection. Both the IPD model and the binaural cue selection are described in [15]. In the following only the conceptually relevant aspects are briefly reviewed.

Most importantly, the signals were analyzed in 23 auditory filters in the range of 200 Hz to 5.0 kHz. Considering the human limit to binaurally exploit fine-structure information above ~1.4 kHz, the fine-structure filter is only implemented in the 12 lowest auditory filters below 1.4 kHz. Envelope IPDs are derived from all 23 filters, but are not exploited in the current study.

A problem occurring especially for fine-structure IPDs in filters above 700 Hz is that their corresponding ITDs do no longer cover the whole range of possible interaural delays, resulting in an ambiguity of direction. Inspired by psychoacoustic findings such as time–intensity trading—for instance, [33]—the sign of the ILD is employed here to extend the unambiguous range of IPDs from $[-\pi, \ \pi]$ to $[-2\pi, \ 2\pi]$. Accordingly, the frequency range for unambiguous fine-structure IPD-to-azimuth mapping is extended from ~700 to 1400 Hz. IPD-to-azimuth mapping itself is performed with a previously learned mapping function.

As argued in [15], the IPD model does not rely on cross-correlation, and, thus, interaural coherence (IC) is not directly assessable. However, Goupell and Hartmann [22] have shown that the temporal fluctuations of the interaural functions are possibly an even better measure for psychoacoustic decorrelation sensitivity. Therefore, in the IPD model, the IPD fluctuations are directly accessible and are specified in the form of the interaural vector strength (IVS). The IVS was used to derive a filter mask which consists of a binary weighting of the interaural parameters based on a threshold value of $\text{IVS}_0 = 0.98$.

By processing each of these high-coherence segments as a single event called *glimpse*, a sparse representation of the binaural features is generated from the median value of the azimuth estimation of this segment. If the IVS constantly exceeds $\text{IVS}_0$ for more than 20 ms, a new glimpse is assigned from the same segment. Depending

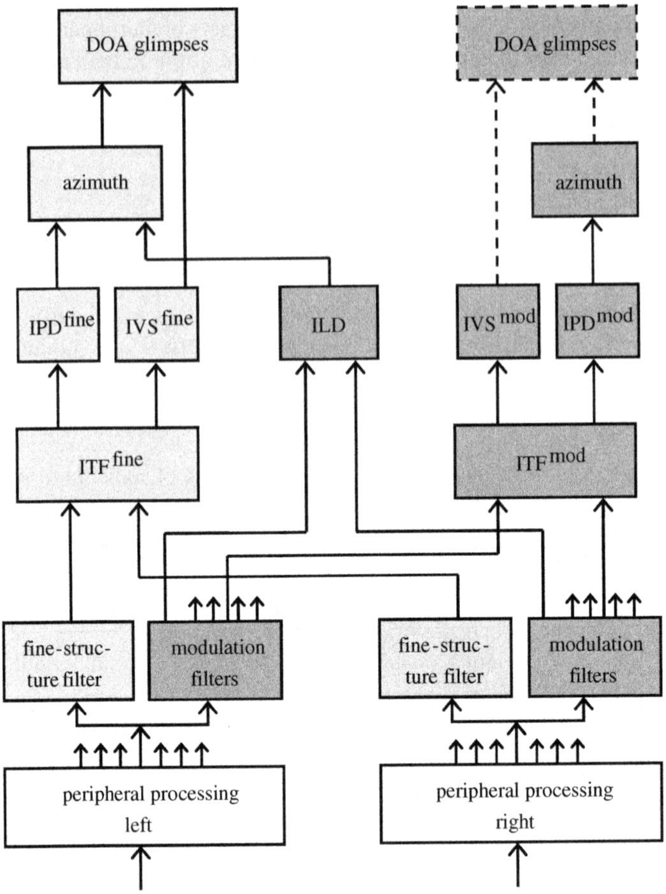


**Fig. 4** Processing stages of the IPD model from [15]. Peripheral processing splits the input signal up into 23 frequency channels from 200 to 5000 Hz. Only one of these channels is drawn for the further processing blocks. IPDs and interaural vector strength (IVS) are derived from one fine-structure and from several modulation filters. Fine-structure information is only derived in the 12 lowest frequency channels from 200 to 1400 Hz. In addition, the ILD is derived at the output of an envelope low-pass filter. The azimuth is derived from the IPDs with a previously stored frequency dependent mapping function. For fine-structure channels from 700 to 1400 Hz additional ILD information is employed to unwrap the IPD—see main text. Those azimuth estimates that occur during IVS $> 0.98$, result in so-called *glimpses*, which represent expectedly salient and sparse estimates of the direction of arrival. Within the current study only glimpses from fine-structure channels are considered

on the application it is decided whether the segments are grouped together to form a glimpse, or not. Here, for DOA estimation of stationary sounds, Sect. 4.3, it is not necessary, while it is highly beneficial for tracking applications such as the tracking of moving speakers, Sect. 4.4, in order to reduce the computational load.

## *4.2 Multi-Channel Speech Material*

This section describes the multi-channel speech material used for the experiments. The first and second subsection describe the monaural speech corpus and the generation of the spatial multi-channel signals, respectively. The simulation of moving sources is presented in the last subsection.

### Speech Data

The speech data used for the experiments consists of sentences produced by ten speakers—four male, six female. The syntactical structure and the vocabulary were adapted from the Oldenburg Sentence Test (OLSA) [67], where each sentence contains five words with ten alternatives for each word and a syntax that follows the pattern <name><verb><number><adjective><object>, which results in a vocabulary size of 50 words. The original recordings with a sampling rate of 44.1 kHz were downsampled to 16 kHz and concatenated—using three sentences from the same speaker. This resulted in sentences with a mean duration of 6.44 s, suitable for speaker tracking. For ASR experiments, the speech material was split into training and test sets with a total duration of 30 and 88 min, respectively. With this amount of speech data, a good ASR performance can be expected in relatively clean acoustics, whereas the estimation of acoustic models from noisy observations usually requires a larger database even for a relatively small vocabulary. Hence, the experiments presented in this chapter concentrate on the performance with one competing, moving speaker. The generation of training and test material is based on processing with a beamformer and is described in more detail in Sect. 4.6.

### Generation of Multi-Channel Signals

Spatially localized and diffuse sound sources are simulated using a database of head-related impulse responses, the HRIR database, which features impulse responses recorded with three microphones from each of two behind-the-ear (BTE) hearing aids attached to left and the right ear and two in-ear microphones. The HRIRs used in this study are a subset of the database described in [29]: Anechoic free-field HRIRs from the frontal horizontal half-plane measured at a distance of 3 m between microphones and loudspeaker were selected. The HRIRs from the database were measured with a 5° resolution for the azimuth angles, which was interpolated to obtain a 0.5° resolution. The coordinate system is illustrated in Fig. 5.

### Moving Speakers

The signals used throughout the experiments contain data of two moving speakers without interfering noise sources. Initial and final speaker positions were randomly drawn from a $-90°$ to $+90°$ azimuth interval, which represents the valid azimuth

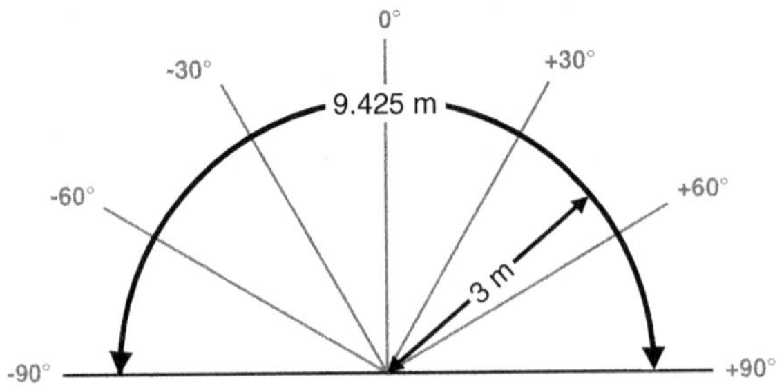


**Fig. 5** Available azimuth range of the generated signals

range of the binaural model. The speakers moved linearly from the start to the end point for the duration of the respective stimulus and crossed their tracks with a 50 % probability. A frame-wise processing scheme was employed by applying 64 ms Hann windows with 50 % overlap and convolving each time frame with the respective HRIR. Since a source separation cannot be performed with a beamformer when the signals come from the same direction, boundary conditions were defined that guaranteed an average angle difference of at least 10°. Additionally, the minimal distance between the start and end points was set to 10 and 20° for non-crossing and crossing speakers, respectively.

## *4.3 Statistical Analysis of Binaural Features*

As demonstrated in [15], the IPD model can be employed to localize several concurrent speakers. While the model suffers stronger from reverberation than normal hearing human listeners, its accuracy and performance is very good in free field multi speaker conditions. Even three speakers in noise at −6 dB SNR with same frequency characteristics as speech were robustly localized. While the number of speakers was only increased up to five in this previous study, Fig. 6 shows that even 6 concurrent speakers can be localized by analyzing the azimuth distribution in the 12 fine-structure channels over a few seconds. The time course of the azimuth estimate of an exemplary channel, $f_c = 1000$ Hz, is plotted in Panel (b). It can be seen that the estimate quickly oscillates between the six speaker positions. Over ten groups of glimpses per second indicate robust DOA estimates while transition periods that can contain any azimuth value are reliably suppressed by the IVS filter.[2]

[2] A demo folder containing the file *exp_spille2013* used to run the IPD model and to generate Fig. 6 is available in the *AMToolbox* [56].

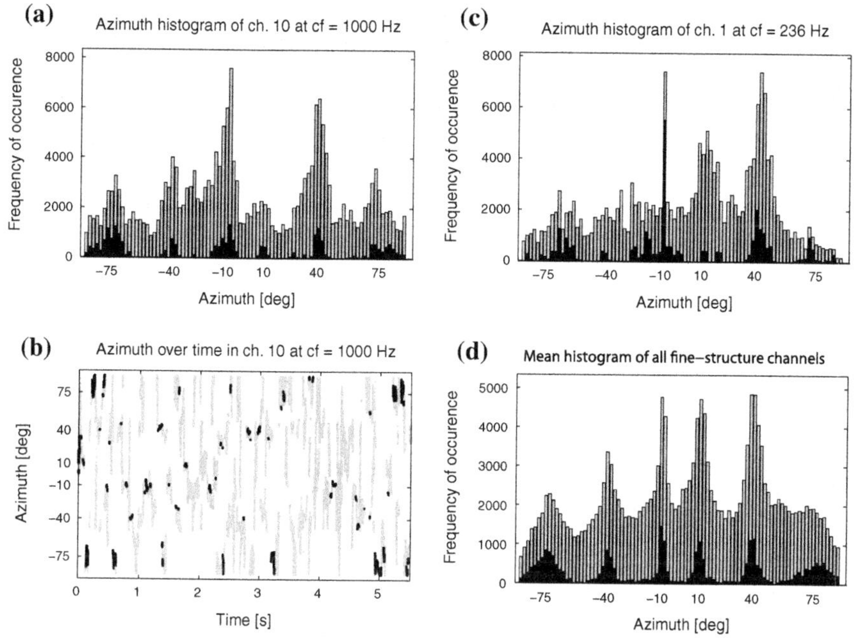


**Fig. 6** Model output for six simultaneously speaking stationary speakers at −75, −45, −10, +10, +45, and +75°. The speech duration was 5.5 s. *Grey color* indicates all DOA estimates without IVS filtering, *black color* indicates DOA estimates with IVS filtering, $IVS_0 = 0.98$. **a** Azimuth histogram of the fine-structure channel centered at $f_c = 1000$ Hz. **b** Time course of the azimuth estimation for the same channel and input signal as in panel (a). **c** Azimuth histogram of the fine-structure channel centered at $f_c = 236$ Hz with the same format and input signal as in (a). **d** Mean azimuth histogram of the twelve fine-structure channels. Same format and same input signal as in (a). It can be seen that the position of the speaker at +75° can only be determined with IVS filtering

## 4.4 Tracking Superposed Speakers

In the framework of the current application, the knowledge of the speaker positions is used to steer a beamformer that enhances the selected speaker by spatial filtering of the six BTE-microphone signals, that is, a binaural multi-channel beamformer. A tracking algorithm for multiple-speaker conditions was already implemented in [15], but only as a proof of concept. Here a more elaborate version is presented and its precision in several two-speaker scenarios is demonstrated.

### Particle Filters and Monte-Carlo Data Association

The main challenge in the tracking of multiple targets is the mapping from observations—in this case, of DOA glimpses—to a specific target, which is a prerequisite for the actual tracking. In this chapter, an algorithm provided by

Särkkä et al. [55] is applied to solve this problem.[3] The main idea of the algorithm is to split up the problem into two parts—so-called *Rao-Blackwellization*. First, the posterior distribution of the data association is calculated using a equential-importance resampling SIR, particle-filtering algorithm. Second, the single targets are tracked by an extended Kalman filter that depends on the data associations. Rao-Blackwellization exploits the fact that it is often possible to calculate the filtering equations in closed form. This leads to estimators with less variance compared to the method using particle filtering alone [9]. For more details of the algorithms see [25, 55].

### Application to Speaker Tracking

The tracking toolbox described in the previous section was applied to tracking the speakers from the DOA *glimpses* given by the IPD model. To apply the filter to the signals, the so-called dynamic model and the measurement model have to be defined. The dynamic model defines the temporal dynamics of the system and implements the block *state prediction* of Fig. 2. The state $x$ of the system is determined by the actual position of the target, $\alpha$, and velocity, $v$. $x$ is a vector consisting of the elements $\alpha$ and $v$. The dynamic model is then given by

$$x_k = A_{k-1} x_{k-1} + q_{k-1}. \tag{1}$$

The matrix, $A$, is the transition matrix of the dynamic model and reflects the dynamics of the system. In this case it is given as

$$A_k = \begin{pmatrix} 1 & \Delta t_k \\ 0 & 1 \end{pmatrix}, \tag{2}$$

where $\Delta t_k = t_{k+1} - t_k$ is the time step between two states of the system. This means that the system's state at time step $k$ is a linear progression of the system at time $k-1$ with constant speed plus some process noise, $q_k$, which is introduced to account for uncertainties in the system's development

$$x_k = \begin{pmatrix} \alpha_{k-1} + \Delta t_{k-1}\, v_{k-1} \\ v_{k-1} \end{pmatrix} + q_{k-1}. \tag{3}$$

The process noise is assumed to be a multivariate Gaussian with zero mean and covariance matrix

$$q_k = \begin{pmatrix} \frac{1}{3}\Delta t_k^3 & \frac{1}{2}\Delta t_k^2 \\ \frac{1}{2}\Delta t_k^2 & \Delta t_k \end{pmatrix} q_f, \tag{4}$$

which is calculated using the previously mentioned toolbox [24]. $q_f$ is a process-noise factor that was set to 0.1 in this case.The prior distribution of the state $x_0$ (see

[3] The algorithm is part of a Matlab-Toolbox provided by [25].

block *initialization of states* in Fig. 2), is also a multivariate Gaussian of the form

$$x_0 \sim N(m_0, P_0),$$

where $m_0$ denotes the prior mean of the state and $P_0$ its prior covariance matrix containing the variances of the system's position and velocity that is set to

$$P_0 = \begin{pmatrix} 50 & 0 \\ 0 & 15 \end{pmatrix},$$

In other words, the actual position $\alpha$ has a variance of 50 $\deg^2$ and the variance of the velocity is 15 $\mathrm{m}^2/\mathrm{s}^2$. The Kalman filter predicts the mean and the covariance of the state using the prior values together with the transition matrix, $A$, and the covariance matrix of the process noise, $q$. The equations for the predicted mean $\overline{m_k}$ and the predicted covariance $\overline{P_k}$ are as follows,

$$\begin{aligned} \overline{m_k} &= A_{k-1} m_{k-1} \\ \overline{P_k} &= A_{k-1} P_{k-1} A_{k-1}^T + q_{k-1}. \end{aligned} \tag{5}$$

Note that the process noise is only used for predicting the new covariance matrix. During the update step—block *weight update* in Fig. 2—these predictions are updated using the actual measurement, that is, *glimpses*, at time step $k$ as well as the measurement model which describes the relation between the measurement and the state of the system. The measurement model is given by

$$y_k = H_k x_k + r_k, \tag{6}$$

where $y_k$ is the actual measurement at time $k$, $H_k$ is the measurement model matrix and $r_k$ is the Gaussian measurement noise, $r_k \sim N(0, R)$. In the measurement model used here, only the position of the target is measured. This measurement can be corrupted by some noise reflecting the variance of the DOA estimation. Thus, the measurement model matrix, $H$, and the noise variance, $R$, are given by

$$H = (1 \quad 0) \; R = 50 \deg^2.$$

As the glimpses are sparse and occur with varying distance in time, the choice of the sampling interval is crucial. A sampling frequency equal to $1/\Delta t$ was chosen for the tracking algorithm and each glimpse was assigned to the nearest sampling point at this sampling rate. Glimpses were sampled at the original rate of the speech material. In the very rare cases that more than one glimpse fell in one bin, all but one glimpse were discarded.

Several sampling frequencies were tested and the minimum median-squared-error of the tracking was derived. For this, a dataset consisting of 71 sentences was used. A final sampling frequency of 500 Hz was chosen based on the results in Table 1.

**Table 1** Median-squared-errors and their roots for the different sampling frequencies

| Sampling frequency | Median-squared-error | Root median-squared-error |
|---|---|---|
| 50 | 6.6615 | 2.5810 |
| 100 | 2.8361 | 1.6841 |
| 200 | 2.1954 | 1.4817 |
| 400 | 1.9353 | 1.3912 |
| 500 | 1.4857 | 1.2189 |
| 1000 | 1.5297 | 1.2368 |
| 1600 | 1.8244 | 1.3507 |

## Speaker Tracking

The particle filter was initialized with a set of 20 particles using a known starting position of the first speaker, that is, the location variable of the first target was set to the position for all particles. The location variable of the second target was altered for each particle in equidistant steps throughout the whole azimuth range. Initial velocities were set randomly between $\pm 2\,\mathrm{m/s}$ for each target in each particle. The covariance matrix was equal for both targets and was set to $P_0$ as above.

If no glimpse is observed at time step $t$, the update step of the Kalman filter was skipped for this time step and the prediction was made based on the internal particle states. The range of the predicted angles was limited to the interval $[-90, 90]$ by setting all predictions outside that range to $-90°$ or $90°$, respectively.

Figure 7 presents two exemplary tracking results. The figure shows that the particle filter is able to track speakers even when they cross tracks, *left panel*. The tracking algorithm was evaluated by calculating the root median-squared error for each of the 9 data sets. On average the error was below 1.5°.

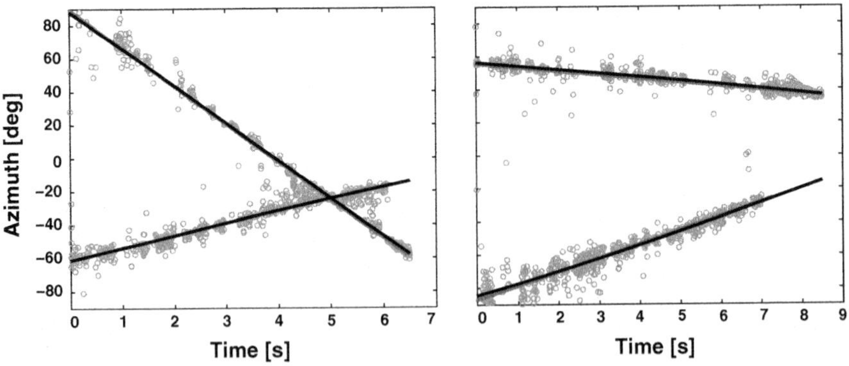


**Fig. 7** Tracking results of a two-speaker scenario. *Light-grey circles* represent the glimpses produced by the binaural model—see text. *Dark-grey lines* represent the real azimuth angles of the speakers. *Solid black lines* show the smoothed estimates obtained by tracking

## 4.5 Steerable Beamformer for Source Selection

In the proposed application, a position estimate for both the target and concurrent speaker are required to control the beamformer parameters to either enhance the speech of a certain speaker or block out a concurrent speaker, thereby increasing the overall signal-to-noise ratio and subsequently lower the word error rates of an automatic speech recognizer. The beamformer employed here is a super-directive beamformer based on the minimum-variance distortionless-response principle [12] that used the six BTE microphone inputs jointly—three channels left and three channel from the right ear. In general, it suppresses the noise coming from all directions while not affecting the speech of the desired speaker. Additionally, the beamformer strongly suppresses the speech of the concurrent speaker which, in this setup, is has a higher impact than the enhancement of the desired source. Let $W$ be the matrix containing the frequency domain filter coefficients of the beamformer, $d_1$ and $d_2$ the vectors containing the transfer functions to the microphones of speakers one and two respectively and $\Phi_{VV}$ the noise power-spectral density, PSD, matrix. Then, the following minimization problem has to be solved,

$$\min_{W} \; W^H \Phi_{VV} W, \qquad \text{with } W^H d_1 = 1 \text{ and } W^H d_2 = 0. \tag{7}$$

The solution to this is the minimum-variance distortionless-response beamformer [3]. The transfer functions in vectors $d_1$ and $d_2$ result from the impulse responses that are chosen based on the angle estimation of the tracking algorithm. The coherence matrix which is required to solve (7) is also estimated using the impulse responses used for generating the signals. Note that relying on the true impulse responses implies the use of *a-priori* knowledge not available in a real-world application, for which the impulse responses need to be estimated. The beamforming by itself therefore represents an upper bound, and will be extended to be used with estimated impulse responses in future work. However, since the IPD model, the tracking algorithm and the ASR system do not use such *a-priori* knowledge in reflecting realistic conditions, and robust methods for estimation of impulse responses exist, the results should still be transferable to real-world applications.

## 4.6 ASR System

### Feature Extraction and Classifier

The benefits of the proposed processing chain for speech processing are analyzed by performing automatic speech recognition (ASR) on the output signals of the beamformer. The ASR system consists of a feature extraction and a classification stage.

The features extracted from speech should represent the information required to transcribe the spoken message and ideally suppress unwanted signal components.

For the experiments, the feature type most commonly applied in ASR, namely, *Mel-frequency cepstral coefficients* (MFCCs) [13] was chosen. These features effectively encode the smoothed short-time Fourier transform (STFT) magnitude, which is computed every 10 ms using overlapping analysis windows of 25 ms duration. Each frame of the STFT is processed by a mel-filterbank that approximates the frequency sensitivity of the human ear, compressed with the logarithm and transformed to cepstral parameters using a discrete cosine transformation. By selecting twelve lower cepstral coefficients, only the coarse spectral structure is retained. By adding an energy value and calculating an estimate for the first and second derivative, the so-called delta and double-delta features, to include some information about temporal dynamics on the feature level, 39-dimensional feature vectors were finally obtained.

The feature vectors are used without normalization to train and test the Hidden-Markov model (HMM) classifier, which has been set up as word model with each word of the vocabulary corresponding to a single HMM. During testing, the likelihoods of each HMM generating the observed sequence of feature vectors are compared and the word with the highest likelihood is selected. A grammar reflecting the fixed syntax of OLSA sentences is used to ensure a transcription with a valid OLSA sentence structure, in particular the following, <name><verb><number> adjective><object>, repeated three times due to the concatenation of sentences. The HMM used ten states per word model and six Gaussians per mixture and was implemented using the Hidden-Markov Toolkit (HTK) decribed in [72].

## Training and Test Material

ASR training was carried out using sentences with one moving speaker, which were processed with the beamformer. The steering vectors of the beamformer were set to the true azimuth angles of the desired speaker instead of using the output of the complete processing chain including DOA estimation. This resulted in signals containing some beamforming artifacts, that is, the classifier was able to adapt to the resulting feature distortions and still carried the relevant information to create proper word models. The effects of speaker-dependent (SD) versus speaker-independent (SI) recognition was investigated by creating two training sets with the test speaker being either included in the training data, SD, or excluded from training, SI. The original data contained 71 long sentences each of which was used several times for the simulation of moving speakers, thereby increasing the amount of training material. Each sentence was processed four times with random start and end positions of the speakers, which resulted in 284 training sentences or 30 min, repectively, for the SD system and approximately 250 training sentences or 27 min for the SI system.

For *testing*, signals with two moving speakers were processed by the complete chain depicted in Fig. 3, one being the target source and the other one the suppressed source, and the recognition rate for the words uttered by the target speaker was obtained. To increase the number of test items, each speaker was selected as the target speaker once and the training/testing procedure was carried out ten times. As for the training set, the original 71 sentences were used for movement simulation several times to further increase the number of test items and hence the significance of the results. For testing, a factor of 11 was chosen due to computational constraints. This resulted in a total number of 781 sentences or 88 min with randomised start and end positions for two speakers.

## 4.7 ASR Results

When using the complete processing chain that included the DOA estimation, tracking, beamforming, and ASR, a word-recognition rate (WRR) of 88.4 % was obtained for the speaker-dependent ASR system. When using a speaker-independent system, a word-recognition rate of 72.6 % was achieved. The data presented in the following were obtained with the speaker-dependent ASR system. When the ASR system cannot operate on beamformed signals, but is limited to speech that was converted to mono signals by selecting one of the eight channels from the behind-the-ear or in-ear recordings, the average WRR was 29.4 %. The variations of WRRs between channels were relatively small, ranging from 28.1 to 30.8 %. When the best channel for each sentence was selected, that is, the channel that resulted in the highest WRR for that specific sentence to simulate the best performance when limited to one channel, the average WRR was increased to 38.8 %.

It is interesting to note that the WRRs were very similar when analyzing crossing and non-crossing speaker tracks separately, namely, 88.3 and 88.4 %, respectively. An analysis of the average separation of speakers in ° showed that the overall accuracy was nearly constant for spatial distances ranging from 40 to 100°—Fig. 8a—but will definitely drop down for smaller distances. The average distance was, of course, significantly higher for non-crossing speakers, namely, 64.9°, than for crossing speak-

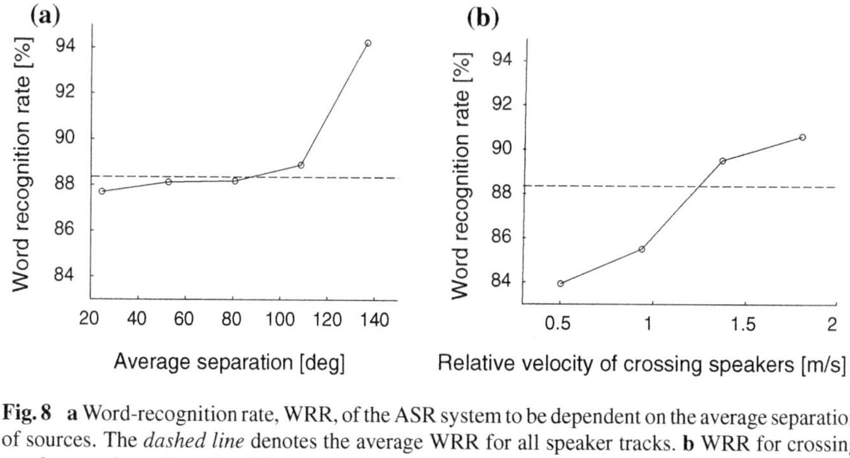


**Fig. 8** **a** Word-recognition rate, WRR, of the ASR system to be dependent on the average separation of sources. The *dashed line* denotes the average WRR for all speaker tracks. **b** WRR for crossing speakers to depend on the difference speed of competing speakers. The *dashed black line* shows the recognition rate that was obtained for crossing speaker tracks

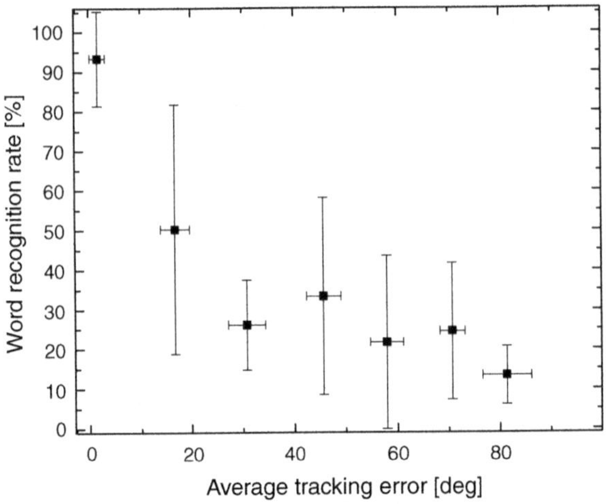


**Fig. 9** Word-recognition rate versus average tracking error. The range of the tracking error was devided in equidistant bins. Data points denote the mean tracking error and the mean word-recognition rate. *Error bars* show the corresponding standard deviations

ers, 41.0°. But, due to the constraints in the case of crossing speakers, the average distance was high enough to not reduce the WRR. The parameter determining the average distance of crossing speakers is the relative velocity of speakers, where high relative velocities correspond to short durations of spatially close speakers—Fig. 8b. The constraints for starting and end positions that were chosen for crossing speakers resulted in an average relative speed difference of 1.2 m/s. Hence, the comparable high WRRs for crossing tracks can be attributed to the high relative velocity of speakers ensured by the contraints in signal generation.

The word-recognition rate also depends strongly on the localization accuracy: The overall localization accuracy was quantified by calculating the average tracking error, which is the root median squared error between the smoothed tracking estimates and the real azimuth angles of the speakers—see Table 1. Figure 9 shows that the WRR is highly dependent on the average tracking error where higher traking errors cause significantly lower WRRs.

## 5 Summary and Conclusions

This study provided an overview of computational auditory scene analysis based on binaural information and its application to a speech recognition task. The usability of the IPD model in automated speech processing was demonstrated by performing a DOA estimation for stationary and moving speakers. For the moving-speaker scenario, it was also shown that the binaural model enables efficient tracking and greatly increases the performance of an automatic speech recognition system in situations with one interfering speaker. The word-recognition rate (WRR) was increased from 30.8 to 88.4 %, which shows the potential of integrating models of binaural hearing into speech processing systems. It remains to be seen if this performance gain in anechoic conditions can be validated in real-world scenarios, that is, in acoustic conditions with strong reverberation, several localized noise sources embedded in a 3D-environment compared to the 2D simulation presented here or with a changing number of speakers. Follow-up studies are suggested that explore a combination of a binaural model, a tracking system and beamforming for other problems in speech and hearing research, such as speaker identification, speaker diarization or the improvement of noise reduction in hearing aids.

**Acknowledgments** Supported by the DFG—SFB/TRR 31 *The active auditory system*, URL: http://www.uni-oldenburg.de/sfbtr31. The authors would like to thank M. Klein-Hennig for casting the IPD model code in the AMToolbox format, D. Marquardt and G. Coleman for their contributions to the beamforming algorithm, M. R. Schädler for sharing the code of the OLSA recognition system, H. Kayser for support with the HRIR database, and two anonymous reviewers for constructive suggestions.

## References

1. M. S. Arulampalam, S. Maskell, N. Gordon, and T. Clapp. A tutorial on particle filters for online nonlinear / non-Gaussian bayesian tracking. *IEEE Trans*. Signal Process., 50:174–188, 2002.
2. R. Beutelmann and T. Brand. Prediction of speech intelligibility in spatial noise and reverberation for normal-hearing and hearing-impaired listeners. *J. Acoust. Soc. Am.*, 120:331–342, 2006.
3. J. Bitzer and K. U. Simmer. Superdirective microphone arrays. In M. Brandstein and D. Ward, editors, *Microphone Arrays*, chapter 2. Springer, 2001.
4. A. Brand, O. Behrend, T. Marquardt, D. McAlpine, and B. Grothe. Precise inhibition is essential for microsecond interaural time difference coding. *Nature*, 417:543–547, 2002.
5. J. Breebaart, S. van de Par, and A. Kohlrausch. Binaural processing model based on contralateral inhibition. I. Model structure. *J. Acoust. Soc. Am.*, 110:1074–1088, 2001.
6. A. S. Bregman. *Auditory scene analysis: The perceptual organization of sound.* MIT Press, 1990.
7. K. O. Bushara, T. Hanakawa, I. Immisch, K. Toma, K. Kansaku, and M. Hallett. Neural correlates of cross-modal binding. *Nat. Neurosci.*, 6:190–195, 2003.
8. C. E. Carr and M. Konishi. Axonal delay lines for time measurement in the owl's brainstem. *Proc. Natl. Acad. Sci. U. S. A.*, 85:8311–8315, 1988.
9. G. Casella and C. Robert. Rao-Blackwellisation of sampling schemes. *Biometrika*, 83:81–94, 1996.
10. H. Christensen, N. M. N. Ma, S. N. Wrigley, and J. Barker. A speech fragment approach to localising multiple speakers in reverberant environments. In *IEEE ICASSP*, 2009.
11. M. Cooke. Glimpsing speech. *Journal of Phonetics*, 31:579–584, 2003.
12. H. Cox, R. Zeskind, and M. Owen. Robust adaptive beamforming. *IEEE Trans. Acoust., Speech, Signal Process.*, 35:1365–1376, 1987.
13. S. B. Davis and P. Mermelstein. Comparison of parametric representations for monosyllabic word recognition in continuously spoken sentences. *IEEE Trans. Acoust., Speech, Signal Process.*, 28:357–366,1980.

14. M. Dietz, S. D. Ewert, and V. Hohmann. Lateralization of stimuli with independent fine-structure and envelope-based temporal disparities. *J. Acoust. Soc. Am.*, 125:1622–1635, 2009.
15. M. Dietz, S. D. Ewert, and V. Hohmann. Auditory model based direction estimation of concurrent speakers from binaural signals. *Speech Commun.*, 53:592–605, 2011.
16. M. Dietz, S. D. Ewert, and V. Hohmann. Lateralization based on interaural differences in the second-order amplitude modulator. *J. Acoust. Soc. Am.*, 131:398–408, 2012.
17. M. Dietz, S. D. Ewert, V. Hohmann, and B. Kollmeier. Coding of temporally fluctuating interaural timing disparities in a binaural processing model based on phase differences. *Brain Res.*, 1220:234–245, 2008.
18. M. Dietz, T. Marquardt, D. Greenberg, D. McAlpine. The influence of the envelope waveform on binaural tuning of neurons in the inferior colliculus and its relation to binaural perception. In B. C. J. Moore, R. Patterson, I. M. Winter, R. P. Carlyon, H. E. Gockel, editors, *Basic Aspects of Hearing: Physiology and Perception*, chapter 25. Springer, New York, 2013.
19. A. Doucet, N. de Freitas, and N. Gordon. An introduction to sequential Monte Carlo methods. In A. Doucet, N. de Freitas, and N. Gordon, editors, *Sequential Monte Carlo Methods in Practice*. Springer, 2001.
20. C. Faller and J. Merimaa. Source localization in complex listening situations: Selection of binaural cues based on interaural coherence. *J. Acoust. Soc. Am.*, 116:3075–3089, 2004.
21. K. Friston and S. Kiebel. Cortical circuits for perceptual inference. *Neural Networks*, 22:1093–1104, 2009.
22. M. J. Goupell and W. M. Hartmann. Interaural fluctuations and the detection of interaural incoherence: Bandwidth effects. *J. Acoust. Soc. Am.*, 119:3971–3986, 2006.
23. S. Harding, J. P. Barker, and G. J. Brown. Mask estimation for missing data speech recognition based on statistics of binaural interaction. *IEEE T. Audio. Speech.*, 14:58–67, 2006.
24. J. Hartikainen and S. Särkkä. Optimal filtering with Kalman filters and smoothersa Manual for Matlab toolbox EKF/UKF. Technical report, Department of Biomedical Engineering and Computational Science, Helsinki University of Technology, 2008.
25. J. Hartikainen and S. Särkkä. RBMCDAbox-Matlab tooolbox of rao-blackwellized data association particle filters. Technical report, Department of Biomedical Engineering and Computational Science, Helsinki University of Technology, 2008.
26. H. Hermansky. Perceptual linear predictive (PLP) analysis of speech. *J. Acoust. Soc. Am.*, 87:1738–1752, 1990.
27. V. Hohmann. Frequency analysis and synthesis using a Gammatone filterbank. *Acta Acustica united with Acustica*, 88:433–442, 2002.
28. L. a. Jeffress. A place theory of sound localization. *J. Comp. Physiol. Psychol.*, 41:35–39, 1948.
29. H. Kayser, S. D. Ewert, J. Anemüller, T. Rohdenburg, V. Hohmann, and B. Kollmeier. Database of multichannel in-ear and behind-the-ear head-related and binaural room impulse responses. *EURASIP Journal on Advances in Signal Processing*, 2009:298605, 2009.
30. M. Klein-Hennig, M. Dietz, V. Hohmann, and S. D. Ewert. The influence of different segments of the ongoing envelope on sensitivity to interaural time delays. *J. Acoust. Soc. Am.*, 129:3856–3872, 2011.
31. M. Kleinschmidt. Methods for capturing spectro-temporal modulations in automatic speech recognition. *Acta Acustica united with Acustica*, 88:416–422, 2002.
32. D. Kolossa, F. Astudillo, A. Abad, S. Zeiler, R. Saeidi, P. Mowlaee, R. Martin. CHiME challenge : Approaches to robustness using beamforming and uncertainty-of-observation techniques. *Int. Workshop on Machine Listening in Multisource, Environments*, 1:6–11, 2011.
33. A.-G. Lang and A. Buchner. Relative influence of interaural time and intensity differences on lateralization is modulated by attention to one or the other cue: 500-Hz sine tones. *J. Acoust. Soc. Am.*, 126:2536–2542, 2009.
34. N. Le Goff, J. Buchholz, and T. Dau. Modeling localization of complex sounds in the impaired and aided impaired auditory system. In J. Blauert, editor, *The technology of binaural listening*, chapter 5. Springer, Berlin-Heidelberg-New York NY, 2013.
35. W. Lindemann. Extension of a binaural cross-correlation model by contralateral inhibition. I. Simulation of lateralization for stationary signals. *J. Acoust. Soc. Am.*, 80:1608–1622, 1986.

36. R. F. Lyon. A computational model of binaural localization and separation. In *IEEE ICASSP*, volume 8, pages 1148–1151, 1983.
37. T. May, S. Van De Par, and A. Kohlrausch. A probabilistic model for robust localization based on a binaural auditory front-end. *IEEE T. Audio. Speech.*, 19:1–13, 2011.
38. T. May, S. Van De Par, and A. Kohlrausch. A binaural scene analyzer for joint localization and recognition of speakers in the presence of interfering noise sources and reverberation. *IEEE T. Audio. Speech.*, 20:1–15, 2012.
39. T. May, S. Van De Par, and A. Kohlrausch. Noise-robust speaker recognition combining missing data techniques and universal background modeling. *IEEE T. Audio. Speech.*, 20:108–121, 2012.
40. T. May, S. van de Par, and A. Kohlrausch. Binaural localization and detection of speakers in complex acoustic scenes. In J. Blauert, editor, *The technology of binaural listening*, chapter 15. Springer, Berlin-Heidelberg-New York NY, 2013.
41. D. McAlpine and B. Grothe. Sound localization and delay lines-do mammals fit the model? *Trends Neurosci.*, 26:347–350, 2003.
42. D. McAlpine, D. Jiang, and a. R. Palmer. A neural code for low-frequency sound localization in mammals. *Nat. Neurosci.*, 4:396–401, 2001.
43. J. Nix and V. Hohmann. Sound source localization in real sound fields based on empirical statistics of interaural parameters. *J. Acoust. Soc. Am.*, 119:463–479, 2006.
44. J. Nix and V. Hohmann. Combined estimation of spectral envelopes and sound source direction of concurrent voices by multidimensional statistical filtering. *IEEE T. Audio. Speech.*, 15:995–1008, 2007.
45. B. Opitz, A. Mecklinger, A. D. Friederici, and D. Y. Von Cramon. The functional neuroanatomy of novelty processing: integrating ERP and fMRI results. *Cereb. Cortex*, 9:379–391, 1999.
46. B. Osnes, K. Hugdahl, and K. Specht. Effective connectivity analysis demonstrates involvement of premotor cortex during speech perception. *Neuroimage*, 54:2437–2445, 2011.
47. P. Paavilainen, M. Jaramillo, R. Näätänen, and I. Winkler. Neuronal populations in the human brain extracting invariant relationships from acoustic variance. *Neurosci. Lett.*, 265:179–182, 1999.
48. K. Palomäki and G. J. Brown. A computational model of binaural speech recognition: Role of across-frequency vs. within-frequency processing and internal noise. *Speech Commun.*, 53:924–940, 2011.
49. K. J. Palomäki, G. J. Brown, and D. Wang. A binaural processor for missing data speech recognition in the presence of noise and small-room reverberation. *Speech Commun.*, 43:361–378, 2004.
50. D. P. Phillips. A perceptual architecture for sound lateralization in man. *Hear. Res.*, 238:124–132, 2008.
51. V. Pulkki and T. Hirvonen. Functional count-comparison model for binaural decoding. *Acta Acustica united with Acustica*, 95:883–900, 2009.
52. L. Rayleigh. On our perception of sound direction. *Philos. Mag.*, 13:214–232, 1907.
53. H. Riedel and B. Kollmeier. Interaural delay-dependent changes in the binaural difference potential of the human auditory brain stem response. *Hear. Res.*, 218:5–19, 2006.
54. N. Roman, D. Wang, and G. J. Brown. Speech segregation based on sound localization. *J. Acoust. Soc. Am.*, 114:2236–2252, 2003.
55. S. Särkkä, A. Vehtari, and J. Lampinen. Rao-Blackwellized particle filter for multiple target tracking. *Information Fusion*, 8:2–15, 2007.
56. P. Søndergaard and P. Majdak. The auditory-modeling toolbox.In J. Blauert, editor, *The technology of binaural listening*, chapter 2. Springer, Berlin-Heidelberg-New York NY, 2013.
57. S. Spors and H. Wierstorf. Evaluation of perceptual properties of phase-mode beamforming in the context of data-based binaural synthesis. In 5th *International Symposium on Communications Control and Signal Processing (ISCCSP)*, 2012, pages 1–4, 2012.
58. R. Stern and N. Morgan. Hearing is believing: Biologically-inspired feature extraction for robust automatic speech recognition. *IEEE Signal Processing Magazine*, 29:34–43, 2012.

59. R. Stern, A. Zeiberg, and C. Trahiotis. Lateralization of complex binaural stimuli: A weighted-image model. *J. Acoust. Soc. Am.*, 84:156–165, 1988.
60. R. M. Stern and H. S. Colburn. Theory of binaural interaction based in auditory-nerve data. IV. A model for subjective lateral position. *J. Acoust. Soc. Am.*, 64:127–140, 1978.
61. S. K. Thompson, K. von Kriegstein, A. Deane-Pratt, T. Marquardt, R. Deichmann, T. D. Griffiths, and D. McAlpine. Representation of interaural time delay in the human auditory midbrain. *Nat. Neurosci.*, 9:1096–1098, 2006.
62. S. P. Thompson. On binaural audition. *Philos. Mag.*, 4:274–276, 1877.
63. S. P. Thompson.On the function of the two ears in the perception of space. *Philos. Mag.*, 13:406–416, 1882.
64. M. van der Heijden and C. Trahiotis. Masking with interaurally delayed stimuli: the use of "internal" delays in binaural detection. *J. Acoust. Soc. Am.*, 105:388–399, 1999.
65. G. von Békésy. Zur Theorie des Hörens. Über das Richtungshören bei einer Zeitdifferenz oder Lautstärkenungleichheit der beiderseitigen Schalleinwirkungen. *Phys. Z.*, 31:824–835, 1930.
66. C. Wacongne, J. P. Changeux, and S. Dehaene. A neuronal model of predictive coding accounting for the mismatch negativity. *J. Neurosci.*, 32:3665–3678, 2012.
67. K. C. Wagener and T. Brand. Sentence intelligibility in noise for listeners with normal hearing and hearing impairment: influence of measurement procedure and masking parameters. *Int. J. Audiol.*, 44:144–156, 2005.
68. D. Wang and G. J. Brown. *Computational Auditory Scene Analysis: Principles, Algorithms, and Applications*. Wiley-IEEE Press, 2006.
69. S. Wilson, A. Saygin, M. Sereno, and M. Iacoboni. Listening to speech activates motor areas involved in speech production. *Nat. Neurosci.*, 7:701–702, 2004.
70. I. Winkler. Interpreting the Mismatch Negativity. *J. Psychophysiol.*, 21:147–163, 2007.
71. J. Woodruff and D. Wang. Binaural localization of multiple sources in reverberant and noisy environments. *IEEE T. Audio. Speech.*, 20:1503–1512, 2012.
72. S. Young, G. Evermann, D. Kershaw, G. Moore, J. Odell, D. Ollason, D. Povey, V. Valtchev, and P. Woodland. The HTK book. *Cambridge University Engineering Department*, 3, 2002.

# Extracting Sound-Source-Distance Information from Binaural Signals

**E. Georganti, T. May, S. van de Par and J. Mourjopoulos**

## 1 Introduction

Judgement of the distance from nearby objects is very important in human's everyday life. Normal-hearing people can easily determine the approximate distance between themselves and a sound source by using only auditory cues, even when visual information is not available.

For example, pedestrians can estimate the distance from a car approaching from behind by listening only to the sound of the car's noise or its horn. They can determine how close is a bee that is flying around their head or how far is a dog barking and running towards them. Similarly, the distance from speakers or other sound sources can be identified within rooms, even when multiple sound reflections and reverberation are present. In all such situations, auditory distance perception could prepare a human to take precaution actions.

During the last years, the problem of distance estimation has attracted the attention of scientists and several studies exist in the field. These studies can be divided into two main categories,

- Perceptual studies that aim to understand the auditory mechanisms that underlie distance judgements by humans

E. Georganti (✉) · J. Mourjopoulos
Audio and Acoustic Technology Group, Electrical and Computer Engineering Department, University of Patras, Patras, Greece
e-mail: egeorganti@upatras.gr

T. May
Centre for Applied Hearing Research, Department of Electrical Engineering, Technical University of Denmark, Kgs. Lyngby, Denmark

S. van de Par
Institute of Physics,
University of Oldenburg, Oldenburg, Germany

J. Blauert (ed.), *The Technology of Binaural Listening*, Modern Acoustics and Signal Processing, DOI: 10.1007/978-3-642-37762-4_7,

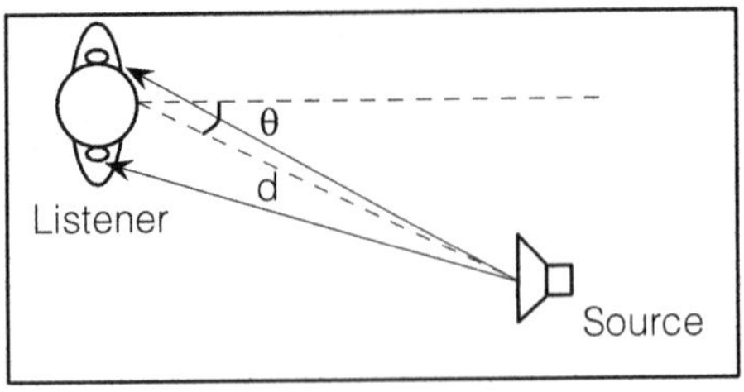


**Fig. 1** The problem of sound-source-distance estimation in rooms. A listener can estimate the line-of-sight distance, $d$, from a single source located at any angle, $\theta$, inside a reverberant space utilising the binaural information

- Studies providing computational methods for the distance-estimation problem, usually based on the findings of such perceptual studies

Detection methods for the actual distance between the source and receiver can be advantageous for various audio and speech applications such as denoising or dereverberation methods [58], intelligent hearing-aid devices [23], auditory scene analysis [50, 66] and hands-free communication systems [22, 24, 44] as often the relative distances between the microphones and the sound source are generally not known.

Figure 1 illustrates a typical example for the problem of distance estimation for a human listener situated in a room where there is one sound source present. The listener utilising binaural perceptual mechanisms can estimate the line-of-sight distance, $d$, between her/him and the sound source based on the signals that arrive to his/her ears. Similarly, a computational distance-estimation method attempts to determine this distance by employing distance-dependent parameters [37, 57, 62]. In practical situations, additive noise and the presence of other simultaneous sound sources may hinter such a task. Here, the problem of distance estimation under room reverberation will be considered in detail, but the effect of other possible interferences is beyond the scope of this chapter.

In Sect. 2, a brief literature overview concerning findings related to the auditory distance perception is given. In Sect. 3, several acoustical parameters that vary with distance in reverberant spaces are presented, providing insight to the reader of the parameters that could potentially assist distance detection techniques. Finally, in Sects. 4 and 5, several existing distance-estimation techniques are presented.

## 2 Auditory-Distance Perception

In this section, in order to provide some further insight to the reader on the distance-judgement ability of humans in enclosed spaces, the main points of several related studies [14, 32, 42, 69] are summarized. Although, there are many studies on distance perception performance [3, 14, 29, 30, 32, 42, 63, 64, 69], it is not completely clear yet what are the exact cues that provide the relevant information for distance estimation since it appears that auditory distance perception depends on a combination of multiple acoustic and non-acoustic cues [32, 69]. Factors known to influence distance perception include the stimulus spectral content or envelope, a priori knowledge of the stimulus-presentation level, azimuthal location of the source, sound reflections from the environment, and visual information about candidate sound sources in the environment [52, 69]. Early studies of anechoic localization show that specific binaural cues convey some distance information to listeners when sources are near the listener's head [11]. However, more recent studies provide hints that perception of the distance of nearby sources in reverberant environments is largely driven by monaural rather than binaural cues [3, 32, 55].

One of the main findings related to auditory distance perception is that human listeners tend to overestimate small distances and underestimate large distances [69] in reverberant rooms. More specifically, people tend to overestimate distances up to approximately 2 m and underestimate distances beyond that. This is in agreement with earlier studies of distance perception [9, 40, 41, 43] and might be related to the *auditory horizon* that represents the maximum perceived distance [65]. Perceived distance depends also on the reverberation time of the room and distance judgements are more accurate in a reverberant space than in an anechoic space [9, 41].

In [69], the relationship given by (1) was established showing a compressive power function between the geometrical sound-source distance, $r$, and the perceived distance, $r'$, that is,

$$r' = kr^{\alpha}, \tag{1}$$

where the constant $k$ and the exponent $\alpha$ are fitted parameters of the power function. The fitted values for $a$ ranged from approximately 0.15 to 0.7 and $k$ was typically close to one. Note here that the $a$ and $k$ values were extracted using an extended dataset taken from 21 relevant studies for various stimulus presentation techniques, source signals and acoustic environments.

In the next section, several acoustical parameters that depend on distance will be presented. Such acoustical factors are present in either the monoaural or binaural signals perceived by humans and could potentially provide distance information for computational systems.

# 3 Distance-Dependent Acoustic Parameters

## *3.1 Sound-Pressure Level*

Sound-pressure level is one of the main parameters that vary with distance and probably provides distance information to humans [40]. As the distance increases the sound pressure decreases, and in free-field conditions this decrease obeys an inverse law—$1/r$. This leads to a 6 dB loss with each doubling of distance in an open-space for point sources. Here, it should be also stated that at distances beyond 15 m, the high-frequency components of the travelling sound waves are further attenuated by the absorbing properties of air [5]. This frequency-dependent attenuation is additive

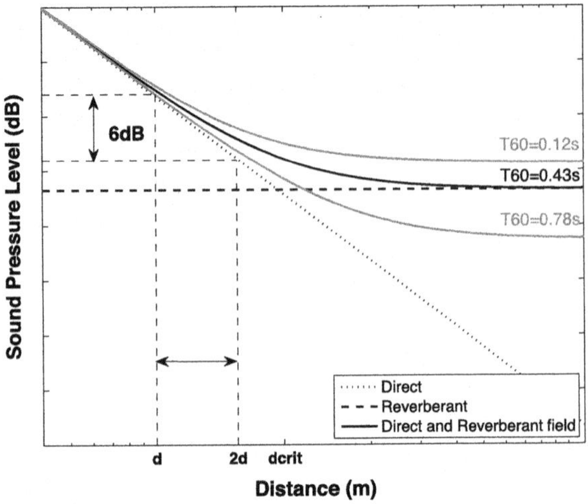


**Fig. 2** Sound-pressure level as a function of the distance between the source and receiver. Beyond the critical distance, $d_{\text{crit}}$, the reverberant field becomes dominant

to the aforementioned inverse law. In the case of enclosed spaces, where the free-field assumptions do not hold, the decrease of the sound-pressure level depends on the acoustic properties of the room and especially its ability to absorb the reverberant energy. Hence, this decrease of sound-pressure level is usually smaller when compared to that in an open-space, due to the presence of early reflections and reverberation.

Figure 2 depicts this well-known dependency of sound-pressure level on distance for three different rooms. The black solid line corresponds to a room that has a reverberation time of 0.43 s. For this case the direct sound-pressure level, the reverberant-field sound-pressure levels and the critical distance, $d_{crit}$, of the room are also indicated on the figure. Here, the 6 dB loss with each doubling of distance is also evident, according to the theoretical predictions [34]. Beyond the critical distance [34], the sound-pressure level does not follow the inverse law due to the presence of reverberation. Similarly, the grey curves correspond to two other rooms that have reverberation times of 0.12 and 0.78 s, respectively.

## 3.2 Direct-to-Reverberant Ratio

From the previous section it is clear that for enclosed spaces, the direct-to-reverberant ratio, DRR, describes the amount of acoustic power due to the direct path [34] and is given by

$$D_{\mathrm{r}} = 10 \log_{10} \frac{\int_0^T h^2(t)}{\int_T^\infty h^2(t)}, \tag{2}$$

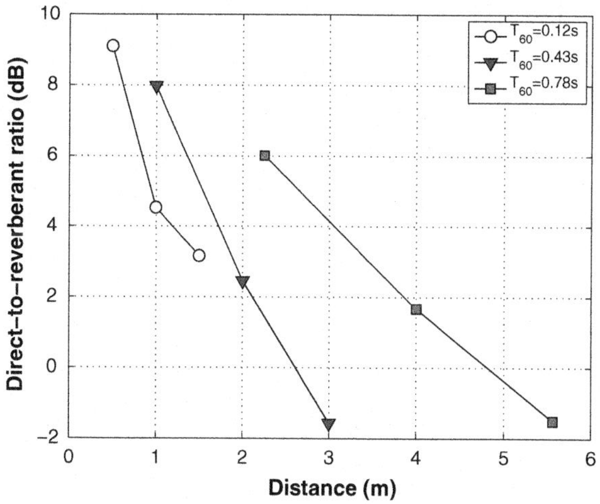


**Fig. 3** RIR–DRR as a function of distance for three different rooms

where $h(t)$ is the room impulse response, RIR, and $T$ is the duration of the direct sound—typically 5 ms. The DRR decreases monotonously as the source/receiver distance increases. This can be seen in Fig. 3, where the DRR is plotted as a function of distance for the three rooms having different reverberation-time values—as illustrated in Fig. 2. For this example, the DRR has been extracted from the corresponding single-channel RIRs measured at different source/receiver distances within the three rooms. Assuming that the head orientation is *facing* the source, the two binaural responses will also exhibit similar dependence on distance since the acoustic paths from source to the two ears will be almost identical.

However, the DRR values obtained from the binaural room impulse responses, BRIRs, for different distance and head orientation angles will vary for each ear. Figure 4 depicts the DRR as a function of distance when extracted from the left and right BRIRs and for various orientation angles in the horizontal plane inside a room that has reverberation time of 0.86 s. It can be seen that for symmetrical positions with respect to the line-of-sight, that is, zero degrees, the DRR values obtained are the same for the left and right BRIRs. For the rest of the orientation angles, there is an interdependency between the DRR values obtained from the left and the right BRIRs, leading to an increase in the left DRR and a decrease in the right DRR when the left ear is *facing* the sound source and vice versa.

## 3.3 Spectral Parameters

It is well known that in enclosed spaces, the magnitude spectra of the received signals are also affected with an increase of distance [27, 53, 54]. Figure 5 depicts typical

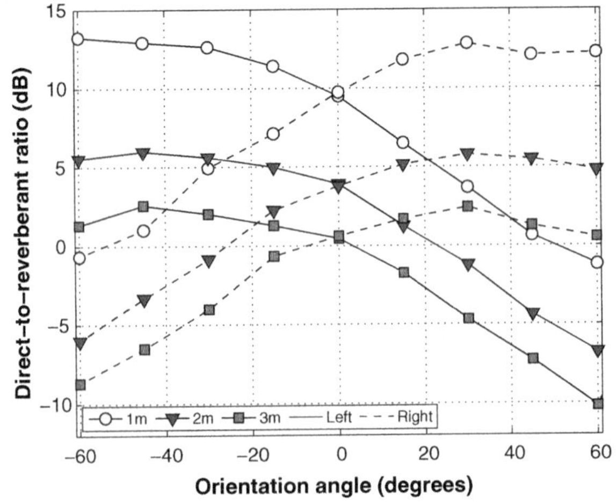


**Fig. 4** BRIR–DRR as a function of distance and head orientation angle for a room with reverberation time of 0.86 s. The *solid* and *dashed lines* correspond to the left and right-channel BRIR–DRR, respectively. When the left ear is *facing* the sound source, the left BRIR–DRR presents higher values than the right BRIR–DRR and vice versa

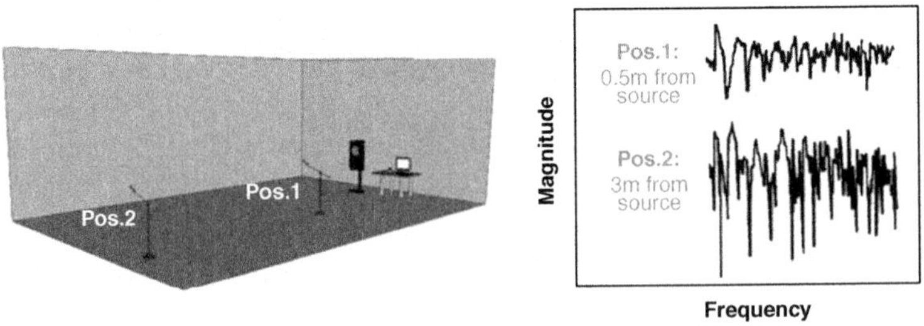


**Fig. 5** Effect of distance increase on the magnitude spectra obtained from the RTFs, measured at two different positions within a room. The spectral values for the distant position, *pos. 2*, deviate more, and the peaks and dips are more evident as compared to *pos. 1*

magnitude spectra obtained from the room transfer functions, RTFs[1] measured at two different positions within a room.

It is evident that for the farther position, *pos. 2*, the spectral values deviate more, and the peaks and dips are more pronounced compared to the spectral peaks and dips of *pos. 1*. Such changes of the magnitude spectra are properly described by specific statistical quantities. Schroeder in 1954 [53] introduced the spectral standard deviation of RTFs, and proved that it increases with distance from the source. This increase is present up to the critical distance [34] of the room, where the spectral standard deviation begins to converge to 5.57 dB.

[1] The term *room transfer function*, refers to the frequency-domain representation of the room impulse response.

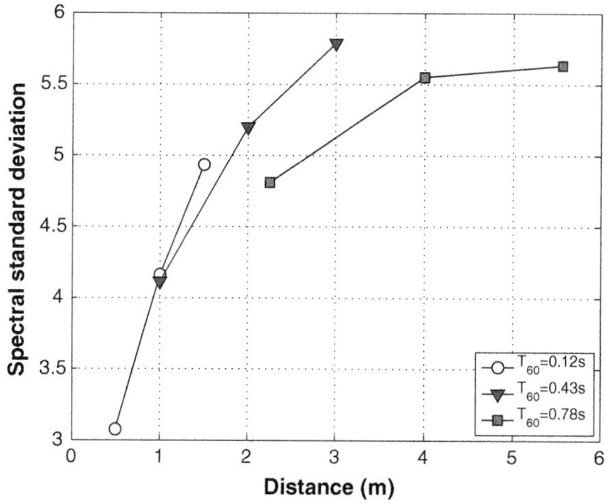


**Fig. 6** RTF spectral standard deviation as a function of distance for three different rooms

Figure 6 shows the spectral standard deviation extracted from the single-channel RTFs as a function of distance for the three rooms of the earlier examples, having different reverberation-time values. More information on the calculation methodology can be found in [20]. The spectral standard deviation shows an increase with distance that is in agreement with the effect observed in Fig. 5.

Similarly, the spectral standard deviation can be extracted from binaural room transfer functions, BRTFs.[2] Figure 7 depicts the BRTFs spectral standard-deviation values as a function of distance when extracted from the left and right BRTFs and for various orientation angles—horizontal plane. It can be seen that for zero degrees the standard-deviation values are the same for the left and right BRTFs. For the rest of the orientation angles, there is an interdependency between the standard-deviation values obtained from the left and the right BRTFs, leading to decreased left-BRTF standard-deviation values and increased right-BRTF standard-deviation values when the left ear is *facing* the sound source and vice versa. Figure 7 follows similar trends in the interdependency of the left and right standard-deviation values as in Fig. 4, although the BRTF standard-deviation values increase with distance instead of decreasing as in the case of the DRR parameter.

## Spectral Standard Deviation and the DRR

The examples shown in Figs. 3 and 6 illustrate the interdependency between the DRR and the spectral standard deviation of RTFs. This has been analysed in [53, 54] and it has been shown that at distances far from the sound source the standard

[2] The term *binaural room transfer function* refers to the frequency-domain representation of the binaural room impulse response.

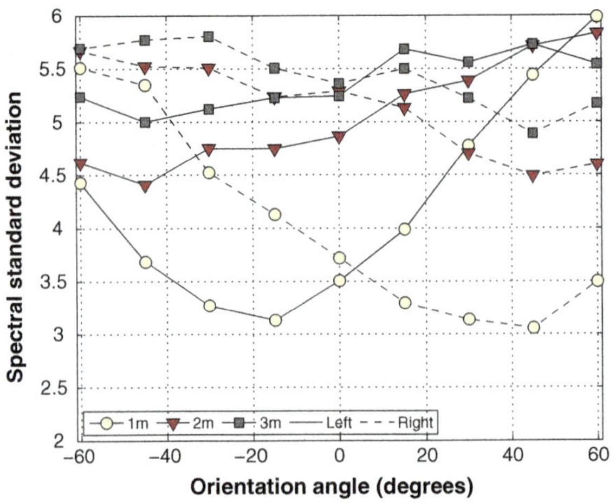


**Fig. 7** BRTF spectral standard deviation as a function of distance and head orientation angle. The *solid* and *dashed lines* correspond to the left and right-channel BRIR–DRR, respectively. When the left ear is *facing* the sound source, the left BRTF standard-deviation value presents lower values than the right BRTF and vice versa

deviation of the RTF magnitude is 5.57 dB. Based on these findings, in the works of Jetzt [27] and Diestel [15] it has been shown that the probability distribution of the RTF sound pressure is related to the DRR. Ebeling [17] proposed a relationship for the dependence of the normalized RTF standard deviation, $\sigma_{nrm}$, on the DRR, as

$$\sigma_{nrm} = \frac{\sqrt{1 + 2 \cdot D_r}}{1 + D_r}, \tag{3}$$

where $D_r$ is the DRR and $\sigma_{nrm}$ is the normalized RTF standard deviation. This simple analytical formula describes the dependence of the normalized fluctuations of the total RTF sound pressure on the DRR. Theoretically, (3) holds for any measured RTF at a certain source/receiver position within the room.

## *3.4 Binaural Parameters*

### Binaural Cues

The human auditory system is capable of analyzing the spatial characteristics, that is, distance, reverberation and orientation angle, etc., of complex acoustic and reverberant environments by exploiting the interaural differences between the signals received at the two ears [5].These binaural cues are the interaural time differences, ITDs, interaural level differences, ILDs, and interaural coherence, IC. ILD is the

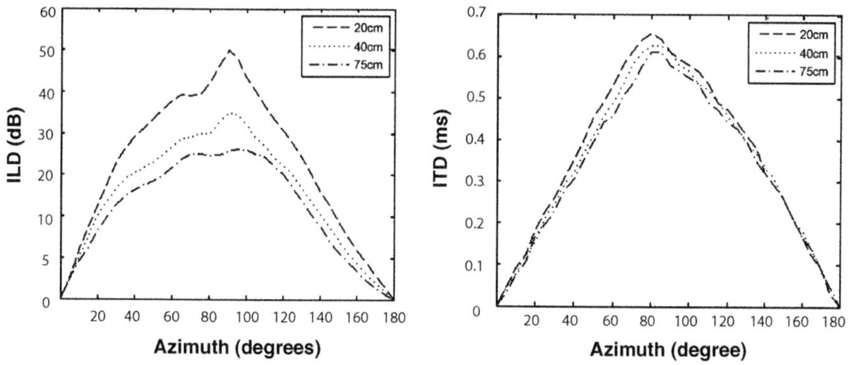


**Fig. 8** *Left* Histogram of ILD values. *Right* Histogram of the ITD values as a function of the azimuth angle for different distances from the source. The obtained values are based on measurements taken in an anechoic chamber—figures taken from [47]

level difference between the sound arriving at the two ears, ITD is the difference in the arrival time of a sound between the two ears and IC expresses the strength of correlation between the left and right-ear signals.These binaural cues are affected by the reverberant energy as a function of the listener's location in a room and the source location relative to the listener [25, 56, 60]. More specifically, reverberant energy decreases the magnitude of ILDs and this effect depends on the actual location of the listener in the room [56]. In addition, the ITD fluctuations across time depend on the amount of reverberation on the signal [60], whereas the IC is directly related to the DRR [6, 33].

On the left side of Fig. 8 an histogram of the ILD values as a function of the azimuth angle measured in an anechoic chamber can be seen—after [47]. It is evident that as the distance increases, the histogram reaches a lower maximum value and becomes more flat. A similar but less pronounced effect can be observed on the measured ITD values—see right side of Fig. 8 [47]. Note here that Fig. 8 is based on measurements taken in an anechoic chamber using a sampling frequency of 65536 Hz, although a similar effect is expected for signals generated inside rooms. In such a case, increase of the distance leads to increased fluctuations in ILDs and ITDs. More specifically, at close distances—depending on angle, binaural cues, specifically ILDs, will have a larger dynamic range that may affect their distributions across frequency. This overall effect of distance on the obtained ILD and ITD values can be potentially captured with the use of various statistical quantities such as the kurtosis, the skewness, etc.

### Differential Spectral Standard Deviation

Here, a distance-dependent feature that can be extracted from binaural signals is described. Let us assume an anechoic signal, $s(t)$, that is reproduced at a certain position within the room and is binaurally recorded using a pair of microphones attached to the ears of a manikin. The left and the right-ear recordings, denoted by $x_\mathrm{l}(t)$ and $x_\mathrm{r}(t)$, can be written as

$$s(t) * h_\mathrm{l}(t) = x_\mathrm{l}(t) \tag{4}$$
$$s(t) * h_\mathrm{r}(t) = x_\mathrm{r}(t), \tag{5}$$

where $h_\mathrm{l}(t)$ and $h_\mathrm{r}(t)$ are the corresponding left and right-ear BRIRs at the corresponding source/receiver positions. A more detailed description of the left and right-ear BRIRs can be found in [59].

The magnitude spectra of the binaural signals, $X_\mathrm{l}(\omega)$ and $X_\mathrm{r}(\omega)$ in dB are given as follows,

$$X_\mathrm{l}^{\mathrm{dB}}(\omega) = 20\log_{10}\left[|X_\mathrm{l}(\omega)|\right], \tag{6}$$
$$X_\mathrm{r}^{\mathrm{dB}}(\omega) = 20\log_{10}\left[|X_\mathrm{r}(\omega)|\right], \tag{7}$$

where $\omega$ corresponds to the frequency index. Applying the Fourier transform to (4) and (5) and using (6) and (7), it can be written

$$X_l^{\mathrm{dB}}(\omega) = S^{\mathrm{dB}}(\omega) + H_l^{\mathrm{dB}}(\omega), \tag{8}$$
$$X_r^{\mathrm{dB}}(\omega) = S^{\mathrm{dB}}(\omega) + H_r^{\mathrm{dB}}(\omega), \tag{9}$$

Denoting the long term, namely, 2-s-magnitude-spectrum difference of the two signals between the two ears as

$$\Delta_X^{\mathrm{dB}}(\omega) = X_l^{\mathrm{dB}}(\omega) - X_r^{\mathrm{dB}}(\omega). \tag{10}$$

then, the standard deviation, $\sigma_x^{jf}$, of the difference of the binaural left and right spectra, $\Delta_X^{\mathrm{dB}}(\omega)$, can be calculated as

$$\sigma_x^{jf} = \left[\frac{1}{n_f - n_j + 1}\sum_{\omega=n_j}^{n_f}\left[\Delta_X^{\mathrm{dB}}(\omega) - \mu_x^{jf}\right]^2\right]^{\frac{1}{2}}, \tag{11}$$

where $n_j$ and $n_f$ define the frequency range of interest, $\omega$ is the discrete-frequency bin, and $\mu_{jf}$ is the spectral-magnitude mean for the specific frequency band, given by

$$\mu_x^{jf} = \frac{1}{n_f - n_j + 1}\sum_{\omega=n_j}^{n_f}\Delta_X^{\mathrm{dB}}(\omega). \tag{12}$$

From now on, in this chapter this differential spectrum $\sigma_x^{jf}$ will be referred to as BSMD–STD—binaural spectral magnitude difference standard deviation.

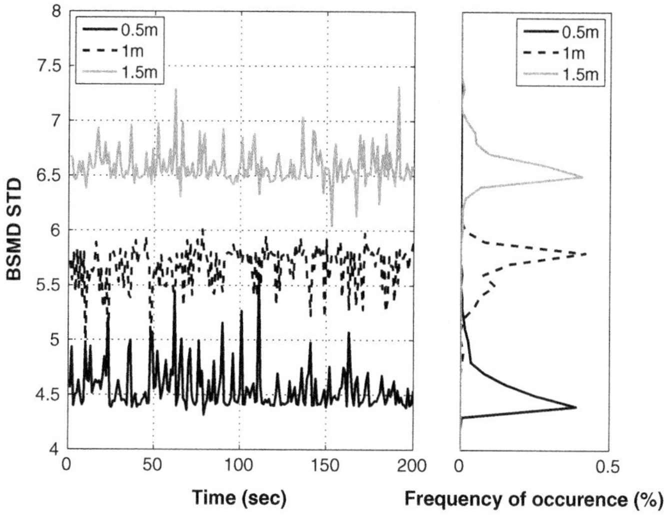


**Fig. 9** *Left* BSMD–STD extracted from speech signals recorded at different source/receiver distances in a small room with a reverberation time of $T_{60} = 0.12$ s. *Right* The corresponding histogram of the extracted BSMD–STD values

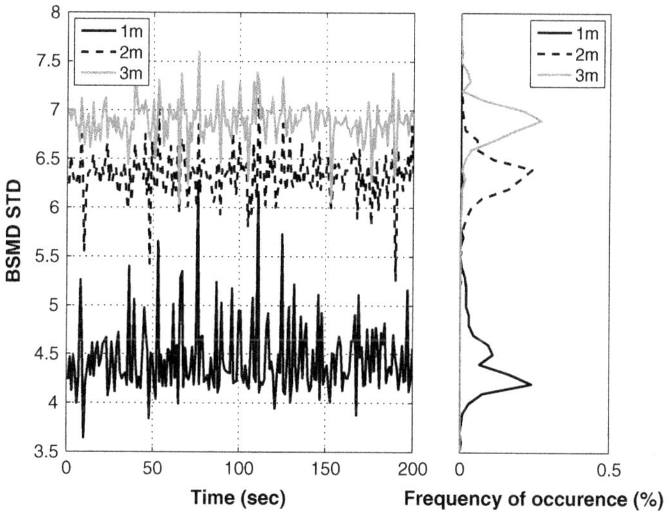


**Fig. 10** *Left* BSMD–STD extracted from speech signals recorded at different source/receiver distances in room with a reverberation time of $T_{60} = 0.89$ s. *Right* The corresponding histogram of the extracted BSMD–STD values

Figures 9 and 10 show such typical results for the BSMD–STD extracted from speech signals as a function of time and the corresponding histograms for the same function. For these examples, the rooms have a reverberation time of 0.12 and 0.89 s respectively and the measurements are taken at different source/receiver distances.

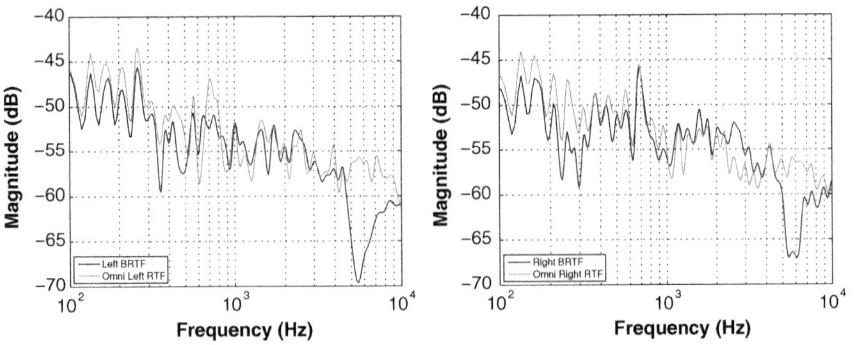


**Fig. 11** *Left* Left-channel BRTF and corresponding RTF. *Right* Right-channel BRTF and corresponding RTF. The BRTFs are measured at a source/receiver distance of 1 m using a manikin and the RTFs are measured at the same positions using an omni-directional microphone

It is evident that the feature values present a clear distance-dependent behaviour and there is small overlap among the three distance classes. This can be explained by the fact that the BSMD–STD feature is related to the spectral standard deviation of the actual RTFs that is highly distance-dependent—see also Sect. 3.3 [19].

Figure 11 shows the left and right BRTFs measured at a source/receiver distance of 1 m using a manikin and the RTFs measured at the same positions using omni-directional microphones. For this example, the orientation angle was 0° and the reverberation time of the room was 0.78 s. It is evident that the BRTF spectrum is mainly affected by the presence of the head and pinnae for the higher frequencies, that is, 4–10 kHz, whereas up to 2 kHz the BRTF spectrum resembles the spectrum of the corresponding RTF.

# 4 Literature Overview on Distance-Estimation Methods

It is now of interest to examine whether it is possible to estimate the sound source distance computationally by exploiting such parameters. In this section, an overview of the existing sound-source-distance-estimation methods is given.

A usual approach to such source-localization and distance detection tasks is to use a microphone array and to perform time delay estimation, TDE, for instance using the generalized cross-correlation, GCC algorithm [31]. The angle of arrival can be calculated from the TDE and applying the triangulation rule can lead to the bearing estimation. This basic bearing-estimation process forms the foundation of most of the source-localization techniques, but the triangulation approach has the disadvantage that the distance cannot be much larger than the maximum distance between the microphones that are used. For this reason, many algorithms exist that solve the problem from a different theoretical perspective [2]. Lately, research work on the localization problem has been undertaken using binaural signals [36, 62].

These methods utilize acoustical parameters and perceptual cues as discussed in Sect. 3. Such listeners' abilities to determine source distance under reverberant conditions have been extensively studied [1, 8, 13, 40, 41, 43, 46, 51, 67–69] and they have initiated novel techniques for the localization problem, especially for distance estimation using only two sensors [37, 57, 61, 62].

Some of the proposed methods for distance estimation in rooms [26, 33, 37] try to estimate the DRR, which seems to be one of the most important parameters for such task. The DRR can be typically extracted from measured RIRs, but in practice these functions are not always available since intrusive measurements within the rooms are required. For this reason, several methods have been developed recently that can *blindly* estimate the DRR from the reverberant signals.

Lu et al. in [37] extracts the DRR from binaural signals by segregating the energy arriving from the estimated direction of the direct source from that arriving from other directions, assuming that reverberant components result in a spatially diffuse field. The DRR estimation method proposed by Lu et al. is based on the equalization-cancellation, EC, concept proposed by Durlach in [16] in order to explain the reduction of the masking ability in the presence of noise. According to this EC concept, equalization is used to align the coherent components of a binaural signal optimally such that after cancellation, the coherent components, which are associated with the direct sound, are removed. Thus, after cancellation only the reverberant part is present and its level can be estimated and be used to also estimate the direct sound level. Figure 12 depicts the block diagram of the method of the EC-based DRR estimation method proposed by Lu et al. [37]. Initially, the binaural signals are passed through a 32-channel Gammatone filterbank. Then, the cross-correlation between the corresponding filtered signals is computed in order to identify the angular position of the direct signal. In parallel, the EC block takes place, and the direct-energy component $E_{\text{direct}}$ is estimated using the azimuthal information of the source localizer. The direct-energy component is also used to select the j-th direct source power. The DRR, $D_{\text{r}}$, is then estimated as the ratio of the direct-to-reverberant energy, the latter computed as the residual of total signal energy, $S$, after subtraction of the direct-energy component, $E_{\text{direct},j}$, as

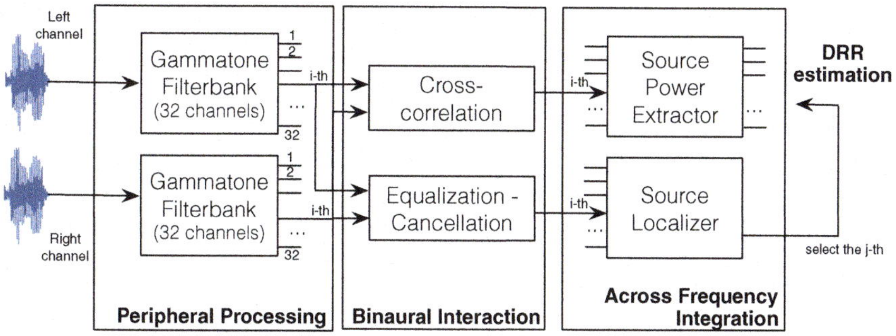


**Fig. 12** Schematic diagram of the EC–DRR method proposed by Lu et al. [37]

$$D_{\mathrm{r}} = \frac{E_{\mathrm{direct},j}}{S - E_{\mathrm{direct},j}}. \tag{13}$$

This method can be also used as a distance estimator, although it is designed to operate for distances beyond 2 m and requires knowledge of the reverberation time of the room.

Hioka et al. [26] has introduced a method for the DRR estimation that uses a direct and reverberant sound spatial correlation-matrix model. This method can be utilized for measuring the absolute distance, but only for a restricted close range of distances. It is mainly designed for microphone arrays, which usually consist of more than two microphones and where the impact of the *head-related transfer function*, HRTF, is not considered.

Recently, a method for the DRR estimation has been proposed by Kuster in [33], where an analytical relationship between the DRR and the coherence, $\gamma^2_{\mathrm{pp}}$, obtained by two pressure signals has been derived analytically as

$$\gamma^2_{\mathrm{pp}} = \frac{D^2_{\mathrm{r}}}{(1 + D_{\mathrm{r}})^2}. \tag{14}$$

Note here, that (14) has been described by Bloom and Caine in [7]—without providing the derivation of the equation. The method proposed by Kuster does not require any special measurement signals or training data and the mean DRR estimation error is found to be higher for low frequencies and lower for higher frequencies. However, this method is not designed specifically as a distance estimator, thus the calculation of the absolute distance from the DRR requires knowledge of either the critical distance or the reverberation time and volume of the room.

Other methods, mainly designed for absolute distance estimation, use features related to the DRR and have been proposed in [49, 57] and [62]. In [57], Smaragdis uses the cross-spectra of the signals recorded by two microphones in order to recognize the position of a sound source. More specifically, the logarithmic ratio of the Fourier transforms of the left and right signals is used for the estimation of distance. By taking the complex logarithm

$$R'(\omega) = \log \frac{X_l(\omega)}{X_r(\omega)}, \tag{15}$$

then the real and imaginary parts are used as features for distance detection, that is,

$$\Re\{R'(\omega)\} = \log \frac{|X_l(\omega)|}{|X_r(\omega)|}, \tag{16}$$

$$\Im\{R'(\omega)\} = \angle X_l(\omega) \cdot X^*_r(\omega). \tag{17}$$

The block diagram of the method developed by Smaragdis [57] can be seen in Fig. 13. An improved distance estimation method has been later proposed by

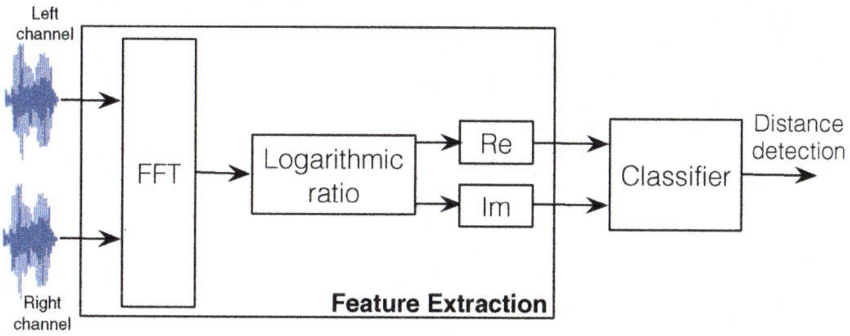


**Fig. 13** Schematic diagram of the distance detection method proposed by Smaragdis [57]

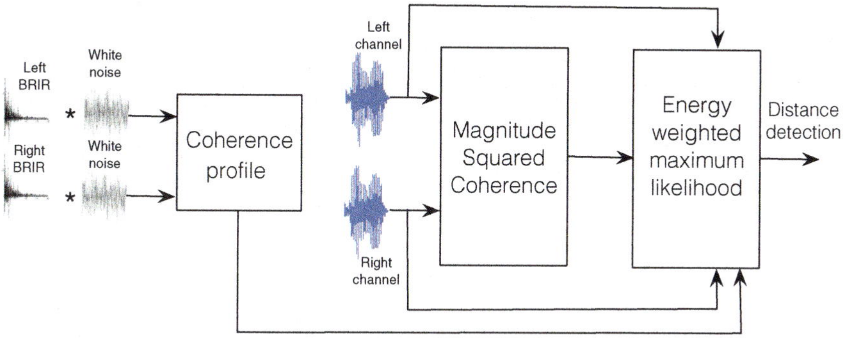


**Fig. 14** Schematic diagram of the distance detection method proposed by Vesa [62]

Vesa [62] in order to account for the positions that have the same azimuth angle using the magnitude squared coherence between the left and right ear as a feature for the training of a Gaussian maximum-likelihood scheme for distance detection. Figure 14 depicts the block diagram of the method developed by Vesa [62]. Initially, a coherence profile is calculated by convolving the BRIRs with white noise. The coherence profile is then used along with the extracted magnitude-squared coherence from the binaural signals in order to estimate the energy-weighted maximum likelihoods. Then, the mutual order of the likelihood magnitudes is used to determine the most likely distance. The method performs very well in rooms when trained and tested in the same positions within a room. However, the method is less successful when swapping the training data between different positions within the same room.

Another distance estimation method has been developed in [49], where several binaural features such as the ILDs, the ITDs, the sound amplitude and the spectral characteristics are used for the distance estimation. The method is trained and tested in a specific acoustic environment. Knowledge of the actual orientation angle increases the performance of the method. In [18], a distance-estimation method based on single-channel inputs has been presented, based on several temporal and spectral statistical features.

More recently [19], a method for distance detection from binaural signals has been reported. The method does not require a priori knowledge of the RIR, the reverberation time or any other acoustical parameter. However, it requires training within the rooms under examination and relies on a set of features extracted from the reverberant binaural signals. These features are incorporated into a classification framework based on GMMs. This method will be presented in more detail in Sect. 5.

# 5 Binaural-Statistics-Based Distance-Detection Method

## *5.1 Features Dependent on Sound-Source Distance*

The distance-detection method [19] presented in detail here, is extending an earlier similar work based on single-channel inputs [18]. It utilizes a large set of statistical features derived from the binaural parameters discussed in Sect. 3.4. Figure 15 presents the block diagram of this binaural distance detection method. The method consists of three stages,

- Binaural feature extraction
- Feature selection
- Classification

In this section, the first block of the method that consists of the binaural-feature-extraction procedure is described. More details on the blocks of the features selection and the classifier will be provided in Sect. 5.2.

The first feature employed by the distance-detection method is the BSMD–STD feature as has already been described in Sect. 3.4—see also Fig. 9. In addition to this, other binaural features are also employed by the method and will be described here.

For this method, an auditory front-end that models the peripheral processing stages of the human auditory system [39] is used. First, the acoustic signal is split

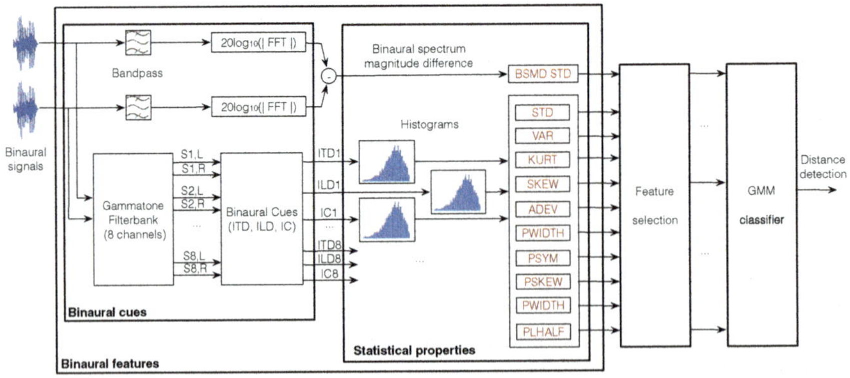


**Fig. 15** Block diagram for the distance estimation method presented in [19] showing the binaural feature extraction, feature selection and classification procedures

into eight auditory channels with center frequencies equally spaced on the equivalent rectangular bandwidth, ERB, scale using a fourth-order Gammatone filterbank. More specifically, phase-compensated Gammatone filters are employed in order to synchronize the analysis of binaural cues across all frequency channels at a given time [10]. Then, the neural-transduction process in the inner hair cells is approximated by halfwave-rectification and square-root compression. Based on this representation, the binaural cues of ITD, ILD and IC are estimated for short frames of 20 ms for each Gammatone channel. Based on the binaural analysis of the auditory frond-end, a set of features is extracted to capture the statistical properties of the estimated binaural cues—see Fig. 15. Here, it should be noted that since the ILDs and ITDs inherently dependent on the sound source direction, the statistical quantities are chosen in such a way so that they do not reflect this direction-dependent behavior. These statistical quantities are

*Standard Deviation,* STD

The standard deviation extracted from a binaural cue is defined here as

$$\sigma^i_{\mathrm{cue}} = \sqrt{\frac{1}{N}\sum_{k=1}^{N}\left(B^{i,k}_{\mathrm{cue}} - \bar{B}^i_{\mathrm{cue}}\right)^2}, \tag{18}$$

where $i$ is the index of the Gammatone channel, $k$ indexes the frame, $N$ is the total number of frames and $\bar{B}^i_{\mathrm{cue}}$ is the average of $B^{i,k}_{\mathrm{cue}}$ over $N$ frames given by

$$\bar{B}^i_{\mathrm{cue}} = \frac{1}{N}\sum_{k=1}^{N} B^{i,k}_{\mathrm{cue}}. \tag{19}$$

*Variance,* VAR

The variance extracted from a binaural cue is given by

$$\mathcal{V}^i_{\mathrm{cue}} = \frac{1}{N}\sum_{k=1}^{N}\left(B^{i,k}_{\mathrm{cue}} - \bar{B}^i_{\mathrm{cue}}\right)^2. \tag{20}$$

*Kurtosis,* KURT

The kurtosis extracted from a binaural cue is given by [21]

$$\Gamma_{\text{cue}}^{i} = \frac{\frac{1}{N}\sum_{k=1}^{N}\left(B_{\text{cue}}^{i,k} - \bar{B}_{\text{cue}}^{i}\right)^{4}}{\left[\frac{1}{N}\sum_{k=1}^{N}\left(B_{\text{cue}}^{i,k} - \bar{B}_{\text{cue}}^{i}\right)^{2}\right]^{2}}. \tag{21}$$

*Skewness,* SKEW

The skewness extracted from a binaural cue is given by [21]

$$\Phi_{\text{cue}}^{i} = \frac{\frac{1}{N}\sum_{k=1}^{N}\left(B_{\text{cue}}^{i,k} - \bar{B}_{\text{cue}}^{i}\right)^{3}}{\left[\frac{1}{N}\sum_{k=1}^{N}\left(B_{\text{cue}}^{i,k} - \bar{B}_{\text{cue}}^{i}\right)^{2}\right]^{\frac{3}{2}}}. \tag{22}$$

*Average Deviation,* ADEV

The average deviation extracted from a binaural cue is given by

$$\Omega_{\text{cue}}^{i} = \sqrt{\frac{1}{N}\sum_{k=1}^{N}|B_{\text{cue}}^{i,k} - \bar{B}_{\text{cue}}^{i}|}. \tag{23}$$

*Percentile Width,* PWIDTH

Ludvigsen [38] proposed a set of statistical features based on the percentile properties and their relationships. These features have been used for sound classification and more details can be found in [12]. The m-th percentile $P_{\text{cue}}^{m,i}$ of a binaural cue for the $i$th Gammatone channel is that value of the binaural cue that corresponds to a cumulative frequency of $Nm/100$, where $N$ is the total number of frames. The width of the histogram is well described by the distance between the $P_{\text{cue}}^{90,i}$ and the $P_{\text{cue}}^{10,i}$ percentile as

$$\Xi_{cue}^{i} = P_{\text{cue}}^{90,i} - P_{\text{cue}}^{10,i}. \tag{24}$$

*Percentile Symmetry,* PSYM

The symmetry of an histogram can be investigated by looking at the difference of the percentiles as

$$\Upsilon_{cue}^{i} = (P_{\text{cue}}^{90,i} - P_{\text{cue}}^{50,i}) - (P_{\text{cue}}^{50,i} - P_{\text{cue}}^{10,i}). \tag{25}$$

This property is near zero for symmetrical distributions, positive for left-sided distributions, and negative for right-sided distributions.

*Percentile Skewness,* PSKEW

The percentile skewness is defined as the difference between the 50 % percentile and the median as

$$\Theta^{i}_{cue} = P^{50,i}_{\text{cue}} - \frac{P^{90,i}_{\text{cue}} + P^{10,i}_{\text{cue}}}{2}. \tag{26}$$

For asymmetrical distributions the difference between the 50 % percentile and the approximated median should be large, for symmetrical distributions approximately zero.

*Percentile Kurtosis,* PKURT

The percentile kurtosis corresponds to the approximation

$$\Psi^{i}_{cue} = \frac{P^{70,i}_{\text{cue}} - P^{30,i}_{\text{cue}}}{2(P^{90,i}_{\text{cue}} - P^{10,i}_{\text{cue}})}, \tag{27}$$

indicating whether the distribution has a narrow or broad peak.

*Lower Half Percentile,* PLHALF

The lower half percentile feature is expressed as

$$\Lambda^{i}_{cue} = (P^{50,i}_{\text{cue}} - P^{30,i}_{\text{cue}}) - (P^{30,i}_{\text{cue}} - P^{10,i}_{\text{cue}}) \tag{28}$$

and indicates whether the histogram is right-sided by encoding the relations between the lower and upper half, namely, below and above $P^{30,i}_{\text{cue}}$, of the lower half of the total distribution, that is, below $P^{50,i}_{\text{cue}}$.

## 5.2 Method Description

The distance-detection method [19] relies on the set of binaural features described in Sect. 5.1. The method employs a feature-selection algorithm and uses GMMs in order to model and evaluate the degree of distance-dependency of the features. The complete procedure of the distance-detection method can be seen in Fig. 15.

### Feature-Selection Algorithm

A feature-selection algorithm is used in order to extract the most relevant and beneficial features for the distance-detection task. Feature-selection algorithms are widely used in pattern recognition and machine learning to identify subsets of features that are relevant for a particular classification task. Here, the minimal-redundancy-maximal-relevance criterion, mRMR [45], is used in order to select features that correlate the strongest with the class labels—here distance classes. The method is able to select the top features based on a user definition of the selected feature number and this has the practical advantage that only a subset rather than the full set of features is computed, which can lead to improved classification performance when a limited amount of training data is available.

Some more details of the procedure regarding the feature selection algorithm will be given next, for instance, the number of features, etc., along with the presentation of typical results.

### Gaussian-Mixture-Models Initialization

Gaussian mixture models, GMMs, can be used to approximate arbitrarily complex distributions and are therefore chosen to model the distance-depending distribution of the extracted features described in Sect. 5.1. The features are used to train the classifier [4, 48] and each discrete distance that should be recognized by the classifier is represented by a GMM. The GMM is initialized using the k-means algorithm [35]. The expectation-maximization, EM, algorithm [4] is used to estimate the set of GMM parameters with a maximum number of 300 iterations. Five Gaussian components and diagonal covariance matrices are used.

### Recordings Database

In order to train and evaluate the particular system, several anechoic speech recordings are used, taken from various recording databases [18]. In total, approximately 60 min of speech recordings are used. 65 % of the recordings dataset is used for the training and 35 % for the testing of the method. In addition, BRIRs are utilized from the Aachen Database [28] and from [62]. The recordings are convolved with the BRIRs measured at different distances between source and receiver in five different rooms having an orientation angle of zero degrees. The volume, reverberation time, $T_{60}$, and the source/receiver distance for each of the rooms can be seen in Table 1. The sampling frequency is 44.1 kHz.

**Table 1** Volume and reverberation time of the rooms

| Room | Volume ($m^3$) | $T_{60}$ (s) | Description |
|---|---|---|---|
| A | 11.9 | 0.12 | Sound booth |
| B | 103 | 0.3 | Listening room |
| C | 194 | 0.63 | Meeting room |
| D | – | 0.86 | Stairway hall |
| E | – | $\approx 5$ | Large hall |

### Feature Extraction

The BSMD–STD feature is extracted from the binaural reverberant signals, which are segmented into relatively long blocks of 1 s, using hanning windowing. For every block of 1 s, the BSMD–STD feature is calculated for various frequency bands. More specifically, the audible frequency range of interest is divided heuristically in four different regions, namely, 0.2–2.5, 2.5–5, 5–10 and 10–20 kHz and the BSMD–STD feature is calculated for these four frequency bands. After several informal tests, the frequency band of 0.2–2.5 kHz was found to contribute mostly to the performance of the method. Thus, it is used for the extraction of the feature. More details on the choice of the specific frequency band can be found in [19].

For the extraction of the binaural cues, ITD, ILD, and IC, the signals are split into eight auditory channels with center frequencies equally spaced on the ERB scale between 50 and $\frac{f_s}{2}$ Hz, using a fourth-order Gammatone filterbank. More specifically, the center frequencies were 50, 293, 746, 1594, 3179, 6144, 11686 and 22050 Hz for the eight channels, respectively. Then, the binaural cues of ILD, ITD and IC are estimated for 20 ms frames with 50 % overlap for each Gammatone channel. From the binaural cues, the statistical features are computed also for the longer blocks of 1 s duration, whereby one long block is composed of 50 frames of 20 ms. Figure 15 illustrates this procedure.

In total, 240 features, that is, 3 cues $\times$ 8 channels $\times$ 10 features, are extracted from the binaural cues. This yields a total set of 241 available features with the addition of the feature BSMD–STD, for one frequency band, namely, 0.2–2.5 kHz.

## 5.3 *Method Performance*

The method described in the previous sections is evaluated for various test cases. First, the performance of the method is evaluated using the BSMD–STD feature within the five different rooms for three distance classes—coarse detection. Then, the method is also tested for seven distance classes—fine detection—using only the BSMD–STD feature and when using the additional statistical features selected by the mRMR feature-selection algorithm. From these tests, the effect of the number of selected features on the distance detection performance can be deduced.

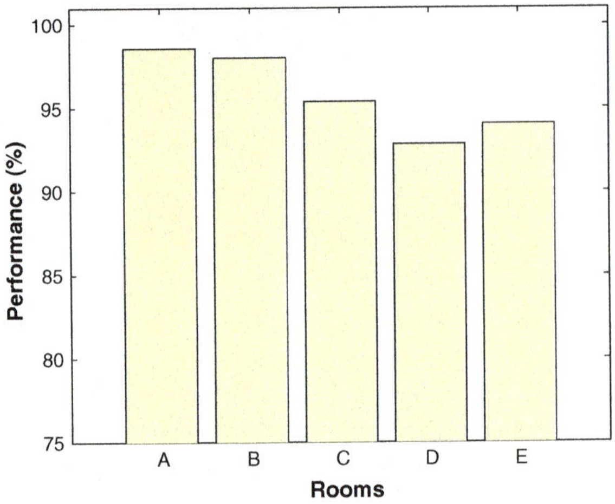


**Fig. 16** Performance of the method [19] when using the BSMD–STD feature for rooms A-E for speech signals and for three distance classes—coarse distance detection

The method is initially tested using only the BSMD–STD feature in rooms A, B, C, D and E and for three different distance classes illustrating its performance for coarse distance detection tasks. For rooms B, C, D and E these distance classes are 1, 2 and 3 m, and for the smallest room, A, the distance classes are 0.5, 1 and 1.5 m since distance measurements above 2 m are not available due to the small size of the room. The performance of the method across all rooms can be seen in Fig. 16. It is evident that the method employing the BSMD–STD feature alone achieves high performance being always higher than 90 % for all rooms for such a coarse distance detection task. For the larger and more reverberant rooms, it can be seen that there is a decrease in the method performance of approximately 5 %.

### Fine Distance Detection

The method using the BSMD–STD feature alone can successfully identify the correct distance with performance rates being higher than 90 % within all rooms for three distance classes. However, it is of interest to examine whether the method using the same feature can perform equally well when there are requirements for higher distance resolution. Here, the method is evaluated for seven distance classes—fine distance detection—across three different rooms. The specific tested distances can be seen in Table 2. The results for rooms B, C and E when using only the BSMD–

**Table 2** Test cases—fine distance detection

| Room | Distances (m) |
|---|---|
| B | 1, 1.5, 2, 2.5, 3, 3.5, 4 |
| C | 1, 1.5, 2, 2.5, 3, 3.5, 4 |
| E | 1, 2, 3, 5, 10, 15, 20 |

Room B — Mean performance: 73%

| | 1m | 1.5m | 2m | 2.5m | 3m | 3.5m | 4m |
|---|---|---|---|---|---|---|---|
| 1m | 99 | 1 | 0 | 0 | 0 | 0 | 0 |
| 1.5m | 1 | 79 | 18 | 2 | 0 | 0 | 0 |
| 2m | 0 | 14 | 77 | 8 | 1 | 0 | 0 |
| 2.5m | 0 | 1 | 8 | 73 | 4 | 1 | 13 |
| 3m | 0 | 0 | 0 | 5 | 54 | 20 | 21 |
| 3.5m | 0 | 0 | 0 | 1 | 13 | 83 | 3 |
| 4m | 0 | 0 | 0 | 9 | 30 | 15 | 46 |

Room C — Mean performance: 74.4%

| | 1m | 1.5m | 2m | 2.5m | 3m | 3.5m | 4m |
|---|---|---|---|---|---|---|---|
| 1m | 95 | 4 | 1 | 0 | 0 | 0 | 0 |
| 1.5m | 3 | 68 | 24 | 4 | 1 | 0 | 0 |
| 2m | 1 | 18 | 71 | 9 | 1 | 0 | 0 |
| 2.5m | 0 | 1 | 9 | 68 | 19 | 3 | 0 |
| 3m | 0 | 0 | 2 | 20 | 61 | 16 | 1 |
| 3.5m | 0 | 0 | 1 | 4 | 18 | 67 | 10 |
| 4m | 0 | 0 | 0 | 0 | 0 | 9 | 91 |

Room E — Mean performance: 63.6%

| | 1m | 2m | 3m | 5m | 10m | 15m | 20m |
|---|---|---|---|---|---|---|---|
| 1m | 90 | 0 | 9 | 1 | 0 | 0 | 0 |
| 2m | 0 | 34 | 0 | 21 | 0 | 4 | 41 |
| 3m | 4 | 0 | 74 | 1 | 17 | 3 | 1 |
| 5m | 0 | 14 | 0 | 77 | 0 | 0 | 9 |
| 10m | 0 | 1 | 9 | 1 | 54 | 34 | 1 |
| 15m | 0 | 2 | 1 | 1 | 18 | 74 | 4 |
| 20m | 0 | 35 | 0 | 17 | 0 | 6 | 42 |

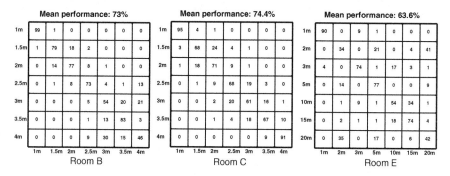

**Fig. 17** Confusion matrices indicating the performance of the method [19] when using the BSMD–STD feature. The *rows* represent the actual classes and the *columns* the predicted ones—fine distance detection

STD feature are also presented in form of confusion matrices in Fig. 17. The rows represent the actual distance classes and the columns the predicted ones. The mean performance rates can be seen on top of the confusion matrices. It is evident that the performance now decreases for all three rooms as compared to the coarse distance estimation results for only three distance classes, as can be seen in Fig. 16. More specifically, when the method is tested for three distance classes the performance for rooms B, C and E is 95.4, 98.1 and 94 % and for seven distance classes it decreases to 73, 74.4 and 63.6 %, respectively.

Since the BSMD–STD feature is not sufficient for higher distance resolution, the additional binaural statistical features are also used in order to increase the performance of the method. Using the mRMR [45], the most relevant and beneficial features for the distance detection task are selected—compare Sect. 5.2.

The results for increasing the number of features can be seen in Fig. 18. It is evident that the addition of more features further increases the performance of the method. In room B, which is the less reverberant, the performance increases by adding more features, up to the number of ten features. Similarly, in room C the method increases up to the number of nine features and room E presents the highest performance for eleven features. In all three rooms, a steep increase of the performance is achieved for adding two features to the first top feature.

A list with the selected features by the mRMR method for each room can be found in [19]. It is worthy of note that the BSMD–STD feature is chosen by the mRMR method as the top feature among the 241 features for all three rooms and thus appears to be particularly important for the distance-detection task.

## Adaptability of the Method Across Different Rooms

The previously described results require training and testing within the same room, but it is of interest to examine whether the method is able to generalize to unknown

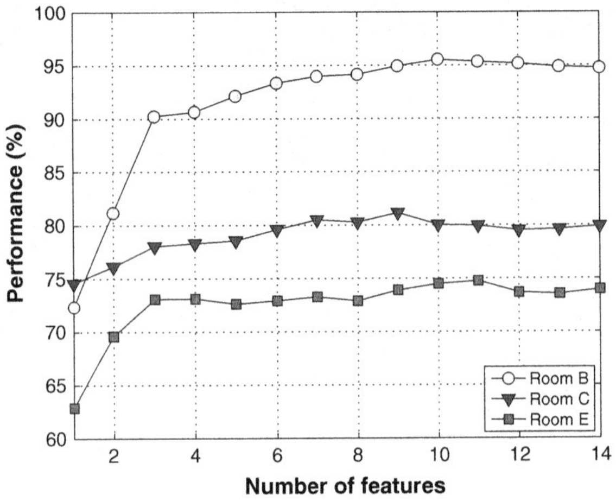


**Fig. 18** Performance of the method as a function of the number of features used in the classifier for seven distance classes—fine distance detection. The first feature is always the BSMD–STD

**Table 3** Test cases for three distance classes—coarse distance detection

| Training | Testing | Performance (%) |
|---|---|---|
| Room B | Room C | 95.4 |
| Room C | Room B | 95.9 |
| Room C | Room D | 70.5 |
| Room D | Room C | 77.6 |
| Room C | Room E | 44.4 |
| Room E | Room C | 44.4 |
| Room D | Room E | 52.9 |
| Room E | Room D | 40.1 |

rooms without prior training within the specific rooms, that is, training in one of the rooms and tested in another one. For this reason, some additional experiments are conducted, where various combinations of rooms are examined. The room combinations for training and testing can be seen in Table 3. The first column contains the room where the method is trained and the second column contains the room where the method is tested. Finally, in the third column the performance for each room combination can be seen when the method is tested for three different classes, namely, 1, 2 and 3 m and when using the BSMD–STD feature. It can be seen that the highest performance is achieved for the combination of rooms B and C, which have a small difference in their reverberation time of approximately 0.2 s. However, for rooms C and D, having a difference in their reverberation time of approximately 0.3 s, the performance is lower, namely, 70 %. The combination of rooms, E and D, E and C, cannot lead to high performance rates. This might be explained by the fact that there is a substantial difference in their corresponding reverberation times.

Hence, it can be concluded that for crude distance classification task, the method could potentially perform well into unknown acoustical room environments, if the model is trained with reverberation characteristics that are *expected* in the testing phase. Extending the training of the model with a large collection of different rooms could potentially further increase the performance of the method.

# 6 Conclusions

In this chapter, the problem of distance estimation by computational methods utilizing binaural cues has been discussed. The main points of several studies related to the auditory distance perception have been summarized, and several acoustical parameters that depend on the source/receiver distance have been presented—especially for the case of sound generated inside reverberant rooms. Various distance-estimation techniques have been described, most of them being based on acoustical parameters depending on physical distance and/or perceptual cues.

It has been shown that the DRR energy in the received signals is often used as the most appropriate feature for distance detection. For example, a method appropriate for distance detection for distances beyond 2 m was introduced in [37], based on the approximation of the DRR—noting that an alternative method that estimates DRR can be found in [33]. In general, it can be deduced that the distance can be easily estimated when the DRR and the critical distance of the room are known. The problem of distance detection can be also simplified by using microphone arrays, which usually consist of more than two microphones and an example employing such information can be found in [26]. Among the *blind*-estimation techniques, the method presented by Vesa in [62] appears to outperform all existing techniques, especially since it has been designed specifically as a distance-estimator technique. This method performed especially well when training and testing took place at the same positions within the room, but it appeared to be sensitive to the actual training and testing positions.

A new method, recently developed by the authors of this chapter, relies on several statistical features extracted from binaural signals. It has been presented in more detail in [19]. The features were incorporated into a classification framework based on GMMs. This work extended earlier work of the authors designed for single-input signals [18] and introduced a novel distance-dependent feature, the binaural spectral-magnitude-difference standard deviation, BSMD–STD. This feature turned out to be particularly effective in the distance-detection tasks. Along with the BSMD–STD, additional binaural features were employed by the method, being extracted by use of an auditory front-end that models the peripheral processing of the human auditory system [39].

It has been shown that the BSMD–STD alone could successfully predict the distance class with high performance rates, that is, above 90 %, when tested in five different acoustical environments and coarse distance detection—three classes—was sufficient. For a finer distance detection—seven classes—a decrease of the performance of the method was observed. To compensate for this decrease, additional binaural features were employed based on a particular feature-selection algorithm. Incorporating these additional features increased the performance of the method further. As a result, the method was able to recognize distances reliably, when trained and tested in the same rooms. The method was also found capable of generalizing and performing well in unknown acoustical environments, at least for crude distance-classification tasks with three classes. The only requirement then was that the training and testing environment should have a reverberation time of similar order.

From the previous discussion it is evident that the variation of the acoustical environments makes *blind* computational distance estimation a challenging task, especially when it is not possible for the computational systems to be trained in the specific room environments where they are finally tested. Yet, the method presented in [19] appears to be quite robust to the various acoustical environments when crude distance estimation is sufficient, however, an approximate estimate of the reverberation time is required.

In conclusion, binaural cues contain distance-dependent information that can effectively be utilised by computational systems. However, it is not yet completely clear which are the actual cues that humans exploit when estimating sound-source distance estimation. Further research work is required for a better understanding of the various acoustical parameters that vary with distance, and which of these parameters are the perceptually relevant ones. The use of statistical quantities, such as statistical moments, could further assist research in this direction, as they can potentially provide alternative descriptions of well-known acoustical parameters and might reveal additional distance-dependent information.

**Acknowledgments** The authors would like to thank the authors of [47] and S. Vesa for offering the database employed also in [19]. Further, thanks are due to two anonymous reviewers for most valuable comments. This research has been co-financed by the European Union through the European Social Fund, ESF, and the Greek national funds through its Operational Program *Education and Lifelong Learning* of the National Strategic Reference Framework, NSRF. Research Funding Program: *Heracleitus II: Investing-in-Knowledge Society through the European Social Fund.*

## References

1. D. R. Begault. Perceptual effects of synthetic reverberation on three-dimensional audio systems. *J. Audio Eng. Soc.*, 40:895–904, 1992.
2. J. Benesty, J. Chen, and Y. Huang, editors. *Microphone Array Signal Process*. Springer Berlin Heidelberg, March 2008.
3. A. Bidart and M. Lavandier.Do we need two ears to perceive auditory distance in rooms? In *2nd Pan-American/Iberian Meeting on Acoustics*, Cancun, Mexico, 2010.
4. C. M. Bishop. *Pattern Recognition and Machine Learning (Information Science and Statistics)*. Springer, August 2006.
5. J. Blauert. *Spatial Hearing - The psychophysics of human sound localization*. Cambridge MA: MIT Press, 1996.
6. P. Bloom and G. Cain. Evaluation of two-input speech dereverberation techniques. *IEEE Intl. Conf. Acoust., Speech, Signal Process., ICASSP '82.*, 7:164–167, 1982.

7. P. J. Bloom and G. D. Gain. Evaluation of two-input speech dereverberation techniques. In *IEEE Intl. Conf. Acoust., Speech, Signal Process.*, New York, 1982.
8. A. W. Bronkhorst. Localization of real and virtual sound sources. *J. Audio Eng. Soc.*, 98:2452–2553, 1995.
9. A. W. Bronkhorst and T. Houtgast. Auditory distance perception in rooms. *Nature*, 397:517–520, 1999.
10. G. J. Brown and M. Cooke. Computational auditory scene analysis. *Comput. Speech and Language*, 8:297–336, 1994.
11. D. S. Brungart, N. I. Durlach, and W. M. Rabinowitz. Auditory localization of nearby sources. ii. localization of a broadband source. *J. Acoust. Soc. Amer.*, 106(4):1956–1968, 1999.
12. M. C. Büchler. *Algorithms for sound classification in hearing instruments*. PhD thesis, Swiss Federal Institute of Technology, Zurich, Switzerland, ETH, 2002.
13. R. A. Butler, E. T. Levy, and W. D. Neff. Apparent distance of sounds recorded in echoic and anechoic chambers. *J. Experim. Psychol.*, 6:745–750, 1980.
14. J. Chomyszyn. *Distance of Sound in Reverberant Fields*. PhD thesis, Department of Music, Stanford University, 08/1995 1995.
15. H. G. Diestel. Zur Schallausbreitung in Reflexionsarmen Räumen. *Acustica*, 12:113–118, 1962.
16. N. Durlach. Note on the equalization and cancellation theory of binaural masking level differences. *J. Acoust. Soc. Amer.*, 32:1075–1076, 1960.
17. K. J. Ebeling. Influence of direct sound on the fluctuations of the room spectral response. *J. Acoust. Soc. Am.*, 68(4):1206–1207, 1980.
18. E. Georganti, T. May, S. van de Par, A. Harma, and J. Mourjopoulos. Speaker distance detection using a single microphone. *IEEE Audio, Speech, Language Process.*, 19(7):1949–1961, Sept. 2011.
19. E. Georganti, T. May, S. van de Par, and J. Mourjopoulos. Sound source distance estimation in rooms based on statistical properties of binaural signals. *IEEE Audio, Speech, Language Process.* (in press)
20. E. Georganti, T. Zarouchas, and J. Mourjopoulos. Reverberation analysis via response and signal statistics. In *128th AES convention Proc.*, London, UK, 2010.
21. J. A. Gubner. *Probability and Random Processes for Electrical and Computer Engineers*. Cambridge University Press, 2006.
22. S. Gustafsson, R. Martin, P. Vary. Combined acoustic echo control and noise reduction for hands-free telephony. *Signal Process. - Special issue on acoustic echo and noise, control*, 64(1):21–32, 1998.
23. V. Hamacher, J. Chalupper, J. Eggers, E. Fischer, U. Kornagel, H. Puder, and U. Rass. Signal processing in high-end hearing aids: State of the art, challenges, and future trends. *EURASIP J. Appl. Signal Process.*, 2005:2915–2929, 2005.
24. A. Härmä. Ambient human-to-human communication. In *Handbook of Ambient Intelligence and Smart Environments*, pages 795–823. Springer, 2009.
25. W. Hartmann, B. Rakerd, and A. Koller. Binaural coherence in rooms. *Acta Acust United Ac*, 91(3):451–462, 2005.
26. Y. Hioka, K. Niwa, S. Sakauchi, K. Furuya, and Y. Haneda. Estimating direct-to-reverberant energy ratio using d/r spatial correlation matrix model. *IEEE Audio, Speech, Language Process.*, 19(8):2374–2384, nov. 2011.
27. J. J. Jetzt. Critical distance measurement of rooms from the sound energy spectral response. *J. Acoust. Soc. Am.*, 65:1204–1211, 1979.
28. M. Jeub, M. Schäfer, and P. Vary. A binaural room impulse response database for the evaluation of dereverberation algorithms. In *Intl. Conf. Proc. on Digital Signal Processing (DSP), Santorini, Greece*, 2009.
29. S. Kerber, H. Wittek, and H. Fastl. Ein Anzeigeverfahren für psychoakustische Experimente zur Distanzwahrnehmung. In H. Fastl and M. Fruhmann, editors, *Tagungsband Fortschritte der Akustik - DAGA 05, München*, volume 1, pages 229–230. Berlin, 2005.
30. S. Kerber, H. Wittek, H. Fastl, and G. Theile. Experimental investigations into the distance perception of nearby sound sources: Real vs. WFS virtual nearby sources. In D. Cassereau,

editor, *Proceedings of the 7-th Congrès Français d' Acoustique/30th Deutsche Jahrestagung für Akustik*(CFA/DAGA 04), pages 1041–1042. Strasbourg, France, 2004.
31. C. H. Knapp and G. C. Carter. The generalized correlation method for estimation of time delay. *IEEE Speech Audio Process.*, 24:320–327, August 1976.
32. N. Kopčo and B. G. Shinn-Cunningham. Effect of stimulus spectrum on distance perception for nearby sources. *J. Acoust. Soc. Amer.*, 130(3):1530–1541, 2011.
33. M. Kuster. Estimating the direct-to-reverberant energy ratio from the coherence between coincident pressure and particle velocity. *J. Acoust. Soc. Am.*, 130(6):3781–3787, 2011.
34. H. Kuttruff. *Room Acoustics, 3rd edition*. Elsevier, 1991.
35. S. P. Lloyd. Least squares quantization in PCM. *IEEE Trans. Inf. Theory*, 28:129–137, 1982.
36. Y. C. Lu and M. Cooke. Binaural distance perception based on direct-to-reverberant energy ratio. In *Proc. Intl. Workshop on Acoust. Echo and Noise, Control*, September 2008.
37. Y.-C. Lu and M. Cooke. Binaural estimation of sound source distance via the direct-to-reverberant energy ratio for static and moving sources. *IEEE Audio, Speech, Language Process.*, 18(7):1793–1805, Sept. 2010.
38. C. Ludvigsen. Schaltungsanordnung für die automatische Regelung von Hörhilfsgeräten [An algorithm for an automatic program selection mode]. In *Deutsches Patent Nr. DE43 40817 A1*, 1993.
39. T. May, S. van de Par, and A. Kohlrausch.A probabilistic model for robust localization based on a binaural auditory front-end. *IEEE Audio, Speech, Language Process.*, 19:1–13, 2011.
40. D. H. Mershon and J. N. Bowers. Absolute and relative cues for the auditory perception of egocentric distance. *Perception*, 8:311–322, 1979.
41. D. H. Mershon and E. King. Intensity and reverberation as factors in auditory perception of egocentric distance. *Perception & Psychophysics*, 18(6):409–415, 1975.
42. S. Nielsen. Distance perception in hearing. Master's thesis, Aalborg University, Aalborg, Denmark, 05/1991 1991.
43. S. H. Nielsen. Auditory distance perception in different rooms. *J. Audio Eng. Soc.*, 41:755–770, 1993.
44. S. Oh, V. Viswanathan, and P. Papamichalis. Hands-free voice communication in an automobile with a microphone array. In *IEEE Intl. Conf. Acoust., Speech, Signal Process.*, volume 1, pages 281–284, Los Alamitos, CA, USA, 1992.
45. H. Peng, F. Long, and C. Ding. Feature selection based on mutual information criteria of max-dependency, max-relevance, and min-redundancy. *IEEE Pattern Anal. Mach. Intell.*, 27(8):1226–1238, Aug. 2005.
46. J. W. Philbeck and M. D. H.Knowledge about typical source output influences perceived auditory distance. *J. Audio Eng. Soc.*, 111:1980–1983, 2000.
47. T. Qu, Z. Xiao, M. Gong, Y. Huang, X. Li, and X. Wu. Distance-dependent head-related transfer functions measured with high spatial resolution using a spark gap. *IEEE Audio, Speech, Language Process.*, 17(6):1124–1132, Aug. 2009.
48. D. A. Reynolds and R. C. Rose. Robust text-independent speaker identification using gaussian mixture speaker models. *IEEE Audio, Speech, Language Process.*, 3(1):72–83, 1995.
49. T. Rodemann.A study on distance estimation in binaural localization. In *IEEE Intl. Conf. Intel. Robots, Systems*, Taipei, Taiwan, 2010.
50. D. F. Rosenthal and H. G. Okuno, editors. *Computational auditory scene analysis*. Lawrence Erlbaum Associates Inc., Mahwah, New Jersey, 1998.
51. N. Sakamoto, T. Gotoh, and Y. Kimura. On "out-of-head localization" in headphone listening. *J. Audio Eng. Soc.*, 24:710–716, 1976.
52. S. G. Santarelli, N. Kopčo, and B. G. Shinn-Cunningham. Distance judgments of nearby sources in a reverberant room: Effects of stimulus envelope. *J. Acoust. Soc. Amer.*, 107(5):2822–2822, 2000.
53. M. Schroeder. Die Statistischen Parameter der Frequenzkurven von grossen Räumen (in german). *Acustica*, (4):594–600, 1954.
54. M. R. Schroeder. Statistical parameters of the frequency response curves of large rooms. *J. Audio Eng. Soc*, 35(5):299–306, 1987.

55. B. Shinn-Cunningham. Localizing sound in rooms. In *Proc. ACM SIGRAPH and EUROGRAPHICS Campfire: Acoustic Rendering for Virtual Environments*, pages 17–22, May 2001.
56. B. G. Shinn-Cunningham, N. Kopčo, and T. J. Martin. Localizing nearby sound sources in a classroom: Binaural room impulse responses. *J. Acoust. Soc. Amer.*, 117(5):3100–3115, 2005.
57. P. Smaragdis and P. Boufounos. Position and trajectory learning for microphone arrays. *IEEE Audio, Speech, Language Process.*, 15:358–368, 2007.
58. A. Tsilfidis. Signal processing methods for enhancing speech and music signals in reverberant environments. *PhD Thesis, University of Patras*, 2011.
59. A. Tsilfidis, A. Westerman, J. Buchholz, E. Georganti, and J. Mourjopoulos. Binaural dereverberation. In J. Blauert, editor, *The technology of binaural listening*, chapter 14. Springer, Berlin-Heidelberg-New York NY, 2013.
60. J. van Dorp Schuitman. *Auditory modelling for assessing room acoustics*. PhD thesis, Technical University of Delft, the Netherlands, 2011.
61. S. Vesa. Sound source distance learning based on binaural signals. In *Proc. 2007 Workshop on Applicat. of Signal Process., Audio, Acoust. (WASPAA 2007)*, pages 271–274, 2007.
62. S. Vesa. Binaural sound source distance learning in rooms. *IEEE Audio, Speech, Language Process.*, 17:1498–1507, 2009.
63. F. Völk. Psychoakustische Experimente zur Distanz mittels Wellenfeldsynthese erzeugter Hörereignisse. In *Tagungsband Fortschritte der Akustik, DAGA 2010*, pages 1065–1066, Berlin, 2010.
64. F. Völk, U. Mühlbauer, and H. Fastl. Minimum audible distance (MAD) by the example of wave field synthesis. In *Tagungsband Fortschritte der Akustik, DAGA 2012*, pages 319–320, Darmstadt, 2012.
65. G. von Békésy. The moon illusion and similar auditory phenomena. *Am. J. Psychol.*, 111:1832–1846, 2002.
66. D. Wang and G. J. Brown, editors. *Computational auditory scene analysis: Principles, Algorithms, and Applications*. Wiley-IEEE, October 2006.
67. P. Zahorik. Assessing auditory distance perception using virtual acoustics. *J. Acoust. Soc. Amer.*, 111:1832–1846, 2002.
68. P. Zahorik. Direct-to-reverberant energy ratio sensitivity. *J. Acoust. Soc. Amer.*, 112(5):2110–2117, 2002.
69. P. Zahorik, S. D. Brungart, and W. A. Bronkhorst. Auditory distance perception in humans: A summary of past and present research. *Acta Acust United Ac*, 91:409–420, May/June 2005.

# A Binaural Model that Analyses Acoustic Spaces and Stereophonic Reproduction Systems by Utilizing Head Rotations

**J. Braasch, S. Clapp, A. Parks, T. Pastore and N. Xiang**

## 1 Introduction

The *ceteris-paribus* assumption[1] is a fundamental concept in science. It builds on fixing all experimental parameters except for a number of test parameters to study the effect of parameter changes on the outcome of the experiment. Further, science builds on acceptable approximations and simplifications of a real world problem to allow the application of known methods and procedures to solve it. In this context, *acceptable* means that the problem is simplified to some degree but not altered to such an extent that it will lead to solutions that no longer help to solve the real-world problem. Once the solution is found, models and test paradigms are refined to understand the real-world situation in greater detail.

The concept of a room impulse response is a good example of this approach. By defining the response of a room between a designated source with stationary directivity pattern and position—for instance, an omni-directional point source and a stationary receiver such as an omni-directional microphone-type receiver—the complex system can be described by a simple two-port system with the source signal as the input and the receiver as its output. Using the two-port approach, standard engineering solutions become accessible to describe the system. The system can be extended to three-ports by replacing the microphone by a binaural manikin that is a model of a human head. Now, the system has two outputs in the form of two ear signals and, in an anechoic room, the transfer functions between the sound source and

[1] A Latin term meaning that all other factors are held unchanged—literally: with other things the same.

J. Braasch (✉) · S. Clapp · A. Parks · T. Pastore · N. Xiang
Graduate Program in Architectural Acoustics,
School of Architecture, Rensselaer Polytechnic Institute, Troy, NY, USA
e-mail: braasj@rpi.edu

J. Braasch · S. Clapp · A. Parks · T. Pastore
Rensselaer Polytechnic Institute, Troy, NY, USA

J. Blauert (ed.), *The Technology of Binaural Listening*, Modern Acoustics and Signal Processing, DOI: 10.1007/978-3-642-37762-4_8,

the two ears are described by the head-related transfer functions, in which the spatial position of the sound source relative to the head is encoded via frequency-dependent interaural time and level differences as well as spectral cues.

While the three-port approach has its advantages and enabled many psychoacoustic findings, it represents a very passive view of spatial hearing—one where the listeners do not move to explore their acoustic environment. Most binaural models describing psychoacoustic phenomena follow the traditional assumption of a fixed listener position by feeding the model with ear signals that correspond to a fixed head position and orientation and then analyzing how the binaural system responds to the given cues. In this chapter, the current stationary approach will be extended to a model that can simulate basic head rotations to examine the advantages of active listening in the context of auditory scene analysis. This model will be tested based on its replication of the human ability to resolve front/back confusions using head rotations. It will be further tested for how closely its performance resembles that of humans in a diffuse field and on the effect of early reflections on perceived auditory source width.

The binaural model described here is part of a larger project that seeks to develop a *creative artificially-intuitive and reasoning agent*, CAIRA, capable of improvising music [11, 12, 29]. CAIRA's architecture, which is shown in Fig. 1, is comprised of several stages:

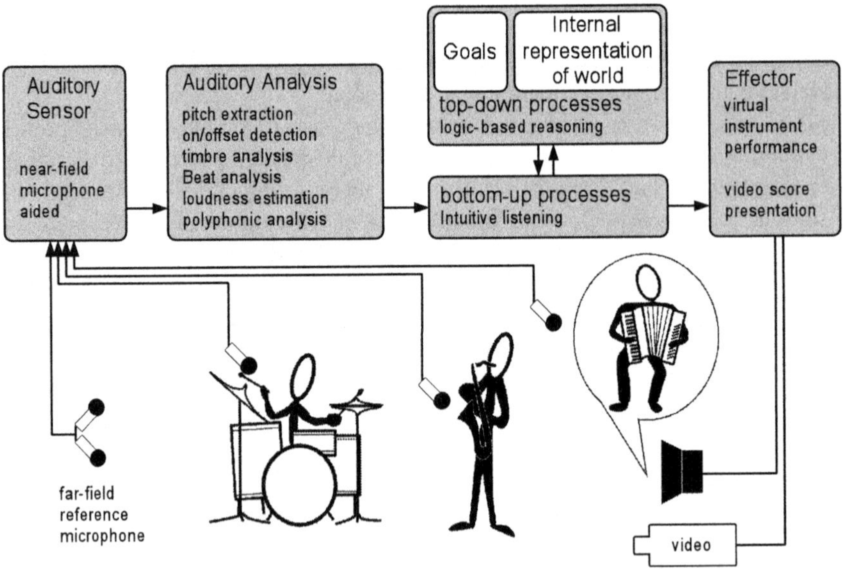


**Fig. 1** Schematic of the creative artificially-intuitive and reasoning agent CAIRA

- To perform computational auditory scene analysis
- To use machine learning to extract and recognize musical textures
- To build on logic-based reasoning to understand music and operate on a symbolic level
- To employ genetic algorithms to create material for musical performance in live response to other musicians.

Currently, the agent has two different sensors to capture the other musicians' sounds. While it analyses sound using an attached binaural head, it also captures the sounds of the individual instruments using closely-positioned microphones. The latter approach allows the system to operate on individual auditory streams for each instrument, to overcome a problem that has not yet been solved in *computational auditory-scene analysis*, CASA, namely the segregation of individual instruments from a mixed microphone signal. Capturing the individual instruments with closely-positioned microphones is a necessary compromise until there are more powerful binaural CASA systems capable of extracting information from complex scenarios. Switching from a passive system that is forced into reading the acoustic world from a fixed perspective to one that can actively explore environments through head rotations is an important step in this direction. This chapter will only scratch the surface of what this paradigm shift could potentially achieve.

A good example of how humans benefit from head rotations while listening is their ability to resolve front/back confusions that can occur because the interaural cues, which are typically stronger than spectral cues, are often ambiguous for the same lateral position in the front or rear hemisphere. A number of psychoacoustical studies have quantified the improvement human listeners experience when using head rotations to determine whether a sound source is located in the front or back.

Most investigations on spatial hearing with natural sound sources use a head-related coordinate system to describe the location of sound sources respective to the head position such as the one shown in Fig. 2. In this coordinate system, the interaural axis intersects the upper margins of the entrances to the left and right ear canals. The origin of the coordinate system is positioned on the interaural axis, halfway between the entrances to the ear canals. The horizontal plane is defined by the interaural axis and the lower margins of the eye sockets, while the frontal plane lies orthogonal to the horizontal plane, intersecting the interaural axis. The median plane is orthogonal to both the horizontal and frontal planes.

The position of a sound source is described using polar coordinates, azimuth, $\alpha$, elevation, $\delta$, and distance, $d$. If $\delta$ is zero and $d$ is positive, the sound source moves anti-clockwise through the horizontal plane with increasing $\alpha$. At $\alpha = 0°$ and $\delta = 0°$, the sound source is directly in front of the listener, intersecting the horizontal and median planes.

A simple geometrical model, like the one by Hornbostel and Wertheimer [30], reveals that interaural time differences, ITDs, of the same magnitude trace hyperbolas in the horizontal plane. At greater distances the shell of a cone in three-dimensional space is apparent. These are the so-called *cones of confusion*. Hence, there exist

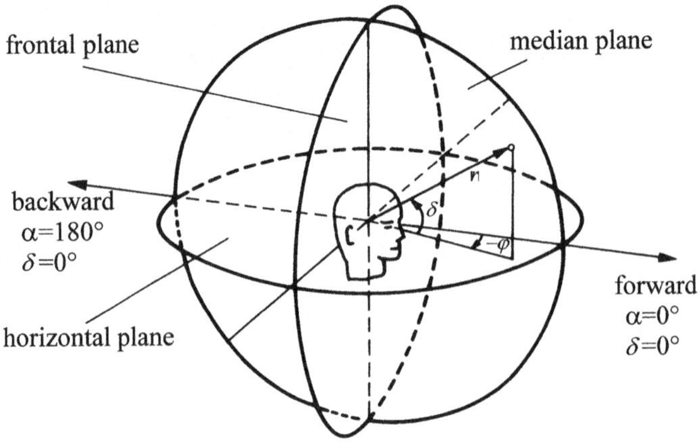


**Fig. 2** Head-related coordinate system [5]

multiple positions with identical ITDs, and while they are reliable cues to determine the left–right lateralization of a sound source, the ambiguities described by the cones of confusion often make it difficult to determine the elevation and front/back directivity of a sound source in case that head movements are not utilized.

Wenzel et al. [31] reported a front/back confusion error of 19 % for free-field gaussian distributed broadband-noise stimuli as an average over 16 listeners and 24 directions with different positions in 3-dimensional space. The average error increased to 31 % when the same listeners localized the stimuli in a virtual auditory environment using non-individual head-related transfer functions. Begault and Wenzel [2] found similar results for speech signals. An average error of 29 % was reported for five different lateral angles. The error for the 0°/180° directions was found to be higher then the error over multiple directions with an average value of 40 %.

A study by Perrett and Noble [24] is an ideal test case for the model analysis in this chapter, because the authors instructed their test participants precisely how to move their heads while listening to the test stimuli. The aim of the investigation was to determine the effect of head movements on listeners' ability to discriminate between front and back directions. In one of the test conditions, the listeners where asked to continuously move their head between $-30°$ and $+30°$ azimuth. In the control condition, the listeners were asked to keep their heads motionless during the stimulus presentations. The listeners' heads where not fixed in position during this condition, but monitored using a head-harness-mounted laser pointer for self monitoring and a Polhemus Isotrak-II head tracker. In a further test condition, the authors also inserted 3-cm-long open plastic tubes into the listeners' ears to circumvent the function of the pinnae using a method originally proposed by Fisher and Freedman [17]. For the motionless conditions, Perrett and Noble's [24] data reveal that for the non-distorted condition—that is, no tubes—the front-back confusion error was approximately 15 % for a 3-s broadband white-noise signal and 35 % for a low-pass-filtered

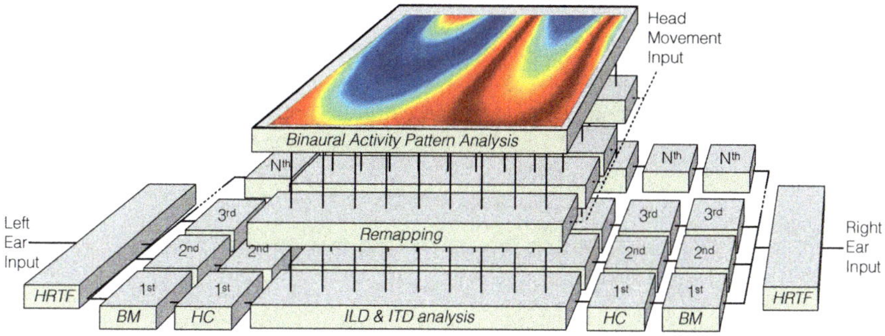


**Fig. 3** General model structure of the binaural localization model utilizing head rotations with the following components. *HRTF*: External-ear simulation/HRTF filtering. *BM*: Basilar membrane/bandpass filtering. *HC*: Hair cell/halfwave rectification. *ITD & ILD analysis*: Interaural time-difference-cue extraction/interaural cross correlation and interaural-level-difference-cue analysis with EI-cells. *Remapping* to azimuth angles with head-rotation compensation. *Binaural-activity-pattern analysis* to estimate the sound-source positions

white-noise signal with a cut-off frequency of 1 or 2 kHz. In the open-tube condition, the results for the low-passed stimuli did not change significantly, but the error for the broadband signal went up to 35 %. With head movements, these errors decreased to 0 % for the non-distorted conditions and 0.6 % for the open-tube conditions.

The initial idea of the model proposed here was to design a prototype model that can utilize head rotations to improve localization performance. The data by Perrett and Noble [24] was chosen as a test bed to evaluate the performance of the model, which will be outlined in the next section (Fig. 3).

## 2 Model Structure

### Periphery

The model is based on the interaural cross-correlation method introduced by Sayers and Cherry [26] to estimate ITDs. The basic model structure is similar to the one proposed by Braasch [9]. In line with earlier approaches by Blauert and Cobben [6] and Stern and Colburn [27], it also builds on the integration of the auditory periphery. The transformations from the sound sources to the eardrums and the influence of the outer ear and, occasionally, room reflections, are taken into account by filtering the sounds with HRTFs from a specific direction. Afterwards, the outputs for all sound sources are added together separately for the left and right channels. Basilar-membrane and hair-cell behavior are simulated using a gammatone-filter bank with 36 bands and a simple half-wave rectifier at a sampling frequency of 48 kHz, as described by Patterson et al. [23].

### Cross Correlation

After the half-wave rectification, the normalized interaural cross correlation is computed for each frequency band over the whole target duration as follows,

$$\Psi_{y_{l,r}}(t,\tau) = \frac{\int\limits_{t'=t}^{t+\Delta t} y_l(t'-\tau/2)\cdot y_r(t'+\tau/2)\,dt'}{\sqrt{\int\limits_{t'=t}^{t+\Delta t} y_l^2(t')\,dt' \cdot \int\limits_{t'=t}^{t+\Delta t} y_r^2(t')\,dt'}}, \tag{1}$$

with the internal delay, $\tau$, and the left and right ear signals, $y_l$ and $y_r$. The variable $t$ is the start time of the analysis window and $\Delta t$ its duration. Only the frequency bands # 1–16, covering 23–1559 Hz, are analyzed, reflecting the human auditory system's inability to resolve temporal fine structure at high frequencies, as well as the fact that the time differences in the fine structure of the lower frequencies are dominant—if they are available [32].

### Remapping and Decision Device

Next, the cross-correlation functions will be remapped from interaural time differences to azimuth positions. This is important for the model to be able to predict the spatial position of the auditory event. In addition, this procedure will help to align the estimates for the individual frequency bands as one cannot expect that the interaural time differences are constant across frequency for a given angle of sound-source incidence. An HRTF catalog is analyzed to convert the cross-correlation function's $x$-axis from interaural time differences to the azimuth.[2] After filtering the HRTFs with the gammatone-filter bank, the ITDs for each frequency band and angle are estimated using the interaural cross-correlation, ICC, algorithm of Eq. 1. This frequency-dependent relationship between ITDs and azimuthal angles is used to remap the output of the cross-correlation stage, namely, ICC curves, from a basis of ITDs, $\tau(\alpha, f_i)$, to a basis of azimuth angles in every frequency band, that is,

$$\tau(\alpha, f_i) = g(\mathrm{HRTF}_l, \mathrm{HRTF}_r, f_i), \tag{2}$$
$$= g(\alpha, f_i), \tag{3}$$

with azimuth $\alpha$, elevation $\delta = 0°$, and distance $r = 2$ m. $f_i$ is the center frequency of the bandpass filter and $\mathrm{HRTF}_{l/r} = \mathrm{HRTF}_{l/r}(\alpha, \delta, r)$.

In a subsequent step, the ICC curves, $\psi(\tau, f_i)$, are remapped to a basis of azimuth angles using the following simple `for`-loop in Matlab using a step size of 1°,

[2] The HRTF catalogs used for this investigation were measured at the Institute of Communication Acoustics of the Ruhr-University Bochum, Germany. They were obtained at a resolution of 15° in the horizontal plane and then interpolated to 1° resolution using the spherical spline method—see Hartung et al. [20].

```
for alpha=1:1:360
   psi_rm(alpha,freq)=psi(g(alpha,freq),freq)
end
```

Here, 'psi(tau,freq)' is the original, frequency dependent, interaural cross-correlation function with the internal delay, $\tau$, that is, 'tau'. The function 'g(alpha, freq)' provides the measured 'tau'-value for each azimuth and frequency. Inserting this function as 'tau' input in 'psi' transforms the 'psi'-function into a function of azimuth using the specific Matlab syntax. In the decision device, the average of the remapped ICC functions: 'psi_rm(alpha, freq)' over the frequency bands # 1–16 is calculated, dividing by the number of frequency bands. The model estimates the sound sources at the positions of the local peaks of the averaged ICC function.

Figure 4 shows an example of a sound source in the horizontal plane with an azimuth of 30° for the eighth frequency band. The top-left graph shows the original ICC curve obtained using Eq. 1 as a function of ITD. The graph is rotated by 90° with the ICC on the $x$-axis and ITD on the $y$-axis to demonstrate the remapping procedure. The curve has only one peak at an ITD of 0.45 ms. The top-right graph depicts the relationship between ITD and azimuth for this frequency band. As mentioned previously, the data was obtained by analyzing HRTFs from a human subject. Now, this curve will be used to project every data point of the ICC-vs.-ITD function to an ICC-vs.-azimuth function as shown for a few data points using the straight dotted and dashed-dotted lines. The bottom panel shows the remapped ICC function, which now contains two peaks—one for the frontal hemisphere and one for the rear hemisphere. The two peaks fall together with the points where the cone-of-confusion hyperbolas intersect the horizontal plane for the ITD value of the maximum peak that is shown in the top-left panel.

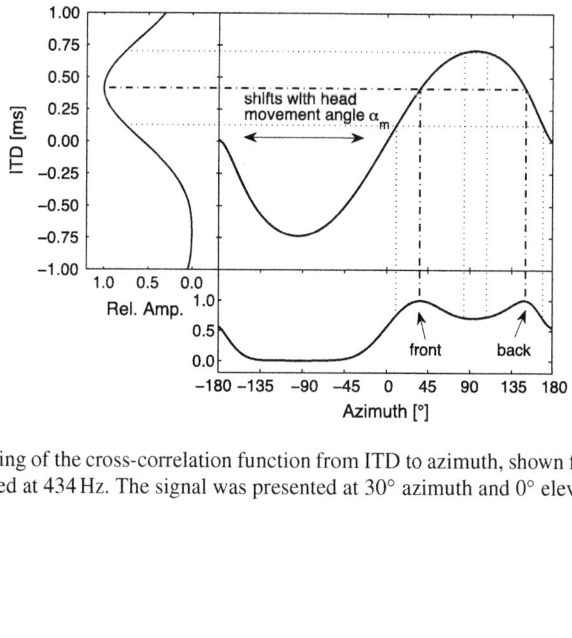


**Fig. 4** Remapping of the cross-correlation function from ITD to azimuth, shown for the frequency band # 8, centered at 434 Hz. The signal was presented at 30° azimuth and 0° elevation

### Integrating Head Rotations

Now it will be assumed that the head rotates to the left while analyzing an incoming sound source from the front. Related to the head, the sound source will now move toward the right. However, if the sound source had been in the rear, the sound source would have moved to the left. This phenomenon will now be used to distinguish between both options: frontal and rear position. For this purpose, a further coordinate system will be introduced, namely, a room-related coordinate system. The fact that human listeners maintain a good sense of the coordinates of a room, as they move through it, motivates this approach. If a stationary head position is considered, the head-related coordinate system is fully sufficient. However, if the head rotates or moves, the description of stationary sound source positions can become challenging, because every sound source starts to move with the changing head position. An easy way to introduce the room-related coordinate system is to define a reference position and orientation of the human head, and then determine that the room-related coordinate system coincides with the head-related coordinate system for the chosen head position.

Consequently, the room- and head-related coordinate systems are identical as long as the head does not move. In this investigation, only head rotations within the horizontal plane are considered and, for this case, the difference between the head-related coordinate system and the room-related coordinate system can be expressed through the head-rotation angle, $\alpha_m$, which converts the room-related azimuth, $\alpha_r$, to the head-related azimuth, $\alpha_h$—see Fig. 5:

$$\alpha_r = \alpha_m + \alpha_h . \tag{4}$$

Given restricted head movement, the origin of both coordinate systems and the elevations are always identical. While the sound-source position changes relative to the head with head rotation, a static sound source will maintain its position in the room-related coordinate system. Using this approach, a further coordinate transformation of the ICC function is executed in the model, this time from a function of head-related azimuth to room-related azimuth. This can be accomplished by rotating the remapping function when the head is moving by $-\alpha_m$ to compensate for the head rotation.

If a physical binaural manikin were to be used with a motorized head in connection with the binaural model, the HRTFs would be automatically adjusted with the rotation of the manikin's head. In the model discussed here, where the manikin or human head is simulated by means of HRTFs, the HRTFs have to be adjusted virtually. Also, the HRTFs have to correspond at every moment in time to the sound-source angle relative to the current head position. This can be achieved with the help of a running-window function, where the sound source is convolved with the current HRTF pair. In this case, a Hanning window of 10 ms duration and a step size of 5 ms is used. The smooth edges of this window will cross-fade the signal, allowing for a smooth transition during the exchange of HRTFs. For each time interval, the model will

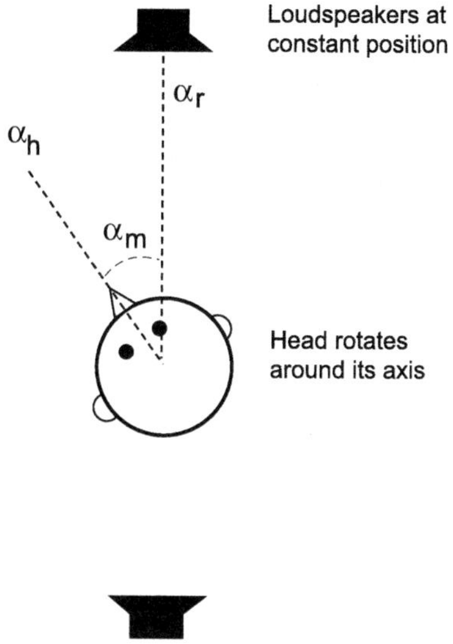


**Fig. 5** Sketch to illustrate the front/back confusion problem. If an ongoing sound source is located in front of the listener who turns his head left, the sound source will move to the right from the perspective of the listener's head. But if the sound source is located in the back, the sound source appears to move to the left for the same head rotation. The variable $\alpha_r$ denotes the azimuth in the room-related coordinate system, here pointing at 0°, $\alpha_h$ is the azimuth in the head-related coordinate system, also pointing at 0°, but for this coordinate system. The third angle, $\alpha_m$, is the head-rotation angle. It indicates by how much the head is turned from the reference head orientation that coincides with the room-related coordinate system

1. Determine the head rotation angle $\alpha_m$ based on a predetermined trajectory
2. Select the HRTFs that correspond to the sound source angle of incidence for the head-related coordinate system
3. Compute the normalized ICC for each frequency as a function of ITD
4. Convert the ICC function to a function of head-related azimuth $\alpha_h$ using the remapping function shown in Fig. 4
5. Circular-shift the remapping function based on the head rotation angle by $-\alpha_m$ to transform the ICC curve into the room-related coordinate system
6. Compute the mean ICC output over all frequency bands
7. Average the ICC outputs over time
8. Estimate the position of the auditory event at the azimuth where the ICC peak has its maximum

The first example is based on a bandpass-filtered white-noise signal with a duration of 70 ms. The signal is positioned at 150° azimuth in the room-related coordinate system. At the beginning of the stimulus presentation, the head is oriented toward the front, that is, $\alpha_h = 0$, and then rotates with a constant angular velocity to the left until it reaches an angle of 90° when the stimulus is turned off. The ICC functions are integrated over the whole stimulus duration. Figure 6 shows the result of the

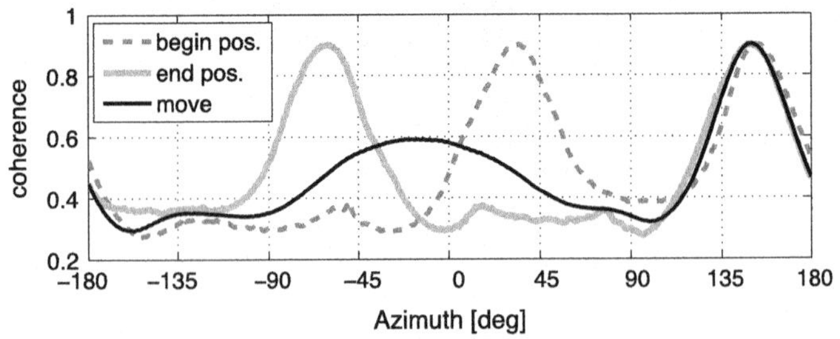


**Fig. 6** Interaural cross-correlation pattern for a sound source at 150° that is presented during a head rotation from $\alpha_m = 0°$ to $45°$. *Dashed line* ICC curve for the initial time window. *Solid gray curve* ICC for the last segment when the head is fully turned. Note that the ICC pattern was shifted in the opposite direction of the head rotation to maintain the true peak position at $\alpha_r = 150°$. *Black curve* Time-averaged ICC curve with the main ICC peak remaining and the secondary ICC partly dissolving

simulation. The initial ICC-vs.-$\alpha_r$ function, the output of step 6 for $\alpha_m = 0°$, is depicted by the dashed gray curve. Here two peaks can clearly be observed, one at $\alpha_{r=h} = 30°$ and the other one at $\alpha_{r=h} = 150°$. At the end of the stimulus presentation—solid gray curve, $t = 70$ ms, $\alpha_m = 45°$—the position of the rear peak, indicating the true sound source location, is preserved at $\alpha_{r \neq h} = 150°$, because the head rotation was compensated for by rotating the remapping function in opposite direction of the head movement. However, in case of the front peak—the front/back confused position—the peak position was counter-compensated for and it rotates twice the value of the head-rotation angle, that is, $\alpha_m = 45°$. The new peak location is shifted by $-90°$ to a new value of $-60°$. The time-averaged curve—solid black line, which shows the output of step 7—demonstrates the model's ability to robustly discriminate between front and rear angles. The secondary peak, the one representing the solution for a frontal-sound source, is now smeared out across azimuth because of the head rotation. Further, its peak height is reduced from 0.9 to 0.6, making it easy to discriminate between front and rear.

Figure 7 shows the same example, but this time the averaged ICC curve, as obtained after step 7, is shown after every time step. In the beginning of the simulation both peaks for the front and back location have equal heights, but over time the peak for the front location smears out and diminishes further. The solid black line shows the position of the true location in the rear, which remains stable over time. The dashed black curve shows the maximum peak position for the confused front position. In the opposite case, where the sound source is located in the front, namely, $\alpha_m = 30°$, the ICC peak for the rear location smears out over the azimuth, while the peak position of the true, frontal position is maintained—see Fig. 8.

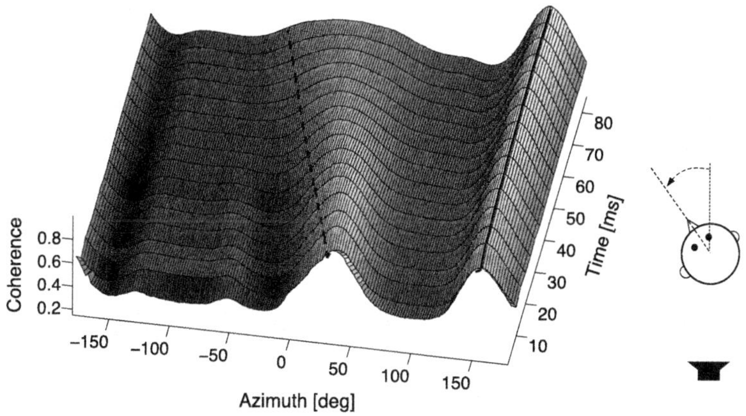


**Fig. 7** Similar to Fig. 6, but this time averaged ICC functions are shown for every time step of 5 ms. *Solid black line* Azimuth of the sound source, $\alpha_r = 150°$. *Dashed black curve* Momentary location of the secondary peak

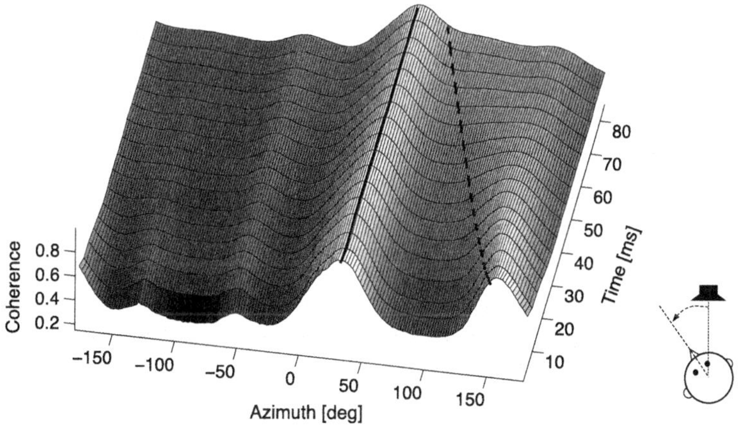


**Fig. 8** Same as Fig. 7 but for a sound-source position of $\alpha_r = 30°$

## *2.1 Processing of Interaural Level Differences*

Interaural level differences, ILDs, are often computed directly from the ratio of the signal powers, $P_{\mathrm{l,r}}$, in both the left and right channels, after the simulation of the auditory periphery. In this case, a physiologically motivated algorithm is utilized that was introduced by Reed and Blum [25]. The algorithm had been implemented into other binaural models with combined ITD and ILD cues [10, 15]. In this model, a cell population of EI cells is simulated. Each cell is tuned to a different ILD as shown in Fig. 9. One possible way to simulate the activity of the EI cells, $E(m, \varphi)$, at a given ILD, $\varphi$, is the following:

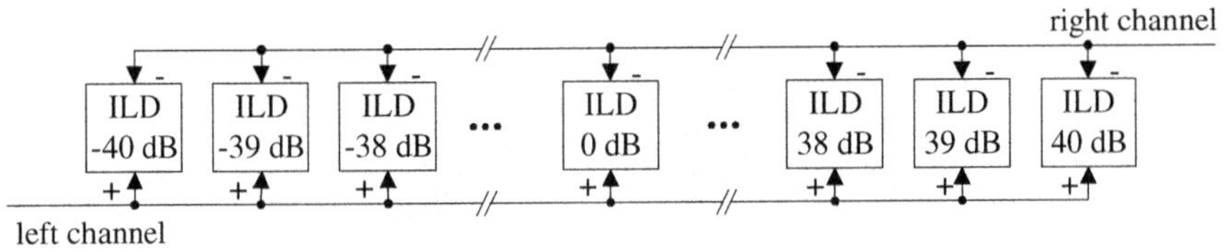


**Fig. 9** Model structure using EI cells

$$E_k(m, \varphi) = \exp\left[\left(10^{\varphi/\mathrm{ILD_{max}}}\sqrt{P_{k,l}(m)} - 10^{-\varphi/\mathrm{ILD_{max}}}\sqrt{P_{k,r}(m)}\right)^2\right], \quad (5)$$

with the time, $m$. $P_{\mathrm{k,l}}(m)$ and $P_{\mathrm{k,r}}(m)$ are the power in the left and right channels and $k$ refers to the $k^{th}$ frequency band. $\mathrm{ILD_{max}}$ is the maximal ILD magnitude that the cells are tuned to. The equation describes an *excitation/inhibition*, EI, process, because the input of the left channel has a positive sign indicating *excitation*, while the sign for the input of the right channel is negative, describing *inhibition*. The terms excitation and inhibition refer to the response of a neuronal cell—an excitatory cell input increases the cell's firing rate, whereas an inhibitory cell input reduces it.

The model is ideal for this study, because it produces an EI-cell-activity-vs.-ILD curve in a way similar to how the interaural cross-correlation function does it. For a given ILD, the EI-cell activity reaches the maximum at the cell that is tuned to this ILD. This property is used to estimate the ILDs in the model. The EI-cell activity decreases with mismatch between the given and the tuned ILD. Using the concept of the remapping function, the EI-cell-activity-vs.-ILD function can be transformed into an EI-cell-activity-vs.-azimuth function, similarly to how the ICC-vs.-internal delay to ICC-vs.-azimuth functions are converted. The transformation map can be derived again from measured HRTFs by calculating the ILDs as a function of frequency and angle of incidence as was done for ITDs in Eq. 2. The map can also be rotated with the head-movement angle, $\alpha_m$, in the same way as it was done for ITDs.

## Simulating Pychophysical Front/Back Confusion Data

The next step of this investigation is to determine how well the model simulates the psychophysical data by Perrett and Noble [24]. In order to simulate front/back discrimination, an optimal detector with internal noise was assumed, which is a central aspect of *Signal Detection Theory*—see, for instance, [19]. In the study presented here, signal detection theory enables the model to predict the front/back direction of the signals by comparing the peak heights of the two cross-correlation peaks of the ICC-vs.-azimuth function and selecting the direction with the higher peak. Internal noise was added to avoid the model always making the same pick for conditions close to the threshold. In this way a psychometric curve was effectively simulated that gradually progresses from chance where the difference between both cues is below detection threshold, up to a 100 % recognition rate, where the difference between the front- and back-peak heights is well above this threshold.

In the present study, internal noise was added after the ICC function was integrated over time and before the decision on the front/back directionality was made—steps 7 and 8 of the itemized list given in Sect. 2. The magnitude of the internal noise was adjusted to data by Gabriel and Colburn [18]. These authors found in their study that the threshold for discriminating two broadband-noise stimuli based on their interaural coherence was 0.04 when the tested coherence values were on the order of one, while a much higher threshold was found when the coherence values were on the order of zero. Since the model compares the height of the two ICC peaks for the front and rear direction and, therefore, effectively analyzes interaural coherence, the first value of 0.04 was chosen for the internal noise, based on the observation that the height of the normalized cross-correlation peaks are close to one. In the practical implementation, a random number is added to each coefficient of the time-averaged cross-correlation function before the height of both functions are compared to each other. The random numbers were drawn from a Gaussian-distributed function with a mean of 0.0 and a standard deviation of 0.04.

The HRTF catalogs from eight human subjects were taken from a previous experiment by Braasch and Hartung [13].[3] The test stimuli were 85 ms broadband white-noise samples, and the model was applied as described in Sect. 2. The sound sources were positioned at an azimuth, $\alpha_r$, of 0° or 180°, and the simulation was repeated 100 times for each direction, using different internal-noise samples and newly-generated broadband white-noise bursts. For the motionless condition, the head rotation angle was kept constant at $\alpha_m = 0°$. For the head-rotation condition, the head-movement angle was computed as

$$\alpha_m(t) = -30 + \frac{4}{s} \cdot t, \tag{6}$$

with the evolved time, $t$ in milliseconds and the duration, $s$, of one step. With a stimulus duration of $t_{max} = 75$ ms and a step size $s$ of 5 ms, the head-rotation angle spans the range of $-30°$ for $t = t_0 = 0\,ms$ to 30° for $t = t_{max} = 75$ ms, during the stimulus presentation. Both the head-movement and the motionless conditions were simulated for a model based on ITD-cue analysis in the frequency bands # 1–16. Further, an alternative version was also applied, using two frequency regions in accordance with Rayleigh's duplex theory [22], namely, low-frequency ITD cues in the frequency bands # 1–16, 23–1559 Hz,—and high-frequency ILD cues in the frequency bands # 17–36, 1805–21164 Hz. The ILDs were processed using EI-activity cells as were described in Sect. 2.1.

The top graph in Fig. 10 shows the results for the ITD model for eight HRTF catalogs from human subjects—one female and seven male. The ITD and ILD remapping functions were individually computed for each HRTF catalog. For the motionless

---

[3] The catalogs were measured in the anechoic room of the Institute of Communication Acoustics of the Ruhr-University Bochum, Germany [13]. The measurement procedure is described in the same study.

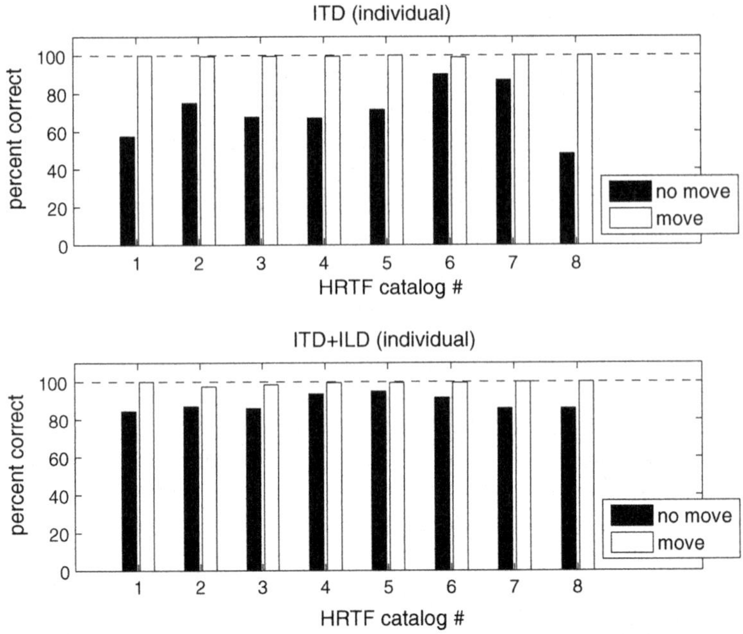


**Fig. 10** Percent correct scores for the binaural model discriminating between the front and back positions of sound sources at 0° and 180° azimuth using eight different HRTF catalogs. *Top graph* Results for the ITD analysis. *Bottom graph* Scores for the combined ITD and ILD analysis. *Black bars* Results without head movement. *White bars* Results with head movement

condition, the percent correct scores vary between 50 %, that is, guessing, for catalog 8 and 90 % for catalog 6. The average error was 30 % compared to 35 % measured by Perrett and Noble for the lowpass noise or open tube conditions. For the head-movement condition the average percent correct scores went up from 70 % to nearly 100 %. These values are reached for all HRTF catalogs. The results are in line with the findings of Perrett and Noble [24], who also found very small errors for the head-movement conditions.

The percent-correct scores improve substantially for the motionless condition if ILDs are taken into consideration as well—Fig. 10, bottom graph. The average percent-correct scores now rise from 70 % for the ITD condition to 89 % for the ITD+ ILD condition. The individual percent-correct scores vary between 85 % and 95 %. If head movements are applied to the model these results improve, yielding 100 percent-correct scores for all HRTF catalogs. Both the values for the motionless and head-movement results are in line with the findings of Perrett and Noble who found error rates of 15 %, that is, 85 % correct, for the motionless, no-tube, broadband condition and no errors with head movements involved.

Next, it will be investigated in more detail how the front/back-confusion-error rates drop with increasing head-rotation angle for the ITD model. For this simulation, the head-rotation angle was computed the following way,

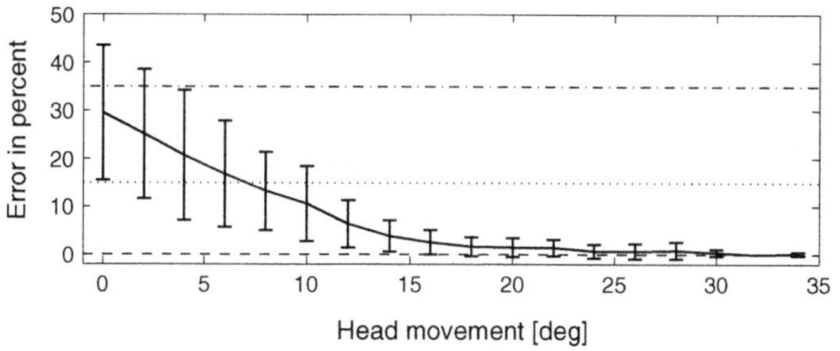


**Fig. 11** Percentage of front/back errors for different head rotation angles that were obtained using the binaural model analysing ITD cues only for HRTF catalog 8

$$\alpha_m(t) = 0 + \frac{x}{s} \cdot t, \tag{7}$$

with the evolved time, $t$, the duration of one step, $s$, and the change in angle per step, $x$. Again, the duration of the signal was 75 ms. Figure 11 shows the results of the simulation. The computed overall head-rotation angles are provided on the x-axis and the front/back-discrimination error is given on the y-axis. Each data point was computed 100× for the $\alpha_r = 0°$ and 100× for the 180° directions using HRTF catalog 8. The mean values are shown and the error bars depict the standard deviation. The value for the motionless condition, that is, 0° head-rotation angle, coincides with the no-movement condition for catalog 8—Fig. 10, top graph. The front-back confusion error decreases almost monotically with increasing head-rotation angle. At an angle of 30°, the front/back confusion errors approximate 0 %, and the 10 % error threshold is reached at an angle of 10°.

In the final test condition, the effect of non-individual HRTF catalogs was examined. This was accomplished by using a different HRTF catalog for computing the remapping functions for ITDs and ILDs than the one used for the convolution of the test signals during the model analysis. In this study, catalog 8 was used for the remapping functions to transform the results of the other seven HRTF catalogs. Figure 12 shows the simulation results for the non-individual HRTF data. Compared to the individual HRTF data for the motionless ITD condition—Fig. 10, top graph, black bars—the average percent-correct scores dropped from 70 % to approximately 50 % when a non-individual remapping function was used—Fig. 12, top graph, black bars. Interestingly, the performance no longer improved if additional ILD cues were considered—Fig. 12, top graph, gray bars. In fact, for HRTF catalog 6 a counter effect was observed where the percent correct score dropped from 50 to 10 % as the ILD cues apparently provided conflicting cues. With head movements—ITD condition shown only—the percent-correct scores reached values at or above 99 %, supporting psychoacoustic findings that head rotations can help to overcome front/back confusion errors when using non-individual HRTF catalogs [3].

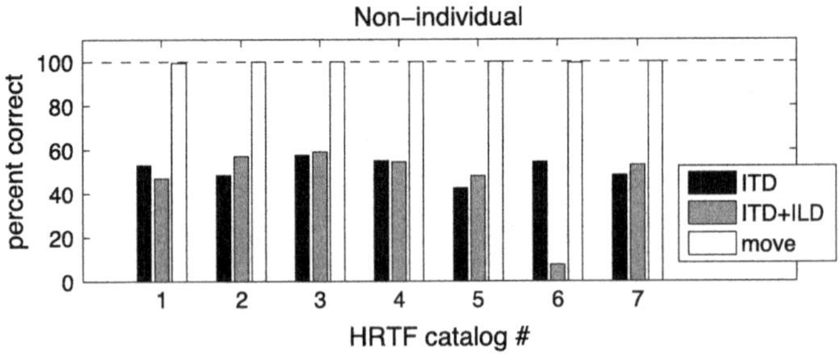


**Fig. 12** Same as Fig. 10 but for non-individual HRTF catalogs, using the eighth catalog as reference set for the remapping procedure. *Black bars* ITD analysis condition. *Gray bars* Combined ITD/ILD data. In both cases no head movements were applied. *White bars* Head movement conditions based on ITD cues

# 3 Head Rotations in the Context of Stereophonic Surround Systems

So far, the role of head movements was only examined in analyzing the direct sound. Now, a number of aspects related to room acoustics will be discussed. In the first case, the effect of head rotations on the perception of a diffuse field, the so-called *listener envelopment*, LEV, [7, 8, 14], will be investigated. The next section deals with the effect of head movements on the perception of early reflections in the context of the *apparent source width*, ASW [1]. An easy way to study the effect of a diffuse sound field is to examine a diffuse field produced by a stereophonic loudspeaker set-up. In a diffuse field of late reverberation, the sound ideally arrives from all directions, but in a stereophonic loudspeaker set-up, the angles of incidence are restricted to the loudspeaker positions. The spatial diffuseness of the sound field increases with the number of distributed loudspeakers. Interaural coherence is often used to predict the perceived spaciousness of a diffuse field, because the coherence decreases with increasing diffuseness [8, 16, 21]. Subsequently, it will be investigated how different loudspeaker configurations affect the prediction of spaciousness with the proposed model when head movements are considered.

In a standard stereo configuration, two loudspeakers are placed in the horizontal plane at azimuth angles of $\pm 30°$ or $\pm 45°$. This of course is a problem, if one needs to present ambient sound from the rear as well, and it was the main reason for the development of surround-sound systems. Transaural systems have attempted to provide surround sound using two loudspeakers by delivering HRTF filtered sounds over loudspeakers after applying an inverse filter to compensate for the transfer function between the actual loudspeakers and the listener. A cross-talk cancellation method avoids the sound for one loudspeaker channel reaching the contralateral ear. However, common transaural systems do not account for head-movements. It will be investigated how a frontal, two-channel loudspeaker system performs with a model that is capable of discriminating front/back directions.

A simple method to simulate a diffuse field is to present statistically independent, that is, uncorrelated, broadband noise signals to each loudspeaker. One-second white-noise bursts were used in this investigation and then either the broadband interaural coherence was calculated or the interaural coherence was computed using the binaural model which included the simulated auditory periphery. Three different loudspeaker configurations were tested:

- A two-loudspeaker configuration with speaker placements at ±45° azimuth
- A four-loudspeaker configuration with speakers placed at ±45° and ±135° azimuth
- An eight-loudspeaker configuration with speaker placements at 0°, ±90°, ±135°, and 180° azimuth

In all three configurations, the loudspeakers were placed in the horizontal plane. For each configuration, the broadband signal for each loudspeaker position was filtered with its corresponding HRTFs. Then, all left and right-ear signals were separately added up to simulate the total sound field at both ears. The simulation was repeated for different head-rotation angles, $\alpha_m$, in steps of 1° from 0° to 360°. It should be noted that, for this simulation, the binaural-cue values were not averaged over different head-movement positions but were presented for discrete positions. The actual head-related HRTFs were selected for each head position. HRTF catalog 8 was used for this simulation.

Figure 13 shows the data for the broadband interaural coherence condition. The black curve shows the results for the two-loudspeaker configuration. When the head is facing toward the front, that is, $\alpha_m = 0°$, the coherence has a value of 0.2. It increases immediately when the head moves left or right. The maxima were found at −90°, with a coherence of 0.6 and +90°, with a coherence of 0.5 for the two conditions where both loudspeakers were located at one side of the head. The variance of interaural coherence decreases if the number of loudspeakers is increased to four units. Now, four maxima, located at the four loudspeaker positions ±45° and ±135° azimuth, can be observed where the coherence reaches values near 0.25. That is a much lower value than found for the two-loudspeaker condition. For most other

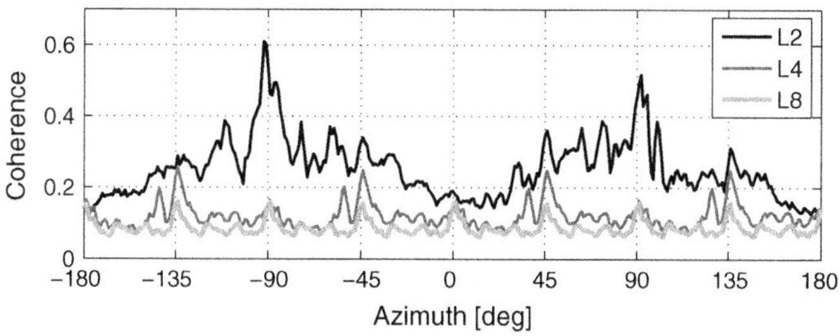


**Fig. 13** Interaural coherence, IC, for a diffuse field simulated with different loudspeaker configurations, L2, L4 and L8, as a function of head rotation angle. The numbers denote the number of loudspeakers involved. The IC was directly computed from the broadband noise signal

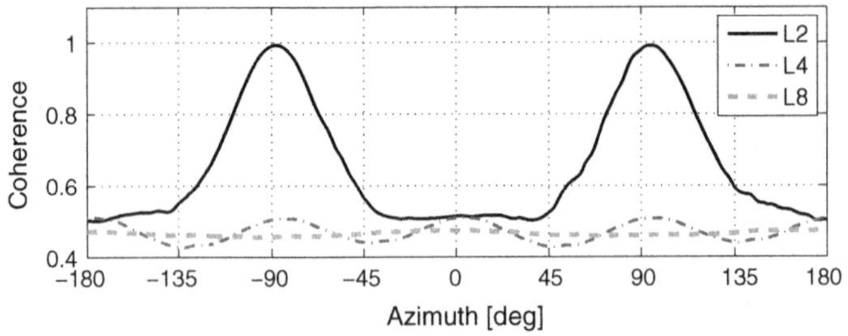


**Fig. 14** Same as Fig. 13, but this time the IC was computed as the average of an auditory-frequency-band analysis for the bands #7–11

angles, the coherence values are on the order of 0.15. Eight maxima are observed for the eight-loudspeaker configuration that always occur when the head is facing one of the eight loudspeakers. However, in this case, the interaural coherence is noticeably lower than for the other two loudspeaker configurations with floor values around 0.1 and peak values of 0.15, corresponding far more to the ideal of a diffuse sound field.

The results change if the coherence is computed from individual frequency bands. Figure 14 shows the results for the binaural model analysis as averaged over frequency bands #7–11. For the two-loudspeaker configuration, the interaural coherence is about 0.5 for a head rotation angle of 0°. This value remains constant for small head rotations, but increases greatly when the head rotation exceeds values of ±40° reaching maxima of one at head rotation angles of $\alpha_m = \pm 90°$. At these angles, both ear signals are fully correlated at the examined frequencies. The curve smooths out when four loudspeakers are used instead of two, oscillating between values of about 0.4 and 0.5 with minima at the loudspeaker positions and peak values in between the loudspeaker positions. In the third loudspeaker configuration, the interaural coherence no longer varies much, but remains fairly constant at a value of just below 0.5.

The next figure shows the variation of the ILD with the head rotation angle $\alpha_m$—see Fig. 15. The values depict the average across frequency bands #16–24. Again,

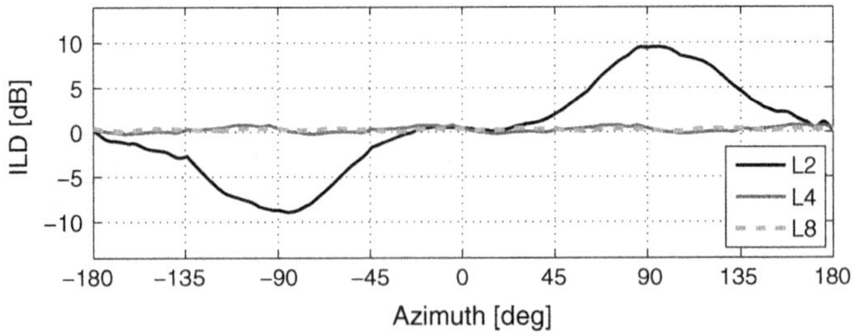


**Fig. 15** Interaural level differences for a diffuse field simulated with different loudspeaker configurations, L2, L4, and L8, as a function of head rotation angle. The numbers denote the number of loudspeakers involved. The ILDs were computed from the auditory band outputs, and the results reflect the average for the bands #16–24, that is, 1559–4605 Hz

the strongest variations can be found for the two-loudspeaker condition with azimuth angles of $\pm 45°$. For small head rotations, the ILDs do not vary much, but at an angle of about $\pm 20°$ the ILD starts to change, exceeding values of $\pm 2$ dB at $\pm 45°$ and reaching maxima of approximately $\pm 8$ dB at a head-rotation angle of $\pm 90°$. These are the two conditions where both loudspeakers are located sideways respective to the orientation of the head. The ILD variations decrease substantially when two rear loudspeakers are added to the two front loudspeakers. Now the level variation is within a range of 1–2 dB. The ILDs are zero for the four head orientations where the head is directly facing one of the four loudspeakers and directly in between two loudspeakers, with magnitude maxima in between these eight magnitude-minima positions. Using the eight-loudspeaker set-up the variation decreases further and the curve can be considered flat within the detection threshold of ILD discrimination, which is in the order of 1 dB. It can be concluded that the interaural coherence, and therefore the predicted perceived spaciousness, is not much affected if the head is kept still in the median plane. However once head movements are executed, the maximum interaural coherence is reduced with an increasing number of loudspeakers. While the diffuseness of the sound field changes substantially with increasing number of loudspeakers, the model simulation shows that the differences can only be experienced when the head is rotated.

# 4 Head Rotations in the Context of Apparent Source Width

The next goal is to investigate the role of head rotations in the context of early room reflections. Here, a tradition developed by researchers to investigate human perception in concert halls is used. Two delayed copies of the direct sound are taken to simulate the first two side reflections. These sideward refections are very important to widen the *apparent source width*, ASW, a perceptual measure that describes how wide a sound source is perceived to be in an enclosed space [1, 16]. Again, interaural coherence is typically used as predictor for apparent source width, which widens with decreasing coherence [21]. ASW is usually estimated from the interaural coherence of the first 80 ms of an impulse response and LEV from the late part. However, since early reflections are generally examined separately from late reverberation, the interaural coherence can be simply calculated from the running signal.

For this study, the delays were set to 20 ms for the left side and 24 ms for the right side. These values are considered to be in the ideal range for the acoustics of concert halls—compare Beranek [4]. Acousticians generally agree that the arrival times need to be slightly different for both side reflections to avoid a symmetrical situation where the sideness of both reflections is perceptually integrated to a center image. The amplitude of both reflections was adjusted to four different values, namely, 0 for direct sound only, and 0.25, 0.33, and 1.0. The reference amplitude for the direct sound was one. The amplitudes for the left and right side reflections always matched. The locations of the side reflections were chosen to be $-45°$ and $+45°$ azimuth, while the direct sound source was positioned at $0°$ azimuth, represented by

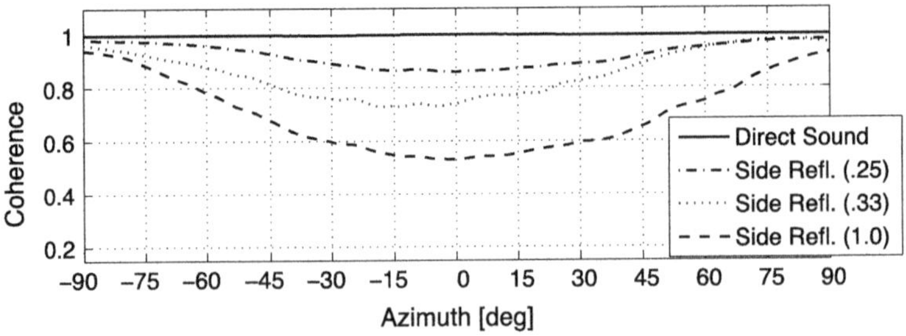


**Fig. 16** Interaural coherence as a measure for apparent source width, analysing a direct sound source in the presence of two reflections as the function of head rotation angle $\alpha_m$. *Solid curve* Results for the direct sound only. *Dashed/dotted curves* Three conditions with reflections of different amplitudes as indicated in the legend

HRTFs corresponding to the chosen directions. All three positions were located in the horizontal plane. A broadband-noise signal served as the sound source. It was convolved with the impulse response created from the direct source position and the two reflections with the given amplitudes and then analyzed using the same binaural model as used in the previous section. Again, different head-rotation angles were tested and the different head rotation angles were simulated by adjusting the HRTFs with head rotation.

Figure 16 shows the results for this simulation for the cumulative coherence values for frequency bands #7–11, that is, 352–738 Hz. It comes as no surprise that the coherence remains constantly at a value of one for the direct-sound-only conditions for all head-rotation angles. For all cases where the head is directly facing the front, the coherence decreases with increasing side-reflection amplitudes, to the values 0.85 with side-reflection amplitudes of 0.25, 0.75 with side-reflection amplitudes of 0.33 and 0.5 with side-reflection amplitudes of 1.0. However, these values increase monotically for all three conditions when the head is turned away from the center position, approximating coherence values of 1.0 when the head is fully turned sideways, that is, $\alpha_m = \pm 90°$. The simulation suggests that the apparent source width is affected by head movements. However, the simulation cannot predict to what extent human listeners can compensate for head movements when judging spatial impression. It thus remains unclear if the ASW varies with head movement or if a general impression is derived over all observation angles (Fig. 17).

## 5 Discussion, Future Directions and Conclusion

The model presented in this chapter is able to demonstrate a number of psychoacoustic effects that build on utilizing head movements, including resolving front/back confusions, analyzing the diffuse field of stereophonic sound reproduction systems and measuring the influence of head movements on the perceived auditory-source width. While the model was intended to be the first step toward a model that can

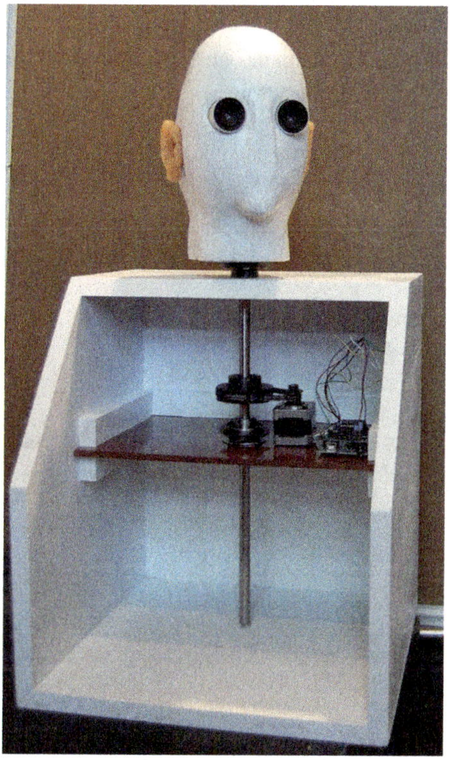

**Fig. 17** Rotating binaural head with stereoscopic vision

simulate active exploration of a sound field using a functionally-plausible model of the human auditory system, it should be pointed out that the model can also work in a passive mode, meaning that the model will produce similar results with its head being rotated on a continuous basis without receiving feedback from higher stages. Nevertheless, it can work equally well with a simple top-down mechanism that turns the head only when the system needs to resolve front/back confusions. This can be done by comparing the cross-correlation peaks for the front and back directions—see Fig. 4—and if the height of both peaks are within a given threshold, which could, for instance, be based on the coherence discrimination data of Gabriel and Colburn [18], the decision device commands the model to turns its head.

The authors are currently in the process of integrating the binaural model with a motorized binaural manikin, so that it can be tested in real environments. For this purpose, a dummy head was constructed that can rotate in the azimuthal plane with a stepper motor. The manikin was built using a rapid prototyping CNC machine to transform a 3-D CAD model of a human head into a dummy head made from medium-density fiberboard [28]. The ears for the head were molded from a human subject out of silicone with small electret Sennheiser KE-4 microphone capsules inserted into the canals. The addition of stereoscopic video capture is to provide visual cues to the binaural manikin, to set markers on individual players and to track their movements, gestures, and spatial arrangement. For cameras, Apple iSight webcams with 0.25" diameter sensors were used and inserted into the eye sockets.

**Acknowledgments** The project reported here is based upon work supported by the National Science Foundation, NSF, under Grant #1002851. The authors would like to thank two anonymous reviewers for their valuable comments and suggestions.

## References

1. M. Barron and A. H. Marshall. Spatial impression due to early lateral reflections in concert halls: the derivation of a physical measure. *J. Sound Vib.*, 77(2):211–232, 1981.
2. D. Begault and E. Wenzel. Headphone localization of speech. *Human Factors*, 35:361–376, 1993.
3. D. Begault, E. Wenzel, and M. Anderson. Direct comparison of the impact of head tracking, reverberation, and individualized head-related transfer functions on the spatial perception of a virtual speech source. *J. Audio Eng. Soc.*, 49:904–916, 2001.
4. L. Beranek. Subjective rank-orderings and acoustical measurements for fifty-eight concert halls. *Acta Acustica united with Acustica*, 89:494–508, 2003.
5. J. Blauert. Spatial Hearing. MIT Press, Cambridge, 1997.
6. J. Blauert and W. Cobben. Some consideration of binaural cross correlation analysis. *Acustica*, 39:96–104, 1978.
7. J. Blauert and W. Lindemann. Auditory spaciousness: Some further psychoacoustic analyses. *J. Acoust. Soc. Am.*, 80:533–542, 1986.
8. J. Blauert and W. Lindemann. Spatial mapping of intracranial auditory events for various degrees of interaural coherence. *J. Acoust. Soc. Am.*, 79:806–813, 1986.
9. J. Braasch. Localization in the presence of a distracter and reverberation in the frontal horizontal plane: II. Model algorithms. *Acta Acustica united with Acustica*, 88(6):956–969, 2002.
10. J. Braasch. Localization in the presence of a distracter and reverberation in the frontal horizontal plane: III. The role of interaural level differences. *Acta Acustica united with Acustica*, 89(4):674–692, 2003.
11. J. Braasch. A cybernetic model approach for free jazz improvisations. *Kybernetes*, 40:972–982, 2011.
12. J. Braasch, S. Bringsjord, C. Kuebler, P. Oliveros, A. Parks, and D. Van Nort. CAIRA - a creative artificially-intuitive and reasoning agent as conductor of telematic music improvisations. In *Proc. 131th Conv. Audio Eng. Soc.*, 2011. Paper Number 8546.
13. J. Braasch and K. Hartung. Localization in the presence of a distracter and reverberation in the frontal horizontal plane. I. Psychoacoustical data. *Acta Acustica/Acustica*, 88, 2002.
14. J. S. Bradley and G. A. Soulodre. The influence of late arriving energy on spatial impression. *J. Acoust. Soc. Am.*, 97:2263–2271, 1995.
15. J. Breebaart, S. van de Par, and A. Kohlrausch. Binaural processing model based on contralateral inhibition. I. Model setup. *J. Acoust. Soc. Am.*, 110:1074–1088, 2001.
16. W. de Villiers Keet. The influence of early lateral reflections on the spatial impression. In *Proc. 6th Intern. Congr. Acoustics, ICA 1968*, pages E-53-E-56, Tokyo, Japan, 1968.
17. H. Fisher and S. Freedman. The role of the pinna in auditory localization. *J. Aud. Res.*, 8:15–26, 1968.
18. K. Gabriel and H. Colburn. Interaural correlation discrimination. I. Bandwidth and level dependence. *J. Acoust. Soc. Am.*, 69:1394–1401, 1981.
19. D. M. Green and J. A. Swets. *Signal Detection Theory and Psychophysics*. Peninsula Publishing, Los Altos, 1989.

20. K. Hartung, J. Braasch, and S. J. Sterbing. Comparison of different methods for the interpolation of head-related transfer functions. In *Audio Eng. Soc.* 16$^{th}$ *Intern. Conf. Spatial Sound, Reproduction*, pages 319–329, 1999.
21. T. Hidaka, L. L. Beranek, and T. Okano. Interaural cross-correlation, lateral fraction, and low- and high-frequency sound levels as measures of acoustical quality in concert halls. *J. Acoust. Soc. Am.*, 98:988–1007, 1995.
22. Lord Rayleigh. On our perception of sound direction. *Phil. Mag.*, 13:214–232, 1907.
23. R. D. Patterson, M. H. Allerhand, and C. Giguère. Time-domain modeling of peripheral auditory processing: A modular architecture and software platform. *J. Acoust. Soc. Am.*, 98:1890–1894, 1995.
24. S. Perrett and W. Noble. The contribution of head motion cues to localization of low-pass noise. *Perception and Psychophysics*, 59:1018–1026, 1997.
25. M. C. Reed and J. J. Blum. A model for the computation and encoding of azimuthal information by the lateral superior olive. *J. Acoust. Soc. Am.*, 88:1442–1453, 1990.
26. B. M. Sayers and E. C. Cherry. Mechanism of binaural fusion in the hearing of speech. *J. Acoust. Soc. Am.*, 29:973–987, 1957.
27. R. M. Stern and H. S. Colburn. Theory of binaural interaction based on auditory-nerve data. IV. A model for subjective lateral position. *J. Acoust. Soc. Am.*, 64:127–140, 1978.
28. M. T. Umile. Design of a binaural and stereoscopic dummy head. Master's thesis, Rensselaer Polytechnic Institute, Troy, New York, 2009.
29. D. Van Nort, P. Oliveros, and J. Braasch. Developing systems for improvisation based on listening. In *Proc. 2010 Intl. Computer Music Conf., ICMC 2010*, New York, NY, 2010.
30. E. M. von Hornbostel and M. Wertheimer. Über die Wahrnehmung der Schallrichtung. Technical report, Sitzungsber. Akad. Wiss., Berlin, 1920.
31. E. M. Wenzel, M. Arruda, D. J. Kistler, and F. L. Wightman. Localization using nonindividualized head-related transfer functions. *J. Acoust. Soc. Am.*, 94:111–123, 1993.
32. F. L. Wightman and D. J. Kistler. The dominant role of low-frequency interaural time differences in sound localization. *J. Acoust. Soc. Am.*, 91:1648–1661, 1992.

# Binaural Systems in Robotics

**S. Argentieri, A. Portello, M. Bernard, P. Danès and B. Gas**

## 1 Introduction

In the seventies, the word *robot* mainly termed a manipulator arm installed in a workshop. It was designed to perform repetitive and/or high-precision tasks, such as pick-and-place, assembly, welding or painting. Its environment was fully controlled with no human around, and its behavior was fully programmed in advance. Since then, robotics has dramatically evolved. Nowadays, a robot is endowed with advanced perception, decision and action capabilities. In essence, it is in interaction with its environment with humans and/or with other robots and is capable of autonomy and adaptation—full or shared. The spectrum of applications has kept broadening, and spans not only manufacturing and supply chain, but also exploration, health—such as in surgery, rehabilitation, assistance—and professional and personal-service robotics as, for instance, in mining, agriculture, transports, monitoring, rescue, guidance, cleaning, assistance, and games. Among the hot topics, one may cite robot deployment in uncontrolled and dynamic environments, *Human–robot interaction*, task-oriented behaviors and networked robotics devices in smart environments, or *ubiquitous robotics* [16]. For the last years, one could observe growing two-way connections between robotics and neurosciences, with the methods, models and experimental achievements of each discipline being a source of inspiration and strengthening for the other one [64].

Perception is a key requirement to robot autonomy, adaptation and self-awareness. Traditionally, a distinction is made between proprioception, that is, the ability for

S. Argentieri · M. Bernard · B. Gas (✉)
Inst. des Systèmes Intelligents et de Robotique,
Univ. Pierre et Marie Curie, National Centre for Scientific Research (CNRS), Paris, France
e-mail: bruno.gas@upmc.fr

A. Portello · P. Danès
Lab. d'Analyse et d'Architecture des Systèmes,
Univ. de Toulouse, Univ. Paul Sabatier, National Centre for Scientific Research (CNRS),
Toulouse, France

J. Blauert (ed.), *The Technology of Binaural Listening*, Modern Acoustics and Signal Processing, DOI: 10.1007/978-3-642-37762-4_9,

a robot to sense its own internal status, for instance, in terms of wheels angular positions/velocities, joint angles, odometry, gyroscope- and exteroception, which provides the robot with information on its environment. Among others, one can cite exteroceptive modalities, such as bumpers for emergency stop, ultrasound/infrared/ laser scanning devices for range sensing, microphones, force sensors, tactile sensors and cameras—be it in the visible, infrared or multispectral range.

So far, the visual modality has undoubtedly received greatest interest. This is due to the richness of the information brought by images and to the high performance, low cost and embeddability of visual sensors. Vision has been used for decades in nearly all kinds of robotic tasks—from control loops and tracking routines to localization, map building, or scene modeling and interpretation. Numerous vision-based functions run nowadays in industry, for instance, non-destructive testing, scene/process monitoring, robot guidance, and other areas of application [5]. Besides, *computer vision* is a discipline by itself that does not take into account the specificities of robotics, such as real time constraints or changes in experimental conditions but, nevertheless, enriches the field.

Like vision, audition is a key sense in humans that plays a fundamental role in language and, consequently, in learning and communication. Quite surprisingly, *robot audition* was identified as a scientific topic of its own only since about the year 2000 [53], though related work existed before as part of bigger projects [12, 28, 30]. The reasons may be cultural as regards the importance of images in our society and also physiological—think of the predominance of vision in primates. More pragmatically, they are also certainly related to the difficult fulfillment of constraints like embeddability, high-performance acquisition, or real time processing. Consequently, while many theoretical results have long been developed in *acoustics* and *signal processing*, the literature on audition for robotics has remained scarce until recently. Fortunately, the timely topics of *cognitive robotics* and Human–robot interaction have promoted the field [25].

In its early days, robot audition benefited from developments in *computational auditory-scene analysis*, CASA [75]. Thereafter, Okuno et al. [63] identified the three main functions that any auditory robot should implement. These are

- Source localization, which may include a tracking system
- Separation of audio flows or source extraction
- Source recognition, which includes but is not limited to automatic speech recognition and can extend to scene interpretation

These functions encompass low-level issues as well as higher-level skills like emotion understanding or acoustic-scene meta-description.

In an attempt to provide a state of the art of binaural systems in robotics, this paper is structured as follows. The paradigms and constraints of robot audition are first summarized. Then the most prominent binaural approaches to canonical low- and high-level auditory functions are presented. After a review of robotics platforms, research projects and hard- and software related to the field, some novel *active* approaches developed by the authors are outlined, namely, combining binaural sensing and robot motion. A conclusion ends the chapter.

## 2 Paradigms and Constraints of Robot Audition

Two main paradigms for robot audition exist in the literature. On the one hand, *microphone arrays* have been used in a lot of applications. Various geometries have been selected, such as a line, a circle, a sphere, or the vertices of a cube. The redundancy in the sensed data is known to improve the acoustic-analysis performance and/or robustness [85]. Specific contributions have been concerned with

- The detection of the number of active sources, for example, through statistical identification [21]
- Source localization, for instance, through beamforming [59] or broadband beamspace MUSIC [2]
- Source extraction, for example, through geometrical source separation [84]
- Online assessment of uncontrolled dynamic environments
- Adaptation of speaker/speech recognition techniques to the robotics context [37]

On the other hand, *binaural approaches* have been developed.[1] These rely on a single pair of microphones that can be in free field, mounted inside an artificial pinna—not necessarily mimicking a human outer ear—and/or placed on a dummy head. From an engineering viewpoint, the possible use of cheap and efficient commercial stereo devices and drivers greatly eases the implementation. However, this simplification may imply an increased computational complexity.

Even though there is no fundamental need to restrict an acoustic sensor to only two microphones, for instance, when advanced data-acquisition and processing units are available, other arguments justify the binaural paradigm. First, robotics can be advocated as a privileged context to the investigation of some aspects of human perception. Indeed, as robots are endowed with locomotion and can incorporate multiple sensory modalities, they constitute a versatile experimental test bed to the validation/refutation of assumptions regarding the sensing structures as well as the processing and cognitive functions in humans. Conversely, these functions can be a source of inspiration for engineering approaches to perception. Last, there is an increasing demand for symbiotic interaction between humans and robots. This may imply the design of humanoid platforms, endowed with bioinspired perception and able to acceptably interact with humans in uncontrolled environments. Important constraints are, however raised by the robotics context, such as

*Embeddability* In the field of array processing, a large antenna involving a high number of microphones is often used. If such a sensor is to be embedded on a mobile robot, then a tradeoff must be handled between its size and, thus, aperture, and potential performances. Binaural sensors do not suffer from this geometrical constraint. Whatever the number of microphones is, the size and power consumption of the data acquisition and processing unit also constitute an important issue.

*Real time* Significantly distinct computation times are sometimes required by algorithms targeting the same goal. In robotics, low-level auditory functions such as

[1] Single-sensor approaches exist, such as [76, 83], but are rarely addressed in the literature.

binaural-cue calculation or source detection/localization must often run within a guaranteed short-time interval. Typically, up to 150 ms are demanded when their output is needed in reflex tasks such as exteroceptive control or people tracking. Specific processing units may have to be designed in order to guarantee real time behavior.

*Environment* Robotics environments are fundamentally dynamic, unpredictable and subject to noise, reverberation, or spurious sound sources. In order to ensure a guaranteed performance, adaptive or robust auditive functions must be designed.

*Sources* Most meaningful sound sources involved in robotics are broadband, with a spectrum spreading over the whole audible frequency bandwidth. This precludes the direct use of narrowband approaches developed elsewhere. Source non-stationarity is also an important issue. In addition, specific methods may be required depending on the source distance. Last, source motion can complexify the processing of the sensed signals, for example, when it breaks their assumed joint stationarity.

*Robot* Robot parts and robot motion generate so-called *self-noise*, or *ego-noise*, which may of course disturb the acoustic perception. Besides, in so-called *barge-in* situations, some sounds emitted intentionally by the robot may be interrupted, for example, by human utterance during a spoken dialog. Hence, they must be filtered-out online for not to damage the analysis of the scene.

Historically, most initial contributions to robot audition took place within the binaural paradigm. However, the results remained mixed when facing wideband non-stationary sources in noisy environments. Nevertheless, the last years have witnessed a renewal of interest for such approaches. A particular focus is put on *active* variations, which, thanks to the coupling of binaural sensing and robot motion, can overcome limitations of their passive counterparts. In computer vision, the coupling between visual perception and behavior has long been envisaged [1, 6]. The usefulness of active processes in hearing is discussed in [17, 53]. Importantly, the increased availability of cheap and accurate head-tracking solutions has given rise to related research amongst the hearing community with the potential of contributing to robot-listening strategies—see, for instance Ref. [11].

## 3 Binaural Auditory Functions in Robotics

As aforementioned, a robot should be able to understand several sources at a time by using its own ears in a daily environment, that is, with permanent background noises, intermittent acoustical events, reverberation, and so on. This challenging environment may be dynamic, due to sound sources moving or changing their acoustic properties. Additionally, the robot ego-noise is part of the problem. Whatever the conditions, most embedded auditory systems in robotics aim at generating an acoustic map of the robot's surroundings in real time. This map can then be used as an input to higher cognitive/decisional layers of the robot's software architecture in order to gaze or

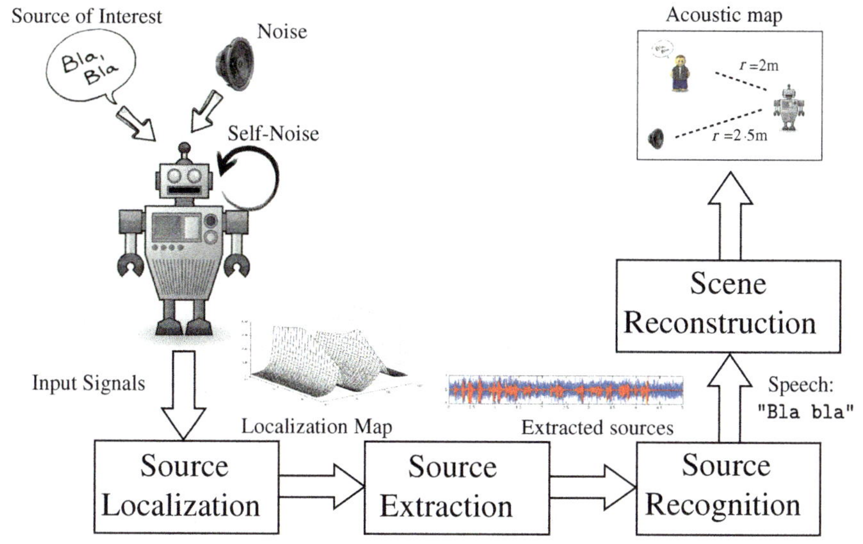


**Fig. 1** Bottom-up working flow representation of classical robot audition systems

move towards the source of interest, answer to an interacting human partner, and other related issues. Such elementary behaviors can of course be enriched by considering low-level-reflex actions or multimodal approaches to artificial sound perception [80] or [81], this volume.

Acoustic maps of the environment are generally obtained along the bottom-up, signal-driven, workflow illustrated in Fig. 1. The successive computational steps are often split into two categories, namely *low-level* and *high-level* auditory functions, which are

*Sound localization* This stage is probably the most important low-level auditory function. Lots of efforts have been made to provide efficient binaural sound localization algorithms suited to robotics.

*Extraction* Once localized, each sound source can be separated so as to provide clean speech signals, noise signals, etc.

*Ego-noise cancellation* Its importance was acknowledged in the early days of active audition. Indeed, the noise of the motors enabling the robot motion may significantly degrade the robot's auditory perception.

*Voice-activity detection, speaker recognition, speech recognition* The need of these functions comes from the fact that most robotics systems are concerned with speech signals for interaction purposes.

*High-level scene description* Finally, all the above extracted information are gathered to accurately describe the acoustic scene.

Elements of this traditional bottom-up approach are reviewed in the following.

### 3.1 Sound Localization

In *binaural robot audition*, the azimuth and elevation of the multiple sources in the environment, and possibly their distance, are inferred from cues extracted from left and right signals. The literature reports the use of binaural cues, namely, *interaural time/phase difference*, ITD/IPD, *interaural level difference*, ILD, and monaural cues, that is, spectral information, and further characteristics, for instance, distance-related, [74]. In [94] various cue-extraction methods are reviewed. Whatever the localization policy, the problem is then to inverse the transformation that relates the spatial source locations to such cues. This requires a-priori knowledge about the propagation of an acoustic wave onto the robot's scatterers. This knowledge is generally captured in a simplified analytical model, or comes as experimental measurements of sound source properties, such as frequency contents, positions and induced auditory cues.

Considering source-azimuth estimation, the first model proposed in robotics was the *auditory epipolar geometry*, AEG [53]. It expresses the interaural phase difference, IPD, as a function of the source azimuth measured with the two microphones in the free field. As AEG does not take into account the shadowing effect of the head on sound propagation, some alternatives were proposed. Among them, one can cite the *revised auditory epipolar geometry*, RAEG [57], inspired by the Woodworth–Schlosberg formula [90]. These models are now commonly used in robotics, but are shown to be not so robust to changes in the environment. For instance, Nakadai et al. [56] showed that the simulated ITD obtained from RAEG is consistent with experimental measurements gathered in an anechoic room in the range of 500–800 Hz. Yet, if the comparison is made in a real robotics environment including reverberation and noises, then the basic models do not fit real-life data anymore.

Another analytical model, based on *scattering theory*, ST, was proposed in [55], considering the scattering induced by a perfectly-spherical head. In comparison with previous models, ST provides a more reliable closed-form model of the IPD as a function of the source azimuth. Of course, room acoustics is still not taken into account, so that the measured IPD remains heavily influenced by the environment. A similar approach was exploited in [27] and experimentally tested in [26] on a spherical plastic head.

In humans, source elevation is inferred from monaural cues taking the form of spectral notches. These are induced by the interference of the direct path of the wave to the concha and its reflections brought by the pinna. They are reproduced in robotics by artificial pinnae in charge of collecting the sound wave to the microphones. Importantly, both the placement of conchas on a robotics dummy head and the shape of pinnae can be optimized to maximize localization sensitivity to the source position. A solution to the first problem was proposed in [79]. The second one related to pinnae design has been more deeply investigated [41, 42, 76]. Yet, so far, both problems remain open. Numerical simulations may be required, which complicates the design. At large, the only emerging rule of thumb consists in designing an irregular or asymmetric pinna shape to provide elevation-dependent spectral cues.

Most of the previous approaches aim at getting closed-form equations of interaural or monaural cues. Such models could also involve the head-related transfer functions, HRTFs, of the robot that is, the transfer functions from the center of the head if it were absent and the two microphones. An HRTF captures all the effects exerted by the robot's body, head and pinnae on the incoming sound waves. In such a way it subsumes all the above head models. As closed-form HRTF expressions are very difficult to obtain for a generic mobile platform, a prior identification step is mandatory. It must be performed in an anechoic room, thus limiting its applicability to robotics. Nevertheless, in the case of well-identified environments, some HRTF-related applications to localization in robotics were developed. For instance, [50] estimated the position of a single talker in a known environment. This work was extended in [37] to simulate a moving talker. In [29] a learning approach is proposed for sound localization based on audio-motor maps that implicitly captures the HRTF. Self-organizing feature maps were also used in [60] to fuse binaural and visual information so as to estimate the 3-D position of a sound source. In the same vein, a multimodal approach to sound localization based on neural networks was proposed in [93], where vision is also combined with auditory cues to estimate the source azimuth. Further, a *gaussian-mixture model*, GMM, approach is proposed in [51], this volume, to evaluate the position of multiple sound sources. Learning approaches thus seem a promising generic tool to adapt an HRTF to various acoustic conditions.

## 3.2 Source Extraction

Once the sound sources have been localized in the robot's environment, the subsequent steps in low-level auditory analysis generally consist in extracting the sound sources of interest. Depending on the authors, source localization and extraction can be inverted or even gathered into a single function. Though binaural extraction of sources was addressed in the early days of robot audition, the number of approaches has remained quite small. One of the most famous solution is the *active direction-pass filter*, ADPF [57]. It works by collecting frequency sub-bands that are presumably linked to the same sound source in order to generate the extracted signal. This fusion of sub-bands is performed only if their associated binaural cues, that is, IPDs/ILDs, are spatially coherent according to a given head model, for instance, AEG or RAEG—see Sect. 3.1—taking into account the better spatial-discrimination ability for sources in front of the robot. This system has been proven to be effective to extract and recognize three simultaneous speakers [58].

One of the main advantages of ADPF is that it requires no a-priori knowledge of the signal contents. The same applies to other recent binaural separation techniques based on localization cues. Well-known approaches based on *independent-component analysis* [70] that can separate two sources at most from the output of a binaural sensor, are then generally overcome. For instance, Weiss et al. [89] proposed to combine binaural cues with higher-level information related to speaker identity in order to separate more than two sources in the presence of reverberation. Another solution, already explored with ADPF, that outperforms beamforming approaches in reverberant scenes, is to derive time-frequency masks on the basis of binaural cues—see for instance [77]. In [23], multiple-speaker joint detection and localization was recasted as a probability-based spatial clustering of audio–visual observations into a common 3-D representation. Cooperative estimates of both the auditory activity and the 3-D position of each object were then deduced from a substantiated variation of the expectation-maximization algorithm. Experiments were conducted with the POPEYE platform—see Sect. 4.4—on single or multiple speakers in the presence of other audio sources.

Whatever the approach, it is still very difficult to extract multiple sources covering the same frequency range. In such cases, some extracted signals may mistakenly come from unrelated sources and may thus present missing or uncertain sections. In the downstream pattern-matching algorithms, for example, for speaker/speech recognition purposes, such problems can be handled within the *missing-feature theory*, MFT [18, 19, 45], which consists in tagging the missing sections by a null confidence weight, as will be shown further down.

### *3.3 Ego-Noise Cancellation*

Two very restrictive solutions to *ego-noise cancellation* are generally proposed for the binaural case. On the one hand, loud enough sources can mask ego-noises during a movement, thus improving the signal-to-noise ratio, SNR, of the perceived signals. On the other hand, stop-and-listen approaches are sometimes used, so as to process sounds while the robot is at rest. Both approaches are unsatisfactory, and recent developments have tried to overcome these limitations.

Canceling the noise originating from the robot can be considered by source separation techniques. But existing studies mainly rely on microphone arrays, which makes them inappropriate for the binaural context. Additionally, as motors are generally placed in the vicinity of the microphones, a diffuse sound field should be used to model the noise, precluding the direct use of standard state-of-the-art approaches to source separation.

One of the first solutions to the specific ego-noise-cancellation problem in robotics was proposed in [54] on the SIG humanoid robot—see Sect. 4.2. It relies on two pairs of microphones, one of them being dedicated to the perception of the interior of the robot. With the help of these inner transducers, the method was able to classify spectral bands as being related either to ego-noise or to the source of interest, but could not suppress the noise from the perceived signals. Other approaches consist in predicting the noise to be emitted on the basis of the generated motion. For instance, Ito et al. [35] did this on a time-frame basis with a neural network. The most promising solution relies on noise patterns. In this vein, joint noise templates related to specific movements were stored offline into a large ego-noise database [32, 62], then were identified online according to the robot motion to be performed, and subsequently subtracted from the perceived signals [31].

## 3.4 Voice-Activity Detection

When considering Human–robot interaction applications, the perceived signals are mainly composed of speech information, non-informative signals, and various types of noise. Efficient speech extraction is necessary to decrease the error rate in high-level auditory analysis. It can be performed by detecting speech segments in the sensed signals, prior to localizing the corresponding sources of interest and spatially filtering them out of the noise. *Voice-activity detection*, VAD, algorithms have often been used, which generally classify signal snippets as either *noise-and-speech* or *only-noise*. Again, multiple solutions to VAD have been proposed in the literature, such as energy feature, zero-cross rate [71] or higher-order statistics [61]. However, very few of them are specifically dedicated to binaural audition and/or suited to robotics. Energy feature can hardly cope with individual differences and dynamic volume changes. Zero-cross rate is better in this respect, but is more sensitive to noise. Statistics show good behavior but their performance decrease in an acoustic environment which shows significant differences with the one used to learn the statistics. As a solution suited to robotics, Kim et al. [37] proposed an enhanced speech detection method that can be used to separate and recognize speech in a noisy home environment. Nevertheless, the detected utterances should take place in front of the robot. Another approach is outlined in [20], inspired by wireless sensor network applications in the context of hearing aids. Therein, a basic energy-based VAD was combined with a cross-correlation based VAD to detect speech in the two signals. But again, the speaker should be uttering in front of the system. Besides, a Bayesian network based integration of audio–visual data for VAD was proposed in [91] as the first layer of an automatic speech recognition system.

## 3.5 Speaker and Speech Recognition

In robotics, speaker and speech recognition are probably the key high-level auditory functions required to perform natural Human–robot interaction. Traditional *automatic speech recognition*, ASR, algorithms are known to be very sensitive to the speech signal-acquisition conditions, such as quality of the microphones, distance to the speaker, environmental noise, and so on. Therefore, considering this problem at large, the robotics context requires a trade-off between large-vocabulary and multiple-speaker applications in the well-known framework of the *cocktail-party problem*.

Environmental noise is probably the most prominent challenge to be faced by ASR systems in robotics. This is probably the reason why most recent studies have focused on noise removal from the speech signals, that is, *speech enhancement*. The aim here is to retrieve ideal acquisition conditions, generally by applying a set of spatial filters which enable the attenuation of noisy sources or echoes in the perceived signal in order to use traditional ASR systems. In this topic, the aforementioned missing features approaches are of particular interest. They are able to cope with additive, possibly non-stationary noise sources [72, 78] by discarding specific regions in the speech spectrogram with low SNRs. Likewise, one can mention *missing-feature compensation* techniques, which are able to accurately estimate the omitted regions of these incomplete spectrograms. These approaches were used in [82] in order to separate two speakers uttering simultaneously from the front of a humanoid robot endowed with a binaural sensor, on the basis of an *independent-component analysis* based source-separation technique. This allows for improving the speech-recognition rates by 15 % with respect to a state-of-the-art-based *hidden Markov model*, HMM, recognition system. Recent developments in the missing-features framework were concerned with adaptive recognition systems, that is, with MFT–ASR approaches which allow to change the weight of those spectrogram sections that are considered damaged [73]. In this domain, Ince et al. [33] exploits an MFT–HMM approach to cancel ego-noise by applying a time-frequency mask. This makes it possible to decrease the contribution of unreliable parts of distorted speech in signals extracted from a microphone array. Such an approach can also be applied in binaural systems.

Compared to speech recognition, speaker recognition has rarely been addressed in robotics. Like speech recognition, it is usually analyzed for the monaural case under similar recording conditions. Initially, the subject was already addressed by means of microphones arrays [44]. For applications to robotics, one can refer to [36], and to the recent preliminary study [95]. The latter paper shows that in a reverberant and noisy environment, the success rate of binaural speaker recognition is much higher than with monaural approaches.

## 3.6 High-Level Scene Description

Several studies have been conducted in order to endow a user with auditory awareness about a complex auditory scene. A basic, though incomplete, solution consists in applying 3-D acoustic spatialization techniques to an extracted and labeled source in such a way that the users sense better where the corresponding message comes from. Intuitive tools exist in our daily lives to obtain awareness on a visual scene, such as overviewing, zooming, scrutinizing, (re)playing at various places, browsing, indexing, etc. In the same vein, a 3-D auditory scene visualizer according to the mantra *"overview first, zoom and filter—and then detail on demand"* was developed on top of face-tracking and auditory functions [40] and integrated into the $\mathcal{HARK}$ robot-audition toolbox—Sect. 4.2. This system has been improved to get better immersive feeling. Though a microphone array is assumed, the underlying concepts extend to binaural techniques.

# 4 Platforms and Research Projects

This section reviews some notable auditory robots and/or research projects. Associated hard- and software dedicated to robot audition are also mentioned as far as available.

## 4.1 Cog

The upper torso humanoid Cog, from MIT, is probably the first platform endowed with audition[2] It was targeted towards the scientific goal of understanding human cognition and the engineering goal of building a general purpose flexible and dextrous robot. Interestingly, the authors pointed out in their manifesto [12] that the first goal implied the study of four essential aspects of human intelligence that, by themselves, involve manufacturing a human-like platform. These topics, discarded in conventional approaches to *artificial intelligence*, were

- *Development* Considering the framework by which humans successfully acquire increasingly more complex skills and competencies
- *Social interaction* Enabling humans to exploit other humans for assistance, teaching, and knowledge
- *Physical embodiment and interaction* Humans use the world itself as a tool for organizing and manipulating knowledge
- *Integration* Humans maximize the efficacy and accuracy of complementary sensory and motor systems

Cog was endowed with an auditory system, comprising two omni-directional microphones mounted on its head, and crude pinnae around them to facilitate localization. The sound acquisition and processing units were standard commercial solutions. Companion development platforms were built, similar in mechanical design to Cog's head with identical computational systems. One of them, oriented towards the investigation of the relationships between vision and audition, complemented the binaural auditory system with one single color camera mounted at the midline of the head. Visual information was used to train a neural network for auditory localization [34].

## 4.2 SIG/SIG2 and HARK

The SIG Project [54] was initiated by Kitano Symbiotic Systems, ERATO and JST Corp., Tokyo. The pioneering program has then been pursued further in collaboration of Kyoto University and the Honda Research Institute.[3] In an effort to understand

[2] http://www.ai.mit.edu/projects/humanoid-robotics-group/cog/

[3] http://winnie.kuis.kyoto-u.ac.jp/SIG/

high-level perceptual functions and their multi-modal integration towards intelligent behavior, an unprecedented focus was put on *computational auditory-scene analysis*, CASA, in robotics. The authors promoted the coupling of audition with behaviors also known as *active audition* for CASA, so as to dynamically focus on specific sources for gathering further multimodal information through active motor control and related means. This approach paved the road to many developments, the first one being ego-noise cancellation.

SIG is an upper-torso humanoid. It has a plastic cover designed to acoustically separate its interior from the external world. It is fitted with a pair of CCD cameras for stereo vision and two pairs of microphones—one in the left and right ears for sound-source localization and the other one inside the cover, mainly for canceling self-motor noise in motion. A second prototype, named SIG2, was designed to solve some problems in SIG, such as the loud self-noise originating from motors, an annoying sound reflection by the body, sound resonance and leakage inside the cover, and the lack of pinnae. This implied changes in the material and actuators, as well as the design of human-shaped ears. Many striking achievements were obtained on SIG/SIG2, such as multiple sound-source localization and tracking from binaural signals while in motion, multiple-speaker tracking by audio–visual integration, human–robot interaction through recognition of simultaneous speech sources.

Subsequently and importantly, array-processing techniques for source localization and source separation were designed and implemented on SIG/SIG2. This gave rise to the *open-source robot-audition toolbox*, $\mathcal{HARK}$,[4] that gathers a comprehensive set of functions enabling computational auditory-scene analysis with any robot, any microphone configuration and various hardware. Within the recent revival of active binaural audition, $\mathcal{HARK}$ has been complemented with a package for binaural processing.

## 4.3 iCub

An open-source platform, comprizing hardware and software, well suited to robot audition is the iCub humanoid robot. iCub has been developed since 2004 within the RobotCub project,[5] and disseminated into more than twenty laboratories. Sized as a 3.5 year-old child, it is endowed with many degrees of freedom and human-like sensory capabilities, including binaural audition. It has also been designed towards research in embodied cognition, including study of cognition from a developmental perspective in order to understand natural and artificial cognitive systems.

Two electret microphones are placed on the surface of its 5-DOF head and plastic reflectors simulate pinnae. The shape of these ears has been kept simple, so as to ease their modeling and production while preserving the most prominent acoustic characteristics of the human ear. To better manage the frequencies of the resonances and notches to be used for vertical localization, a spiral geometry was selected [29].

---

[4] HRI-JP audition for robots with Kyoto University, http://winnie.kuis.kyoto-u.ac.jp/HARK/. In Ariel's Song, The Tempest, from Shakespeare, *hark* is an ancient english word for *listen*.

[5] http://www.icub.org/projects.php

Small asymmetries between right and left pinnae, namely, a rotation of 18°, enable to tell the notches due to the source contents from these due to its spatial location just by comparing the binaural spectral patterns. A supervised learning phase matches the (ITD, ILD, notches)-tuples extracted from binaural snippets with sound-source positions inside audio-motor maps. These maps are then used online to drive the robot by sound. The maps are seamlessly updated using vision to compensate for changes in the HRTFs as imposed by ears and/or environment. Experiments show that the robot can keep the source within sight by sound-based gaze control, with worst-case errors of pan and tilt below 6°.

## 4.4 POP and HUMAVIPS, and Their Associated Platform/Datasets

A recent milestone in robot-audition research is undoubtedly the *perception-on-purpose*, POP, project.[6] This European scientific collaboration in 2006–2008 was oriented towards the understanding and modeling of the interactions between an agent and its physical environment from the biological and computational points of view, concentrating on the perceptual modalities of vision and audition. Aside from a fundamental investigation of cognitive mechanisms of attention on the basis of measures of brain physiology brought about by *functional magnetic-resonance imaging*, fMRI, and *electro/magneto-encephalography*, EEG/MEG, a sound mathematical framework was targeted, enabling a robot to feature purposeful visio–auditive perception by stabilizing bottom-up perception through top-down cognition. A specific focus was put on crossmodal integration of vision and audition along space and time, the design and development of methods and algorithms to coordinate motor activities and sensor observations, the design and thorough evaluation of testable computational models and on the provision of an experimental testbed.

The following achievements can be mentioned. A two-microphone binocular robotic platform, POPEYE, was built [15]. This highly repeatable system can undergo high velocities and accelerations along 4-DOFs, namely, pan, tilt, and the two camera-independent pan angles. POPEYE allows the use of a dummy head for binaural audition but is not fully bio-mimetic since the binaural axis is higher than the stereovision axis. Novel algorithms for real-time robust localisation of sound sources in multisource environments were proposed, based on a fusion of interaural time difference and pitch cues, using source-fragment methods inspired by glimpsing models of speech perception. Active-listening behaviors were defined that can use planned movement to aid auditory perception, namely, head rotation in order to maintain a tracked source in the auditory fovea, judgement of distance by triangulation, and others. As mentioned in Sect. 3.2, an original approach to detection and localization of multiple speakers from audio–visual observations was also developed and experimented on POPEYE [23].

6 http://perception.inrialpes.fr/POP/

The dataset CAVA, which stands for *computational audio–visual analysis of binaural–binocular recordings* [3] was made freely available for non-profit applications. It was recorded from sensors mounted on a person's head in a large variety of audio–visual scenarios, such as multiple speakers participating in an informal meeting, static/dynamic speakers, presence of acoustic noise, and occluded or turning speakers.

The subsequent HUMAVIPS project, *humanoids with auditory and visual abilities in populated spaces*, runs from 2010 to 2013 and concerns multimodal perception within principled models of Human–robot interaction and humanoid behavior.[7] In this project coordinated audio–visual, motor and communication abilities are targeted, enabling a robot to explore a populated space, localize people therein, assess their status and intentions, and then decide to interact with one or two of them by synthesizing an appropriate behavior and engaging a dialog. Such cocktail-party and other social skills are being implemented on an open-source-software platform and experienced on a fully-programmable humanoid robot.

Open-source datasets have also been disseminated in this framework. Two of them have been recorded with the aforementioned POPEYE system. To investigate audio–motor contingencies from a computational point of view and to experiment with new auditory models and techniques for computational auditory-scene analysis, the CAMIL dataset, *computational audio–motor integration through learning*, provides recordings of various motionless sources, like random spectrum sounds, white noise, speech, music, from a still or moving dummy head equipped with a binaural pair of microphones.[8] Over 100 h of recordings have been elaborated, each of them being annotated with the ground-truth pan-and-tilt motor angles undergone by the robot.

Likewise, to benchmark Human–robot interaction algorithms, the RAVEL corpora, *robots with auditory and visual abilities*, provides synchronized binaural auditory and binocular visual recordings by means of a robocentric stable acquisition device in realistic natural indoor environments.[9] It gathers high-quality audio–visual sequences from two microphone pairs and one camera pair in various kinds of scenarios concerning human-solo- action recognition, identification of gestures addressed to the robot, and human–human as well as Human–robot interaction. The scenes may be affected by several kinds of audio and visual interferences and artifacts. To ease the statement of ground truth, the absolute position and utterances of actors in the scene are also recorded by external cross-calibrated and synchronized devices, namely, a commercial 3-D tracking system and four distributed headset microphones.

[7] http://humavips.inrialpes.fr/

[8] http://perception.inrialpes.fr/~Deleforge/CAMIL_Dataset/index.html

[9] http://ravel.humavips.eu

## 4.5 BINAAHR

BINAAHR, *binaural active audition for humanoid robots*, was established as a french–japanese collaboration focused on two accepted concepts of active (binaural) robot audition.[10] On the one hand, a low-level auditory function is said to be active if it combines, and is improved by, the perception and the motor commands of the sensor. On the other hand, a high-level interaction is active if it is bidirectional and involves the robot and multiple parties. The project has contributed to the design of artificial binaural systems, active binaural localization (Sect. 5.2), binaural separation of more than two sources, ego-noise cancellation, binaural speaker recognition, audio–visual speech recognition and other significant issues.

A separate line of research in BINAAHR champions robotics, because of its locomotion and multimodal-sensing capabilities, as a privileged context to investigate *psychology-of-perception* theories of *active human perception*, that is, theories that hypothesize an interweave of human perception and action. In this context, some innovative developments have been tightly connected with the sensorimotor contingency theory [65, 66]. These developments enable the analysis of the sensorimotor flow of a naive agent, be it endowed with hearing only or with both vision and audition, in order to characterize the dimension of the rigid group of the space underlying its input–output relationship, that is, the dimension of its physical space—see Sect. 5.1. Experimental issues concern unitary testing of low-level functions on binaural prototypes as well as the integration of functions on a HRP-2 humanoid robot.

## 4.6 Further Hardware

Further specific hardware suited to the needs of robot audition has been developed with the aim of pushing forward the integration of auditory functions on embeddable autonomous sensors. Corresponding achievements have mainly been oriented towards array processing, partly because off-the-shelf multichannel data-acquisition devices are often unsatisfactory because of limited embeddability, and high cost due to a too high genericity and other reasons. Although suitable commercial stereo devices can be used for binaural acquisition, array processing-oriented hardware may still constitute an inspiration, for example, for computational issues.

The active direction-pass filter (Sect. 3.2, [57]) was integrated in a dedicated reconfigurable processor and could separate a mixture of sounds in real time with good accuracy [43]. A more generic low-cost low-consumption sound card was developed in order to fit the requirements of the ManyEars project, which features an array based source localization, tracking and separation system [52]. This board, named 8-Sounds-USB, performs an eight-channel synchronous audio capture and embeds computational power. Likewise, the *embedded audition for robotics*, EAR, sensor,

[10] http://projects.laas.fr/BINAAHR

based on a fully programmable eight-channel data-acquisition board and a powerful FPGA, has recently been revisited within a *system-on-a-programmable-chip* approach [47], namely, a C/C++ compatible soft-processor has been implemented on the FPGA, together with dedicated hardwired modules such as co-processing units, memory controllers, communication and data acquisition modules. A release suited to binaural audition is under development. Several complex and intensive operations will be hardwired. The device will process data sensed by MEMS microphones and be compatible with standard audio interfaces.

### 4.7 Conclusion

While robot audition is a fairly recent field of research, various solutions have been proposed to cope with the constraints of robotics. In the above, most low- or high-level functions have been reviewed independently, yet many contributions have been considering them jointly—see for instance Ref. [23] in Sect. 3.2 for joint source localization and separation. The same holds for higher-level auditory functions. For instance, it was shown in [92] that the design of a recognition system should take into account a trade-off between the recognition rates and the sensitivity to speaker locations. Last, the order of the successive computation steps involved in a complex auditory task can differ from one author to another. To conclude, no universal strategy is available at this time and the optimal architecture for a CASA system in robotics is still a matter of debate. Some related activities were conducted in Europe, for example, in the context of the research cluster CoTeSys, *cognition for technical systems*. For instance, a multi-modal robotics head, Mask-Bot, was built [67] to feature face animation, speech communication and basic sound localization. A generic comprehensive model of binaural listening that could also be of high interest for robotics is proposed in [10], this volume.

## 5 Active Approaches to Perception: Applications to Binaural Audition in Robotics

Theoretical approaches to perception are many, and some of them show significant divergences. For instance, Marr's celebrated computational approach to visual perception [48], which prevailed in the development of *artificial intelligence*, proposes a viewpoint of passive perception where the representation is predominant and behavioral aspects are overlooked. Nowadays, it is still a debate whether this conception should be traded off for another theory, namely, one that hypothesizes perception and action to be interweaved. The latter viewpoint is usually related to Gibson's theory, which puts forward the active and exploratory nature of perception [24]. Such considerations also apply to robotics, which has for long considered perception as a bottom-up process in the sense that action are results of sensory analysis.

This historical viewpoint on perception is currently being questioned experimentally, all the more since the exploratory abilities of robotics platforms can be exploited to improve analysis and understanding of the environment. In this context, the current section gathers two original contributions of the authors with regard to binaural auditive perception. Both entail an active behavior of the robotic platform but along distinct approaches.

In a first subsection, an active strategy for *auditory-space learning* is proposed together with its application to sound source localization. It relies on a general theoretical approach to perception, grounded in sensorimotor theory. In a second subsection, a stochastic-filtering strategy for *active binaural sound localization* is introduced, where it is shown how the motor commands of a moving binaural sensor can be fused with the auditive perception to actively localize a still or moving intermittent sound source in the presence of false measurements.

## 5.1 Active Hearing for Auditory-Space Learning and Sound-Source Localization

Action in robotics is usually viewed as a high-level process and is mostly used to address problems that cannot be solved by passive strategies alone, such as front-back disambiguation or distance perception. The method proposed here investigates an alternative paradigm, where the action is envisaged at the same level as perception. In this framework, action and perception interact so as to build an internal representation of the auditory space. As a first step, an active hearing process is used during the learning of an auditory-motor map. Next, this map is used for a-priori passive sound localization. In what follows the approach is introduced for azimuthal localization by considering a mobile listener endowed with a binaural auditory system and perceiving a single stationary sound source of random azimuth in a $\pm 90^\circ$ range.

### A Sensorimotor Definition of Source Localization

The present method is grounded in the sensorimotor theory [65, 68], claiming that the brain is initially a naive agent that interacts with the world via an unknown set of afferent and efferent connexions, with no a-priori knowledge about its own motor capacities or the space it is immersed in. The agent therefore extracts this knowledge by analyzing the consequences of its own movements on its sensory perceptions, building a sensorimotor representation of its embodying space. Generally speaking, consider that all the environments, motor states and sensations that an agent can experiment are depicted as the respective manifolds, $\mathcal{E}$, $\mathcal{M}$ and $\mathcal{S}$ [66].

A sensory state, $s \in \mathcal{S}$, is given as a function of the current motor and environment states, $m \in \mathcal{M}$ and $e \in \mathcal{E}$, through a sensorimotor law, $\Phi$, so that $s = \Phi(m, e)$. Here $e$ models the scene acoustics and spatial and spectral properties of the sound source, $m$

models the agent's body configuration, whereas $\Phi$ represents the body-environment interactions and neural processing that gives rise to the sensation, $s$. Moreover the sensory space, $\mathcal{S}$, lies on a low-dimensional manifold whose topology is similar to the embodying space and, consequently, the learning of spatial perception becomes the learning of such a manifold. Such a process has been applied to auditory-space learning using non-linear dimensionality-reduction techniques [4, 22]. Nevertheless the knowledge of this auditory space is not sufficient for sound localization—an association of a percept in this space and a spatial location have still to be done.

Classical localization methods express a source location in terms of angle or range in an Euclidean physical space. As the sensorimotor approach directly links perception and action in an internal representation of space, a spatial position is here directly expressed as a motor state and as such does not implies any notion of space [68]. Given a motor space, $\mathcal{M}$, and an environment state, $e \in \mathcal{E}$, the source localization problem can thus be defined as the estimation of the motor state, $\tilde{m}$, as follows:

$$\tilde{m} = \underset{m \in \mathcal{M}}{\operatorname{argmin}} |\Phi(m, e) - \Phi(m_0, e_0)| , \quad (1)$$

where $|.|$ denotes a distance metric and $\Phi(m_0, e_0)$ represents a reference sensory state that has to be approximated. In the case of sound-source localization, $\Phi(m_0, e_0)$ corresponds to a source localized in front of the listener with the head in the rest position, which is the most obvious case of azimuthal localization.

## Evoked Behavior for Active Hearing

An evoked or reflex behavior is a simple *hard-wired* behavior allowing a naive agent to react to a stimulus. Considering active hearing, a simple behavior enabling head-to-source orientation can be implemented from ILD cues in a simple way as follows. Once a sound is perceived, the agent orients its head toward the loudest side while the ILD is non-zero. Once the behavior is completed, that is, when the ILD reaches 0, the head of the agent has arrived at an orientation facing the sound source. Because the source azimuth is in a range of $\pm 90°$ only, the agent is not exposed to front-back ambiguity—although front-back disambiguation has also been addressed in the literature, for instance, Ref. [9]. This active hearing process allows an *a-posteriori* localization, $\tilde{m}$ being given after motion as the difference between the initial and final motor states. Moreover, the agent's final configuration with the source in front of the head corresponds to the reference sensory state, $\Phi(m_0, e_0)$, as introduced in (1). This reference state, initially unknown, is approximated through successive executions of the behavior. The orientation behavior has been successfully demonstrated on a robotic platform and been extended to phonotaxis, that is, allowing for a reactive approach of the robot towards the source [8].

**Autonomous Online Learning of Sound Localization**

The evoked behavior that links the initial sensory state in $\mathcal{S}$ to the final motor state in $\mathcal{M}$, provides the sensorimotor association required for an a-priori passive localization. Figure 2 shows an auditory space representation after the learning of high dimensional ILD cues from 1000 auditory stimuli. Each point, corresponding to a different sensory state, is associated with its localization estimation, $\tilde{m}$, computed after the orientation behavior.

After learning of such an association, it becomes possible to localize new percepts based on neighborhood relationships. Suppose a new stimulus corresponding to a sensory state, $s \in \mathcal{S}$, perceived by the agent. $s$ is firstly projected in the sensory-space representation and, if this projection has close neighbors—$s_1$ in Fig. 2—its corresponding motor state, $\tilde{m}$, is interpolated from the neighborhood, giving a passive localization estimation. If the projection is outlying in an area with no neighbors — $s_2$ in Fig. 2—this sensory state is not yet represented and $\tilde{m}$ can not be estimated passively. In this case the orienting behavior is executed, giving an active estimation of $\tilde{m}$.

Instead of learning an auditory space representation from a database, an iterative process can be used by mixing the dimensionality reduction with the evoked behavior. This allows therefore the representation to be learned online, experience after experience. Thus each new percept which projection is outlying in the representation is learned ($s_2$ in Fig. 2): its related sensory state is added in an updated representation and is associated with its related active estimation of $\tilde{m}$. Reference [9] provide more details on this learning algorithm and the auditory system used. The authors propose a simulated experiment where a mean localization error of about 1° is reached after an online learning of 200 iterations.

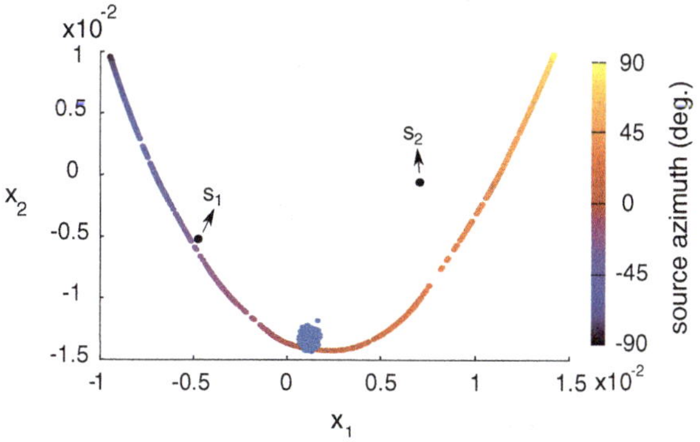


**Fig. 2** Two dimensional manifold of auditory space learned from a set of 1000 sensory states. *Parabolic curve* obtained *before* the orienting behavior—each state corresponding to a sound source of random azimuth in the range of ±90°. *Point cluster close to* $x_1 = 0$: Projections on the manifold of sensory states obtained *after* the orienting behavior, approximating the reference state, $\Phi(m_0, e_0)$. New percepts, such as $s_1$ and $s_2$, can be localized on the manifold as well—see text for details

**Discussion**

The above method has, to be sure, been illustrated for a very simple case, namely, a single stationary sound source in the azimuthal plane. Yet, it seems to be basically suitable for hearing systems in autonomous robotics. In fact, the dimensionality reduction used for the computation of the auditory-space representation allows for unsupervised learning of scene non-linearities, such as reverberation or HRTF filtering. Also, this method requires almost no a-priori knowledge of either the agent or the environment. It mainly depends on the knowledge of the auditory-space representation dimension, typically 2-D or 3-D, and on a dimension-reduction technique robust enough to estimate the non-linear embedding of complex environments in an efficient hard-wired evoked behavior. Active hearing, binaural processing, representation learning and online estimation have the potential to be integrated into a single model that could be applied to more complex problems, thus opening new perspectives for sensorimotor approach to binaural robot audition.

## 5.2 *A Stochastic-Filtering Strategy for Active Binaural Sound Localization*

While the above approach aims at estimating the source position from a binaural sensor with no assumption regarding the environment, this chapter will now be concluded with a strochastic-filtering approach to binaural sound localization from a moving platform. As to this field, reference is due to cite [14, 46], where tracking algorithms based on the particle filtering framework are exploited to detect utterer changes and infer speaker location, respectively. The work presented in this section shows how binaural perception and motor commands of the sensor can be fused to localize an intermittent source in the presence of false measurements.

In the context of binaural audition, sound-source localization relies prominently on *time-delay estimation*, TDE, that is, on an estimation of the arrival-time differences of the sound signals at the two acoustic sensors. The topic of TDE has been widely addressed. In robotics, the most common approach is undoubtedly *generalized cross-correlation*, GCC [38], which consists in cross-correlating truncated and filtered versions of the raw sensed signals and picking the `argmax` of the resulting function. However, given a state vector, $X$, that is, a vector fully characterizing the sensor-to-source relative position, the time delay comes as a nonlinear and noninvertible function, $h$, of $X$. Without any additional information it is not possible to recover the complete state vector from just a time-delay estimate. For instance, consider for simplification, that source and microphones lie on a common plane parallel to the ground, and let the Cartesian coordinates vector, $X = (x, y)^T$, represent the source position in a frame, $(R, e_X, e_Y)$, rigidly linked to the microphone pair, $\{R_l, R_r\}$, with $R_l R = \frac{R_l R_r}{2}$ and $e_Y = \frac{RR_l}{|RR_l|}$. It can be shown that given a time delay, $\tau$, all the pairs, $(x, y)$, satisfying $h(x, y) = \tau$, describe a branch of hyperbola, referred in the literature as cone of confusion. In other words, given a time delay, one cannot locate the true sound source on the associated hyperbola branch. However, with the microphones being mounted on a mobile robot, its motor commands, for instance, translational and rotational velocities, can be fused with audio perception to infer sound localization. Similarly, when the source is moving, prior knowledge about its dynamics can be used. One way to tackle this problem is to use a *Bayesian filtering framework*. In this context and in the presence of relative motion, $X$ is now considered as a discrete hidden-Markov process, characterized by a dynamic equation of the form

$$X_{[k+1]} = f(X_{[k]}, u_{1[k]}, u_{2[k]}) + W_{[k]}. \tag{2}$$

Therein, $X_{[k]}$ is a random vector describing the process $X$ at time step $k$. $u_{1[k]}$ is a deterministic vector gathering information about the robot's motor commands. $u_{2[k]}$ is a vector composed of the source velocities. $u_{2[k]}$ can be deterministic or random, depending on whether the source motion is fully described beforehand or not. $W_{[k]}$ is an additive random noise accounting for uncertainty in the relative motion. At each time, $k$, the time delay measurement, hereafter referred as $Z_{[k]}$, is a memoryless function of the state vector, according to

$$Z_{[k]} = h(X_{[k]}, u_{1[k]}, u_{2[k]}) + V_{[k]}, \tag{3}$$

with $V_{[k]}$ being an additive noise representing the TDE error. Given an initial probability-density function, pdf, of $p(x_{[0]})$ and a sequence of measurements, $z_{[1:k]} \triangleq z_{[1]}, \ldots, z_{[k]}$, considered as samples of $Z_{[1:k]} \triangleq Z_{[1]}, \ldots, Z_{[k]}$, the optimal Bayesian filter consists in the recursive computation of the posterior state probability-density function, that is, $p(x_{[k]}|Z_{[1:k]} = z_{[1:k]})$. When $f, g$ are nonlinear functions and/or $W, V$ are non-gaussian processes, the optimal filter has no closed-form expression. Approximate solutions are thus needed, such as the *extended/unscented Kalman filter*, EKF/UKF, particle filters, PF, or grid-based methods. Whatever the chosen strategy is, certain issues have to be dealt with, such as

*Modeling* The state space model must be defined in such a way that the state vector gathers a minimal set of parameters. For a still source—or a moving source with known velocities with respect to the world—the state vector can be made up with, for example, its Cartesian coordinates in the sensor frame. If the source is moving at unknown speed, an autonomous equation describing the structure of its motion in the world frame must be introduced, whose initial condition and parameters complete the vector $X$ to be estimated. For the localization of human utterers, typically used models are Langevin processes or random walks [13, 86]. However that may be, the mathematical transcription of the prior knowledge of the source dynamics is of crucial importance.

*Consistency and tuning* Generally, the statistics of the dynamic and measurement noise processes are unknown, so they must be hypothesized. Setting a too-high covariance leads to too pessimistic conclusions, while setting them too low may result in an overconfident filter. In general, the noise statistics are set in an ad-hoc manner. However, inconsistency, that is, overconfidence or underconfidence, can arise independently of the noise statistics. Indeed, the approximation of the true-state posterior density function propagated recursively by the filter can become inaccurate when the nonlinearities are not smooth enough and/or the filtering technique is not suited to the model and its parameters. For instance, the runs of basic particle filtering strategies conducted by the authors on simulated and experimental measurements showed that these estimation strategies are not suited to active binaural localization.

*Initialization* When no prior knowledge about the source location is available, one usually set the initial prior $p(x_{[0]})$ as a *flat prior*, that is, a probability distribution with zero-mean and infinite covariance matrix. However, due to the non-linearities involved in the considered source localization problem, the propagation of widely spread distributions often leads to overconfident conclusions. As a solution, the posterior state pdf can be approximated by a *Gaussian mixture*, GM, (GM) whose hypotheses are recursively propagated using a bank of non-interactive filters [7].

*Time-delay extraction* At each time, $k$, the measurement, $Z_{[k]}$, is obtained from a TDE algorithm using audio data collected over a finite time window. Generally, this time window is of short duration for distinct reasons. First, TDE algorithms rely on the hypothesis that the two windowed signals can be regarded as sample sequences of individually and jointly *wide-sense stationary*, WSS, processes. Individual stationarity implies that the source signal is itself WSS, while speech, for instance, cannot be considered as WSS unless the time window is sufficiently short. Joint stationarity implies that the time delay must be approximately constant over the time window. This is of crucial importance when source and sensor move. Classical TDE algorithms do not provide reliable and meaningful estimates if this hypothesis is not satisfied—unless the time-delay variations are specifically taken into account and compensated in the algorithm, like in [39]. Finally, in an embedded application, a cross-correlation cannot have an unreasonable length, due to finite time and space resources.

Because of the environment noise, the non-stationarity of the source and the short duration of TDE windows, the TDE statistics may change significantly over time, namely, if at a considered instant, the SNR and time-bandwidth product, TBP, of the signals are sufficiently high, then the TDE algorithm outputs an accurate estimate of the genuine time delay. If not, the estimate might be unreliable, that is, drawn from a process with large variance or, in the worst case, uncorrelated with the state [87, 88]. Such spurious measurements must be taken into account in order to prevent filter inconsistency/instability. They can be handled in a *hard manner* with an external decision rule that selects, according to some criteria, such as estimated SNR, the measurements that are to be incorporated into the filter—or in a probabilistic way, for example, by probabilistic data association.

Taking all these considerations into account, a filtering strategy was proposed in [69]. It relies on a multiple hypothesis UKF with probabilistic data association and a *source-activity detection*, SAD, system based on *generalized likelihood-ratio test*, GLRT. The results from an experiment conducted in an acoustically prepared room are shown in Fig. 3. Each subfigure represents a time snapshot, the left figure

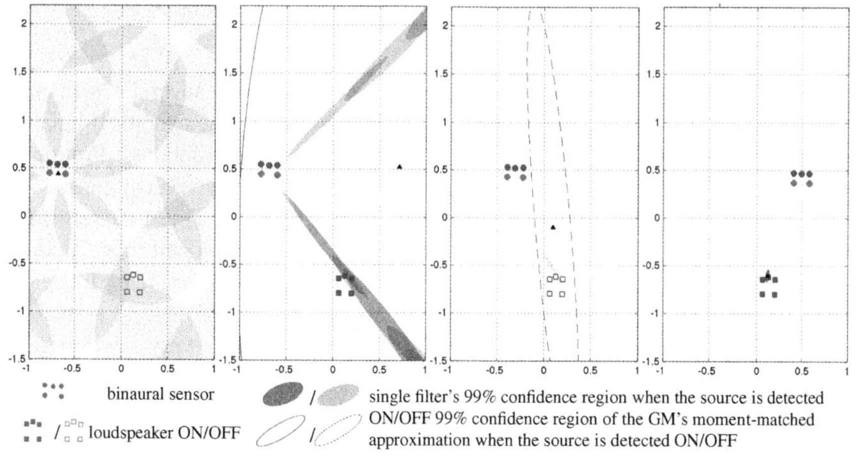


**Fig. 3** Localization results from an experiment conducted in an acoustically prepared room. Each figure represents a time snapshot. *Far left* initial time. *Far right* final time. The results show that the motion of the sensor allowed to disambiguate front and back and to provide information regarding source distance. For more explanations and details see [69]

corresponding to initial time, and the right figure corresponding to final time. At the beginning, the filter is initialized with 24 hypotheses so that the union of the 99 % probability ellipsoids defined from the 24 modes of the initial GM-prior-pdf covers a 4 m-radius circular region around the sensor. In the second snapshot, a part of the hypothesis is spread along the source-to-sensor direction, while another part is spread along the symmetric direction with respect to the $(R_l R_r)$-axis. This behavior depicts that so far there is a large uncertainty on the distance to the source and there is a front-back ambiguity. This originates from the aforementioned time-delay characteristics. In the third snapshot, the loudspeaker is switched off, and the transition from on to off has been detected by the filter. In the fourth snapshot, the loudspeaker is emitting again, and the transition from off to on has been detected correctly. Note that the state pdf is now very sharp around the true-source location. In other words, thanks to motion, front and back have been disambiguated and the distance uncertainty has significantly decreased. The experimental results show that the standard deviation of the errors at the end of the listener's motion is about $\pm 2°$ in terms of azimuth and $\pm 5$ cm in terms of range. However, the good performance of the system is partially due to the favorable experimental conditions, namely, the sensor speeds were precisely known, and the acoustic environment was particularly clean—that is, with only little reverberation and noise. Experiments should be performed in a more realistic environment such as an office, with possibly non-negligible background noise. This is subject to future work.

# 6 Conclusion

This chapter has discussed binaural systems in robotics along several dimensions. After the statement of key constraints raised by this context, the canonical binaural functions underlying the analysis of any auditory scene were detailed. Prominent platforms and research projects were then reviewed. Finally, two recent approaches to binaural audition developed by the authors were presented. These are termed to be *active* because they consider the coupling of binaural sensing and robot motion.

As mentioned before, binaural audition is an attractive paradigm, regarding engineering issues, cross-fertilization between robotics and neurosciences, as well as Human–robot interaction. Though the field is studied by only very few laboratories as compared to binocular vision, it has now reached a certain level of maturity. But there remains ample space for methodological and technological contributions, particularly, to the end of better coping with uncontrolled and dynamic environments. This could then allow to better understand how to use binaural audition as a mechanism for acoustic-scene analysis in general.

To conclude, some broader research areas connected to binaural audition have been mentioned that hopefully bring new researchers to the field. First, the coupling of bottom-up and top-down processes in active audition and, to a larger extent, to active perception, deserves attention. As shown above, two distinct viewpoints have been developed towards purposeful auditory perception. One addresses the definition of top-down feedback from symbolic levels, while the other approaches—including those developed by the current authors—have addressed this topic right at the sensorimotor level. In the authors' opinion, these two lines of research are to be reinforced and must join each other in order to define a comprehensive computational architecture for active analysis of auditory scenes. Some subtopics should finally be mentioned, such as, the definition of binaural audition based control/estimation strategies that explicitly include an exploration goal to collect information about the source location, their interlinking with decision processes, the assimilation of available data over space and time, the generality of such an approach and its ability to tackle multimodality, adaptive approaches to binaural cues extraction and exploitation where auditory cues and algorithms are dynamically changed according to the context, just to name a few. A second fruitful broad research area is the involvement of binaural audition in ubiquitous robotics. The idea here is to outfit rooms with embedded sensors, such as microphone arrays, cameras and/or RFID antennas. These areas would be shared by humans and robots interacting with each other, using binaural audition, vision and maybe, further available modalities. The mobility of the robots enables the possibility of combining their motor commands with binaural perception, not only to improve a *local* binaural function, but also to dynamically reconfigure the global network constituted by the microphone arrays and the binaural heads—possibly including prior knowledge. Some contributions have already been developed in this field on the basis of dynamically reconfigurable microphone arrays [49]. They could constitute a valuable source of inspiration for enhancing binaural robot audition.

**Acknowledgments** This work was conducted within the project *binaural active audition for humanoid robots*, BINAAHR, funded under contract # ANR-09-BLAN-0370-02 by ANR, France, and JST, Japan. The authors would like to thank two anonymous reviewers for valuable suggestions.

## References

1. J. Aloimonos, I. Weiss, and A. Bandyopadhyay. Active vision. *Intl. J. Computer Vision*, 1:333–356, 1988.
2. S. Argentieri and P. Danès. Broadband variations of the MUSIC high-resolution method for sound source localization in robotics. In *IEEE/RSJ Intl. Conf. Intelligent Robots and Systems*, IROS'2007, pages 2009–2014, 2007.
3. E. Arnaud, H. Christensen, Y.-C. Lu, J. Barker, V. Khalidov, M. Hansard, B. Holveck, H. Mathieu, R. Narasimha, E. Taillant, F. Forbes, and R. Horaud. The CAVA corpus: Synchronised stereoscopic and binaural datasets with head movements. In *ACM/IEEE Intl. Conf. Multimodal, Interfaces*, ICMI'08, 2008.
4. M. Aytekin, C. Moss, and J. Simon. A sensorimotor approach to sound localization. *Neural Computation*, 20:603–635, 2008.
5. P. Azad, T. Gockel, R. Dillmann. *Computer Vision: Principles and Practice*. Elektor, Electronics, 2008.
6. R. Bajcsy. Active perception. *Proc. of the IEEE*, 76:966–1005, 1988.
7. Y. Bar-Shalom and X. Li. *Estimation and Tracking: Principles, Techniques and Software*. Artech House, 1993.
8. M. Bernard, S. N'Guyen, P. Pirim, B. Gas, and J.-A. Meyer. Phonotaxis behavior in the artificial rat Psikharpax. In *Intl. Symp. Robotics and Intelligent Sensors*, IRIS'2010, pages 118–122, Nagoya, Japan, 2010.
9. M. Bernard, P. Pirim, A. de Cheveigné, and B. Gas. Sensorimotor learning of sound localization from an auditory evoked behavior. In IEEE *Intl. Conf. Robotics and Automation*, ICRA'2012, pages 91–96, St. Paul, MN, 2012.
10. J. Blauert, D. Kolossa, K. Obermayer, and K. Adiloglu. Further challenges and the road ahead. In J. Blauert, editor, *The technology of binaural listening*, chapter 18. Springer, Berlin-Heidelberg-New York NY, 2013.
11. W. Brimijoin, D. Mc Shefferty, and M. Akeroyd. Undirected head movements of listeners with asymmetrical hearing impairment during a speech-in-noise task.*Hearing Research*, 283:162–8, 2012.
12. R. Brooks, C. Breazeal, N. Marjanović, B. Scassellati, and M. Williamson. The Cog project: Building a humanoid robot. In C. Nehaniv, editor, *Computations for Metaphors, Analogy, and Agents*, volume 1562 of LNCS, pages 52–87. Springer, 1999.
13. Y. Chen and Y. Rui. Real-time speaker tracking using particle filter sensor fusion. *Proc. of the IEEE*, 920:485–494, 2004.
14. H. Christensen and J. Barker. Using location cues to track speaker changes from mobile binaural microphones. In *Interspeech 2009*, Brighton, UK, 2009.
15. H. Christensen, J. Barker, Y.-C. Lu, J. Xavier, R. Caseiro, and H. Arafajo. POPeye: Real-time binaural sound-source localisation on an audio-visual robot head. In *Conf. Natural Computing and Intelligent Robotics*, 2009.
16. Computing Community Consortium. *A roadmap for US robotics. From Internet to Robotics*, 2009. http://www.us-robotics.us/reports/CCC%20Report.pdf.
17. M. Cooke, Y. Lu, Y. Lu, and R. Horaud. Active hearing, active speaking. In *Intl. Symp.* Auditory and Audiological Res., 2007.
18. M. Cooke, A. Morris, and P. Green. Recognizing occluded speech. In *Proceedings of the ESCA Tutorial and Res.arch Worksh. Auditory Basis of Speech Perception*, pages 297–300, Keele University, United Kingdom, 1996.

19. M. Cooke, A. Morris, and P. Green. Missing data techniques for robust speech recognition. In *Intl. Conf. Acoustics, Speech, and Signal Processing*, ICASSP'1997, pages 863–866, Munich, Germany, 1997.
20. B. Cornelis, M. Moonen, and J. Wouters. Binaural voice activity detection for MWF-based noise reduction in binaural hearing aids. In *European Signal Processing Conf.*, EUSIPCO'2011, pages Barcelona, Spain, 2011.
21. P. Danès and J. Bonnal. Information-theoretic detection of broadband sources in a coherent beamspace MUSIC scheme. In *IEEE/RSJ Intl. Conf. Intell. Robots and Systems*, IROS'2010, pages 1976–1981, Taipei, Taiwan, 2010.
22. A. Deleforge and R. Horaud. Learning the direction of a sound source using head motions and spectral features. Technical Report 7529, INRIA, 2011.
23. A. Deleforge and R. Horaud. The Cocktail-Party robot: Sound source separation and localisation with an active binaural head. In *IEEE/ACM Intl. Conf. Human Robot Interaction*, HRI'2012, Boston, MA, 2012.
24. J. Gibson. *The Ecological Approach to Visual Perception*. Erlbaum, 1982.
25. M. Giuliani, C. Lenz, T. Müller, M. Rickert, and A. Knoll. Design principles for safety in human-robot interaction. *Intl. J. Social Robotics*, 2:253–274, 2010.
26. A. Handzel, S. Andersson, M. Gebremichael, and P. Krishnaprasad. A biomimetic apparatus for sound-source localization. In *IEEE Conf. Decision and Control*, CDC'2003, volume 6, pages 5879–5884, Maui, HI, 2003.
27. A. Handzel and P. Krishnaprasad. Biomimetic sound-source localization. *IEEE Sensors J.*, 2:607–616, 2002.
28. S. Hashimoto, S. Narita, H. Kasahara, A. Takanishi, S. Sugano, K. Shirai, T. Kobayashi, H. Takanobu, T. Kurata, K. Fujiwara, T. Matsuno, T. Kawasaki, K. Hoashi. Humanoid robot-development of an information assistant robot, Hadaly. In *IEEE Intl. Worksh. Robot and Human, Communication*, RO-MAN'1997, pages 106–111, 1997.
29. J. Hörnstein, M. Lopes, J. Santos-victor, and F. Lacerda. Sound localization for humanoid robots - building audio-motor maps based on the HRTF. In *IEEE/RSJ Intl. Conf. Intelligent Robots and Systems*, IROS'2006, pages 1170–1176, Beijing, China, 2006.
30. J. Huang, T. Supaongprapa, I. Terakura, F. Wang, N. Ohnishi, and N. Sugie. A model-based sound localization system and its application to robot navigation. *Robotics and Autonomous Syst.*, 270:199–209, 1999.
31. G. Ince, K. Nakadai, T. Rodemann, Y. Hasegawa, H. Tsujino, and J. Imura. Ego noise suppression of a robot using template subtraction. In *IEEE/RSJ Intl. Conf. Intelligent Robots and Systems*, IROS'2009, pages 199–204, Saint Louis, MO, 2009.
32. G. Ince, K. Nakadai, T. Rodemann, J. Imura, K. Nakamura, and H. Nakajima. Incremental learning for ego noise estimation of a robot. In*IEEE/RSJ Intl. Conf. Intelligent Robots and Systems*, IROS'2011, pages 131–136, San Francisco, CA, 2011.
33. G. Ince, K. Nakadai, T. Rodemann, H. Tsujino, and J. Imura. Multi-talker speech recognition under ego-motion noise using missing feature theory. In *IEEE/RSJ Intl. Conf. Intelligent Robots and Systems*, IROS'2010, pages 982–987, Taipei, Taiwan, 2010.
34. R. Irie. Multimodal sensory integration for localization in a humanoid robot. In *IJCAI Worksh. Computational Auditory Scene Analysis*, pages 54–58, Nagoya, Aichi, Japan, 1997.
35. A. Ito, T. Kanayama, M. Suzuki, and S. Makino. Internal noise suppression for speech recognition by small robots. In *Interspeech*'2005, pages 2685–2688, Lisbon, Portugal, 2005.
36. M. Ji, S. Kim, H. Kim, K. Kwak, and Y. Cho. Reliable speaker identification using multiple microphones in ubiquitous robot companion environment. In *IEEE Intl. Conf. Robot & Human Interactive Communication*, RO-MAN'2007, pages 673–677, Jeju Island, Korea, 2007.
37. H.-D. Kim, J. Kim, K. Komatani, T. Ogata, and H. Okuno. Target speech detection and separation for humanoid robots in sparse dialogue with noisy home environments. In *IEEE/RSJ Intl. Conf. Intelligent Robots and Systems*, IROS'2008, pages 1705–1711, Nice, France, 2008.
38. C. Knapp and G. Carter. The generalized correlation method for estimation of time delay. *IEEE Trans. Acoustics, Speech and, Signal Processing*, 24:320–327, 1976.

39. C. Knapp and G. Carter. Time delay estimation in the presence of relative motion. In *IEEE Intl. Conf. Acoustics, Speech, and Signal Processing*, ICASSP'1977, pages 280–283, Storrs, CT, 1977.
40. Y. Kubota, M. Yoshida, K. Komatani, T. Ogata, and H. Okuno. Design and implementation of a 3D auditory scene visualizer: Towards auditory awareness with face tracking. In *IEEE Intl. Symp. Multimedia*, ISM'2008, pages 468–476, Berkeley, CA, 2008.
41. M. Kumon and Y. Noda. Active soft pinnae for robots. In *IEEE/RSJ Intl. Conf. Intelligent Robots and Systems*, IROS'2011, pages 112–117, San Francisco, CA, 2011.
42. M. Kumon, R. Shimoda, and Z. Iwai. Audio servo for robotic systems with pinnae. In *IEEE/RSJ Intl. Conf. Intelligent Robots and Systems*, IROS'2005, pages 885–890, Edmonton, Canada, 2005.
43. S. Kurotaki, N. Suzuki, K. Nakadai, H. Okuno, and H. Amano. Implementation of active direction-pass filter on dynamically reconfigurable processor. In *IEEE/RSJ Intl. Conf. Intelligent Robots and Systems*, IROS'2005, pages 3175–3180, Edmonton, Canada, 2005.
44. Q. Lin, E. E. Jan, and J. Flanagan. Microphone arrays and speaker identification. *IEEE Trans. Speech and Audio Processing*, 2:622–629, 1994.
45. R. Lippmann and B. A. Carlson. Using missing feature theory to actively select features for robust speech recognition with interruptions, filtering, and noise. In *Eurospeech*'1997, pages 863–866, Rhodos, Greece, 1997.
46. Y.-C. Lu and M. Cooke. Motion strategies for binaural localisation of speech sources in azimuth and distance by artificial listeners. *Speech Comm.*, 53:622–642, 2011.
47. V. Lunati, J. Manhès, and P. Danès. A versatile system-on-a-programmable-chip for array processing and binaural robot audition. In*IEEE/RSJ Intl. Conf. Intelligent Robots and Systems*, IROS'2012, pages 998–1003, Vilamoura, Portugal, 2012.
48. D. Marr. Vision: A Computational Investigation into the Human Representation and Processing of *Visual Information*. Feeeman, W.H., 1982.
49. E. Martinson and B. Fransen. Dynamically reconfigurable microphone arrays. In *IEEE Intl. Conf. Robotics and Automation*, ICRA'2011, pages 5636–5641, Shangai, China, 2011.
50. Y. Matsusaka, T. Tojo, S. Kubota, K. Furukawa, D. Tamiya, K. Hayata, Y. Nakano, and T. Kobayashi. Multi-person conversation via multi-modal interface - a robot who communicate with multi-user -. In *Eurospeech*'1999, pages 1723–1726, Budapest, Hungary, 1999.
51. T. May, S. van de Par, and A. Kohlrausch. Binaural localization and detection of speakers in complex acoustic scenes. In J. Blauert, editor, *The Technology of Binaural Listening*, chapter 15. Springer, Berlin-Heidelberg-New York NY, 2013.
52. F. Michaud, C. Côté, D. Létourneau, Y. Brosseau, J.-M. Valin, E. Beaudry, C. Raïevsky, A. Ponchon, P. Moisan, P. Lepage, Y. Morin, F. Gagnon, P. Giguère, M.-A. Roux, S. Caron, P. Frenette, and F. Kabanza. Spartacus attending the 2005 AAAI conference.*Autonomous Robots*, 22:369–383, 2007.
53. K. Nakadai, T. Lourens, H. Okuno, and H. Kitano. Active audition for humanoids. In *Nat. Conf. Artificial Intelligence*, AAAI-2000, pages 832–839, Austin, TX, 2000.
54. K. Nakadai, T. Matsui, H. Okuno, and H. Kitano. Active audition system and humanoid exterior design. In *IEEE/RSJ Intl. Conf. Intelligent Robots and Systems*, IROS'2000, pages 1453–1461, Takamatsu, Japan, 2000.
55. K. Nakadai, D. Matsuura, H. Okuno, and H. Kitano. Applying scattering theory to robot audition system: Robust sound source localization and extraction. In *IEEE/RSJ Intl. Conf. Intelligent Robots and Systems*, IROS'2003, pages 1147–1152, Las Vegas, NV, 2003.
56. K. Nakadai, H. Okuno, and H. Kitano. Epipolar geometry based sound localization and extraction for humanoid audition. In *IEEE/RSJ Intl. Conf. Intelligent Robots and Systems*, IROS'2001, volume 3, pages 1395–1401, Maui, HI, 2001.
57. K. Nakadai, H. Okuno, and H. Kitano. Auditory fovea based speech separation and its application to dialog system. In *IEEE/RSJ Intl. Conf. on Intelligent Robots and Systems*, IROS'2002, volume 2, pages 1320–1325, Lausanne, Switzerland, 2002.
58. K. Nakadai, H. Okuno, and H. Kitano. Robot recognizes three simultaneous speech by active audition. In *IEEE Intl. Conf. Robotics and Automation*, ICRA'2003, volume 1, pages 398–405, Taipei, Taiwan, 2003.

59. H. Nakajima, K. Kikuchi, T. Daigo, Y. Kaneda, K. Nakadai, and Y. Hasegawa. Real-time sound source orientation estimation using a 96 channel microphone array. In *IEEE/RSJ Intl. Conf. Intelligent Robots and Systems*, IROS'2009, pages 676–683, Saint Louis, MO, 2009.
60. H. Nakashima and T. Mukai. 3D sound source localization system based on learning of binaural hearing. In *IEEE Intl. Conf. Systems*, Man and Cybernetics, SMC'2005, pages 3534–3539, Nagoya, Japan, 2005.
61. E. Nemer, R. Goubran, and S. Mahmoud. Robust voice activity detection using higher-order statistics in the LPC residual domain. *IEEE Trans. Speech and Audio Processing*, 9:217–231, 2001.
62. Y. Nishimura, M. Nakano, K. Nakadai, H. Tsujino, and M. Ishizuka. Speech recognition for a robot under its motor noises by selective application of missing feature theory and MLLR. In *ISCA Tutorial and Research Worksh. Statistical and Perceptual Audition*, Pittsburgh, PA, 2006.
63. H. Okuno, T. Ogata, K. Komatani, and K. Nakadai. Computational auditory scene analysis and its application to robot audition. In *IEEE Intl. Conf. Informatics Res. for Development of Knowledge Society Infrastructure*, ICKS'2004, pages 73–80, 2004.
64. J. O'Regan. How to build a robot that is conscious and feels. *Minds and Machines*, pages 117–136, 2012.
65. J. O'Regan and A. Noë. A sensorimotor account of vision and visual consciousness. *Behavioral and brain sciences*, 24:939–1031, 2001.
66. D. Philipona and J. K. O'Regan. Is there something out there? inferring space from sensorimotor dependencies. *Neural Computation*, 15:2029–2049, 2001.
67. B. Pierce, T. Kuratate, A. Maejima, S. Morishima, Y. Matsusaka, M. Durkovic, K. Diepold, and G. Cheng. Development of an integrated multi-modal communication robotic face. In *IEEE Worksh. Advanced Robotics and its Social Impacts*, RSO'2012, pages 101–102, Munich, Germany, 2012.
68. H. Poincaré. L'espace et la géométrie. *Revue de Métaphysique et de Morale*, pages 631–646, 1895.
69. A. Portello, P. Danès, and S. Argentieri. Active binaural localization of intermittent moving sources in the presence of false meaurements. In *IEEE/RSJ Intl. Conf. Intelligent Robots and Systems*, IROS'2012, pages 3294–3299, Vilamoura, Portugal, 2012.
70. R. Prasad, H. Saruwatari, and K. Shikano. Enhancement of speech signals separated from their convolutive mixture by FDICA algorithm. *Digital Signal Processing*, 19:127–133, 2009.
71. L. Rabiner and M. Sambur. An algorithm for determining the endpoints of isolated utterances. *The Bell System Techn. J.*, 54:297–315, 1975.
72. B. Raj, R. Singh, and R. Stern. Inference of missing spectrographic features for robust speech recognition. In *Intl. Conf. Spoken Language Processing*, Sydney, Australia, 1998.
73. B. Raj and R. M. Stern. Missing-feature approaches in speech recognition. *IEEE Signal Processing Mag.*, 22:101–116, 2005.
74. T. Rodemann. A study on distance estimation in binaural sound localization. In *IEEE/RSJ Intl. Conf. Intelligent Robots and Systems*, IROS'2010, pages 425–430, Taipei, Taiwan, 2010.
75. D. Rosenthal and H. Okuno, editors. *Computational Auditory Scene Analysis*. Lawrence Erlbaum Associates, 1997.
76. A. Saxena and A. Ng. Learning sound location from a single microphone. In *IEEE Intl. Conf. Robotics and Automation*, ICRA'2009, pages 1737–1742, Kobe, Japan, 2009.
77. S. Schulz and T. Herfet. Humanoid separation of speech sources in reverberant environments. In *Intl. Symp. Communications*, Control and Signal Processing, ISCCSP'2008, pages 377–382, Brownsville, TX, 2008.
78. M. L. Seltzer, B. Raj, and R. Stern. A Bayesian classifier for spectrographic mask estimation for missing feature speech recognition. *Speech Comm.*, 43:379–393, 2004.
79. A. Skaf and P. Danès. Optimal positioning of a binaural sensor on a humanoid head for sound source localization. In *IEEE Intl. Conf. Humanoid Robots*, Humanoids'2011, pages 165–170, Bled, Slovenia, 2011.
80. D. Sodoyer, B. Rivet, L. Girin, C. Savariaux, J.-L. Schwartz, and C. Jutten. A study of lip movements during spontaneous dialog and its application to voice activity detection. *J. Acoust. Soc. Am.*, 125:1184–1196, 2009.

81. M. Stamm and M. Altinsoy. Employing binaural-proprioceptive interaction in human machine interfaces. In J. Blauert, editor, *The technology of binaural listening*, chapter 17. Springer, Berlin-Heidelberg-New York NY, 2013.
82. R. Takeda, S. Yamamoto, K. Komatani, T. Ogata, and H. Okuno. Missing-feature based speech recognition for two simultaneous speech signals separated by ICA with a pair of humanoid ears. In *IEEE/RSJ Intl. Conf. Intelligent Robots and Systems*, IROS'2006, pages 878–885, Beijing, China, 2006.
83. K. Tanaka, M. Abe, and S. Ando. A novel mechanical cochlea "fishbone" with dual sensor/actuator characteristics. *IEEE/ASME Trans. Mechatronics*, 3:98–105, 1998.
84. J. Valin, J. Rouat, and F. Michaud. Enhanced robot audition based on microphone array source separation with post-filter. In *IEEE/RSJ Intl. Conf. Intelligent Robots and Systems*, IROS'2004, pages 2123–2128, Sendai, Japan, 2004.
85. H. Van Trees. *Optimum Array Processing (Detection, Estimation, and Modulation Theory, Part IV)*. Wiley-Interscience, 2002.
86. D. Ward, E. Lehmann, and R. Williamson. Particle filtering algorithms for tracking an acoustic source in a reverberant environment. *IEEE Trans. Speech and Audio Processing*, 11:826–836, 2003.
87. E. Weinstein and A. Weiss. Fundamental limitations in passive time delay estimation - Part II: Wideband systems. *IEEE Trans. Acoustics, Speech and Signal Processing*, pages 1064–1078, 1984.
88. A. Weiss and E. Weinstein. Fundamental limitations in passive time delay estimation - Part I: Narrowband systems. *IEEE Trans. Acoustics, Speech and Signal Processing*, pages 472–486, 1983.
89. R. Weiss, M. Mandel, and D. Ellis. Combining localization cues and source model constraints for binaural source separation. *Speech Comm.*, 53:606–621, 2011.
90. R. Woodworth and H. Schlosberg. *Experimental Psychology*. Holt, Rinehart and Winston, 3rd edition, 1971.
91. T. Yoshida and K. Nakadai. Two-layered audio-visual speech recognition for robots in noisy environments. In *IEEE/RSJ Intl. Conf. Intelligent Robots and Systems*, IROS'2010, pages 988–993, 2010.
92. K. Youssef, S. Argentieri, and J. Zarader. From monaural to binaural speaker recognition for humanoid robots. In *IEEE/RAS Intl. Conf. Humanoid Robots*, Humanoids'2010, pages 580–586, Nashville, TN, 2010.
93. K. Youssef, S. Argentieri, and J.-L. Zarader. A binaural sound source localization method using auditive cues and vision. In *IEEE Intl. Conf. Acoustics, Speech and Signal Processing*, ICASSP'2012, pages 217–220, Kyoto, Japan, 2012.
94. K. Youssef, S. Argentieri, and J.-L. Zarader. Towards a systematic study of binaural cues. In *IEEE/RSJ Intl. Conf. Intelligent Robots and Systems*, IROS'2012, pages 1004–1009, Vilamoura, Portugal, 2012.
95. K. Youssef, B. Breteau, S. Argentieri, J.-L. Zarader, and Z. Wang. Approaches for automatic speaker recognition in a binaural humanoid context. In *Eur. Symp. Artificial Neural Networks*, Computational Intelligence and Machine Learning, ESANN'2011, pages 411–416, Bruges, Belgium, 2011.

# Binaural Assessment of Multichannel Reproduction

**H. Wierstorf, A. Raake and S. Spors**

## 1 Introduction

Sound reproduction systems have evolved over the years in the direction of including more and more loudspeakers. The goal is to create a sense of auditory *immersion* in the listeners. A first binaural transmission of a concert via telephone was demonstrated in 1881 [9]. Later followed a proposal by Snow and Steinberg [29], aiming at the transmission of entire sound fields. The basic idea of these authors was that a sound field could be captured by an array of microphones and, consequently, be reproduced by replacing the microphones with loudspeakers that are fed with the signals picked up by the microphones. The loudspeaker signals then superimpose and, together, recreate the sound field in a similar manner as described by Huygen's principle [15]. If the number of independent microphones and loudspeakers is restricted to two each, the sound field is recreated correctly only in one specific location, the so-called *sweet spot*. This recording-and-reproduction technique known as stereophony, is still the prominent spatial-audio technique. This is due to its technical simplicity, the wide use of respective audio-mastering chains and the convincing perceptual results. The latter are mainly due to inherent properties of the human auditory system as have extensively been investigated in psychoacoustics—for example, [2].

In parallel to the continuing success of stereophony, the old idea as proposed by Steinberg and Snow has been revisited in recent years. Todays technology allows for the use of several hundreds of loudspeakers, such enabling the reproduction or synthesis of extended sound fields. These novel techniques are termed *sound-field synthesis*, which includes methods like *higher-order ambisonics*, HOA, and *wave-field*

A. Raake (✉) · H. Wierstorf
Assessment of IP-based Applications, Telekom Innovation Laboratories (T-Labs), Technische Universität Berlin, Berlin, Germany
e-mail: alexander.raake@telekom.de

S. Spor
Signal Theory and Digital Signal Processing, Institute of Communications Engineering, Universität Rostock, Rostock, Germany

J. Blauert (ed.), *The Technology of Binaural Listening*, Modern Acoustics and Signal Processing, DOI: 10.1007/978-3-642-37762-4_10,

*synthesis*, WFS. In these methods, the sound field is treated as being spatially-sampled by the loudspeaker array. As these methods target the recreation of a sound field with frequencies up to 20 kHz, the highest audible frequency, loudspeaker spacings below 1 cm are theoretically required to avoid spatial aliasing, which is still impossible. To get around this problem, research in the field has progressed towards the exploitation of psychoacoustics, to the end of synthesizing sound fields with inaudible perceptual errors as compared to natural hearing.

This chapter presents work that investigated how errors in a given sound field, as synthesized with a multichannel loudspeaker array employing the WFS method, influence listeners' localization of virtual sound sources.

Section 2 presents a discussion of what is needed in order to create a convincing reproduction of a given auditory scene, including relevant aspects of the sound-source localization performed by the human auditory system. Then, Sect. 3 provides the required theoretical background regarding WFS and outlines the binaural re-synthesis approach as used here for simulating different WFS setups. In Sect. 4, test results of localization tests with WFS are presented. Finally, in Sect. 5, it is shown how the results of the localization-test can be predicted by means of a model of binaural processing, that is, a so-called *binaural model.*

## 2 Creating a Convincing Auditory Scene

Anyone who deals with sound reproduction should consider how the reproduced scene is perceived by a listener. In the context of this chapter, an auditory scene is considered to consist of different elements, namely, the underlying auditory events that the listener interactively analyzes in terms of the available auditory information and which leads to the formation of auditory objects. According to [17], an auditory event is characterized by its loudness, its pitch, its perceived duration, its timbre, and spaciousness. Here, spaciousness comprises the perceived location and spatial extent associated with the auditory event [2]. Obviously, localization is just one aspect of auditory scenes. In this section, the perception of the entire scene is discussed, specifically addressing the role that localization plays in it.

Usually, in physical terms, the sound field associated to a reproduced sound scene deviates from that of the intended scene. Prior to reproduction, the intended scene is represented in terms of a *recorded* scene in a specific representation format, for example, created from a given recording using a specific set of source models, or modified based on such recordings and models. When such a stored scene representation is provided as input to a given reproduction method, the result is known as a *virtual sound scene*. A sound-reproduction system is required to present acoustic signals to the listeners' ears in such a way that the corresponding auditory scene matches the desired one as closely as possible, that is, the auditory scene as intended by its creator. The focus of the following considerations will be on system properties that enable a perceptually *authentic* or *plausible* reproduction of a sound scene [3]. Here, *authentic* means indiscernible from an explicit or implicit reference, in other words true to the original. *Plausible* means that the perceived features of the reproduced scenes show plausible correspondence with the listener's expectations in the given context, without necessarily being authentic. Perceived features in this context are nameable and quantifiable features of the auditory scene and its elements, such as loudness, timbre, localization, and spatial extent.

The totality of features can be considered as the *character* of the auditory scene—perceived or expected. The expected character is often referred to as internal reference. However, for systems such as used for stereo reproduction, the listening experience itself has led to fixed schemata of perception that are linked with internal references of their own kind. The *quality* of a system as perceived by a listener is considered to be the result of *assessing* perceived features with regard to the desired features, that is, the internal reference. With quality being expressed as unidimensional index, systems can be ranked according to their perceived quality, and quality differences can be measured quantitatively.

No comprehensive overview of sound reproduction technology evaluation is available from the literature. A basic concept for evaluation is described in [23], which focuses mainly on the evaluation of multichannel stereophony-based reproduction, and a auditory-scene-based evaluation paradigm. By evaluating various 5.1 surround setups, it was found that the overall quality is composed of timbral and spatial fidelity. The same shows up in first results for typical WFS setups collected in the current research, namely, the timbral fidelity may be of greater importance for overall quality than spatial fidelity. However, the research on this topic is still at an early stage [38].

Timbral and spatial fidelity are perceptual constructs of multidimensional nature. In order to describe their perceptual dimensions that they are composed of, attribute descriptions have been sought for in different studies, employing verbal description and attribute ratings [6, 14, 39]. Multidimensional analysis methods such as multidimensional scaling or the repertory-grid technique followed by attribute scaling are suitable methods when no a-priory knowledge of the perceptual character of the auditory events associated to the stimuli is available. A relevant example is reported in [13], where so-called focused sources in WFS have been assessed. In the current chapter, a special focus is put on *localization* of sound sources reproduced with WFS. In this context, localization is one of the key features associated with spatial fidelity.

## 2.1 Localization

One basic ability of the human auditory system is the localization of sound. Localization describes the process of assessing the location of auditory events with respect to the positions and other properties of corresponding sound events. Note that the sound events giving rise to one auditory event can be manifold, for example, in the case of classical stereophony the two loudspeakers can create one auditory event at a position located between the two loudspeakers.

For localization, the auditory system evaluates differences between the two ear signals that depend on the position of the sound source. The most prominent cues for localization are interaural time differences, ITDs, and interaural level difference, ILDs [2]. In addition, the auditory system estimates the distance and the vertical position of a sound source by interpreting monaural cues, such as the frequency spectrum resulting from a known type of source signal that is transmitted to the eardrums from a distance or from a specific vertical source position, thus yielding direction-specific filtering due to the shape of the outer ear. However note, that in this chapter, only displacements in the horizontal plane are be addressed and, hence, only horizontally-oriented loudspeaker setups are used. As a consequence, vertical source displacement and distance are not considered in the localization assessment. Consequently, whenever the term localization is used, it refers to angular displacement in the horizontal plane.

For broad-band content, the localization is dominated by ITDs [36] of the spectral components below 1 kHz. Moreover, incident wave fronts arriving within a time-delay of around 1 ms after the first wave front are summed up by the auditory system, a phenomenon known as *summing localization*. Delayed wave fronts arriving later, but no later than approximately 50 ms after the first wave front, have no influence on localization at large, an effect known as the *precedence effect* [20]. Considering the delayed playout of sound by the different loudspeakers of a given multichannel loudspeaker setup, the precedence effect has implications with regard to the perceived directions of virtual sources. These are specific for the particular reproduction method and listening position.

In stereophony, the perceived location of an auditory event is caused by the superposition of the wave fronts coming from the two loudspeakers. For example, at low frequencies level differences between the two loudspeakers transform into a corresponding interaural time difference at a central position between the loudspeakers, the so-called *sweet spot*. For positions outside of sweet spot, the superposition is impaired, and the closest loudspeaker dominates localization. This is visualized in Fig. 1. The arrows in the figure point towards the location of the auditory event that a listener perceives when placed at the position of the arrow. The gray-shades of the arrow indicate the deviation from the intended direction, which is, in this example, given by the virtual source to be located in the middle between the loudspeakers, that is, at the $x$, $y$-point (0, 0) m. The sweet spot is indicated by the position of the listener placed at (0, 2) m. Calculation of the directions of the individual arrows was performed with the binaural model as described in Sect. 5. Figure 1 is provided for illustrative purposes only. For an overview of methods to predict the sweet spot in stereophony see, for example, [22].

In sound-field synthesis, the physical or authentic reproduction of a given sound field is intended. However, due to the limited number of loudspeakers and respective spatial aliasing, this is only possible up to a specific frequency for a given listening position. Due to this limitation, the localization of a virtual source reproduced with WFS depends on the position of the listener, and on the loudspeaker array configuration. For determining the best possible system layout in practise, it will be helpful to provide a model that predicts localization in the synthesis area of WFS, like the presentation as depicted in Fig. 1 for two-channel stereophony. In the further course of this chapter the development of such a tool will be described. Thereby an

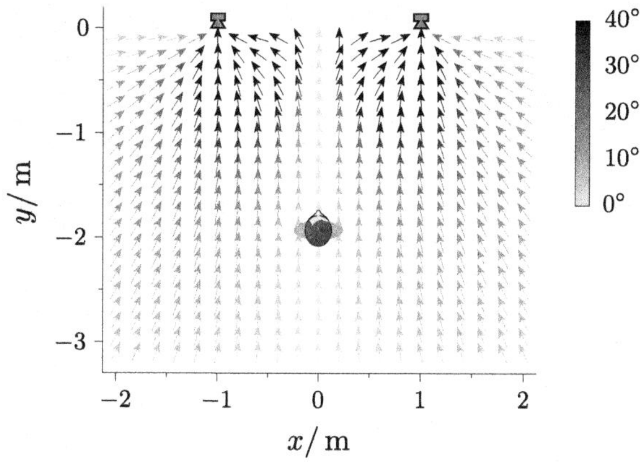


**Fig. 1** Sketch of the *sweet spot* phenomenon in stereophony. The *arrows* point into the direction of where the auditory event of a listener appears, if he/she sits at the position of the *arrow*. Increasing *gray-shades* of the *arrows* indicate the deviation from the intended direction which, in this case, is right in the *middle* between the two loudspeakers

existing binaural model will be modified to produce the output needed here, namely, for an application to localization performance analysis. Yet, to be able to specify such a model, the theory of WFS needs to be shortly revisited.

# 3 Wave-Field Synthesis

*Wave-field synthesis*, WFS, is a sound-field synthesis method that targets physically accurate synthesis of sound fields over an extended listening or synthesis area, respectively. WFS was formulated in the eighties for linear loudspeaker arrays [1]. In the following, a formulation of WFS is presented that is embedded into the more general framework of sound-field synthesis. Furthermore, restrictions regarding a 2-dimensional only loudspeaker setup are discussed, as well as the usage of loudspeakers with a given fixed inter-loudspeaker spacing. At the end of this section, WFS theory is discussed by means of an example.

## 3.1 Physical Fundamentals

The sound pressure, $P(\mathbf{x}, \omega)$, at the position, $\mathbf{x}$, synthesized by a weighted distribution of monopole sources located on the surface, $\partial V$, of an open area, $V \subset \mathbb{R}^3$, is given as the *single-layer potential*

$$P(\mathbf{x}, \omega) = \oint_{\partial V} G(\mathbf{x}|\mathbf{x}_0, \omega) D(\mathbf{x}_0, \omega)\, dA(\mathbf{x}_0), \tag{1}$$

where $G(\mathbf{x}|\mathbf{x}_0, \omega)$ denotes the sound field of a monopole source located at $\mathbf{x}_0 \in \partial V$. $D(\mathbf{x}_0, \omega)$ is its weight, usually referred to as the *driving signal*. The geometry of the problem is illustrated in Fig. 2.

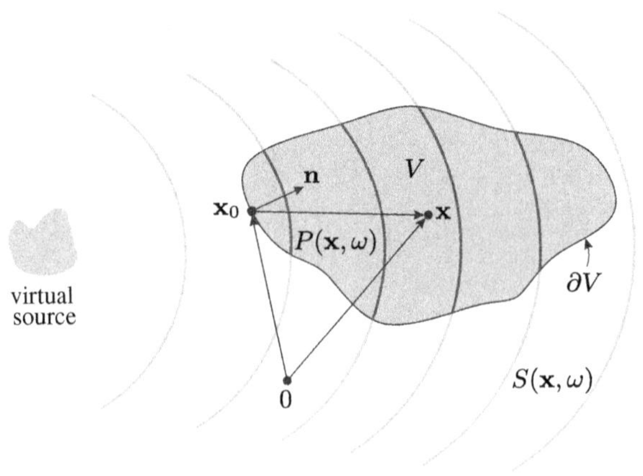


**Fig. 2** Illustration of the geometry used to discuss the physical bases of sound-field synthesis and single-layer potential (1)

In sound-field synthesis, the monopole sources are referred to as secondary sources. Under free-field conditions, their sound field, $G(\mathbf{x}|\mathbf{x}_0, \omega)$, is given by the three-dimensional Green's function [37]. The task is to find the appropriate driving signals, $D(\mathbf{x}_0, \omega)$, for the synthesis of a virtual source, $P(\mathbf{x}, \omega) = S(\mathbf{x}, \omega)$, within $V$. It has been shown that the integral Eq. (1) can be solved under certain reasonable conditions [10].

## 3.2 Solution of the Single-Layer Potential for WFS

The single-layer potential (1) satisfies the homogeneous Helmholtz equation both in the interior and exterior regions, $V$ and $\bar{V} := \mathbb{R}^3 \setminus (V \cup \partial V)$. If $D(\mathbf{x}_0, \omega)$ is continuous, its pressure value, $P(\mathbf{x}, \omega)$, is continuous when approaching the surface, $\partial V$, from the inside and outside. Due to the presence of the secondary sources at the surface, $\partial V$, the gradient of $P(\mathbf{x}, \omega)$ is discontinuous when approaching the surface. As a consequence, $\partial V$ can be interpreted as the boundary of a scattering object with Dirichlet boundary conditions, hence as an object with low acoustic impedance. Considering this *equivalent scattering problem*, the driving signal is given as follows [11].

$$D(\mathbf{x}_0, \omega) = \partial_{\mathbf{n}} P(\mathbf{x}_0, \omega) + \partial_{-\mathbf{n}} P(\mathbf{x}_0, \omega), \tag{2}$$

where $\partial_{\mathbf{n}} := \langle \nabla, \mathbf{n} \rangle$ is the directional gradient in direction $\mathbf{n}$. Acoustic scattering problems can be solved analytically for simple geometries of the surface $\partial V$, such as spheres or planes.

The solution for an infinite planar boundary, $\partial V$, is of special interest. For this specialized geometry and Dirichlet boundary conditions, the driving function is given as

$$D(\mathbf{x}_0, \omega) = 2\partial_{\mathbf{n}} S(\mathbf{x}_0, \omega), \tag{3}$$

since the scattered pressure is the geometrically mirrored interior pressure given by the virtual-source model, $P(\mathbf{x}, \omega) = S(\mathbf{x}, \omega)$, for $\mathbf{x} \in V$. The integral equation resulting from introducing (3) into (1) for a planar boundary, $\partial V$, is known as *Rayleigh's first integral equation*.

An approximation of the solution for planar boundaries can be found by applying the *Kirchoff approximation* [7]. Here, it is assumed that a bent surface can be approximated by a set of small planar surfaces for which (3) holds, locally. In general, this will be the case if the wave length is much smaller than the size of a planar surface patch, hence, for high frequencies. In addition, the only part of the surface that is active is the one which is illuminated from the incident field of the virtual source. This also implies that only convex surfaces can be used to avoid contributions from outside of the listening area, $V$, to re-enter. The outlined principle can be formulated by introducing a window function $w(\mathbf{x}_0)$ into (3), namely,

$$P(\mathbf{x}, \omega) \approx \oint_{\partial V} G(\mathbf{x}|\mathbf{x}_0, \omega) \underbrace{2w(\mathbf{x}_0)\partial_{\mathbf{n}} S(\mathbf{x}_0, \omega)}_{D(\mathbf{x}_0, \omega)} \, dA(\mathbf{x}_0), \tag{4}$$

where $w(\mathbf{x}_0)$ describes a window function for the selection of the active secondary sources, according to the criterion given above. Equation (4) constitutes an approximation of the Rayleigh integral that forms the basis for WFS-type sound reproduction methods.

## 3.3 Virtual-Source Models

In WFS, sound fields can be described by using source models to calculate the driving function. The source model is given as $S(\mathbf{x}, \omega)$, and with (3), the driving function can be calculated. Two common source models are point sources and plane waves. For example, point sources can be used to represent the sound field of a human speaker, whereas plane waves could represent room reflections.

The source model for a point source located at $\mathbf{x}_\text{s}$ is given as

$$S(\mathbf{x}, \omega) = \hat{S}(\omega) \, \frac{e^{-i\frac{\omega}{c}|\mathbf{x}-\mathbf{x}_\text{s}|}}{|\mathbf{x} - \mathbf{x}_\text{s}|}, \tag{5}$$

where $\hat{S}$ is the temporal spectrum of the source signal $\hat{s}(t)$.

The source model for a plane wave with a propagation direction of $\mathbf{n}_\text{s}$ is given as

$$S(\mathbf{x}, \omega) = \hat{S}(\omega) \, e^{-i\frac{\omega}{c}\mathbf{n}_\text{s}\mathbf{x}}. \tag{6}$$

### 3.4 2.5-Dimensional Reproduction

Loudspeaker arrays are often arranged within a two-dimensional space, for example as a linear or circular array. From a theoretical point of view, the characteristics of the secondary sources in such setups should conform to the two-dimensional free-field Green's function. Its sound field can be interpreted as the field produced by a line source. Loudspeakers exhibiting the properties of acoustic line sources are not practical. Real loudspeakers have properties similar to a point source. In this case three-dimensional free-field Green's functions are used as secondary sources for the reproduction in a plane, which results in a dimensionality mismatch. Therefore, such methods are often termed *2.5-dimensional synthesis* techniques. It is well known from WFS, that 2.5-dimensional reproduction techniques suffer from artifacts [30]. Amplitude deviations are most prominent.

### 3.5 Loudspeakers as Secondary Sources

Theoretically, when an infinitely-long continuous secondary source distribution is used, no errors other than an amplitude mismatch due to 2.5-dimensional synthesis are expected in the sound field.

However, such a continuous distribution cannot be implemented in practice, because a finite number of loudspeakers has to be used. This results in a *spatial sampling* and *spatial truncation* of the secondary source distribution [28, 30]. In principle, both can be described in terms of diffraction theory—see for example [4]. Unfortunately, as a consequence of the size of loudspeaker arrays and the large range of wave lengths in sound as compared to light, most of the assumptions made to solve diffraction problems in optics are not valid in acoustics. To present some of the basic properties for truncated and sampled secondary source distributions, simulations of the sound field are made and interpreted in terms of basic diffraction theory, where possible.

#### Spatial Sampling

The spatial sampling, which is equivalent to the diffraction by a grating, only has consequences for frequencies greater than the aliasing frequency

$$f_{\text{al}} \geq \frac{c}{2\Delta x_0}, \tag{7}$$

where $\Delta x_0$ describes the spacing between the secondary sources [27]. In general, the aliasing frequency is dependent on the listening position $\mathbf{x}$—compare [28, Eq. 5.17].

For the sound field of a virtual source, the spatial aliasing adds additional wave fronts to the signal. This can be explained as follows. Every single loudspeaker is sending a signal according to (3). If no spatial aliasing occurs, the signals cancel each other out in the listening area, with the exception of the intended wave front. In the case of spatial aliasing and for frequencies above the aliasing frequency, the cancellation does not occur, and several additional wave fronts reach a given listener position, following the intended wave front. The additional wave fronts also add energy to the signal.

### Truncation

The spatial truncation of the loudspeaker array leads to further restrictions. Obviously, the listening area becomes smaller when a smaller array is used.

Another problem is that a smaller loudspeaker array introduces diffraction in the sound field. The loudspeaker array can be seen as a single slit that causes a diffraction of the sound field propagating through it. This can be described in a way equivalent to the phenomenon of edge waves as shown by Sommerfeld and Rubinowicz—see [4] for a summary. The edge waves are two additional spherical waves originating from the edges of the array, which can be softened by applying a tapering window [31].

## *3.6 Example*

For the simulations shown in Fig. 3, a circular loudspeaker array is assumed with a diameter of 3 m, consisting of 56 loudspeakers, which results in a loudspeaker spacing of $\Delta x_0 = 0.17$ m. Note that a circular array constitutes a 2.5-dimensional scenario.

Figure 3 illustrates the reproduced wave field for two different frequencies of the virtual plane wave, and its spatio-temporal impulse response. For 1 kHz, the reproduced wave field shows no obvious artifacts. However, some inaccuracies can be observed close to the secondary sources. This is due to the approximations applied for the derivation of the driving function in WFS. For plane waves with the frequencies of 2 and 5 kHz sampling artifacts are visible, and rather evenly distributed over the listening area. The amplitude decay in the synthesized plane wave due to the 2.5-dimensional approach is clearly visible in Fig. 3a.

The impulse response depicted in Fig. 3d shows that WFS reconstructs the first wave front well, with prominent artifacts following behind. The artifacts consist of additional wave fronts coming from the single loudspeakers. These additional wave fronts would vanish for a loudspeaker array with a loudspeaker spacing smaller than $\lambda_{\min}/2$, where $\lambda_{\min}$ is the smallest wavelength to be reproduced.

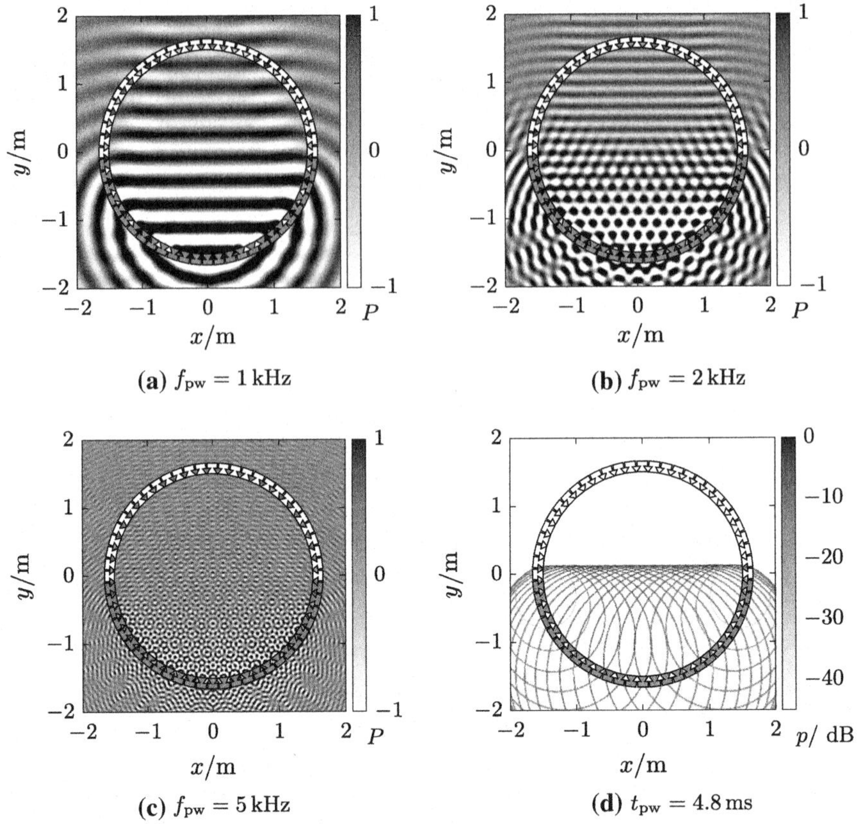


**Fig. 3** Snapshot of sound fields synthesized by 2.5-dimensional WFS using a *circular array*, $R = 1.50\,\text{m}$, with 56 loudspeakers. The virtual source constitutes a plane wave with an incidence angle of $-90°$ and the frequency $f_{pw}$. The *gray shades* denote the acoustic pressure, the active loudspeakers are *filled*. **a–c** Snapshot of $P(\mathbf{x}, \omega)$. **d** Snapshots of broad-band $p(\mathbf{x}, t)$

# 4 Localization Measurement with Regard to Wave-Field Synthesis

## *4.1 Binaural Synthesis*

As discussed in the last section, spatial aliasing depends on the listening position and the loudspeaker array. In a listening test targeting localization assessment, it is not sufficient anymore to test only one position and one loudspeaker setup. Instead, different listener positions have to be investigated, and different types of loudspeaker setups must be applied, switching configurations more or less instantaneously and without disturbing the listener.

In practice, a real-life physical setup can be approximated by applying dynamic binaural synthesis to simulate the ear signals for the listeners for all needed conditions. Dynamic binaural synthesis simulates a loudspeaker by convolving the *head-related transfer functions*, HRTFs with an intended audio signal, which is played back to a listener via headphones. Simultaneously, the orientation of the head of the listener is tracked, and the HRTFs are exchanged according to the head orientation of the listener. With this dynamic handling included, results from the literature show that the localization performance for a virtual source is equal to the case of a real loudspeaker, provided that individual HRTFs are used. For non-individual HRTFs, the performance can be slightly impaired, and an individual correction of the ITD may be necessary [19]. For the case of a real loudspeaker, the localization performance of listeners in the horizontal plane lies around 1°–2°. Note that this number only holds for a source located in front of the listener. For sources to the sides of the listener, localization performance can get as bad as 30°, due to the fact that the ITD changes only little for positions to the side of the head. An accuracy of around 1° sets some requirements on the experimental setup. One has to ensure that the employed setup introduces a measurement error that is smaller than the error expected in terms of human localization performance. This is especially difficult for the acquisition of the perceived direction based on the indications/judgments collected from the test listener. A review of different techniques and their advantages and drawbacks can be found in [21, 25].

In general, the WFS simulation based on binaural synthesis can be implemented as follows. For each listener position and individual loudspeaker of the WFS system, a dedicated set of HRTFs is used. The ear signals are constructed from the loudspeaker-driving signals, which are convolved with the respective *head-related impulse responses*, HRIRs, and then superimposed. For the tests and respective setups considered in this chapter, the *SoundScape Renderer* has been used as the framework for implementation [12], as well as the *Sound-Field Synthesis Toolbox* [34]. To simulate loudspeaker setups that deviate from the set of HRIRs, which are typically measured with a loudspeaker at a given radial distance from the dummy head, for different angular loudspeaker positions, the HRIRs are extrapolated using delay and attenuation, according to the propagation delay and respective distance-related attenuation.

## *4.2 Verification of Pointing Method and Dynamic Binaural Synthesis*

For the localization indication, it has been decided to use a method where the listeners have to turn their heads to the direction of the auditory event during sound presentation. This has the advantage that the listener is directly facing the virtual source, a region where the localization performance is at its best. If the listeners point their heads in the direction of the auditory event, an estimation error of the sources at the side will occur, due to an interaction with the motor system. In other words, listeners

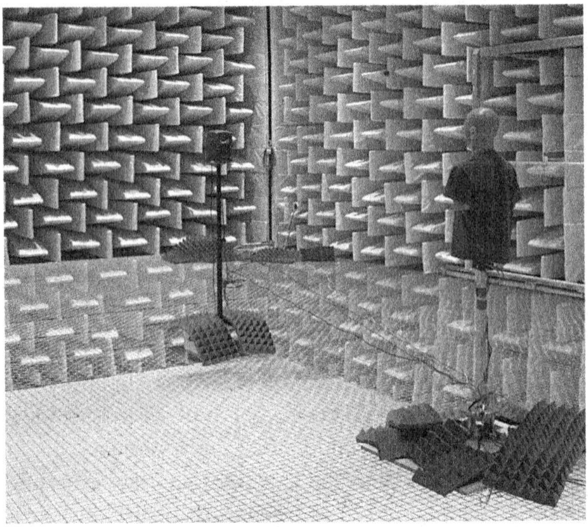

**Fig. 4** Measurement of HRTFs with a dummy head in an anechoic chamber [35]

do not turn their heads sufficiently far as to indicate the real location. This can be overcome by adding a visual pointer that indicates to the listeners where their noses are pointing at [18].

Before investigating the localization in WFS, a pre-study was conducted [35], where the performance of the pointing method was verified, and it was studied whether the dynamic binaural synthesis introduces errors to the localization of a source. For the binaural synthesis, non-individual HRTFs were used that had been measured with a KEMAR dummy head in an anechoic chamber, as shown in Fig. 4 [32].

For the pre-study, the listeners were seated in an acoustically damped listening room, 1.5 m in front of a loudspeaker array, with an acoustically transparent curtain in between. Eleven of the 19 loudspeakers of the array were used as real sources and also simulated via the dynamic binaural synthesis. The listeners were seated on a heavy chair and were wearing open headphones, AKG K601, both for the loudspeaker and the headphone presentation. A laser pointer and the head tracker sensor, Polhemus Fastrack, were mounted onto the headphones. A visual mark on the curtain was used to calibrate the head-tracker setup at the beginning of each test run. For each trial, the listener was presented with a Gaussian white-noise train, consisting of periods of 700 ms noise and 300 ms silence. The experimenter instructed the listener to look towards the perceived source and to hit a key when the intended direction was correctly indicated by the laser. The conditions in terms of virtual-source directions and loudspeaker-versus-headphone presentation were randomized. The setup is shown in Fig. 5.

Eleven listeners participated in the experiment, and every condition was repeated five times. Figure 6 shows the deviation between the direction of the auditory event

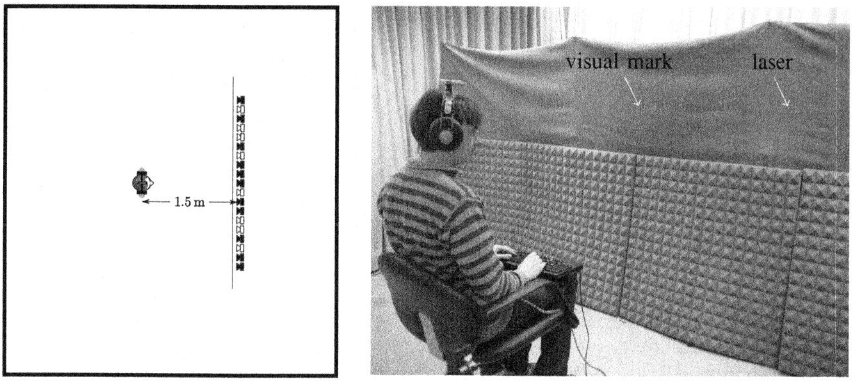


**Fig. 5** Sketch of the experimental setup (*left*) and picture of a listener during the experiment (*right*), [35]. Only the filled loudspeakers were used in the first experiment. The light in the room was dimmed during all experiments

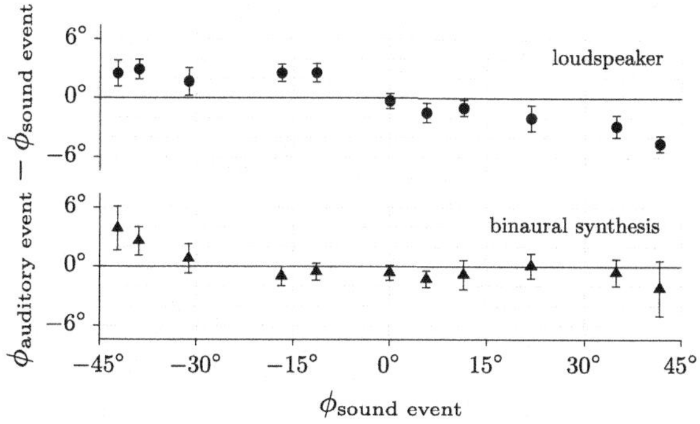


**Fig. 6** Deviation $\Delta\phi$ between the position of the auditory event and the position of the sound source. The mean across all listeners and the 95 % confidence intervals are indicated. *Top row* Real loudspeakers. *Bottom row* Binaural synthesis

and the sound event for every single loudspeaker. It can be seen that there are only slight differences between the binaural simulation using headphones and the localization of the noise coming from the real loudspeakers. The mean absolute deviation, $\Delta\phi$, of the direction of the auditory event compared to the position of the sound event together with its confidence interval is $2.4^\circ \pm 0.3^\circ$ for the real loudspeakers and $2.0^\circ \pm 0.4^\circ$ for the binaural synthesis. In both cases, the mean deviation gets higher for sources more than $30^\circ$ to the side of the listener. For these conditions, the position of the auditory event is underestimated and pulled towards the center. To avoid this kind of error in the examination of localization in WFS, only virtual-source positions within the range of $-30^\circ$ to $30^\circ$ are be considered in the following. The only differences between simulation and the loudspeakers can be found in the localization blur for individual listeners. The mean standard deviation for a given position is $2.2^\circ \pm 0.2^\circ$ for the loudspeakers and $3.8^\circ \pm 0.3^\circ$ for the binaural synthe-

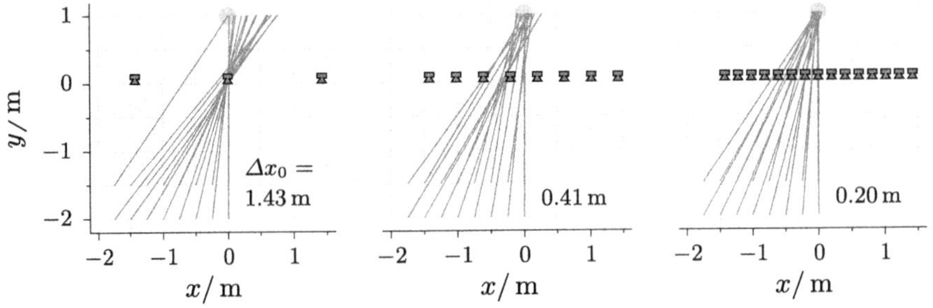


**Fig. 7** Average directions that listeners were looking at from the 16 different listener positions evaluated, [33]. Results for the three different loudspeaker spacings. The *gray point* above the loudspeaker array indicates the intended virtual-source position

sis conditions. This is most likely due to higher ease of localization when listening to the real loudspeaker—an interesting issue discussed further in [35].

## 4.3 Localization Results for Wave-Field Synthesis

The same setup as presented in Sect. 4.2 and shown in Fig. 5 was employed. This time, a virtual source located at $\mathbf{x}_\text{s} = (0, 1)$ m was presented via the loudspeaker array, now driven by WFS. Following the descriptions above, the loudspeaker array was simulated using dynamic binaural synthesis. It had a length of 2.85 m, and consisted of 3, 8, or 15 loudspeakers, translating to a loudspeaker spacing of 1.43, 0.41, and 0.20 m. Like in the pre-test, the listeners were seated in the heavy chair in front of the curtain. Now, however, different listener positions of the listeners were introduced via binaural synthesis. The positions were at $x = -1.75$ m up to 0 m in steps of 0.25 m, with $y = -2$ m, and $y = -1.5$ m, leading to a total of 16 positions—compare Fig. 7.

Figure 7 summarizes the results. A line goes from every position of the listener to the direction where the corresponding auditory event was perceived, taking the average over all listeners. The gray point indicates the position of the virtual point source. As can be seen from this figure, the loudspeaker array with 15 loudspeakers leads to high localization accuracy. The intended position of the auditory event is reached with a deviation of only 1.8°. For the arrays with eight and three loudspeakers, the deviations are 2.7° and 6.6°, respectively. For the array with three loudspeakers, a systematic deviation of the perceived direction towards the loudspeaker at (0, 0) m can be observed for all positions except one. For all three array geometries, the mean error is slightly smaller for the listener positions with $y = 2$ m than for that with $y = 1.5$ m. The results for every single position are presented in Fig. 10.

In WFS, the aliasing frequency determines the cut-off frequency up to which the sound field is synthesized correctly. For localization, mainly the frequency content below 1 kHz is important. The aliasing frequencies for the three loudspeaker arrays are 120, 418, 903 Hz, starting from the array with a spacing of 1.43 m between the loudspeakers. Further, the aliasing frequency is position-dependent and can be higher for certain positions. According to the results for the 15-loudspeaker array, a loudspeaker spacing of around 20 cm seems to be sufficient to yield unimpaired localization for the entire range of listener positions. For a central listening position, a similar result was obtained in other experiments [28, 30, 38]. It was further discovered with the measuring method used here that even for spacings twice as large, the localization is only impaired by 1°. For larger spacings, the behavior tends more towards a stereophonic setup, that is, showing a sweet spot and localization towards the loudspeaker nearest to the sweet spot.

In the next section, a binaural model will be extended to enable predictions of the localization test results found for WFS. It is shown how the model can be used to predict localization maps for the entire listening area, going beyond the set of tested conditions.

# 5 Predicting Localization in Wave-Field Synthesis

In this section, a binaural model will be extended to enable predictions of the localization test results found for WFS. It is shown how the model can be used to predict localization maps for the entire listening area, going beyond the set of tested conditions.

An important difference of WFS in comparison to stereophony is the feature of uniform localization across an extended listening area. This is in clear contrast to the confined sweet spot of stereophonic systems. The sweet spot phenomenon was illustrated in Fig. 1. It would be of advantage to be able to predict such *localization maps* for further loudspeaker setups and reproduction methods as well, for example, for multichannel loudspeaker arrays and WFS. To this end the binaural model after Dietz [8] was modified and extended to be able to predict the direction of the auditory events for any pair of given ear signals. Specifically, the same ear signals were used as input signals to the binaural model, as have been synthesized for the listening tests by means of binaural synthesis—see Sect. 4.

In the following, the predictions from the binaural model are compared to the actual localization data as obtained in the listening tests. Given that the model provides localization predictions that agree with the listening-test data, it can be used to create localization maps for setups other than those that have been investigated perceptually.

## *5.1 Modelling the Direction of the Auditory Event*

Binaural auditory models as outlined in [16], this volume, typically process the signals present at the right and left ear canal. For example, the model developed by Dietz [8] provides as its output a set of interaural arrival-time-difference values, ITDs, namely, one for every auditory filter. For the prediction of the direction of an auditory event, the ITD values have to be transformed into azimuth values that

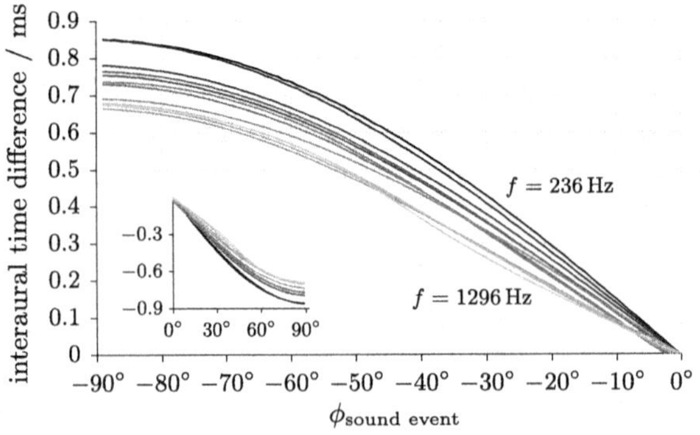


**Fig. 8** Lookup table for ITD values and corresponding sound-event directions, shown for the first twelve auditory filters. Data derived with the binaural model of Dietz [8]

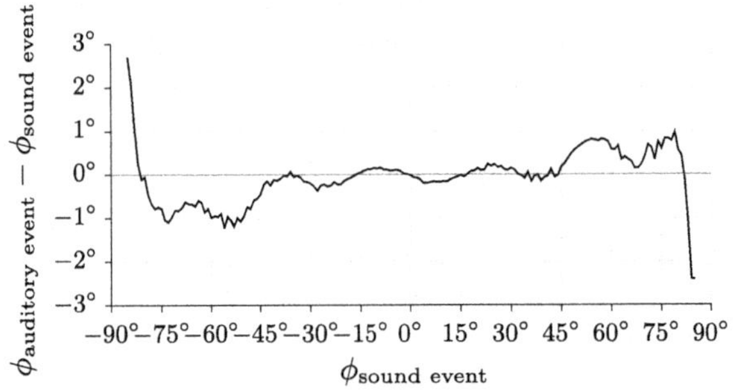


**Fig. 9** Deviation of the predicted direction of auditory events from the direction of corresponding sound events

describe the direction of the auditory events. This can be accomplished by means of a lookup table of ITD values and corresponding angles [8].

In the study presented here, such a table was created by convolving a 1-s-long white-noise signal with head-related impulse responses from the same database as has been used for the listening tests presented in Sect. 4. The database has a resolution of 1°. The convolved signals were fitted to match the input format of the binaural model and stored. The result for the first twelve auditory filters are shown in Fig. 8.

For the prediction of the perceived direction belonging to a given stimulus, the binaural model first calculates the ITD values. Then, for each of the twelve auditory channels, the ITD value is transformed into an angle by use of the lookup table—Fig. 8. If the absolute ITD value in an auditory channel turns out to be larger than the natural limit of 1 ms, this channel is disregarded in the following step. Afterwards, the median value across all angles is taken as the predicted direction. If the angle in an auditory filter differs by more than 30° from the median, it is considered an outlier and skipped, and the median is re-calculated.

In order to test whether the predictions depend on the actual method used for determining the look-up table, the head-related impulse responses from the same HRTF database were convolved with another white-noise signal and again fitted to match the model input format. Figure 9 shows the deviation between the predicted direction of the auditory event and the direction of the sound sources for this case. Only for angles of more than $\pm 80^\circ$, the deviation exceeds a value of $1.2^\circ$. The deviation is is due to the decreasing slope of the ITD for large angles—compare Fig. 8. This effect makes it more difficult to achieve proper fit of the ITDs and their corresponding azimuths.

## 5.2 Verification of Prediction

The modified binaural model can now be used to *predict* the direction of an auditory event. In this part of the study, the model prediction performance were analyzed in view of the localization results of Sect. 4. Due to limitations of the binaural model used here—for example, the precedence effect is not included—it might well be that it fails to properly predict localization in more complex sound fields, such as those synthesized with WFS. To check on this, the predictions that the model renders for the setups that have been investigated by the listening tests—compare Sect. 4—have been analyzed. See Fig. 10 regarding the results obtained in the localization test and the corresponding model predictions. Open symbols denote a listener distance of 2 m to the loudspeaker array, filled symbols a distance of 1.5 m. The model predictions are presented as dashed lines for the 2 m case and solid lines for 1.5 m case.

For most of the configurations, the model predictions are in agreement with the directions perceived by the listeners. Only for positions far to the side some deviations of up to $7^\circ$ are visible. The overall prediction error of the model is of $1.3^\circ$, ranging from $1.0^\circ$ for the array with 15 loudspeakers to $2.0^\circ$ for the array with three loudspeakers. These results indicate that the model is able to predict the perceived direction of a virtual source in WFS almost independently from the listener position and the array geometry.

## 5.3 Localization Maps

With the method as presented in the previous sections, it is now possible to create a localization map similar to the one shown in Fig. 1. To this end, the ear signals for each intended listener position and loudspeaker array are simulated via binaural synthesis. Then these signals are fed into the binaural model, which delivers the predicted direction for the respective auditory event. In the following, the procedure is illustrated with two different loudspeaker setups. The first one is the same setup as used for the WFS localization test—compare Sect. 4—however, additional listener

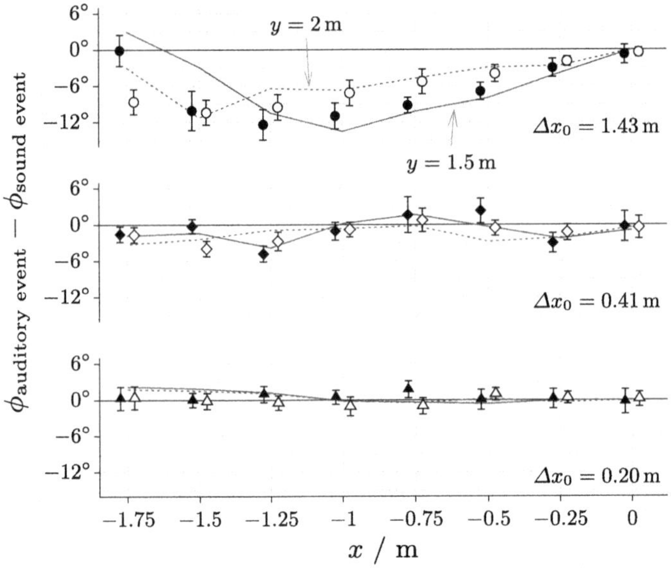


**Fig. 10** Means and 95 % confidence intervals of localization errors in WFS dependent on the listening position and the loudspeaker spacing. *Open symbols* Listener positions at $y = 2\,\mathrm{m}$. *Closed symbols* Listener positions at $y = 1.5\,\mathrm{m}$. The *lines* denote the model predictions. *Solid line* $y = 1.5\,\mathrm{m}$. *Dashed line* $y = 2\,\mathrm{m}$

positions. The second one is a circular loudspeaker array that is installed in the authors' laboratory.

The following virtual sources were chosen: (i) a point source located either at the center of the array or one meter behind it, (ii) a plane wave traveling into the listening area vertically to the loudspeaker array. Both cases were be evaluated separately, because of the expected differences in localization, that is, a point source stays at its position when if the listener moves around in the listening area, but a plane wave moves with the listener.

The resulting localization maps can be presented in the form of arrows pointing into the direction of the auditory event as in Fig. 1. Alternatively, a color can be assigned to each position, denoting the deviation of the perceived direction from the intended one. The latter format renders a better resolution.

### Linear Loudspeaker Array

Figure 11 shows localization maps for the three different linear loudspeaker setups as also used in the listening tests of Sect. 4. The first array consists of three loudspeakers with a spacing of 1.43 m between them, the second of eight loudspeakers with a spacing of 0.41 m, and the third of 15 loudspeakers with a spacing of 0.20 m. The localization maps presented at the top of the figure show a sampling of the listening

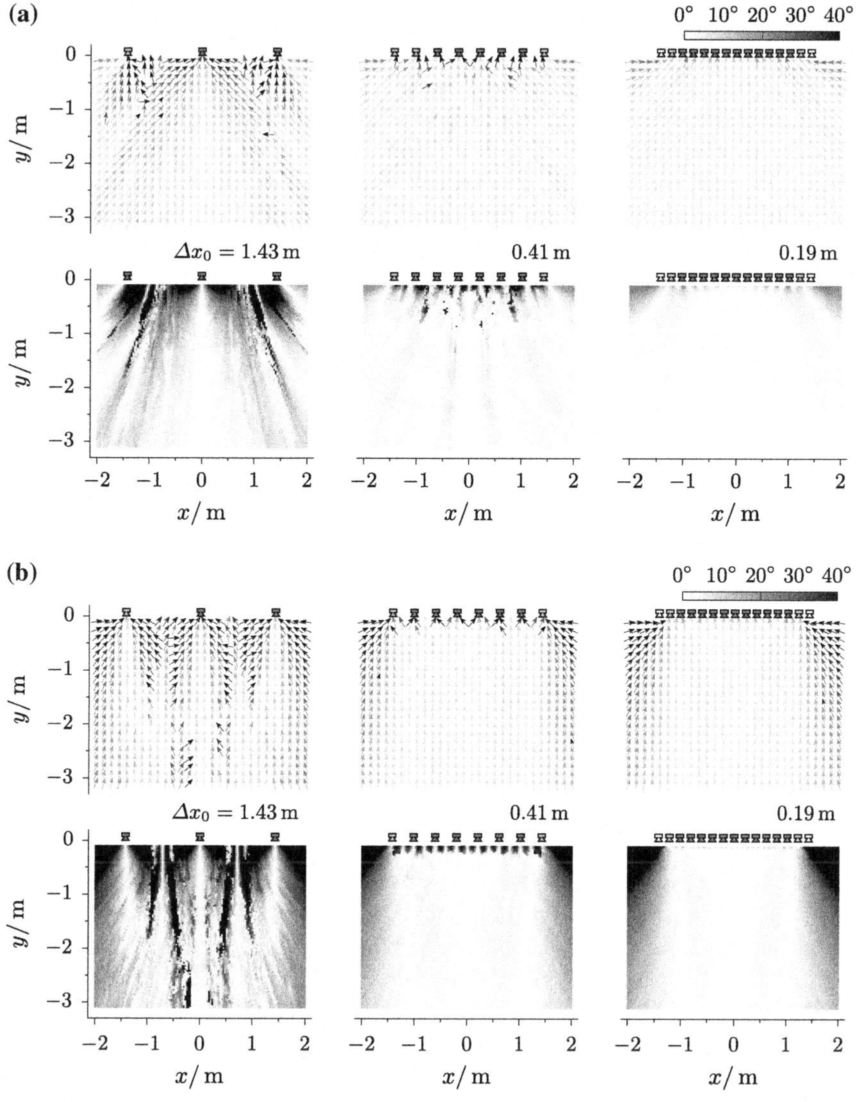


**Fig. 11** Localization maps for a linear loudspeaker array driven by WFS for (**a**) a virtual point source, (**b**) a plane wave. The *arrows* point into the direction of where the auditory event of a listener appears, if he/she sits at the position of the *arrow*. The *gray-shades* indicate the deviation from the intended direction

area of $21 \times 21$ points. The arrows indicate the predicted direction in which the auditory event is predicted to appear as seen from the respective listening position. The localization maps presented at the bottom of the figure show a sampling of the listening area of $135 \times 135$ points. The gray-shades of the points indicate the absolute deviation between the predicted direction of the auditory event and the prospective direction of the virtual sound event. Absolute deviation values are clipped at 40°.

In Fig. 11a, the sound event corresponds to a point source located at (0, 1) m. The same source configuration was used for the listening experiment in Sect. 4. The predicted results fit very well with the results from the experiment, as already shown in Sect. 5. For a spacing of 0.20 m, a large region with no deviations can be seen across the listening area. Only towards the edges of the loudspeaker array are large deviations between intended and predicted directions visible. For loudspeaker arrays with fewer loudspeakers, the deviations of the direction are spread across the listening area, but are worse in the close to the loudspeakers. This is obviously a general trend for all arrays. The larger the distance of the listener to the array in $y$-direction, the smaller is the intended direction from the predicted perceived one.

In Fig. 11b, the sound event is a plane wave impinging parallel to the $y$-direction onto the listener. The pattern of results is similar to the one for the point source, but the deviations are larger for the case of only three loudspeakers. This is mainly due to the fact that the auditory event is bound towards the single loudspeaker which, in the case of a plane wave, leads to larger deviations in the whole listening area. For the loudspeaker array with 15 loudspeakers, the deviation-free region is slightly smaller as for the point source. Deviations due to the edges of the array are more visible.

### Circular Loudspeaker Array

In addition to the linear loudspeaker arrays used in the listening experiment, localization maps were derived for circular arrays with a geometry similar to the one available in the authors' laboratory at the TU Berlin. Three configurations were considered, consisting of 14, 28, or 56 loudspeakers. These numbers correspond to an inter-loudspeaker spacing of 0.67, 0.34, and 0.17 m, respectively. All three configurations have a diameter of 3 m. Again, a point source located 1 m behind the array, and a plane wave traveling parallel to the $y$-direction were used as virtual sources.

The results are shown in Fig. 12. They are very similar to the case of a linear loudspeaker array. For the plane wave, the deviations increase with increasing listener distance to the loudspeaker array, as was also observed for the linear array, but to a smaller degree.

For listener positions in the near-field of the loudspeakers, the predicted direction of the auditory event deviates in most cases toward the direction of the corresponding loudspeaker. This seems to be a plausible result, but it should be mentioned that the model used in the current study is not optimally prepared for the near-field case. Particularly, a HRTF dataset with a distance of 3 m between source and dummy head has been used. It is well known from literature that for distances under 1 m, the interaural level differences, ILDs, vary with distance [5, 32]. Hence, the model predictions could probably be enhanced for the small-distance cases by using appropriate HRTF datasets.

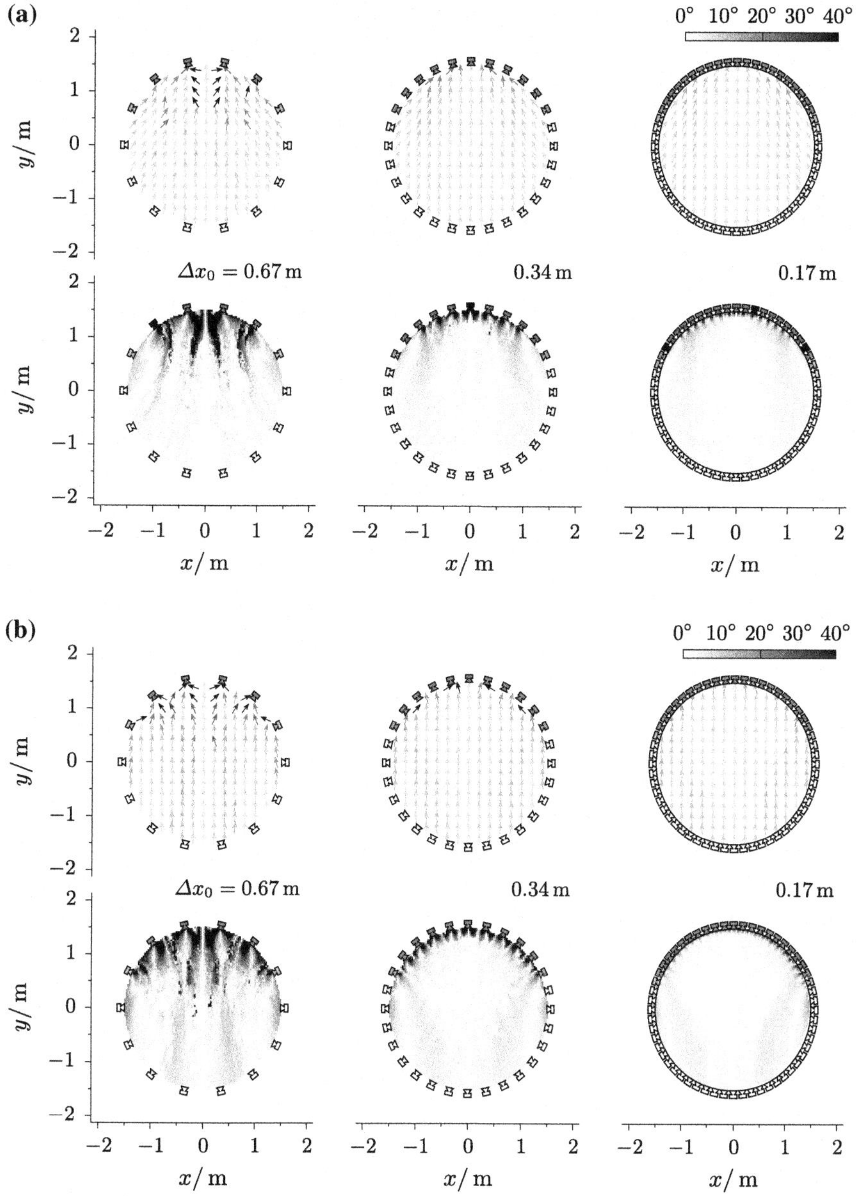


**Fig. 12** Localization maps for a circular loudspeaker array driven by WFS for (**a**) a virtual point source, (**b**) a plane wave. The *arrows* point into the direction of where the auditory event of a listener appears, if he/she sits at the position of the *arrow*. The *gray-shades* indicate the deviation from the intended direction

## 6 Conclusion

In sound-field synthesis such as *wave-field synthesis*, WFS, it is of great interest to evaluate perceptual dimensions such as localization and/or coloration not only at one listener position but also for an extended listening area. As concerns localization, this chapter has provided relevant results along these lines by employing binaural synthesis. This method was applied to generate the ear signals at each listener position for headphone presentation. This approach allows further to feed these ear signals into an auditory model which then predicts the localization at all simulated listening positions. To achieve this, the binaural model by Dietz has been extended by a stage that transforms the interaural time differences provided by the model into azimuths corresponding to the sound-source directions. By combining these predicted angular positions with the binaural simulations, localization maps for different loudspeaker setups and WFS configurations were predicted. With an accompanying listening test, the model results for linear loudspeaker arrays were verified. The results showed that the localization in WFS is not distorted as long as the inter-loudspeaker spacing is below 0.2 m. For larger spacings, small deviations between the intended and perceived source locations occur. In practice, one has to specify the localization accuracy that is needed for the intended application of a given WFS system. The predicted localization maps are a valuable aid when planning the task-required loudspeaker setup. However, for practical applications of WFS, it is not only the localization accuracy which is important. WFS may also be affected by the localization blur, for example, indicated by the standard deviation of the localization. In order to investigate the localization blur via binaural synthesis, one has to account for the localization blur as already contributed by binaural synthesis [35]. Further, beside these spatial-fidelity features, coloration or timbral fidelity is of high relevance, as reported by Rumsey [24], who found by comparison of different stereophonic 5.1 surround setups that the overall quality is composed of timbral and spatial fidelity, whereby, according to Rumsey, timbral fidelity explained approximately 70 % of the variance the overall quality ratings while spatial fidelity explained only 30 %. Hence, to provide further components of a model of integral WFS quality, ongoing work by the authors addresses the prediction of coloration resulting from different WFS system and listener configurations.

### Materials

The algorithm of the binaural model is included in the *AMToolbox* described in [26], this volume. The function for the prediction of the direction of the auditory event is `estimate_azimuth`. In addition, all other software tools and data are also available as open source items. The *Sound Field Synthesis Toolbox* [34], which was used to generate the binaural simulation for WFS, can be downloaded from https://dev.qu.tu-berlin.de/projects/sfs-toolbox/files. The version used in this chapter is 0.2.1. The HRTF data set [32] is part of a larger set available for download from https://dev.qu.tu-berlin.de/projects/measurements/wiki/2010-11-kemar-anechoic. The set that has been used here is the one with a distance of 3 m. The *SoundScape Renderer* [12] that was employed as the convolution engine for the dynamic binaural synthesis is available as open source as well. It can be downloaded from https://dev.qu.tu-berlin.de/projects/ssr/files.

**Acknowledgement**

The authors are obliged to M. Geier, who is the driving force behind the development of the SoundScape Renderer, and who helped with the setup of the experiments. Thanks are also due to two anonymous reviewers for their valuable input. This work was funded by DFG–RA 2044/1-1.

## References

1. A. Berkhout. A holographic approach to acoustic control. *J. Audio Eng. Soc.*, 36:977–995, 1988.
2. J. Blauert. *Spatial Hearing*. The MIT Press, 1997.
3. J. Blauert and U. Jekosch. Concepts Behind Sound Quality: Some Basic Considerations. In *Proc. 32nd Intl. Congr. Expos. Noise, Control*, 2003.
4. M. Born, E. Wolf, A. Bhatia, P. Clemmow, D. Gabor, A. Stokes, A. Taylor, P. Wayman, and W. Wilkock. *Principles of Optics*. Cambridge University Press, 1999.
5. D. S. Brungart and W. M. Rabinowitz. Auditory localization of nearby sources. Head-related transfer functions. *J. Acoust. Soc. Am.*, 106:1465–79, 1999.
6. S. Choisel and F. Wickelmaier. Evaluation of multichannel reproduced sound: Scaling auditory attributes underlying listener preference. *J. Acoust. Soc. Am.*, 121:388, 2007.
7. D. Colton and R. Kress. *Integral Equation Methods in Scattering Theory*. Wiley, New York, 1983.
8. M. Dietz, S. D. Ewert, and V. Hohmann. Auditory model based direction estimation of concurrent speakers from binaural signals. *Speech Commun.*, 53:592–605, 2011.
9. T. du Moncel. The Telephone at the Paris Opera. *Scientific America*, pp. 422–23, 1881.
10. F. M. Fazi. *Sound Field Reproduction*. PhD thesis, University of Southampton, 2010.
11. F. M. Fazi, P. A. Nelson, and R. Potthast. Analogies and differences between three methods for sound field reproduction. In *Proc. Ambis. Sym.*, 2009.
12. M. Geier, J. Ahrens, and S. Spors. The SoundScape Renderer: A Unified Spatial Audio Reproduction Framework for Arbitrary Rendering Methods. In *Proc. 124th Conv. Audio Eng. Soc.*, 2008.
13. M. Geier, H. Wierstorf, J. Ahrens, I. Wechsung, A. Raake, and S. Spors. Perceptual Evaluation of Focused Sources in Wave Field Synthesis. In *Proc. 128th Conv. Audio Eng. Soc.*, 2010.
14. C. Guastavino and B. F. G. Katz. Perceptual evaluation of multi-dimensional spatial audio reproduction. *J. Acoust. Soc. Am.*, 116(2):1105, 2004.
15. C. Huygens. *Treatise on Light (English translation by S. P. Thompson)*. Macmillan & Co, London, 1912.
16. A. Kohlrausch, J. Braasch, D. Kolossa, and J. Blauert. An introduction to binaural processing. In J. Blauert, editor, *The technology of binaural listening*, chapter 1. Springer, Berlin-Heidelberg-New York NY, 2013.
17. T. Letowski. Sound Quality Assessment: Concepts and Criteria. In *Proc. 89th Conv. Audio Eng. Soc.*, 1989.

18. J. Lewald, G. J. Dörrscheidt, and W. H. Ehrenstein. Sound localization with eccentric head position. *Behav. Brain Res.*, 108(2):105–25, 2000.
19. A. Lindau, J. Estrella, and S. Weinzierl. Individualization of dynamic binaural synthesis by real time manipulation of the ITD. In *Proc. 128th Conv. Audio Eng. Soc.*, 2010.
20. R. Y. Litovsky, H. S. Colburn, W. A. Yost, and S. J. Guzman. The precedence effect. *J. Acoust. Soc. Am.*, 106:1633–54, 1999.
21. P. Majdak, B. Laback, M. Goupell, and M. Mihocic. The Accuracy of Localizing Virtual Sound Sources: Effects of Pointing Method and Visual Environment. In *Proc. 124th Conv. Audio Eng. Soc.*, 2008.
22. S. Merchel and S. Groth. Adaptively Adjusting the Stereophonic Sweet Spot to the Listeners Position. *J. Audio Eng. Soc.*, 58:809–817, 2010.
23. F. Rumsey. Spatial Quality Evaluation for Reproduced Sound: Terminology, Meaning, and a Scene-Based Paradigm. *J. Audio Eng. Soc.*, 50:651–666, 2002.
24. F. Rumsey, S. Zielinski, R. Kassier, and S. Bech. On the relative importance of spatial and timbral fidelities in judgments of degraded multichannel audio quality. *J. Acoust. Soc. Am.*, 118:968–976, 2005.
25. B. U. Seeber. *Untersuchung der auditiven Lokalisation mit einer Lichtzeigermethode*. PhD thesis, 2003.
26. P. Søndergaard and P. Majdak. The auditory modeling toolbox. In J. Blauert, editor, *The technology of binaural listening*, chapter 2. Springer, Berlin-Heidelberg-New York NY, 2013.
27. S. Spors and J. Ahrens. Spatial Sampling Artifacts of Wave Field Synthesis for the Reproduction of Virtual Point Sources. In *Proc. 126th Conv. Audio Eng. Soc.*, 2009.
28. E. Start. *Direct Sound Enhancement by Wave Field Synthesis*. PhD thesis, Technische Universiteit Delft, 1997.
29. J. Steinberg and W. B. Snow. Symposium on wire transmission of symphonic music and its reproduction in auditory perspective: Physical Factors. *AT&T Tech. J.*, 13:245–258, 1934.
30. E. Verheijen. *Sound Reproduction by Wave Field Synthesis*. PhD thesis, Technische Universiteit Delft, 1997.
31. P. Vogel. *Application of Wave Field Synthesis in Room Acoustics*. PhD thesis, Technische Universiteit Delft, 1993.
32. H. Wierstorf, M. Geier, A. Raake, and S. Spors. A Free Database of Head-Related Impulse Response Measurements in the Horizontal Plane with Multiple Distances. In *Proc. 130th Conv. Audio Eng. Soc.*, 2011.
33. H. Wierstorf, A. Raake, and S. Spors. Localization of a virtual point source within the listening area for Wave Field Synthesis. In *Proc. 133rd Conv. Audio Eng. Soc.*, 2012.
34. H. Wierstorf and S. Spors. Sound Field Synthesis Toolbox. In *Proc. 132nd Conv. Audio Eng. Soc.*, 2012.
35. H. Wierstorf, S. Spors, and A. Raake. Perception and evaluation of sound fields. In *Proc. 59th Open Sem. Acoust.*, 2012.
36. F. L. Wightman and D. J. Kistler. The dominant role of low-frequency interaural time differences in sound localization. *J. Acoust. Soc. Am.*, 91(3):1648–61, 1992.
37. E. G. Williams. *Fourier Acoustics*. Academic Press, San Diego, 1999.
38. H. Wittek. *Perceptual differences between wavefield synthesis and stereophony*. PhD thesis, University of Surrey, 2007.
39. N. Zacharov and K. Koivuniemi. Audio descriptive analysis & mapping of spatial sound displays. In *Proc. 7th Intl. Conf. Audit. Display*, pages 95–104, 2001.

# Optimization of Binaural Algorithms for Maximum Predicted Speech Intelligibility

A. Schlesinger and Chr. Luther

## 1 The Speech-in-Noise Problem and Solutions

Binaural processing is a central auditory process that takes a vital role in enriched and complex communication tasks. For instance, the normal hearing of a young person binaurally unmasks speech-in-noise, that is, improves the signal-to-noise ratio, SNR, by about 10 dB when a continuous noise source with the long-term spectrum of speech rotates from frontal position, where the target speech is located, to the side. However, elderly people suffering from presbycusis—and that is the majority of hearing-impaired people—experience only a benefit of 2–3 dB in the same binaural comparison [8]. In addition, if the continuous noise source is substituted by a competing voice, young listeners with healthy hearing generally gain another 3–4 dB advantage for lateral noise positions, and even show an advantage of 7 dB when the competing voice source collapses with the target voice in the frontal direction. On the contrary, elderly people suffering from presbycusis are not able to benefit from glimpsing into spectro-temporal regions in which the target signal prevails over the interference signal. This deficit is predominantly caused by the elevated hearing threshold and reduced temporal acuity [8, 12]. In total, peripheral and concomitant central deficits of old people with presbycusis amount to an SNR difference of 5–15 dB with respect to young listeners with healthy hearing [8]. The severity of the problem is even more striking if one considers the well-known fact that 1 dB of SNR change corresponds to 15–20 % of absolute speech-intelligibility change at the threshold of understanding mono-syllable meaningless words by 50 %, which is taken as the speech-reception threshold, SRT, in noise. For a full compensation of the individual hearing loss in noise, the SRT difference between the hearing impaired and normal hearing has to be offset.

---

A. Schlesinger (✉) · Chr. Luther
Institute of Communication Acoustics,
Ruhr-University Bochum, Bochum, Germany
e-mail: anton.schlesinger@rub.de

J. Blauert (ed.), *The Technology of Binaural Listening*, Modern Acoustics and Signal Processing, DOI: 10.1007/978-3-642-37762-4_11,

The enhancement of speech intelligibility is a difficult problem. After many decades of pioneering research it can be summarized that, primarily, algorithms that exploit the spatial diversity by a spatial sampling provide a solution to the problem—see, for instance, [14]. These algorithms are known as multichannel filters. Their unifying objective is to enhance the target speech either by a direct enhancement of the target signal or, implicitly, by a suppression of the noise. Popular multichannel filters are the well-known beamformers. Until now the beamforming filters pose the most robust and practically efficient solution to the speech-in-noise problem. There are different variants of beamforming filters. A powerful variant is the *minimum-variance distortionless response*, MVDR, beamformer that allows for a high and frequency-independent improvement of the SNR. The *generalized side-lobe canceler*, GSC, framework, which is an adaptive method to calculate the optimal filters instantaneously, represents a further advancement of beamformers. The method is powerful in coherent noise conditions, but interference suppression in more complex conditions is generally reduced to the gain that is provided by the underlying fixed processing scheme [13]. Recent implementations extend the GSC processing over two ears or bilaterally head-worn arrays—for example, [14].

Another class of multi-channel filters aims to decompose the input into independent signals. This class is known as *blind source separation*, BSS, approach [14, 44]. If the underlying—in realistic environments generally highly underdetermined—problem can be solved, these filters have shown to be very effective.

The third well-known class of multichannel filters is the *multi-channel Wiener filter*, which draws upon the statistical description of the signal mixture and the mean-square-error criterion. An efficient version of this filter is based on the auditory principles of directional sound perception, hence, the computational mimicry of *auditory scene analysis*, ASA. Directional sound perception is strongly driven by the binaural interaction process as well as the exploitation of the head shadow effect. The ability for unraveling a speech mixture by imitating these functions algorithmically has been successfully demonstrated.

Binaural interaction is an auditory process that can be thought of as the neural execution of cross-correlating the signals of the left and right ear. The most popular model that explains many phenomena of the binaural interaction process is the cross-correlation-based coincidence detector for sound localization by Jeffress [18]. The application of the cross-correlation in order to assess time or equivalently phase differences between the ears, known as *interaural time differences* and *interaural phase differences*, ITD and IPD, respectively, is therefore fundamental to most binaural ASA processors. An alternative technique—and widely viewed as a physiological more plausible approach—to mimic the binaural interaction process represents the equalization-cancellation model of Durlach [9]. Its implementation is, however, relatively difficult and the advantage over the cross-correlation model in the speech enhancement task has not yet been fully shown—see, for example, [23]. The binaural advantage derives a further benefit—in a non-additive manner—from the head shadow effect [6]. The advantage is given by the fact that the ear away from, that is, contralateral to the interferer has a higher SNR than the ipsilateral ear. The effect is quantified by the *interaural level differences*, ILD, that different sources evoke relative to the orientation of the head. Both, IPD and ILD can be readily derived from the interaural transfer function, which holds for almost every sound direction a unique identifier in free-field acoustics.[1] Definitely, it is the application of the ILD that gives binaural processors of source localization and noise suppression an advantage over multi-sensor arrays, which generally do not offer significant location-based amplitude differences among the transducers.

After fundamental endeavors of binaural ASA-based speech enhancement showing promise—see, for example. the pioneering works of [11, 21]—the field recently gained a new impulse with the introduction of statistical models that simulate parts of the auditory pattern-driven processing of binaural cues [15, 29, 41].

The overview of the above-mentioned research lines allows for a second categorization of the speech-enhancement algorithms into approaches that have shown to be suited for the suppression of diffuse noise fields, whereby fixed beamforming filters are generally unmatched to this task, and into approaches that are better suited to the suppression of coherent-noise interference, such as, for example, BSS- and ASA-methods. The combination of these classes of algorithms has been pursued in several works—see, for example, [14, 32]—and was formalized in the fundamental account of Simmer et al. [37] with the factorization of the *minimum mean-square error*, MMSE, solution into an MVDR-beamformer and a single-channel Wiener post-filter . The motivation of this chapter originates from the same intent, by combining bilaterally applied beamforming front-ends with binaural ASA post-filters for the purpose of a higher overall speech intelligibility gain in noise.

Pivotal to these experiments as well as to the studies on binaural core algorithms and classification models are the statistics of directional parameters in noise. So far, binaural algorithms of speech enhancement are often applied without a thorough understanding of the signal power dispersion of multiple sources in different feature spaces and the manner in which binaural parameters change in noise. The following section works toward this understanding by considering the statistics of binaural parameters of the fine-structure and the envelope of a speech source in noise. Given this statistical insight, the aim is to answer why binaural source segregation succeeds in some circumstances and fails in others.

[1] For a particular direction of sound incidence, the interaural transfer function is defined as the quotient of the corresponding head-related transfer functions at each ear. There is also a running interaural amplitude-modulation transfer function for a particular sound direction, which is equivalent to the quotient of the corresponding amplitude-modulation transfer functions of each ear. While the former is rooted in binaural differences of the fine-structure of the waveform, the latter is caused by interaural differences of the envelope. Both types of transfer functions are approximately independent of each other.

## 2 Binaural Statistics

When two signals are added, their waveform *probability-density functions*, PDF, undergo a convolution. As a consequence of additive noise or reverberation, a similar observation is made for interaural parameters, when the signals of an auditory scene lack disjointness in the time-frequency domain.[2] Hence, an undesirable implication for ASA-based source grouping is given by the manner in which the binaural parameters of the target signal are subjected to the nature and strength of the interference. Whereas the binaural detection of the first wave-front can be well exploited for stable localization purposes, a continuous source segregation has to deal with these noise-induced directional source-label alterations in a scene-dependent and pattern-based fashion.

Recent studies quantified the influence of additive noise on binaural parameters at the human head [29], a head mannequin, and at the output of different head-mounted beamforming front-ends [35]. In both studies the binaural parameters were calculated across auditory filters and per short frames of 16 ms length. In the following, the main findings are summarized.

- The free-field statistics of binaural parameters closely reflect the possible exploitation of binaural cues in the auditory scene analysis. The fine-structure IPD parameter is an unequivocal location feature in the lower frequency regime, which is upper-bounded by the spatial Nyquist limit. The fine-structure as well as the envelope-based ILD parameters are each a directional parameter throughout the entire spectrum in an-echoic conditions, although, at low frequencies the level differences are small. The IPD of the envelope represents a meaningful location feature throughout the entire spectrum in an-echoic conditions too. However, at low modulation frequencies of the envelope—approximately modulation frequencies below 100 Hz—the robust computation of the interaural differences is numerically hampered.
- Binaural parameters strongly fluctuate in frequency regions in which the noise level is close to the signal level or higher. The binaural parameters are not equally affected by a signal degradation though. The fine-structure IPD and the envelope ILD proved to be the most reliable location features in noise. The envelope IPD has shown to be the most sensitive feature to additive noise. As regards the first two moments, the ILD mean shifts to the median plane when the SNR decreases. At the same time, the standard deviation increases significantly. A similar observation can be made with the fine-structure IPD. However, even though the standard deviation builds up considerably at lower SNRs, its mean is more resilient to interference.
- The standard deviation of binaural parameters for lateral sources is generally higher than for sources in the front and the back of the head. Hence, a dependence of the statistics of interaural parameters on direction has been demonstrated.

[2] Disjointness between two signals in the time-frequency domain, for instance, the signal $\delta_j(d, m)$ and the signal $\delta_{j'}(d, m)$ can be expressed as $\delta_j(d, m)\delta_{j'}(d, m) \approx 0, \forall j' \neq j$, where $d$ and $m$ denote the frequency and time index, respectively.

- The signal and the noise type are less important to the statistics than the SNR in the analyzed frequency band.
- If beamformers are applied as a front-end, the SNR at the ear-level, that is, after the beamformer, determines the statistics of the binaural parameters, which are—except for a different mean as a function of sound incidence—quantitatively on par with the statistics of binaural parameters at the output of a head mannequin.
- Directional hearing aids alter the front-back ambiguity of natural binaural cues. This ambiguity is, in three dimensions, well-known as the *cone-of-confusion* artifact for narrow-band sounds. Furthermore, the directional hearing aids demonstrated a positive effect on the IPD of the carrier, especially at low frequencies due to the increased distance between the receivers. The ILD of the carrier and the envelope, on the other hand, revealed to be compressed by the directional processing of the front-end.

The list above considers binaural statistics in additive noise, even though a real-world speech-in-noise problem generally consists of additive and convolutional distortions. Therefore, in the following, a statistical analysis of binaural parameters of the fine-structure in reverberation is conducted.

The setup was as follows. Clean female speech was separately convolved with *head-related transfer functions*, HRTFs, and *binaural room impulse responses*, BRIR, for different directions of sound incidence. The recording of the BRIRs took place in a laboratory with variable acoustics in the Bochum institute. The reverberation time in one setting was 0.4 and 0.9 s in the other one, both averaged over 1/3 octave-bands. The critical distances were, respectively, 1.15 and 0.75 m. Using a swept sine technique, the transfer functions were measured at steps of 5 or 30° in the horizontal plane, for the HRTFs and the BRIRs, respectively, with 0° aligned with the median plane and a counterclockwise sense of orientation. The source was at a distance of 2 m from the head mannequin, a Head Acoustics type HMS II.3.[3] For calculating interaural parameters on a short-time basis, the method of Nix and Hohmann was adopted [29]. It comprises a *short-time Fourier transform*, STFT, framework and a subsequent averaging of the *power spectral densities*, PSD, over bands of the auditory bandwidths, before calculating IPD and ILD. In the present work, the binaural analysis stage of the binaural fine-structure algorithm of Sect. 4—including a subsequent averaging of the PSDs over bands of the critical bandwidth [2]—was applied.[4]

The results of a statistical analysis of the binaural fine-structure parameters under the influence of reverberation are given in Fig. 1. As can be easily seen, for both the IPD and the ILD the degradation of directional distinctness grows with the length of the reverberation time. Mean and standard deviation of both parameters are equally affected over a wide frequency range.

---

[3] The same recording chain as well as the same source-receiver distance were applied during the HRTF measurements in the anechoic chamber of the Bochum institute.

[4] Frame length and step size had a length of 16 and 8 ms, respectively. This way they corresponded to the settings in [29, 35].

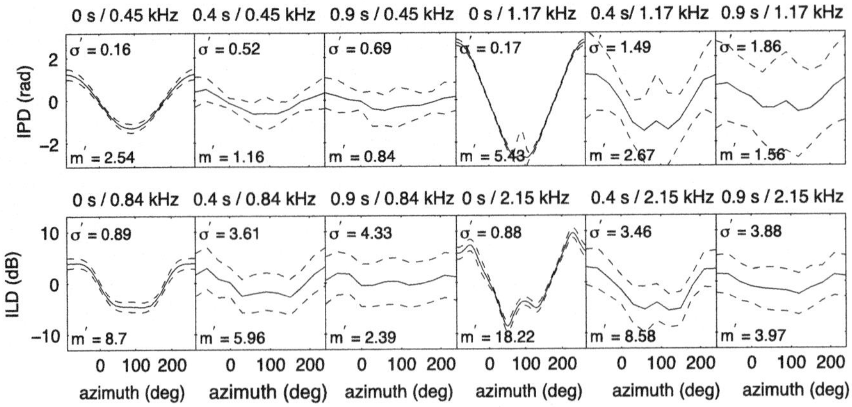


**Fig. 1** Mean (—) and standard deviation (- - -) of the conditional PDFs $P\,(ILD\,|\,azimuth)$ and $P\,(IPD\,|\,azimuth)$, derived from the fine-structure IPD and ILD at different critical bands and in three degrees of reverberation. Center frequencies and reverberation times are given in the titles. The index $m'$ denotes the maximum range of the mean as a function of source azimuth. $\sigma'$ specifies the standard deviation of the mean across all angles

In view of the intended speech enhancement, the binaural statistics of the fine-structure and the envelope in diffuse additive-noise conditions as well as the fine-structure in convolutional noise conditions yield discouraging results. The statistical insight thereby explains why speech processors that have only access to a subset of binaural parameters degrade in diffuse-noise conditions much more than the human hearing does. The challenge for these speech processors remains in the optimal activation of binaural parameters in a pattern-based fashion as was proposed by Harding et al. [15]. It will be applied in the following section. The model hearing process, on the contrary, combines many more cues in demanding circumstances and can, for example, discount equivocal cues of binaural disparity in favor of timbre and modulation. Future ASA-based systems might advance in the source segregation problem through a combination of multiple cue-based classification strategies as well as a weighted scene-dependent activation of these.

## 3 Classical Binaural ASA Algorithms Revisited

Various designs of binaural ASA algorithms for speech enhancement exist. The binaural speech processors can be categorized into three basic groups that use different interaural parameters in the noise-suppression task. Figure 2 depicts those algorithms, which are explained in the following.

Most binaural speech processors originate from the algorithm of Gaik and Lindemann [11], which will, in the following, be named *carrier-level-phase*, CLP, algorithm. This speech processor accomplishes a bilateral frequency decomposition

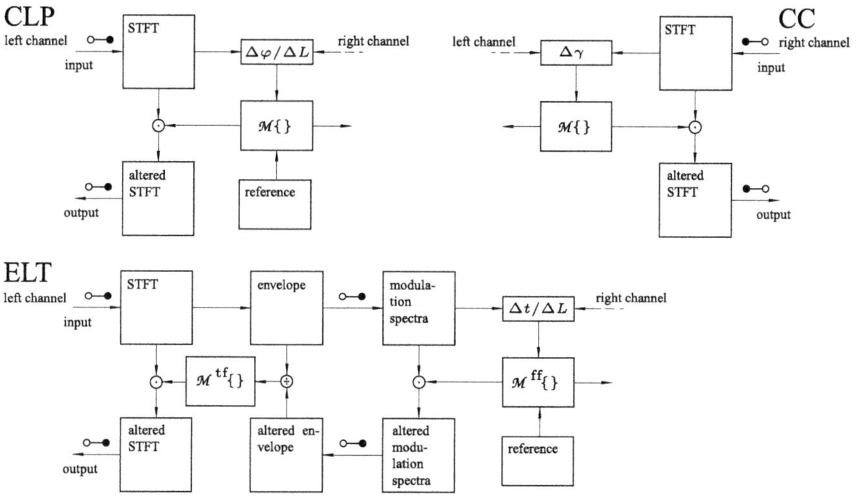


**Fig. 2** Block diagrams of the model-based binaural-speech enhancement processors considered in this section. CLP refers to the processor of Gaik and Lindemann [11], CC to the algorithm of Allen et al. [1], and ELT to that of Kollmeier and Koch [21]. As the binaural algorithms are symmetric around their binaural stages, only one side is drawn. The *barbells* indicate a Fourier transformation, or its inverse. $\Delta\varphi$, $\Delta L$, $\Delta t$ and $\Delta\gamma$ denote, IPD, ILD and ITD, respectively, and the magnitude-squared-coherence at zero lag. $\mathcal{M}$s symbolize transform-domain amplitude-weighting masks

and then calculates the IPDs and ILDs from the fine-structure in each frequency band, in order to employ these parameters subsequently as a directional source feature in an amplitude weighted noise-suppression process. For this purpose, the binaural parameters of the input are compared to the reference parameters of a previously defined *listening* direction. Depending on the deviation from the target direction, a gain between one, that is, target present, and close to zero, that is, target absent, is established. The comparison with the reference is time-variant and executed per frequency band. When thinking of the gain function as a function of time and frequency, one generally speaks of a filter mask—see $\mathcal{M}$s in Fig. 2. A second class of binaural ASA algorithms adopts the concept of the multi-channel spatial coherence algorithm of Allen et al. [1], in the following named *carrier-coherence*, CC, algorithm.

Based on a primitive-grouping scheme, this algorithm exploits the binaural waveform coherence of the fine-structure in bands as a means to suppress diffuse sound. Accordingly, the weighting gain is proportional to the coherence. A third well-known binaural speech processor filters the signal in a conjoint carrier-and-modulation-frequency domain—generally known as amplitude-modulation spectrum—and was developed by Kollmeier and Koch [21]. The noise suppression method of this algorithm is based on ITDs and ILDs of the waveform envelope. These are analyzed in the range of the fundamental frequency of speech. The algorithm is in the following called *envelope-level-time*, ELT, algorithm. Above speech processors offer a binaural output that, when applied in hearing aids, was shown to add to the audiological benefit [14, 28].

In consideration of recent advancements in the field of binaural speech enhancement, the present study undertook an update and a revision of these three binaural speech processors and, subsequently, downstreamed them to a set of binaural front-ends. These front-ends are an artificial head, the ITA head of the RWTH Aachen, a *behind-the-ear*, BTE, hearing aid, GN ReSound type Canta 470-D with and without directional processing, and the *hearing glasses*, HG, of Varibel Innovations BV in two directivity modes.

Throughout the following study, each algorithm was kept conform with the respective initial conceptual design. As regards the implementation, the algorithms are based on a direct FFT/IFFT analysis/synthesis method. Consequently, noise-suppression is applied in the STFT-domain representation of the signal mixture. This offers a high degree of disjointness of speech signals, despite their broadband character. The sampling frequency of each algorithm was set to 16 kHz. Based on an empirical study of separability of a source mixture in the binaural modulation domain of algorithm ELT, the frame length of the—first—STFT was determined. It resulted in a Hanning-weighted segmentation of frames of 256 bins, that is, 16 ms length, and a subsequent zero padding with 256 bins. The adjustment allows for an analysis of modulation frequencies up to 500 Hz in the amplitude-modulation spectrum. The frame shift of the modulation filter stage of algorithm ELT was set to 128 bins, that is, 8 ms length. For good comparability, the STFT settings of algorithm CC and CLP were chosen correspondingly.

In earlier implementations, the lookup tables of binaural parameters—in Fig. 2 referred to as *reference*—often consist of heuristically defined limits that had been found in idealized acoustic situations. As an improvement, Harding et al. introduced the *Bayesian classifier* for estimating soft-gains in binaural filtering [15]. This method was adopted here for the calculation of histogram-based weighting functions in the CLP and ELT algorithms. In contrast, the CC algorithm was based on the standard primitive-grouping scheme, using the non-directional magnitude-squared-coherence at zero lag as a noise classifier.

Subsequent to the basic preparation of binaural ASA algorithms, the central question of this section amounts to the benefit as provided by a particular ASA speech processor with a certain binaural front-end in a specific environment. This implies extensive testing and poses the question in which way speech intelligibility can efficiently be assessed. Furthermore, a constant assessment is required throughout the design of binaural speech processors and the finer optimization of algorithmic parameters such as, for example, the balanced application of directional parameters in the source-separation process. For this purpose, several intelligibility measures for the assessment of binaural and non-linearly processed speech were designed and subjectively evaluated [35]. Based on these and other studies—compare, for instance, with [24]—the I3 measure of Kates and Arehart was selected [20]. Furthermore, the I3 was extended with a better-ear-decision stage, operating per frame of 1.6 s length and per critical band, with a frame shift of 800 ms—the definition is given in [2]—to account for the dominant binaural effect, namely, the head shadow. Using this instrument, any perceptual evaluation of the ASA algorithms is relinquished in this chapter in favor of a broad instrumental analysis. The distortion associated with varying gain-functions was shown to be very similar to center clipping [35].[5] In a listening test of center-clipped speech material, the I3 showed a correlation of more than 90 % with perception [36].

Another challenge is constituted by the optimization of binaural ASA algorithms. Each algorithm possesses a set of algorithmic parameters that need to be tuned to a particular scene and a particular front-end, similar to the facilitation and the cue trading observed in the model hearing process. In this section, a *genetic algorithm* is applied for this complex optimization task. The presentation of the respective algorithmic details is beyond the scope of the synoptic nature of this section. A detailed report on the algorithmic details, their optimization and the assessment is given in [35].

After now having defined the basics of the three binaural algorithms, the review continues in the following way. In the next section, the probabilistic binaural pattern-driven source separation approach is presented. Subsequently, the results of the genetic optimization are summarized. Finally, the binaural speech processors are applied and assessed at the output of several front-ends and in different environments.

## 3.1 Pattern-Driven Source Separation

This section covers the establishment of pattern-driven weighting functions as are used in the speech processors ELT and CLP for the attenuation of noise. A learning-based pattern-driven decision process in the domain of interaural parameters can be built with a Bayesian classification method. In the vanguard of connecting a classifier with a binaural speech processor is, among others, the work of Roman et al., who derived a *maximum a-posteriori*, MAP, classifier from joint ILD-ITD features, that is, bivariate statistics, to estimate a binary mask for noise suppression [33]. The approach was further developed by Harding et al. on the basis of soft-gains for de-noising an automatic speech recognition system [15]. While these works employ histogram-based lookup tables, other researchers estimate the distribution parameters before calculating the filter gains—see, for instance, [26]. In the following, it is solely focussed on histogram-based lookup tables. Later in this chapter, this method is compared with the parametric statistical-filtering approach.

To calculate the noise-suppression mask from the posterior estimation, the short-time binaural parameters are considered to be the output of a stochastic process with $\Delta_{d,m}$ being a time-variant feature of a binaural parameter—or a set of binaural parameters—at every frequency index, $d$, and with realizations in, for example, the range of $\pm\pi$ for the IPD. The underlying stochastic process needs to be estimated in order to generate the required prior distributions, specifically,

[5] The I3 has originally been developed to predict the effects of *additive noise, peak clipping* and *center clipping* on speech intelligibility [20].

the distributions of a binaural parameter—or set—of sound from all directions,[6] $P_d(\Delta_{d,m}) = \sum_\phi P_d(\Delta_{d,m}|\phi)P_d(\phi)$, and the distributions of the joint probability when only the target is present, $P_d(\Delta_{d,m}|\phi_t)P_d(\phi_t)$. Here, $\phi$ is the source azimuth. $\phi_t \in \mathbb{T}$ denotes a set of target directions of the direct sound and room reflections. Consequently, the posterior estimate of the target with a binaural feature or set of these, being given at each time-frequency bin, is found via

$$P_d(\phi_t|\Delta_{d,m}) = \frac{P_d(\Delta_{d,m}|\phi_t)P_d(\phi_t)}{\sum_\phi P_d(\Delta_{d,m}|\phi)P_d(\phi)}. \tag{1}$$

Harding et al. have shown that this equation can be evaluated by the quotient of two histograms as follows,

$$P_d(\phi_t|\Delta_{d,m}) \approx \begin{cases} \frac{H_d^t(\Delta_{d,m})}{H_d^a(\Delta_{d,m})}, \text{ if} & H_d^a(\Delta_{d,m}) > \zeta \\ 0, & \text{otherwise} \end{cases}, \tag{2}$$

where $H_d^t$ and $H_d^a$ are the histograms of the labeled target signal and the histograms of the speech-plus-noise mixture, respectively [15]. $\zeta$ is a threshold to prevent faulty estimations from insufficient statistical data and numerical noise. Consequently, after the division of these distributions, the filter gain can be read from a look-up table by using $\Delta_{d,m}$.

In comparison to binaural speech processors using primitive grouping schemes, the approach taken here constitutes a leap in terms of simplicity and efficiency by applying the statistics of binaural parameters through supervised learning. For the training of the classifier, that is, the collection of the histograms of (2), Harding et al. proposed the ideal binary mask, $\mathcal{M}^b$, as a means to label the data [15]. This preprocessing is adopted in the present section, and reproduced in detail in the final part of this chapter.

## 3.2 Optimization of Binaural Speech Processors

The development of speech-enhancement processors faces an increasing algorithmic complexity that makes the optimization challenging. In fact, a deterministic search for an optimum performance is often not possible, as an exhaustive enumeration of a multidimensional search space demands—even for relatively small problems—an impractical computational effort. At the expense of accuracy, stochastic search algorithms reduce the calculation effort. In general, this trade-off tremendously lessens the time of convergence and yields sufficiently good solutions.

The present work employed a *genetic algorithm*, GA, for the optimization of the three algorithms of this section. To that end, the Genetic Algorithms for Optimization

[6] Throughout this chapter, the target source and the interferers were arranged in the horizontal plane.

Toolbox of Houck et al. [16] was applied with the default settings and the better-ear-I3 measure served as the cost-function. In order to obtain an indication of the optimization complexity, the reproducibility of GA solutions was tested by running several GA optimizations for each setup. Additionally, the GA-optimized parameter sets were applied in changing acoustic environments, which yielded an indication of the robustness and generalizability of a certain solution.

It is not within the scope of the current chapter to reproduce the results of the genetic optimization in detail. A full report on these optimization results can be found in [35]. At large, the GA optimization of algorithmic parameter sets produces practical and psychoacoustically grounded solutions. While the efficiency, in other words, fast convergence to optimal solutions, of the GA procedure is a consequence of the *survival of the fittest* strategy, the regularity of the solutions is a product of the interplay of ASA algorithms in the improvement and prediction of speech intelligibility.

Moreover, the application of an optimization scheme to the holistic framework of ASA-based improvement and ASA-based assessment of speech intelligibility provides solution strategies that may underly the ranking of low-level cues in the model hearing process. By way of example, for front-end and scene combinations in which the carrier ILD is significantly included in the directional weighting process, the ILD never gained more than half of the algorithmic weight in the entire filtering process. By trend, these results correspond to recent psychoacoustic tests regarding the trading of binaural cues in noise [31].

## 3.3 Assessment of Parameter-Optimized ASA Algorithms

This section presents the assessment of the parameter-optimized binaural speech processors at the output of different front-ends under varying acoustical conditions.

### Preparation

In order to employ pattern-driven weighting functions, posterior lookup-tables were generated for the speech processors CLP and ELT at different front-ends. The stochastic processes were approximated with time series of training speech material that comprised a length of two minutes and speech of both genders—composed of concatenated and RMS-equalized utterances. The sentence material was taken from the TNO-SRT database [40]. Male speakers made up 2/3 of the material. This proportion was equally applied for target and masker sources. The training mixtures comprised twelve different spatial configurations. While the target was fixed at $0^\circ$ in all mixtures, the interfering speaker resided in each histogram binning process at one of the angles of $-90, -50, -30, -20, -10, -5, 5, 10, 20, 30, 50$ or $90^\circ$. As regards the canteen setup, that—among other scenes—is assessed in the following section, the interfering speakers were each added to a random time section of the canteen recording at 0 dB SNR. After generating $H_d^{\mathrm{t}}$ and $H_d^{\mathrm{a}}$ for each spatial arrangement, the twelve histograms of each set were summed up.

Although it is an artificial concept, the SNR is determined at the ear-level, including the RMS average between the ears. The rationale is given by the strong correspondence between the statistics of binaural parameters and the SNR at the ear-level. Even though this excludes the directionality of the beamforming front-ends, it allows for a better comparability with omni-directional receivers and among the three binaural speech processors. Genuine applications using the approach proposed here, should reflect the directional level dependence of the binaural front-end in the *a-priori* information.

As regards the parameter optimization of the binaural speech processors, several GA optimization runs at 0 dB SNR were performed using the respective *a-posteriori* weighting functions and speech-in-noise tokens of 15 s length, with a male/female proportion of 2/3. In order to incorporate the stochastic nature of the GA solution in the subsequent assessment, an arbitrary GA-optimized parameter set out of several GA-optimized sets was chosen.

The speech material used to predict the speech-intelligibility enhancement consisted of 45 s of concatenated sentences, for both the target and coherent noise signals, each with a male/female proportion of 2/3. All sentences were RMS equalized and pauses, defined as the RMS level of −50 dB in frames of 10 ms relative to the overall RMS level, were excluded by applying a *voice-activity detection*, VAD, method.

## Assessment Results

The performance of the speech processors in a canteen environment at different mixing SNRs is presented in the following. The respective front-end/post-processor parameter sets of the GA optimization were held constant during the assessment in an SNR range between −10 and 10 dB with increments of 5 dB. Figure 3 presents the results.

Around 0 dB, the differential improvement of speech intelligibility is in the range of 5 to 15 % at the output of algorithm CC and CLP. No improvement is gained with algorithm ELT, which is essentially switched off by an increased compression value of the weighting function—an algorithmic parameter tuned in the GA optimization. Comparing the upper and lower plots in Fig. 3, no marked difference is found between the output of the artificial head and the hearing glasses in the low-directivity mode.[7] None of the post-processors generates a decrease of speech intelligibility, as assessed with the better-ear-I3 measure. This is an important result. If it was possible to classify a scene correctly and, consequently, choose appropriate parameter sets and lookup tables, binaural speech processors may run in a wide range of SNR conditions without detrimental effects on speech intelligibility.

[7] The hearing-aid program *low* of the hearing glasses provides a benefit of 4.4 dB as assessed with the directivity-index method of the ANSI S3.35-2005 standard [4]. In the mode *high* the hearing aid offers an improvement of the directivity index of 7.2 dB.

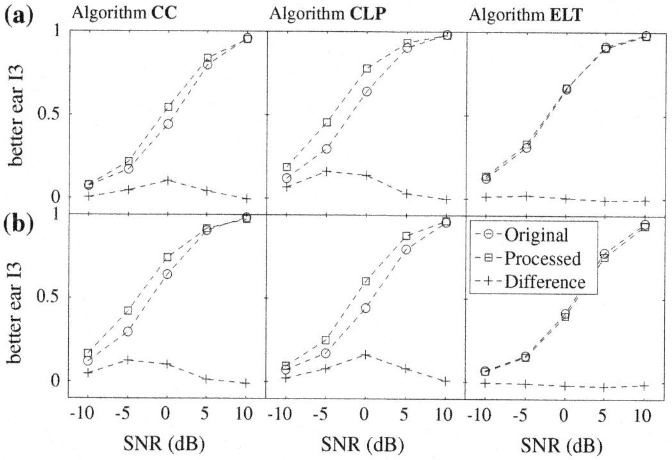


**Fig. 3** Assessment of the binaural speech processors in a canteen environment at varying mixing SNRs and at the output of two front-ends with the better-ear-I3. The target speaker resides at 0 °. **a** ITA head. **b** Hearing glasses in the low-directivity mode

While the previous scene was highly diffuse, the following experiment deals with speech-intelligibility improvement in coherent-interference conditions.[8] In a preliminary analysis, algorithm CC was found to be only applicable for the suppression of incoherent noise. This is obvious, considering that the binaural waveform coherence at zero lag, that is, the weighting function of algorithm CC, is a poor indicator for distinguishing between frontal and lateral sources [35]. Therefore, algorithm CC is not included in the following analysis.

In the first part of the experiment, the CLP processor was applied at the output of different front-ends for the attenuation of two interferers, a first one, N1, at 90 ° and a second one, N2, at varying azimuths between −180 ° and 180 ° with increments of 5 °. The mixing SNRs were adjusted between −10 and 10 dB at increments of 5 dB. The results were interpolated and are given in the contour plots of Fig. 4.

To a first approximation, the benefit of the CLP post-processor is primarily determined by the ear-level SNR. The optimal working point lies at an SNR range of approximately −5 to −10 dB. The highest intelligibility gains are achieved if the interferers collapse at a lateral position of 90 °, such that both interferers share one direction in the binaural domain. Furthermore, mainly low self noise of the hearing aid supports the efficient application of the post-processor. If this requirement is violated, the benefit reduces considerably. This effect is seen for the BTE in the directional gradient processing mode, that generally possesses a significant internal noise level.

[8] The term *coherent-interference condition* was used here to describe an interfering sound source under an-echoic conditions.

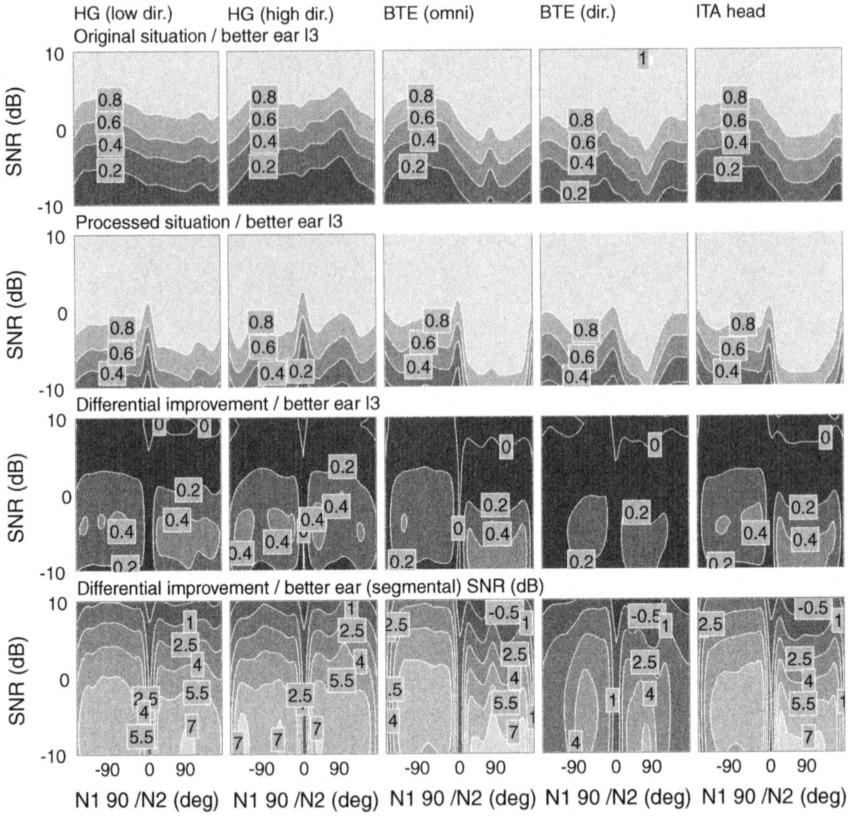


**Fig. 4** Prediction of speech intelligibility before and after applying the CLP processor at the output of several front-ends in the presence of two coherent interferers—using the better-ear-I3. The target speaker is at 0° in all conditions. The *bottom row* gives the segmental SNR, equal to the speech-distortion ratio defined in [20]—with the here proposed *better-ear* computation

In the second part of this simulation experiment, the ELT processor is assessed throughout different coherent interference conditions at the output of the hearing glasses in the low directivitiy mode and the ITA head. Figure 5 depicts the corresponding results—see the subscripts for the spatial arrangements. The comparison with the results of algorithm CLP shows a lower gain of speech intelligibility. If there is only one interferer to be suppressed—the results are presented on the right-hand side in Fig. 5—the disjointness of the sources in the signal domains of algorithm ELT is increased and speech intelligibility recovers substantially.

## Discussion

The assessment delivered the important result that binaural ASA speech processors are approximately independent of the directionality of the front-end. Albeit only

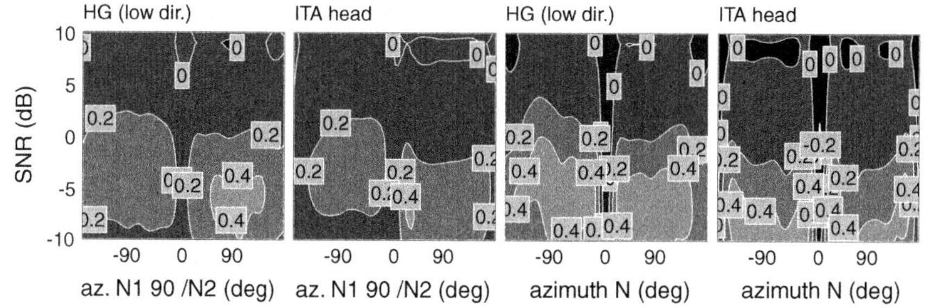


**Fig. 5** Differential speech-intelligibility improvement, assessed with the better-ear-I3, of processor ELT for two front-ends with one and two coherent interferers—see *bottom labels*. The target speaker is fixed at 0°

being grounded in the applied instrumental measure, this confirms the expectation that the SNR gain of a post-processor adds to the SNR improvement of a directional front-end [37]. A high internal noise level, however, was shown to detract from the power of binaural noise suppression.

Comparison of the three binaural speech processors demonstrated the dominance of the CLP algorithm in diffuse- and in coherent-noise conditions. Mainly due to the noise sensitivity of binaural phase differences of the envelope, as shown in [35], algorithm ELT is only beneficial in highly coherent noise conditions. This outcome is in contrast to the results of Kollmeier and Koch [21]. Their implementation of algorithm ELT was audiologically assessed and gained a small improvement of an estimated differential SNR of 2 dB in diffuse-noise conditions at negative mixing SNRs. However, Kollmeier and Koch did not attain an improvement of speech intelligibility in coherent interference conditions. In consideration of the results given here, the contrary outcome might be the result of the second weighting function of algorithm ELT. This weighting function is based on the standard deviation of binaural parameters, and attenuates the filter gain at time-frequency bins where binaural parameters deviate from the predefined listening direction. It was not possible to verify this result with the current implementation, what is probably due to coarser resolution of the modulation domain, caused by a decreased frame-length in the analysis. This reduced frequency resolution is, however, required for a binaural envelope-based suppression of lateral interference. For a deeper study of algorithm ELT see [35].

The adaption processes, that is, the probability-based pattern-driven weighting method and the genetic intelligibility optimization, turned out to produce viable and robust solutions for a great variety of front-ends and speech-in-noise problems. In this respect, the present work advances toward a more comprehensive ASA approach.

Overall, the study gives a revision of the three classical binaural ASA speech processors. Considerable improvements of instrumentally-predicted speech intelligibility were achieved in coherent interference conditions. When applied as a post-processor, the algorithmic benefit showed to be widely independent of the directionality of the front-end, which justifies a combination with beamformers for an increased intelligibility gain in noise. Across coherent and incoherent speech-in-noise problems, algorithm CLP distinctly outperformed the competitors in this comparison. For that reason, this processor will be further analyzed in terms of statistical training methods and models.

# 4 Classification Strategies

Whereas univariate distributions of interaural parameters underly many classical implementations of binaural algorithms, Harding et al. showed that binaural noise suppression in automatic speech recognition can benefit from the application of bivariate distributions [15]. Therefore, in this section, the statistical model of algorithm CLP undergoes a closer inspection. Beginning with the formalization of a simplified version of algorithm CLP, univariate and bivariate distributions will be compared on the basis of the speech-intelligibility gain across a wide range of noise conditions. Further, the training setup of the classifier will be studied in order to find an optimum in terms of the generalization to unseen data.

In a second study, the probabilistic lookup tables are replaced with a parametric *Gaussian mixture model*, GMM. Parametric models possess a series of advantages, such as arbitrary scaling to multivariate feature spaces, less memory space and the efficient exploitation of localization data [5, 25].

## *4.1 Histogram-Based Statistical Filtering*

The histogram-based statistical model for the pattern-driven source separation possesses a set of training parameters that influence the speech intelligibility gain as well as the generalization to unseen speech-in-noise conditions. In the following, these training parameters are under consideration. As a means to focus on the influence of the statistical model, the study applies a stripped-down version of the CLP processor. The simplification involves a reduced set of algorithmic core parameters and their default adjustment. Hence, no stochastic optimization is pursued. Therefore, the implementation reported here differs in parts from the implementation of algorithm CLP in the previous section. It follows a formalization of the algorithm for improved readability of the results.

### Algorithm

Let $s^{\ell}(n)$ be the band-limited signal-plus-noise mixture at the sampling frequency, $f_{\mathrm{s}}$, at the left input of algorithm CLP—compare Fig. 2. Using a Hanning window of $N_{\mathrm{H}}$ samples length, the signal is partitioned into overlapping frames with a frame shift $\Delta p$. Subsequently, the STFT representation of the signal is calculated with an FFT of length $N_{\mathrm{FFT}}$, as

$$S^{\ell}_{d,m} = \sum_{n=0}^{N_{\mathrm{FFT}}-1} s^{\ell}_{m\Delta p+n} h_n e^{-j2\pi n \frac{d}{N_{\mathrm{FFT}}}}, \tag{3}$$

where $d$, $m$ and $h$ are the frequency index, the frame index and the Hanning window, respectively. The signal-plus-noise mixture at the right input is calculated in the same way, which results in $S^{r}_{d,m}$. Thereafter, the power spectral densities, PSD, are estimated. This is done through a modulus and recursive first-order filtering method, known as the Welch method [42]. For the signals at the left and right ear, the Welch method is computed as

$$\begin{bmatrix} \Phi^{\ell}_{d,m} \\ \Phi^{r}_{d,m} \end{bmatrix} = \alpha \begin{bmatrix} \Phi^{\ell}_{d,m-1} \\ \Phi^{r}_{d,m-1} \end{bmatrix} + (1-\alpha) \begin{bmatrix} |S^{\ell}_{d,m}|^2 \\ |S^{r}_{d,m}|^2 \end{bmatrix}, \tag{4}$$

where the smoothing factor, $\alpha$, is calculated as $\alpha = \exp(-\Delta p/(\tau f_{\mathrm{s}}))$, with $\tau$ being the time constant. Furthermore, the cross-power spectral density is calculated as

$$\Phi^{\ell r}_{d,m} = \alpha \Phi^{\ell r}_{d,m-1} + (1-\alpha) S^{\ell}_{d,m} \bar{S^{r}}_{d,m}, \tag{5}$$

in order to infer binaural phase differences. Here, $\bar{S^{r}}$ is the complex conjugate of $S^{r}$. Subsequently, the IPD is computed as

$$\Delta\varphi_{d,m} = \angle \Phi^{\ell r}_{d,m}, \tag{6}$$

and the ILD is found via

$$\Delta L_{d,m} = 10 \log_{10} \frac{\Phi^{\ell}_{d,m}}{\Phi^{r}_{d,m}}. \tag{7}$$

As introduced with (1) and (2), noise suppression in algorithm CLP is based on the *a-posteriori* estimate of the target, given a binaural feature—or feature vector, see below—at each time-frequency atom. Consequently, a soft-mask of probabilities is multiplied with the original STFT-signal

$$\begin{bmatrix} \check{S}^{\ell}_{d,m} \\ \check{S}^{r}_{d,m} \end{bmatrix} = \max\left(P_d(\phi_{\mathrm{t}}|\Delta_{d,m}),\, A\right) \begin{bmatrix} |S^{\ell}_{d,m}| e^{j\angle S^{\ell}_{d,m}} \\ |S^{r}_{d,m}| e^{j\angle S^{r}_{d,m}} \end{bmatrix}, \tag{8}$$

where $A$ is a flooring parameter that allows for balancing the enhancement/distortion trade-off. The check mark over the STFT representation denotes the noise-suppressed signal mixture. The original phase is left unchanged, and, as a last processing step, the waveform of the noise-suppressed signal is reconstructed through an inverse STFT.

In the following, univariate and bivariate distributions are considered. Beginning with the univariate distribution, the classification feature at each time and frequency bin is defined as

$$\Delta_{d,m} = \begin{cases} \Delta\varphi_{d,m}, \text{ if} & d < d_x \\ \Delta L_{d,m}, \text{ otherwise } d \geq d_x \end{cases}, \tag{9}$$

**Table 1** Core parameters of algorithm CLP and training parameters of the histogram-based filtering method

| | | | | |
|---|---|---|---|---|
| *Algorithmic core parameters* | | | | |
| $f_s$ | $N_H$ | $N_{FFT}$ | $\Delta p$ | $\tau$ |
| 16 kHz | 512 | 512 | 128 | 8 ms |
| *Training parameters* | | | | |
| $A$ | $\zeta$ | $SNR_h$ | $\xi$ | $d_x$ |
| 0.02 | [1, 5, 8, 15, 30] | [−5, 0, 5] dB | [−5, 0, 5] dB + $SNR_h$ | [1, 8] kHz |

*Square brackets* denote parameter ranges for which the speech processor is tested

where $d_x$ is the crossover frequency between the operational ranges of the IPD and the ILD parameter. In case of a bivariate distribution of both directional fine-structure parameters, the feature vector is defined as

$$\Delta_{d,m} = \left[\Delta\varphi_{d,m}\ \Delta L_{d,m}\right]. \tag{10}$$

For calculating the *a-posteriori* probability of target presence per time frame and frequency band, as defined in (2), feature histograms for the target speech as well as for the speech-plus-noise mixture need to be generated. As a means to train the classifier in a supervised fashion, Harding et al. suggested the following ideal binary mask definition for labeling the data, namely,

$$\mathcal{M}^{b}_{d,m} = \begin{cases} 1, \text{ if} & 10\log_{10}\frac{\Phi^{s}_{d,m}}{\Phi^{n}_{d,m}} > \xi \\ 0, \text{ otherwise} & \end{cases}, \tag{11}$$

where $\Phi^s$ and $\Phi^n$ are the PSD of the speech signal and the noise signal, respectively, and $\xi$ is the local SNR threshold that categorizes speech and noise [15]. $\xi$ is added—hence relative—to the histogram mixing SNR of the binning process, which is denoted $SNR_h$. Due to the offline mixing process, the separated target and noise signals are accessible after their relative level adjustment. One-channel signals of speech and noise are generated by summing both ear signals, prior to the processing with the STFT and the Welch method, used to calculate $\Phi^s$ and $\Phi^n$.

Accordingly, the ideal binary mask is applied to isolate binaural features that correspond to dominant PSDs of the target signal. These directional parameters, the training features, are binned into the target histograms $H^{t}_{d}$. In this manner, interaural deviation from free-field as a consequence of the noise are taken into account. Histograms of the noisy mixture, $H^{a}_{d}$, on the other hand, are directly binned from the mix of binaural training features.

Univariate histograms of the IPD were designed to have each 500 bins in the range of $\pm\pi$. Correspondingly, univariate histograms of the ILD were arranged to have each 500 bins in the range of −40 to 40 dB. Using the same parameter ranges, bivariate histograms were sampled with a grid of 100 × 100 bins.

The parameters of algorithm CLP—in the stripped-down version—are collected in Table 1. The STFT parameter settings represent optimal disjointness for speech mixtures in the STFT domain [43]. A $d_x$ of 1 kHz can be considered a reasonable choice for separating the working ranges of IPD and ILD. Because of the $2\pi$ periodicity, the IPD becomes an ambiguous directional feature above this frequency limit. Conversely, below 1 kHz, the ILD is a weak indicator of direction, particularly in reverberation. The adjustments of $A$ as well as the histogram bin sizes are based on experiment and observation.

Four central parameters of the estimation process of the distributions are studied in the following. These are the $\mathrm{SNR_h}$ at which the probabilistic model is trained, the local SNR criterion, $\xi$, at which the signals are differentiated into the target and the noise signal, the histogram threshold, $\zeta$, of $H_d^{\mathrm{a}}(\Delta_{d,m})$—see (2)—and the crossover frequency, $d_x$, between either binaural parameters in the univariate model.

## Training and Evaluation

For applying real-world reverberation acting on the target signal as well as on the interference signal, a moderately diffuse sound scene with a reverberation time of 0.4 s was created with the setup described in Sect. 2. Whereas the speech signals of target and interference were convolved with the pre-recorded BRIRs for the first experiment, a single-loudspeaker playback at 300° of a high-quality internet-streamed radio broadcast was additionally recorded in the same setup, and used as an interferer in the second experiment. The speech material used for the training and the assessment consisted of male speakers of the TNO-SRT corpus, speaking phonetically balanced sentences in Dutch, English, French and German [40]. Each speaker was applied once during the following experiments. Except for the directional variation of a particular speech signal within the training and evaluation sets, no overlap existed between the target and interference signals, nor between the training and evaluation sets. The recording of a commercial radio broadcast featured Alpine pop music and partly Austrian conversational speech, always accompanied by background music.

The training of the probabilistic model was carried out with speech mixtures that consisted of one target source and one interfering source. Whereas the target was assigned to the direction of 0°, the interference resided for each of the speech-in-speech mixtures at one of the angular positions: −120, −90, −60, −30, −30, −30, 30, 30, 30, 60, 90 or 120°. The threefold repetition at ±30° was chosen to increase the histogram counts around the median plane. After generating $H_d^{\mathrm{t}}$ and $H_d^{\mathrm{a}}$ for each spatial arrangement, the twelve histograms of each set were summed up. The signals of the target and the interference consisted each of three male speakers and had each a length of 300 s. Using an RMS-based VAD, signal portions in frames of 10 ms smaller than −20 dB relative to the overall RMS were discarded. By applying a relatively high VAD-threshold, spurious background energy from the speaker recordings, that may lead to a lack of disjointness in the time-frequency domain, was partly excluded from the estimation process.

Throughout the evaluation of the first experiment, speech material of four speakers was applied for each the target speech and the lateral interference. The two tokens had each a length of 200 s. Again the RMS-based VAD was applied to exclude longer speech pauses. Therefore, the threshold was set to −30 dB. For each speaker/interference combination, the interference speaker was consecutively set to one of the angular positions: 30, 60, 90 or 150°. The SNRs for assessment ranged from −15 to 15 dB in steps of 5 dB.

The second experiment evaluated the generalizability of the training stage, by replacing the speech interferer with the aforementioned radio broadcast recording at 300°. Target speech of four speakers was mixed with randomly chosen sections of the radio recording. The length of each evaluation token was again set to a length of 200 s, prior to the VAD—the threshold equaled −30 dB. Speech enhancement was predicted at SNRs ranging from −15 to 15 dB, in steps of 5 dB. The signal content of the chosen radio station led to less disjointness between the target signal and the interference and, consequently, made the speech-in-noise problem more difficult when viewed from the local SNR-level.

## Results

The following evaluation is primarily based on the previously introduced better-ear-I3 measure. Because of the limited accuracy with which such an instrumental measure reflects absolute perceptual speech intelligibility[9] as well as for the sake of brevity, only the differential intelligibility changes with respect to the original speech-plus-noise mixtures are given. In addition, as there was no instrumental measure for assessing speech enhancement processors in the presence of reverberation available, the degraded and the enhanced signals are compared to the corresponding reverberated target signals. This approach is supported by the fact that a reverberation time of 0.4 s lowers speech intelligibility by only a small margin [17]. As a means to query the outcomes of the better-ear-I3 measure, the better-ear-STOI measure [38] and the mean-ear-Q3 method [19] were furthermore applied, whereby the latter is a quality measure. The *better-ear* extension of the STOI is based on a maximum speech-intelligibility decision per time frame and frequency band. As regards the *mean-ear* extension of Q3, the quality counts per frequency band and time frame are averaged across both channels. The definitions of frame length and bandwidth are for each of the instrumental measures conform with the original proposals found in literature.

Figure 6 depicts the speech-enhancement predictions when applying the bivariate distributions of IPD and ILD and a histogram threshold, $\zeta$, of one. There are three obvious inferences. First, a considerable improvement of speech intelligibility is

[9] In order to provide for an instrumental measure of absolute perceptual speech intelligibility, a perceptual intelligibility test with the applied speech material as well as a subsequent fitting of the instrumental results to the perceptual recognition scores with a logistic function need to be executed. Therefore, *un-fitted* instrumental measures merely quantify trends.

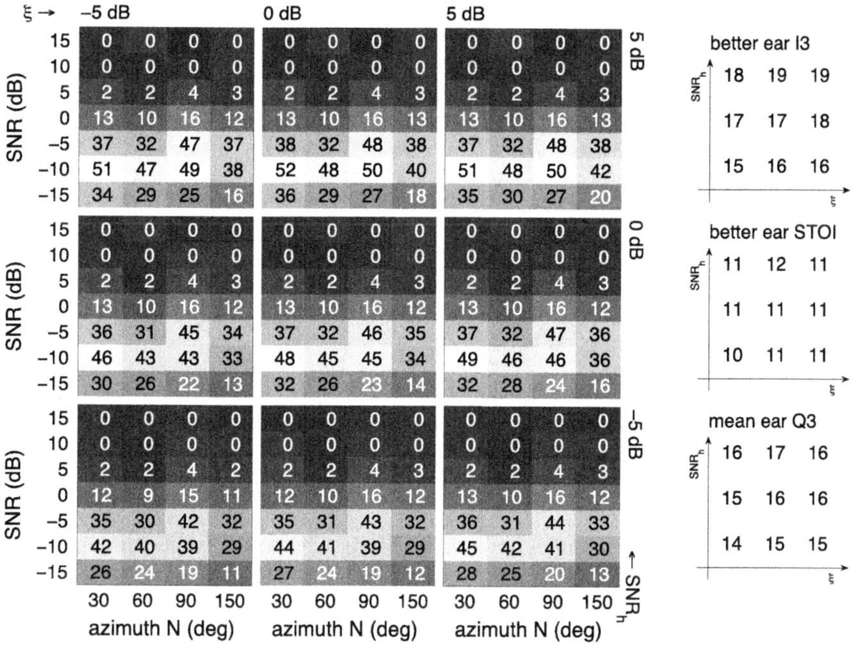


**Fig. 6** Results of the 1st experiment. *Left panel* the *gray coded* $3 \times 3$ matrix gives the differential speech-intelligibility improvement predicted with the better-ear-I3 in %, using bivariate distributions of IPD and ILD in processor CLP with $\zeta = 1$. Target speaker is fixed at 0°. Each of the nine subplots reflect different noise conditions, characterized by mixing SNR and interferer, N, angle. The training parameters $\mathrm{SNR_h}$ and $\xi$ of the statistical model set the nine subplots apart. *Right panel* mean differential-intelligibility changes in % for the better-ear-I3, the better-ear-STOI and the mean-ear-Q3. Summaries are derived from the respective $3 \times 3$ subplot matrices

observed at mixing SNRs of −10 to −5 dB. Second, no decline of intelligibility arises at higher mixing SNRs. Third, the training parameters $\mathrm{SNR_h}$ and $\xi$ show a moderate influence on the performance. Nevertheless, a consistent optimum is found with the three instrumental measures at an $\mathrm{SNR_h}$ of 5 dB and an $\xi$ of 0 dB.

Next, the reverberated speech-in-radio mixture is evaluated in order to test the generalizability of algorithm CLP including the *un-fitted* bivariate histograms with a $\zeta$ set to one. Figure 7 gives the results. As compared to experiment one, above inferences remain valid. Due to the more difficult speech-in-noise problem, the optimal point of operation of the speech processor shows to be shifted by about 5 dB toward higher SNRs. As such, the differential intelligibility improvement is reduced at a mixing SNR of −10 dB by about 10 to 20 %, but is increased at higher mixing SNRs. One the whole, an optimum in noise suppression is again reached with the training parameters $\mathrm{SNR_h} = 5$ dB and $\xi = 0$ dB.

In order to compare the bivariate model with the univariate analogue as well as to quantify the influence of parameter $\zeta$, the mean differential improvement in percent was calculated for the three instrumental measures. Each mean per $\mathrm{SNR_h}$ and

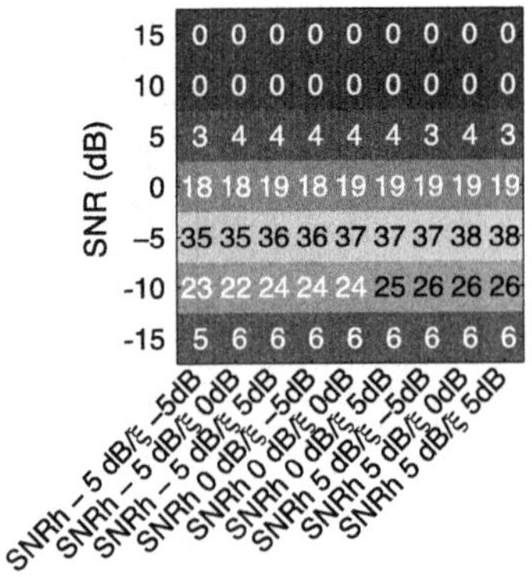


**Fig. 7** Results of the 2nd experiment. Differential speech intelligibility improvement as assessed with the better-ear-I3 in %, using bivariate distributions of IPD and ILD in the CLP processor with $\zeta = 1$. The target speaker is fixed at 0°

$\xi$ includes all interference angles and mixing SNRs. The column of the three subplots on the right-hand side in Fig. 6 already gave a part of the outcomes of this averaging process. In Fig. 8, the resulting mean values of each averaging process are compared statistically while varying $\zeta$ and the lookup histograms. Furthermore, the subfigure at the right-hand side of Fig. 8 juxtaposes the statistical analysis of the mean differential improvement in the 2nd experiment. As can be seen, the bivariate model outperforms the univariate model with a $d_x$ of 1 kHz as well as the univariate model with a $d_x$ of 8 kHz, that is, when only the IPD is applied. The comparison between the univariate models shows that the exclusive application of the IPD results in an advantage in experiment one. However, this outcome is detected as an over-fitting in experiment two, in which the combined application of IPD and ILD leads to an improved noise suppression for the univariate model. Further, the spread of the mean differential improvement, hence, the impact of varying the training parameters $SNR_h$and $\xi$, is influenced by the choice of parameter $\zeta$. Looking at the two experiments, this dependency reveals to be not monotonous. Nevertheless, optimal performance can be isolated for both experiments, with $\zeta = 1$ for the bivariate model and $\zeta =$ 15 for both univariate models.

## Conclusion

The intention of the preceding experiments on noise suppression was to find most beneficial statistical models and optimal parameters sets for training the classifiers. At the same time, a set of training parameters were held constant during the experiments. These are basically the histogram resolution, the spatial sound scene arrangements during the training and the VAD thresholds. This fact needs to be considered throughout the following conclusions.

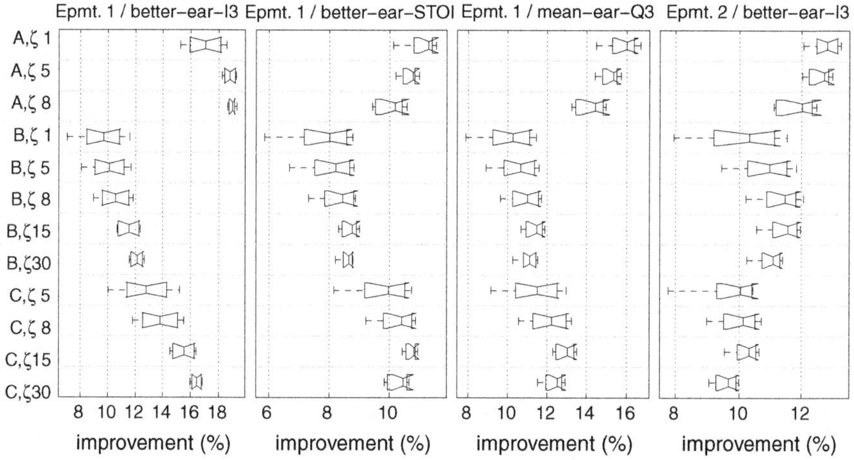


**Fig. 8** Results of the statistical analysis of the mean differential improvement of the 1st and 2nd experiments—see headings. (*A*) Bivariate model. (*B*) Univariate model with a $d_x$ of 1 kHz. (*C*) Univariate model with a $d_x$ of 8 kHz. *Central marks*: Medians. *Box edges* 25th and 75th percentiles. *Whiskers* most extreme values. *Notches* comparison intervals. If intervals do not overlap, the corresponding medians are significantly different at the 5 % level [39]

As was shown across a large number of speech-in-noise conditions, bivariate distributions of binaural features offer an advantage over univariate distributions in terms of noise suppression. No marked decline in performance was found for speech interference at un-trained noise directions as, for instance, presented with the 150° noise condition in the first experiment. When disjointness diminishes, as was rendered with the broad-band music interference of the second experiment, the mean performance of noise suppression drops by only a few percent—assessed with the better-ear-I3. Consequently, the bivariate distribution as a lookup table in algorithm CLP can be recommended. Optimal training parameters were given. The second important result is that binaural speech enhancement processors are applicable in real-world environments, featuring reverberation and continuously active non-speech backgrounds.

## 4.2 Parametric Statistical Filtering

To avoid using histograms to approximate multivariate PDFs, what becomes increasingly impractical for more than 2-dimensional data, it is possible to model these parametrically. A widespread approach to such parametric modeling is the *Gaussian-mixture model*, GMM, sometimes also referred to as *mixture of Gaussians*, MoG. The GMM approximates a PDF as a weighted sum of multivariate Gaussian normal distributions. The estimate of the multivariate PDF of an arbitrary random vector,

$\Delta_m$, as a weighted sum of Gaussian distributions can be defined as

$$\tilde{P}(\phi_t|\Delta_m) = \sum_{c=1}^{C} p_c \cdot \mathcal{N}(\mu_c, \Sigma_c), \quad (12)$$

$$\text{with} \sum_{c=1}^{C} p_c = 1. \quad (13)$$

The model is parameterized by the mixture weights, $p_c$, the component count, $C$, the mean vectors, $\mu_c$, and covariance matrices $\Sigma_c$, that can be estimated to fit a given set of training data through an *expectation-maximization* algorithm. The method is, for example, implemented in the function `gmdistribution.fit` of the MATLAB Statistics Toolbox [39].

Using such a model, it is possible to perform Bayesian classification based on a larger number of localization features than with histogram-based lookup tables—with reasonable use of computational resources. Here, the GMM-based classifier was implemented as part of a generic binaural speech-enhancement framework. It focussed on the possibility of freely-configuring sets of localization features to be employed for filtering.[10] To simplify the implementation and its generality, the STFT filter bank center frequencies at which binaural features are observed, are also treated as part of the feature vector.

The approximation accuracy of GMMs is easily scaleable with the model complexity as determined by the component count, $C$, of the GMM approximation. In theory, an arbitrarily exact approximation of the training-data distribution is possible provided that a sufficiently high component count is chosen—even if the real probability distribution that created the data is not a mixture of Gaussians. In practice, however, high component counts impose several problems. Besides an increasing computational load, high component counts may lead to an over-adaption to the training-data set, which reduces the generalizability of the model. Moreover, with increasing component count, numerical problems may arise during parameter estimation, depending on the statistical properties of the training data. That said, it is crucial for a good modeling accuracy that the data is well conditioned to yield statistical-distribution properties that can be sufficiently modeled even with very few components. For example, GMMs greatly benefit from rather compactly distributed input data that increases the *Gaussianity* of the data and thus reduces the model complexity needed to achieve a certain level of accuracy. The choice of the ITD instead of the IPD is such a measure to improve feature-distribution properties, since

---

[10] The current section is an excerpt from the diploma thesis of Ch. Luther, *Speech intelligibility enhancement based on multivariate models of binaural interaction*, Ruhr-University Bochum, 2012. Contact the author to obtain a copy.

the ITD varies much less over different frequency regions and thus leads to a more compact clustering of observations. Also, the ranges of the individual features should be approximately in the same order of magnitude and preferably not discretely be valued in order to avoid numerical instability of the EM process [39].

### Possible Implementations

The generality of the GMM approach enables us to implement a variety of algorithmic concepts just by defining the set of features to be employed for the classification task, along with the decision whether to process STFT or modulation-spectral signal representations. For a comparison with the results from the previous section, algorithm CLP is analyzed as the basic approach for interaural fine-structure based speech enhancement in conjunction with a parametric statistical model. Therefore and herewithin, the algorithm is referred to as algorithm CLPp. The classification algorithm derives its decision from a feature vector consisting of the fine-structure ILD and ITD values and the STFT center frequency at which these features were observed. The parameters of the STFT framework are given in Table 1.

As a second implementation, an enhanced-CLPp algorithm was developed that is augmented by the frequency-dependent features of the interaural coherence at zero lag. Several studies successfully applied interaural coherence as a selection criterion to improve source localization in complex listening situations [7, 10]. In an attempt to exploit this advantage for speech enhancement, the possible benefits of considering the interaural coherence along with fine-structure ILDs and ITDs were investigated. In this way, the decision of whether a signal component belongs to the target or the interference would not only be based on the ILD and ITD features alone but also on the interaural coherence as a measure for how reliable these features are.

For both implementations, the PDFs of target and interference feature vectors were estimated using the `gmdistribution.fit` function from the MATLAB Statistics Toolbox with a component count of 8. This component count is the result of preliminary tests carried out to find out a good compromise between modeling accuracy and computational resources needed during model estimation.

### Evaluation and Conclusion

The parametric-filtering approach was applied to a subset of the speech-in-noise conditions that are presented in the preceding section. Figure 9 gives the predicted differential speech-intelligibility improvements of the two speech processors.

Overall, the results show that algorithm CLPp is in most of the tested conditions inferior to its non-parametric counterpart with bivariate lookup tables. Due to the reduced accuracy of the GMM approximation, this constitutes no surprise. Generally, the GMM can be considered a smoothed PDF approximation. Thus it will adapt less exactly to its training data but, in turn, interpolate more smoothly than histograms. Therefore, it is generally expected that the parametric model will, within the bounds of the model complexity, adapt to situations that differ considerably from the training

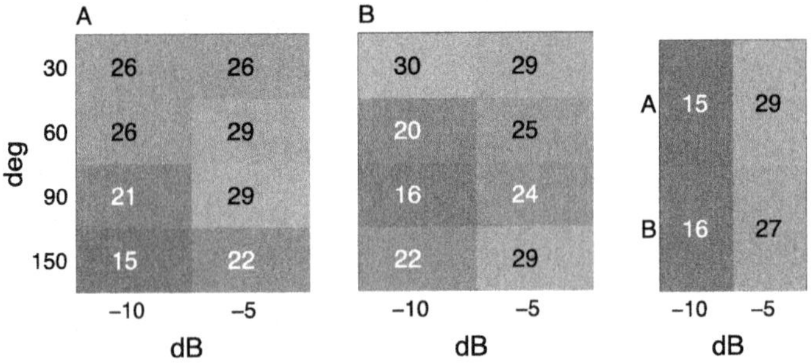


**Fig. 9** Differential speech intelligibility improvement in %, assessed with the better-ear-I3, using GMM-approximated feature distributions. (*A*) CLPp. (*B*) Enhanced-CLPp. *Left* results for scenes with one interfering speaker and reverberation. *Right* radio as the interferer and reverberation. Throughout the simulations, the target speaker is at 0°

conditions. The results found for the attenuation of the radio interference in the second experiment confirm this expectation.

No consistent benefit is seen for the coherence-augmented enhanced-CLPp processor. The result is in line with the conclusion of Sect. 3, which showed that algorithm CC is not suitable for improving speech intelligibility, mainly because of its inability to distinguish between different directions. In addition, the combination of interaural coherence with time and level differences suffers from the fact that interaural coherence offers only a small range—residing close to unity—of high information content. Once the coherence decreases as is the case in slight reverberation and, more so, close or beyond the critical distance, the feature merely introduces fuzziness and cannot contribute to the classification task [27].

Listening to the output reveals a high-pass character of the enhanced-CLPp algorithm. Including coherence as a feature obviously suppresses incoherent low-frequency energy of the target and the interference. This effect can be used as an instrument to improve speech quality. In terms of speech intelligibility, the reader is furthermore reminded that, for the reasons given above in this section, the output of the processor was compared to the reverberated target signal. Therefore, due to the stochastic nature of the short-time binaural coherence, it can be assumed that reverberation in the target signal was partly classified as noise and thus suppressed. Therefore, the instrumentally measured intelligibility declined accordingly. Consequently, more appropriate means have to be applied to study the algorithm further. Peissig, for example, gained encouraging results in listening tests with a similar algorithm [30].

The most powerful speech intelligibility-enhancement-strategies as presented in this chapter, are anticipated to be suitable for wearers of cochlear implants, since non-linear processing artifacts should have little influence on the final low-frequent envelope and modulation-based coding strategies in cochlear implants. Patients with less-severe hearing deficits should generally benefit from rigorous speech enhancement in difficult noise conditions too. However, when speech quality is to be maximized, the parametric filtering method was shown to produce good results. As was audiologically demonstrated, a quality improvement can also result in a moderate but significant speech-intelligibility improvement when the subjects are confronted with meaningful speech material [34].

## Conclusions and Outlook

In this chapter, standard binaural speech processors for improving speech intelligibility in noise were investigated. In the initial study on binaural statistics, interaural temporal parameters of the fine-structure were found to be the most resilient directional features in noise. Interaural level differences of the fine-structure and envelope behave widely equal in noise. The subsequent application of directional features in noise suppression substantiated these results. In coherent as well as diffuse-noise conditions, the binaural fine-structure algorithm clearly surpassed speech processors that employ the binaural envelope or the binaural coherence for noise classification.

Superdirective beamforming filters constitute today's most powerful solution to the speech-in-noise problem. However, for economic and cosmetic reasons, microphone arrays are not unconditionally allowed to extend to dimensions that are physically optimal. Post-filters as, for example, binaural speech processors, can be applied to enhance intelligibility further. Using two commercially-available hearing aids with four beamforming modes, it has been shown that binaural speech processors can be applied as post-processors. It was further demonstrated that the SNR gain of the post-filter is generally additive to the SNR gain of the beamformer.

Further, different statistical models and training parameters were studied with respect to intelligibility using the binaural fine-structure algorithm. In a comparison with univariate models, it was found that speech enhancement benefits from bivariate distributions in trained and untrained conditions. The subsequent application of a multivariate-parametric model could not attain an advantage over histogram-based methods. Yet, the parametric approach produces a high-quality output—along with a considerable noise suppression—and, therefore, enables well-balanced speech enhancement.

Although instrumental measures for rendering the intelligibility of binaural and non-linearly processed speech signals are likely to replace listening tests in the future, the work of this chapter was based on preliminary measures that show only a fair amount of correlation with perception. Therefore, the results of this chapter remain unconfirmed in terms of perception. Considering the present work as a catalogue of binaural filtering techniques, it will be a future task to test promising algorithmic solutions and parameter sets against perception.

The human performance to unravel a speech-in-noise mixture is yet unattainable for computational approaches. In order to match up to this superior functioning, it is likely that the computational mimicry will discard the artificial reconstruction of the target waveform. Recent developments in the field of automatic speech recognition prepare the grounds for more elaborate ways of speech enhancement. Novel automatic speech recognition systems combine low-level and high-level features in order to form hypotheses for accessing portions of clean target speech from databases. As regards instantaneous speech enhancement, such methods are beyond reach in the coming years. As a next step, an obvious sophistication of the algorithms presented herein is the application of a multitude of low-level features, that is, besides the directional parameters one can think of spectral markers of speakers and backgrounds, and time-dependent hierarchical statistical models. Universal concepts underlying such a development were theoretically proposed and already implemented [3, 22, 41].

**Acknowledgments** The authors gratefully acknowledge their indebtedness to two anonymous reviewers for helpful advice. Part of this work has been supported by the Dutch Technology Foundation STW-project # DTF.7459.

## References

1. J. B. Allen, D. A. Berkley, and J. Blauert. Multimicrophone signal-processing technique to remove room reverberation from speech signals. *J. Acoust. Soc. Am.,* 62:912–915, 1977.
2. ANSI/ASA. American national standard methods for calculation of the speech intelligibility index. Technical report, Am. Nat. Standards of the Acoust. Soc. Am., S3.5-1997 (R2007).
3. J. Blauert. Epistemological bases of binaural perception—a constructivists' approach. In *Forum Acusticum 2011,* Aalborg, Denmark, 2011.
4. M. M. Boone. Directivity measurements on a highly directive hearing aid: The hearing glasses. In *AES 120th Conv., Paris, France,* 2006.
5. M. M. Boone, R. C. G. Opdam, and A. Schlesinger. Downstream speech enhancement in a low directivity binaural hearing aid. In *Proc, 20th Intl. Congr. Acoust., ICA 2010,* Sydney, Australia, 2010.
6. A. W. Bronkhorst. The cocktail party phenomenon: a review of research on speech intelligibility in multiple-talker conditions. *Acta Acust,/Acustica,* 86:117–128, 2000.
7. M. Dietz, S. D. Ewert, V. Hohmann. Auditory model based direction estimation of concurrent speakers from binaural signals. it, *Speech Communication,* 53:592–605, 2011.
8. A. J. Duquesnoy. Effect of a single interfering noise or speech source upon the binaural sentence intelligibility of aged persons. *J. Acoust. Soc. Am.,* 74:739–743, 1983.
9. N. I. Durlach and H. S. Colburn. *Handbook of Perception,* volume 4, chapter Binaural phenomena. New York: Academic Press, 1978.
10. C. Faller and J. Merimaa. Source localization in complex listening situations: Selection of binaural cues based on interaural coherence. *J. Acoust. Soc. Am.,* 116:3075–3089, 2004.
11. W. Gaik and W. Lindemann. Ein digitales Richtungsfilter, basierend auf der Auswertung interauraler Parameter von Kunstkopfsignalen. In *Fortschr. Akust., DAGA'86,* volume 86, pages 721–724, Oldenburg, Germany, 1986.
12. E. L. J. George. *Factors affecting speech reception in fluctuating noise and reverberation.* PhD thesis, Vrije Universiteit, The Netherlands, 2007.
13. J. Greenberg and P. Zurek. *Microphone arrays: Signal processing techniques and applications,* chapter Microphone-array hearing aids. Springer-Verlag, 2001.
14. V. Hamacher, U. Kornagel, T. Lotter, and H. Puder. *Advances in digital speech transmission,* chapter Binaural signal processing in hearing aids. John Wiley & Sons Ltd., 2008.
15. S. Harding, J. Barker, and G. J. Brown. Mask estimation for missing data speech recognition based on statistics of binaural interaction. *IEEE Trans. Audio, Speech, and Language Processing,* 14:58–67, 2005.

16. C. Houck, J. Joines, and M. Kay. A genetic algorithm for function optimization: a Matlab implementation. *North Carolina State University, Raleigh, NC, Technical Report,* 1995.
17. T. Houtgast and H. J. M. Steeneken. *Past, present and future of the Speech Transmission Index,* chapter The roots of the STI approach, pages 3–11. TNO Human Factors, Soesterberg, The Netherlands, 2002.
18. L. Jeffress. A place theory of sound localization. J. *Comparative and Physiological Psychol.,* 41:35–39, 1948.
19. J. M. Kates and K. H. Arehart. A model of speech intelligibility and quality in hearing aids. In IEEE Worksh. *Applications of Signal Process. to Audio and Acoustics, WASPAA,* pages 53–56, New Paltz, 2005.
20. J. M. Kates and K. H. Arehart. Coherence and the speech intelligibility index. *J. Acoust. Soc. Am.,* 117:2224–2237, 2005.
21. B. Kollmeier and R. Koch. Speech enhancement based on physiological and psychoacoustical models of modulation perception and binaural interaction. *J. Acoust. Soc. Am.,* 95:1593–1602, 1994.
22. D. Kolossa. High-level processing of binaural features. In *Forum Acusticum 2011,* Aalborg, Denmark, 2011.
23. J. Li, S. Sakamoto, S. Hongo, M. Akagi, and Y. Suzuki. Two-stage binaural speech enhancement with Wiener filter for high-quality speech communication. *Speech Comm.,* 53:677–689, 2010.
24. J. Ma, Y. Hu, and P. C. Loizou. Objective measures for predicting speech intelligibility in noisy conditions based on new band-importance functions. *J. Acoust. Soc. Am.,* 125:3387–3405, 2009.
25. N. Madhu. Data-driven mask generation for source separation. In *Int. Symp. Auditory and Audiological Res., ISAAR,* Marienlyst, Denmark, 2009.
26. M. I. Mandel, R. J. Weiss, and D. P. W. Ellis. Model-based expectation maximization source separation and localization. *IEEE Trans. Audio, Speech, and Language Process.,* 53:382–394, 2010.
27. R. Martin. *Microphone arrays: Signal processing techniques and applications,* chapter Small microphone arrays with postfilters for noise and acoustic echo reduction. Springer-Verlag, 2001.
28. I. Merks. *Binaural application of microphone arrays for improved speech intelligibility in a noisy environment.* PhD thesis, Delft University of Technology, The Netherlands, 2000.
29. J. Nix and V. Hohmann. Sound source localization in real sound fields based on empirical statistics of interaural parameters. *J. Acoust. Soc. Am.,* 119:463–479, 2006.
30. J. Peissig. *Binaurale Hörgerätestrategien in komplexen Störschallsituationen (Binaural strategies for hearing aids in complex noise situations).* PhD thesis, Georg-August Universität, Göttingen, Germany, 1992.
31. B. Rakerd and W. M. Hartmann. Localization of sound in rooms. V. Binaural coherence and human sensitivity to interaural time differences in noise. *J. Acoust. Soc. Am.,* 128:3052–3063, 2010.
32. K. Reindl, P. Prokein, E. Fischer, Z. Y., and W. Kellermann. Combining monaural beamforming and blind source separation for binaural speech enhancement in multi-microphone hearing aids. In *ITG-Fachtg. Sprachkommunikation,* Nürnberg, Germany, 2010.
33. N. Roman, D. L. Wang, and G. J. Brown. A classification-based cocktail-party processor. volume 16, Vancover, Canada, 2003.
34. A. Sarampalis, S. Kalluri, B. Edwards, E. Hafter. Objective measures of listening effort: Effects of background noise and noise reduction. *J. Speech, Language, and, Hearing Res.,* 52:1230–1240, 2009.
35. A. Schlesinger. *Binaural model-based speech intelligibility enhancement and assessment in hearing aids.* PhD thesis, Delft University of Technology, The Netherlands, 2012.
36. A. Schlesinger. Transient-based speech transmission index for predicting intelligibility in nonlinear speech enhancement processors. In *IEEE Intl. Conf. Acoustics, Speech and Signal Process., ICASSP,* volume 1, pages 3993–3996, Kyoto, Japan, 2012.

37. K. U. Simmer, J. Bitzer, and C. Marro. *Microphone arrays: Signal processing techniques and applications,* chapter Post-filtering techniques. Springer-Verlag, 2001.
38. C. H. Taal, R. C. Hendriks, R. Heusdens. A short-time objective intelligibility measure for time-frequency weighted noisy speech. In *IEEE Intl. Conf. Acoust., Speech and Signal Process., ICASSP 2010,* pp. 4214–4217, Dallas, United States of, America, 2010.
39. The MathWorks, Inc. *MATLAB R2012a Documentation,* 2012.
40. TNO. *Multilingual database.* TNO Human Factors Research Institute, Soesterberg, The Netherlands, 2000.
41. R. J. Weiss, M. I. Mandel, and D. P. W. Ellis. Combining localization cues and source model constraints for binaural source separation. *Speech Communication,* 53:606–621, 2011.
42. P. Welch. The use of fast Fourier transform for the estimation of power spectra: a method based on time averaging over short, modified periodograms. *IEEE Trans. Audio and Electroacoustics,* 15:70–73, 1967.
43. O. Yilmaz and S. Rickard. Blind separation of speech mixtures via time-frequency masking. *IEEE Trans. Signal Processing,* 52:1830–1847, 2004.
44. Y. Zheng, K. Reindl, and W. Kellermann. BSS for improved interference estimation for blind speech signal extraction with two microphones. In *3rd IEEE Intl. Worksh. Computational Advances in Multi-Sensor Adaptive Process., CAMSAP,* pages 253–256, Aruba, Dutch Antilles, 2009.

# Modeling Sound Localization with Cochlear Implants

**M. Nicoletti, Chr. Wirtz and W. Hemmert**

## 1 Introduction

*Cochlear implants*, CIs, are the most successful neuroprostheses available today, with approximately 219,000 people implanted worldwide—as of December 2010 [60]. Modern CIs often provide good speech intelligibility, but there is room for improvement. Due to the limited number of independent channels [71], the spectral representation of music is less detailed for cochlear-implant users than for normal-hearing subjects, and they perform more poorly in adverse acoustic environments, such as in a cocktail-party scenario with multiple simultaneous sound sources [12, 50]. To sustain communication in such conditions, humans have developed the remarkable ability to focus on a single speaker even within a highly modulated background noise consisting of concurrent speakers and/or additional noise sources. In such scenarios binaural hearing plays a major role. Time and level differences between the right and left ear are exploited by the auditory system to localize the sound sources and segregate the acoustic information focused upon [3].

As CIs were initially developed for unilateral implantation only, they lack some important prerequisites required for precise sound localization. Automatic gain control, AGC, in CIs is required to compress the large dynamic range of acoustic signals to the limited dynamic range available for electric stimulation of the auditory nerve. Automatic gain-control systems in CIs reflect a compromise between conflicting requirements. Actually, they must handle intense transients but at the same time minimize disturbing side effects such as *"breathing"* or *"pumping"* sounds and distortions of the temporal envelope of speech. This conflict can be solved with a dual

M. Nicoletti · Chr. Wirtz · W. Hemmert (✉)
Bio-Inspired Information Processing, Institute of Medical Engineering,
Technische Universität München, Garching, Germany
e-mail: werner.hemmert@tum.de

Chr. Wirtz
MED-EL Deutschland GmbH, Starnberg, Germany

J. Blauert (ed.), *The Technology of Binaural Listening*, Modern Acoustics and Signal Processing, DOI: 10.1007/978-3-642-37762-4_12, 

time constant compression system [78]. However, in contemporary CIs, AGCs in bilateral cochlear implants work independently from each other, which can degrade the coding of interaural level differences ILDs. In addition, the accuracy of temporal coding in CIs clearly does not yet come close to the precision reached in the intact hearing organ. Given the many limitations involved in the artificial electrical stimulation of the auditory nerve, especially the effects of severe channel crosstalk, it is unclear if and which changes of the coding strategies will improve spatial hearing.

To answer these questions, quantitative models have been developed that help better understand the complex mechanisms involved in the electrical excitation of neurons. Pioneering model-based investigations [17, 35, 54, 63, 64] were initiated with the goal to improve CIs. They tried to optimize stimulus parameters like the stimulation frequency, pulse width and shape. The investigation of Motz and Rattay [55] harnessed the models to improve coding strategies implemented in a speech processor. The modeling approaches can be separated in three different categories, namely, point neuron models, multi-compartment models and population models.

- *Point neuron models* try to capture the detailed dynamic properties of the neurons. Motz and Rattay used them to explain neuronal responses to sinusoidal stimuli and current pulses [54, 64]. Dynes [20] extended these models to capture the refractory period of the neurons more precisely. With the introduction of CI-coding strategies with high stimulation rates, it became important to investigate the stochastic behavior of the auditory nerve to electric stimulation [8, 9] and rate-dependent desynchronization effects [69]. Mino et al. [53] captured channel noise by modeling the stochastic open- and closed states of the sodium-ion-channel population with Markov chains. Most recent models aimed to describe the adaptation of the auditory nerve to electric stimulation at high stimulation rates [37, 85].
- *Multi-compartment models* are an extension of point neuron models and were introduced by McNeal [51]. They are important for investigating effects of electrode position and configuration, for instance, monopolar versus bipolar. Multi-compartment models can predict how and where action potentials are elicited in the axon of a neuron [52, 85]. They are also essential for investigating how cell morphology affects their dynamical properties [67, 74], and how the field spread in the cochlea affects the stimulation of neurons along the cochlea [7, 25, 27].
- *Population models* are required to replicate neuronal excitation patterns along the whole cochlea. Therefore, they usually require modeling of thousands of neurons. For electrical stimulation, large populations of neurons are required to investigate rate-intensity functions [9] and neuronal excitation patterns for speech sounds [32]. Due to the large number of modeled neurons, these population models were implemented with computationally less expensive stochastic spike response models. Nevertheless, the increasing computational power of modern computer clusters has enabled us to model also large neuron populations based on biophysically plausible Hodgkin-Huxley-like ion-channel models [58]. In these models refractoriness and spike rate adaptation result from ion-channel dynamics.

Modeling higher levels of neuronal processing is hindered by the large complexity involved and therefore is limited to a few special cases. Basic perceptual properties like intensity perception [8] or forward masking [32] can be captured by deriving a neuronal representation that corresponds to the respective psychophysical data. More cognitive processes can usually not be predicted with models based on single neurons. However, in recent years, machine-learning techniques based on neuronal features [23, 59] were adopted to tackle highly complex tasks such as the prediction of speech understanding in noise.

The investigation reported in this chapter focuses on a similarly complex task, namely, *sound localization*. The ability to localize sound sources in complex listening environments has fascinated researchers over decades. Models developed in the fifties have been extended and improved by many researchers, but their basic concepts remain valid until today. The models can be divided into two basic groups, namely, coincidence models and equalization-&-cancellation models—see [43], this volume.

- *Coincidence models* Jeffress postulated neuronal coincidence detectors fed by two delay lines with signals traveling in opposite directions within each tonotopic center-frequency channel [38]. This model was extended to predict basic ILD sensitivity [11]. The coincidence model was combined with a simplified inner-ear model to model the basilar membrane and inner hair cells, consisting of a filter bank, an automatic gain control, AGC, a low-pass filter, and a rectifier. References [13–15, 76, 77] provide quantitative predictions on binaural interaction. Blauert and his colleagues Lindemann and Gaik [46] used a complementary approach adding ILD sensitivity to the Jeffress model. Later in this chapter this model will be applied to predict sound-source localization.
- *Equalization-&-cancellation models* These models have primarily been developed to predict binaural masking differences [19, 42]. Recently, cancellation features were added to the Jeffress-Colburn model [4–6].

# 2 Modeling Hearing in Cochlear Implant Users

This chapter presents a modular model framework to simulate auditory nerve responses elicited by a cochlear implant. The framework is schematically depicted in Fig. 1. It consists of a speech processor, a model of the electrical field spread caused by the implanted electrode array and a model of auditory nerve fibers along the inner ear.

The model was duplicated for the case of two ears to evaluate binaural interaction—see Sects. 2.5 and 3.1. The nerve responses are then evaluated by *"cognitive"* stages, which, for example, can be an automatic speech-recognition system or a system that estimates the position of a sound source. Every part of the model can be exchanged by a more or less computationally expensive realization, where the complexity required depends on the scientific question.

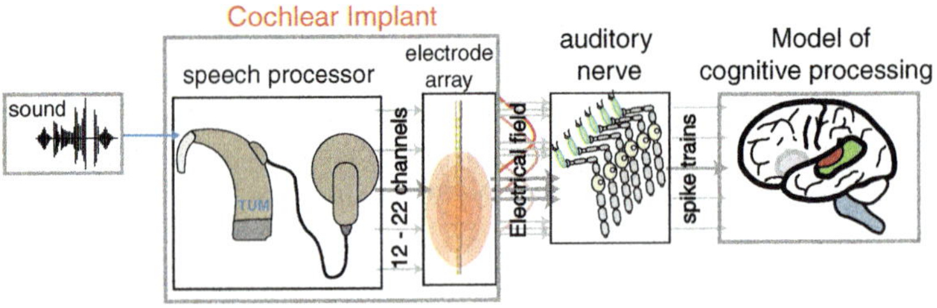


**Fig. 1** Sketch of the model that simulates hearing in cochlear implant users

## 2.1 Speech Processor

In the speech processor input signals, in this model audio-files in .sph and .wav format, are processed by a coding strategy that converts the physical sound signals into electric-current pulse trains—compare Fig. 2. The pulse trains are fed into the inner ear via the electrode array. The coding strategy replaces the processing steps that usually take place in the inner ear and translate the physical sound signal into a

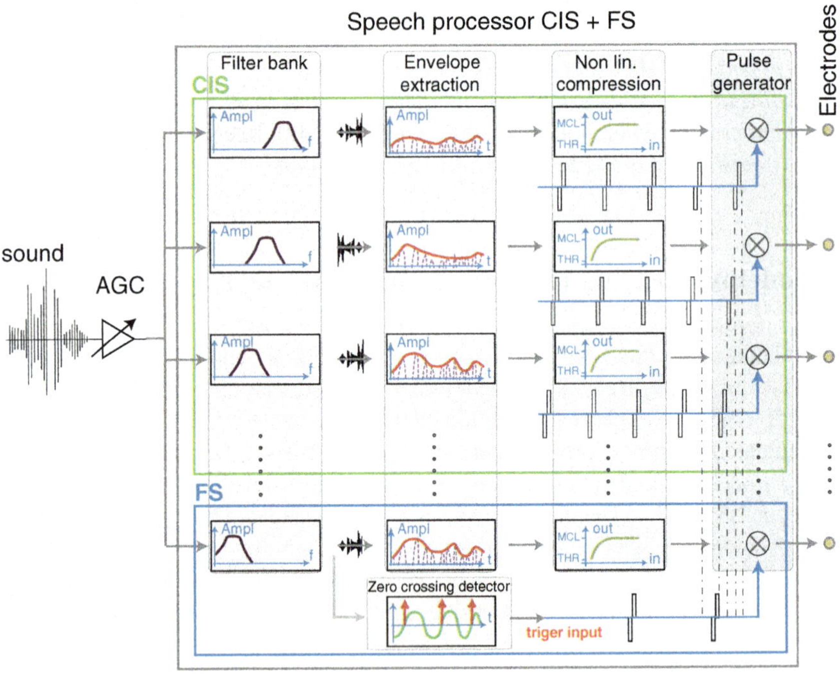


**Fig. 2** Signal processing of a modern coding strategy. Where CIS channels code only the spectral envelopes, FS channels code fine structure information with zero crossing detectors. Electrodes are stimulated with interleaved biphasic, that is, charge-balanced, rectangular pulses

representation that can be processed by the neuronal system. Because of the limited information-transmission capacity between implant and the neural system, coding strategies are predominantly optimized for speech coding. They process a limited frequency range, usually between 100 Hz and 8 kHz in 12–22 frequency bands. The large dynamic range of natural sounds requires effective AGC systems. Whereas many coding strategies were developed by the different manufacturers [87], in this chapter a generic implementation of one of the most successful coding principles, the *continuous interleaved sampling*, CIS, strategy [84] is introduced. In the CIS strategy, the signal is first filtered into frequency bands, then the spectral envelope is extracted. The temporal fine structure is discarded. The spectral envelope is then sampled using biphasic rectangular pulses, which are delivered to one electrode at a time, that is, interleaved.

This implementation applied a dual-time-constant front-end AGC [78], followed by a filter bank. The envelope was extracted in each channel with a Hilbert transformation. The amplitudes of the envelopes are mapped individually for each CI user and electrode between threshold level, THR, and maximum comfort level, MCL. Note that the dynamic range for electrical stimulation is extremely narrow. The difference between THR and MCL is only in the order of 10–20 dB [28, 86]. Mapping is implemented using power-law or logarithmic-compression functions [48]. The small dynamic range that is available for electric stimulation causes another severe limitation, namely, if two electrodes were stimulated simultaneously, their overlapping fields would sum up and cause overstimulation of neurons. The electrodes in CIs are therefore stimulated one after the other, that is, interleaved. Extensions of the CIS concept try to reconstruct the phase locking, as is observed in neuronal responses at low frequencies [41]. In CIs, this can be achieved by implementing a filter bank followed by slope-sensitive zero-crossing detectors—compare Fig. 2—that trigger the stimulation pulses [88].[1] This technique is used in the fine structure coding strategies (FSP, FS4, FS4-p) in the MED-EL MAESTRO cochlear implant system.

In summary, the FS strategies transmit the envelope information with their basal electrodes at high temporal resolution and code additional FS phase information with their most apical electrodes—which is conceptually similar to what happens in the intact inner ear. With this model framework it is now possible to estimate how much of this additional phase information is actually transmitted by the auditory nerve fibers and available for sound localization, and how much of it is corrupted by the limitations of electrical stimulation.

## 2.2 Electrode Model

One of the most severe limitations in modern cochlear implants is imposed by the electrical crosstalk between stimulation electrodes. The CI electrode array is usually

[1] This concept is realized in the MAESTRO cochlear implant system by MED-EL in the lowest-frequency channels, which stimulate the most apical electrodes.

inserted in scala tympani and immersed in perilymph—which has a conductivity of approx. 0.07 kΩmm [30, 66]. The neurons of the auditory nerve are inside the modiolar bone, which has a much higher conductivity of approx. 64 kΩmm [26, 30, 66]. Due to these anatomical constraints, the current spreads predominantly along the cochlear duct [30, 83]. This problem limits the number of independent electrodes to a value of about 7–8 [18, 24, 36] and a single electrode can excite auditory nerve fibers almost along the whole cochlea [30]. The amount of channel crosstalk is dependent on many factors and varies from CI user to CI user [21]. Different methods to measure the spread of excitation provide values between 1 and 4 dB/mm [33, 44, 57, 62]. The electrode array was modeled with 12–22 contacts as electrical point sources. The electrical excitation of a neuron in an electrical field is governed by the *activating function* [64], which is the second derivative of the electrical potential in the direction of the axon.

For a point current source, $I$, in a homogeneous isotropic medium, the activating function can be calculated as

$$\frac{d^2 V_{ex}}{dx^2} = I \frac{\rho}{4\pi} \frac{2x^2 - y^2 - z^2}{\left[x^2 + y^2 + z^2\right]^{5/2}}, \tag{1}$$

where $V_{\text{ex}}$ is the extracellular potential field, $\rho$ the mean conductivity of the surrounding tissue—3 kΩmm [67]. $x$, $y$ and $z$ are the coordinates according to Fig. 3. The value of this function was calculated at a distance $x$ of 500 μm from the electrode in the modiolus, where the electrical stimulation most likely elicits the action potentials in the auditory nerve fibers. Because of the coiling of the cochlea, the current spread can only be solved with three-dimensional models [7, 66, 83]. For simplicity, coiling of the cochlear ducts was neglected and therefore no across-turn stimulation occurred in this model. As the activating function for a homogeneous medium underestimates current spread, which would lead to unrealistic focal stimulation of a neuron population, the activating function was calculated only at the position of the electrode $z = 0$. For the current spread in the $z$ direction an exponential decay with 1 dB/mm was assumed, which was found experimentally [33, 44, 57, 62]—worst-case scenario.

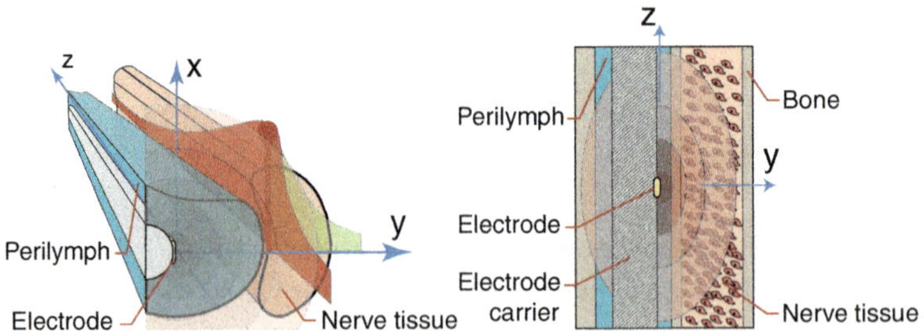


**Fig. 3** Electrical field spread and channel crosstalk of an electrode array in the cochlea

## 2.3 Model of a Single Nerve Fiber

One of the most important steps in this model is the excitation of the auditory nerve. The theory behind this model is reviewed in [29, 49]. A biophysically plausible model was implemented that is based on Hodgkin-Huxley-like ion channels—including hyperpolarization-activated cation channels, HPAC, high-threshold potassium channels, $K_{HT}$, and low-threshold potassium channels, $K_{LT}$. Such ion channels are also found in cochlear nucleus neurons. Due to their large time constants, the auditory nerve exhibits adaptation to electrical stimulation [56]. Conductances and time constants were corrected for a body temperature of 37 °C. The electrical equivalent circuit of the model is shown in Fig. 4.

The equations and parameters for the models are taken from [68] (see also Table 1)—with the units ms and mV. $V_M$ denotes the trans-membrane voltage. The gating variables of the different channels, $x \in \{w; z; n; p; r\}$, are voltage-dependent and they converge with a time constant of $\tau_x$ to their equilibrium value $x_\infty$ as described by the following differential equation.

$$\frac{dx}{dt} = \frac{1}{\tau_x}(x_\infty - x). \tag{2}$$

The behavior of the ion channels is described in the next equations.
(i) Low threshold, $K^+$-channel, $K_{LT}$,

$$i_{KLT} = \bar{g}_{KLT} \cdot w^4 z \cdot (V_m - E_K), \tag{3}$$

$$w_\infty = \left[\frac{1}{1 + exp\left(-\frac{V_m+48}{6}\right)}\right]^{0.25}, \tag{4}$$

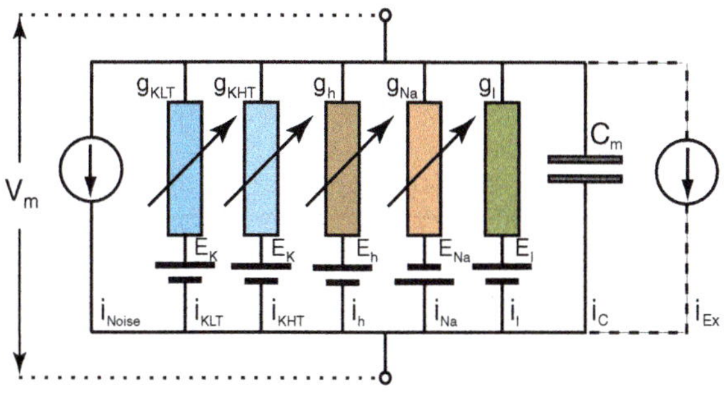


**Fig. 4** Electrical equivalent circuit of a neuron

$$\tau_w = \frac{100}{6 \cdot exp\left(\frac{V_m+60}{6}\right) + 16 \cdot exp\left(-\frac{V_m+60}{45}\right)} + 1.5, \quad (5)$$

$$z_\infty = \frac{1-0.5}{1+exp\left(\frac{V_m+71}{10}\right)} + 0.5, \quad (6)$$

$$\tau_z = \frac{1000}{exp\left(\frac{V_m+60}{20}\right) + exp\left(-\frac{V_m+60}{8}\right)} + 50. \quad (7)$$

(ii) High-threshold, $K^+$-channel $K_{HT}$,

$$i_{KHT} = \bar{g}_{KHT} \cdot (0.85 \cdot n^2 + 0.15 \cdot p) \cdot (V_m - E_K), \quad (8)$$

$$n_\infty = \left[1 + exp\left(-\frac{V_m+15}{5}\right)\right]^{-1/2}, \quad (9)$$

$$\tau_n = \frac{100}{11 \cdot exp\left(\frac{V_m+60}{24}\right) + 21 \cdot exp\left(-\frac{V_m+60}{23}\right)} + 0.7, \quad (10)$$

$$p_\infty = \left[1 + exp\left(-\frac{V_m+23}{6}\right)\right]^{-1}, \quad (11)$$

$$\tau_p = \frac{100}{4 \cdot exp\left(\frac{V_m+60}{32}\right) + 5 \cdot exp\left(-\frac{V_m+60}{22}\right)} + 5. \quad (12)$$

(iii) Hyperpolarization-activated cation current, $i_h$,

$$i_h = \bar{g}_h \cdot r \cdot (V_m - E_h), \quad (13)$$

$$r_\infty = \left[1 + exp\left(-\frac{V_m+76}{7}\right)\right]^{-1}, \quad (14)$$

$$\tau_r = \frac{100{,}000}{237 \cdot exp\left(\frac{V_m+60}{12}\right) + 17 \cdot exp\left(-\frac{V_m+60}{14}\right)} + 25, \quad (15)$$

(iv) Fast Na+ current, $i_{Na}$,

$$i_{Na} = \bar{g}_{Na} \cdot m^3 h \cdot (V_m - E_{Na}), \tag{16}$$

$$m_\infty = \left[1 + exp\left(-\frac{V_m + 32}{7}\right)\right]^{-1}, \tag{17}$$

$$\tau_m = \frac{10}{5 \cdot exp\left(\frac{V_m+60}{18}\right) + 36 \cdot exp\left(-\frac{V_m+60}{25}\right)} + 0.04. \tag{18}$$

The electrical stimulation of neurons in an electric field was analyzed according to Rattay [65] as

$$\tau \frac{dV_m}{dt} = \lambda^2 \frac{d^2 V_m}{dx^2} + \lambda^2 \frac{d^2 V_{ex}}{dx^2} + V_m. \tag{19}$$

In this equation, $\tau = \rho_m c_m$ denotes the time constant of the passive membrane and $\lambda = \sqrt{\rho_m / \rho_a}$ the length constant of an axon. If a long axon is assumed in the field, the term $d^2 V_m / dx^2$ can be neglected if the neuron is at rest. Then the external electrical stimulation acts like a virtual internal current source, which is proportional to $d/4\,\rho_a$, $d^2 V_{Ex}/dx^2$, with axon diameter $d$. Therefore the equation for a section of the axon can be described by the equation

$$C_m \frac{dV_m}{dt} = \frac{d}{4\,\rho_a} \frac{d^2 V_{Ex}}{dx^2} - [i_{KLT} + i_{KHT} + i_h + I_{Na} + i_l]\,. \tag{20}$$

If the analysis is restricted to the compartment, where the action potential is elicited, it is not necessary to solve the equations for all compartments—this would require a computationally intensive multi-compartment model. Instead, it is sufficient to check if this compartment—that is, the compartment where the activating function has its maximum—elicits an action potential. This would then, in the case of a multi-compartment model, propagate along the axon. Following this analysis, it is possible to reduce the model complexity to a single-compartmental model, which allows to calculate the response of a large number of neurons. The nonlinear ion-channel equations are solved in the time domain with the exponential Euler rule [10]. The model presented so far is deterministic and has therefore a fixed, although dynamic, threshold. Recordings from the auditory nerve in laboratory animals show that neurons exhibit a stochastic behavior also for electrical stimulation. This behavior was modeled by including a stochastic current source—compare Fig. 4. Physiological recordings show that a single neuron exhibits a dynamic range in the order of 1–5 dB [72, 73]. In this model a dynamic range of 2.5 dB was modeled by adjusting the current amplitude of the white noise source accordingly. However, the dynamic range of a single neuron is still too small to explain the dynamic range observed in CI users.

**Table 1** Parameters of the auditory nerve fiber model

| Parameter | Value | Description | References |
|---|---|---|---|
| $\rho_m$ | 0.02 [$k\Omega cm^2$] | Specific membrane resistance | [1, 79] |
| $\rho_a$ | 0.14 [$k\Omega cm$] | Specific axial resistance | [1, 81] |
| $E_{Na}$ | 66 [mV] | Reverse potential for sodium | [53, 56] |
| $E_K$ | −88 [mV] | Reverse potential for potassium | [53, 56] |
| $E_h$ | −43 [mV] | Reverse potential for $i_H$ | [53, 56] |
| $E_l$ | −62.5 [mV] | Reverse potential for leak current | |
| $g_{Na}$ | 324 [$mS/cm^2$] | Specific conductance of Na channels | |
| $g_{KHT}$ | 105 [$mS/cm^2$] | Specific conductance of $K_{HT}$ channels | |
| $g_{KLT}$ | 27 [$mS/cm^2$] | Specific conductance of $K_{LT}$ channels | |
| $g_h$ | 16 [$mS/cm^2$] | Specific conductance of HPAC | |
| $g_l$ | 0.006 [$mS/cm^2$] | Specific leak conductance | |
| $c_m$ | 1 [$\mu F/cm^2$] | Specific membrane capacitance | [67] |
| d | $1 \cdot 10^{-4}$–$2 \cdot 10^{-4}$ [cm] | Axon diameter | [45] |

## *2.4 Population Model of the Auditory Nerve: Individual Model for CI Users*

The analysis of the coding of complex sounds like speech requires a large population of neurons along the cochlea. When the stimulation current increases more neurons are excited, which extends the dynamic range for electrical stimulation. Surviving neurons in the spiral ganglion have variations in their axonal diameters, namely, 1.2–2.5 μm [45], and are located at different distances from the stimulating electrode—compare Fig. 5.

Factors which extend the dynamic range of the spiral ganglion neuron, SGN population comprises of the channel noise from an individual neuron and different thresholds of the single fibers due to different axon diameters and the varying distances between electrodes and cells. The dynamic range due to channel noise is about 2.5 dB, variations of the diameter contribute up to 6 dB, and the distance between electrode and SGNs up to 12.5 dB. With appropriate SGN populations, CI users with dynamic ranges between 3 and 21 dB can be modeled. A larger dynamic

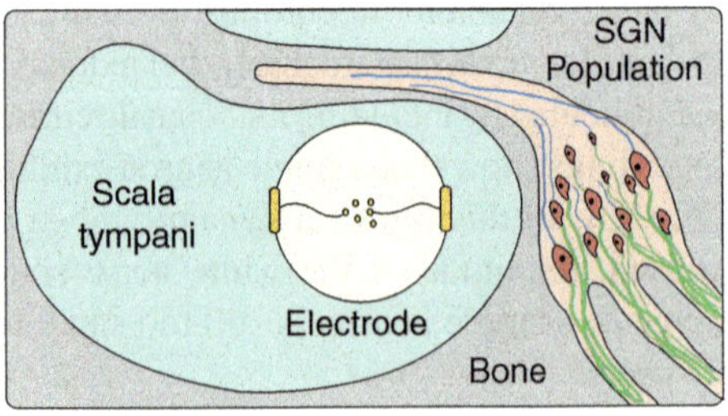


**Fig. 5** Cartoon of the populations composition: SGN with different sizes and distances respect to the electrode

range requires a larger SGN population and in turn longer computing times. The SGN are distributed along the length of the basilar membrane according to cell counts by [22, 75] in hearing impaired subjects. Given that SGNs degenerate further in deaf subjects [2, 40] and limitations in computational power, the model results presented here include up to 6.000 SGNs. SGNs were randomly distributed along the length of the cochlea and also the distance of the cells to the electrode, therefore the population was not uniformly distributed—Fig. 5.

## *2.5 Sound-Localization Model*

For sound localization experiments, left- and right-ear signals were processed with normal-hearing, NH or CI-listening models, which provide auditory-nerve-fiber, ANF, responses for further evaluations. Here the inner ear model from Wang [82] was selected as the NH reference. In the case of CI hearing, the acoustic input was processed by two independent models of speech processors, followed by two models of electrically evoked ANF/SGN responses—see Fig. 6.

Commonly a speech processor uses an AGC to scale the input signal to the limited dynamic range of electrical hearing—see Sect. 2.1. Then a designated coding strategy translates the acoustic input into stimulation patterns for each electrode. Note that there is no common synchronization between the two CIs. As a result, the AGC

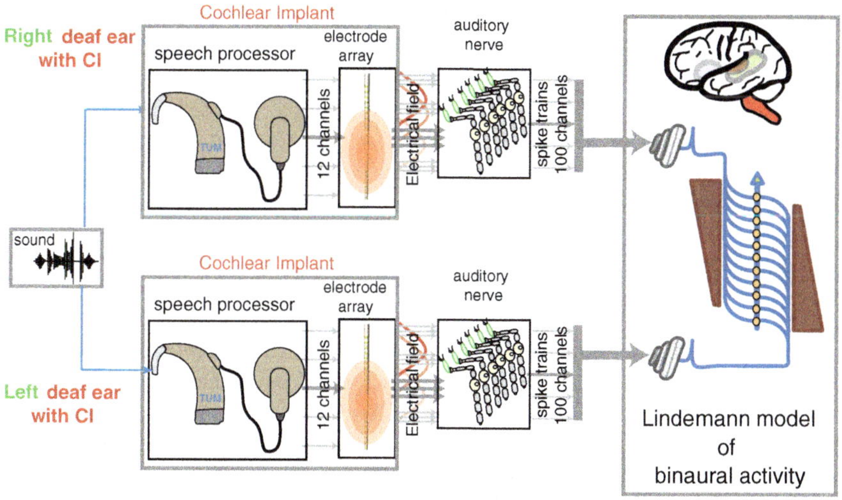


**Fig. 6** Schematic of the used framework. Each channel of a binaural acoustic signal is scaled and analyzed by both speech processors independently and transformed into firing patterns for each of the implant electrodes, according to the used coding strategy. The spike trains of the ANFs are then calculated from the electric field gradients. The Lindemann model, as binaural back end, performs the localization task in one frequency band

and stimulation delivered to the electrodes at the two ears run independent on both implants. Applying independent gains to the left and right CI alters ILD cues in the acoustic signals and unsynchronized firing patterns obscure ITD cues. In the following a best case scenario, where both implants are synchronized will be further investigated.

Lindemann developed a binaural model [46, 47] originally intended for NH listeners. This model is based on the Jeffress model [38] and assumes that coincidence neurons receive input from a tapped delay line from each side of the tonotopic representation of the cochlea. The coincidence detecting neurons are located along the delay lines such that they fire at specific ITDs.

$$l(n) \otimes r(n) = \int_{-\infty}^{+\infty} l(\tau) \cdot r(n+\tau)\, d\tau \tag{21}$$

This process is mathematically described as a cross-correlation function, as defined in Eq. 21. $l(n)$ and $r(n)$ denote the discrete left and right input signals. The cross-correlation output is a value of signal energy as a function of time delay.

The Lindemann model extends the correlation delay line of the Jeffress model by introducing inhibitory elements, which adds ILD sensitivity to the model. This is modeled with attenuation elements along the delay-line. By this arrangement, ILDs are mapped to a corresponding cross-correlation time. The model does not consider any correlation between different frequency bands in the hearing system. The sharpness of the correlation peaks depends on the inhibition parameter. Larger values will sharpen the peaks. In addition, the model features a temporal integration element to stabilize the output for non stationary input signals. For further details see [47].

# 3 Testing the Model

## 3.1 Test Set-Up

For testing purposes a binaural signal generator was implemented (Fig. 7), which provides an acoustic two channel signal carrying ITD and ILD information.

The simulated listening setup consists of a sound source that is circling around the listener's head at 1 Hz. The distance between the two ears was set to 150 mm. An emitted wavefront will reach both ears at different times and thus invoke location-dependent ITDs. ILDs were evoked with a frequency-independent attenuation component, when required. By intention, no head shadow effect was included to control ILD and ITD independently from each other.

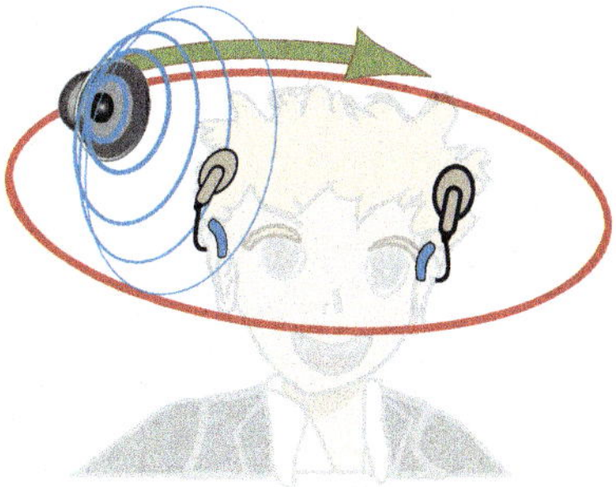

**Fig. 7** Binaural signal generator

## *3.2 Results*

### ANF-Response Patterns

Figure 8 shows smoothed response patterns for 6,000 auditory nerve fibers in response to the spoken utterance /ay/ from the ISOLET database [16]—female speaker fcmc0-A1-t, upper trace, 72.8 dB(A). Smoothing was achieved with a 10 ms Hamming window as low-pass filter.

Figure 9 shows spike patterns with high temporal resolution for the acoustic signals, CIS and FS4. The high-resolution figures for the normal-hearing model show very strong phase locking to the fundamental frequency of approx. 220 Hz of the speech signal—which is not coded at all by the CIS strategy. In the case of FS4 strategy, there are phase-locked responses, which are, however, obscured by additional spikes that are elicited by other nearby electrodes due to electrical crosstalk.

### Lindemann Example with Rectified Bandpass-Filter Input

Adapting the original example from the Lindemann model, a loudspeaker playing a 500 Hz pure tone was circling the listener at a rate of 1 Hz, in a virtual anechoic listening scenario. The signal pre-processing for the original Lindemann NH model consists of a filter bank with half-wave rectification and low-pass filtering to mimic the output of an inner hair cell.

The output of the model is shown in Fig. 10a, which shows that the model localizes the sound source with ease. The signal started with an interaural delay of 0 ms, corresponding to a source location of 0°, and performed two clockwise rotations. The maximum time delay of +0.4 ms was reached at 0.25 and 1.25 s for the right side and −0.4 ms at 0.75 and 1.75 s for the left side, which corresponded to a location at ±90°. The 500 Hz-analysis shows that the model analyzes the simulated time delay correctly. Note that for a 500 Hz sinusoidal input, the cross-correlation time delay has a repetition period of 2 ms. Nevertheless, only delays smaller than 1 ms are considered, as the distance between the ears is only 150 mm.

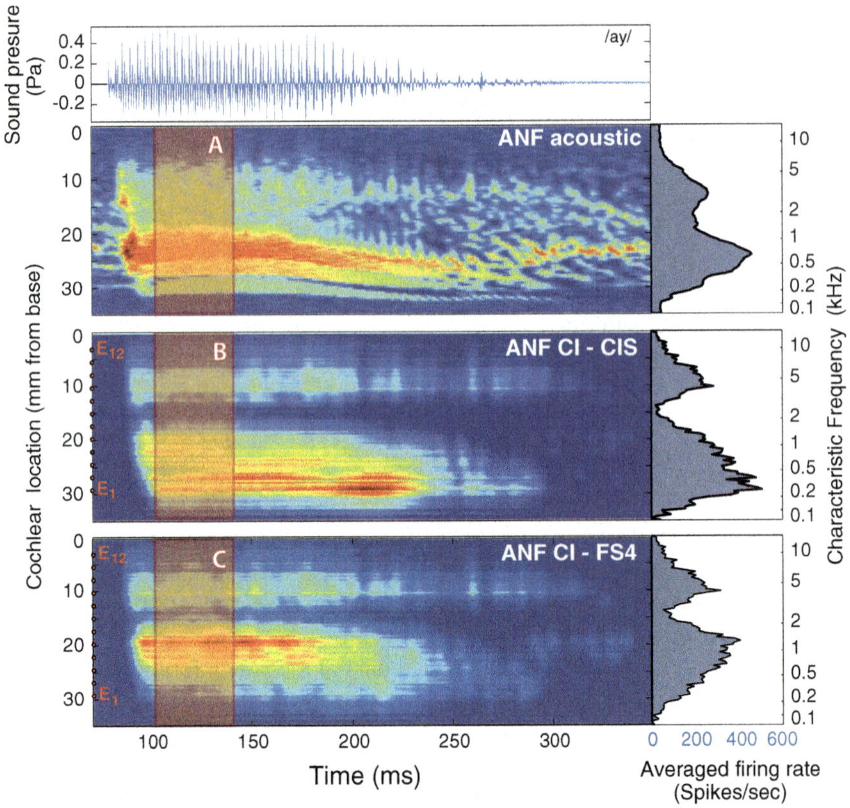


**Fig. 8** Response pattern of auditory nerve fibers in response to the spoken utterance /ay/. *Top panel* acoustic signal. *Second* intact-ear model [82], 60 high-spontaneous-rate ANFs per frequency channel, averaged with a 10 ms Hamming window—*right column* Averaged firing rate over whole utterance. *Third* response to electric stimulation with CIS strategy. *Bottom* same with FS4 strategy. The electrode positions are shown schematically on the *left hand* side. The electrical field spread, here 1 dB/mm, see Sect. 2.2, limits the spatial resolution of electrical stimulation

## Lindemann Model with Spike Count Input

The Lindemann model can be used with spike-count data of the ANF as well. By use of the NH-listener-ANF model as described in Sect. 2.4, the spike response of two ANF populations from the left and right cochleae was calculated and processed by the Lindemann model. Figure 10b shows the Lindemann cross-correlation for the circling source emitting a 500 Hz pure tone, when using spike counts derived from the Wang model—see [82]. The circling can be clearly seen, although the image looks noisier than the original model, what results from the probabilistic nature of the spikes.

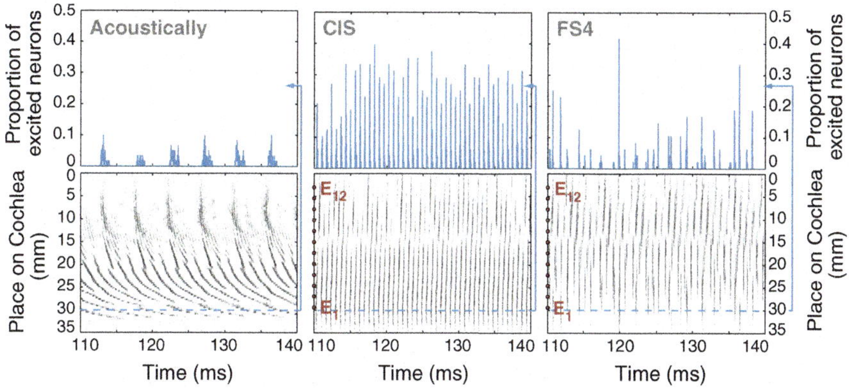


**Fig. 9** Response pattern from the area marked in Fig. 8 at high temporal resolution—0.4 ms time bins. *Left* intact ear. *Middle* implanted ear with CIS, *Right* with FS4 coding strategy. *Upper traces* show the response probabilities of a population of neurons at the position of the most apical electrode in a time bin

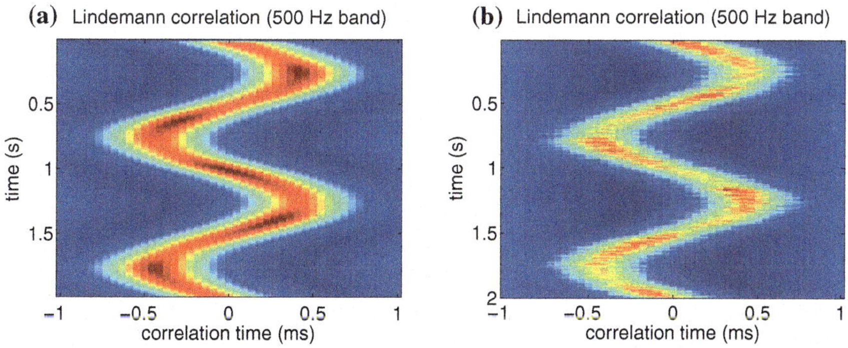


**Fig. 10** Output of the Lindemann cross-correlation for a low-frequency pure tone of 500 Hz, **a** original input, **b** spike-trains derived from an auditory model. A positive/negative Lindemann cross-correlation time-delay indicates a sound-source positioned *right/left* of the median-sagittal plane of the head. A sound source, circling the head once per second leads to a deviation of max. 0.441 ms in the cross-correlation time-delay

As the localization of a pure sine wave is a somewhat artificial example, the model was also tested with speech sounds. Results are illustrated in Fig. 11a. When this sentence is radiated from the moving speaker, it can be well localized using the Lindemann cross-correlation even with ANF spike count inputs simulated with the NH-listener model in the 200-Hz low-frequency region.

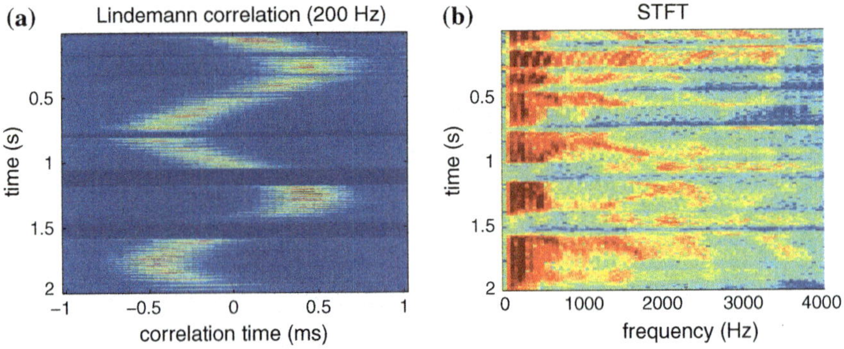


**Fig. 11** NH-listener localization of a moving speaker saying the German sentence "Britta gewann drei schwere Steine". **b** Short-time spectrogram of the acoustic input signal in a 200-Hz band, **a** corresponding Lindemann cross-correlation using ANF spike counts. Note the fricatives /s/ and /sch/ occurring at 1 and 1.4 s in the 200-Hz-band spectrogram, do not provide enough energy for the Lindemann model to localize

## Localization with Cochlear Implants

As sound localization in complex acoustic environments is still poor for most of the CI users, the question arises if—and if yes, how well—today's coding strategies can preserve binaural cues. In the following, two commercial coding strategies are compared, namely, MED-EL's former CIS strategy and their current FS4 strategy.

Speech samples derived from the binaural signal generator were processed with the speech processor and the two coding strategies—see Fig. 7. The ideal assumption was made that both ears, implants and fittings were identical. Before the electrically-evoked spike responses of the ANF is calculated, the electrode-stimulation patterns can be used as inputs to the Lindemann cross-correlation. This is advantageous when comparing coding strategies as no neuronal model is required yet, allowing us to track the point at which the binaural localization cues are compromised.

Figure 12 shows that the Lindemann model fails to localize the speech sample in the case of the CIS strategy, but succeeds in the case of FS4. Therefore, it can be concluded that FS4, with its fine structure channels, preserves the temporal-fine-structure-ITD cues needed for localization, but CIS does not. In the case of the CIS strategy, the Lindemann cross-correlation only outputs values at multiples of 0.6 ms, which is due to the CIS stimulation rate of ≈1,600 pulses per second in each channel. For FS4, the Lindemann correlation shows a time-delay resolution of ≈0.2 ms. This is possible because the FS4 strategy breaks-up the CIS rule and dedicates a higher sampling rate to the FS channels.

The FS4 coding strategy itself preserves localization cues. However, it is still possible that the CI-electrode crosstalk—see Sect. 2.2—deteriorates sound localization cues. For that reason the ANF module was added to the model, and the response for the electrically-evoked hearing was processed further with the Lindemann module. Figure 13 shows results from two different locations, the first next to electrode #2 at

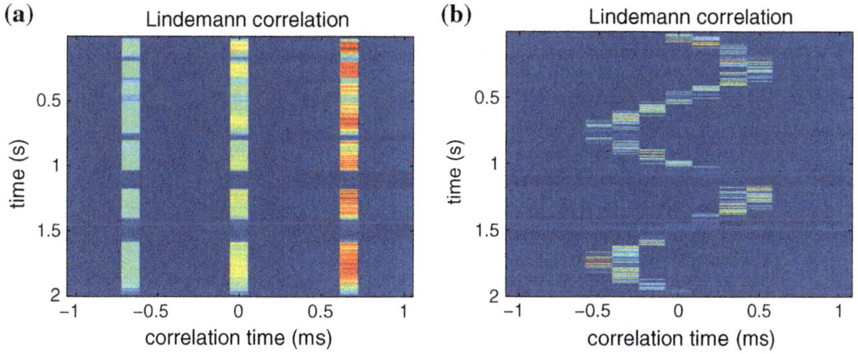


**Fig. 12** Speech-localization comparison of different coding strategies with the German speech sample from Fig. 11b. The output of apical electrode #2 of CIS and FS4 coding strategies is used as input to the Lindemann cross-correlation module. Whereas the model is unable to localize the moving sound source using the CIS strategy, the FS4 clearly manages to provide cues required for sound localization to the CI user. **a** CIS electrode 2. **b** FS4 electrode 2

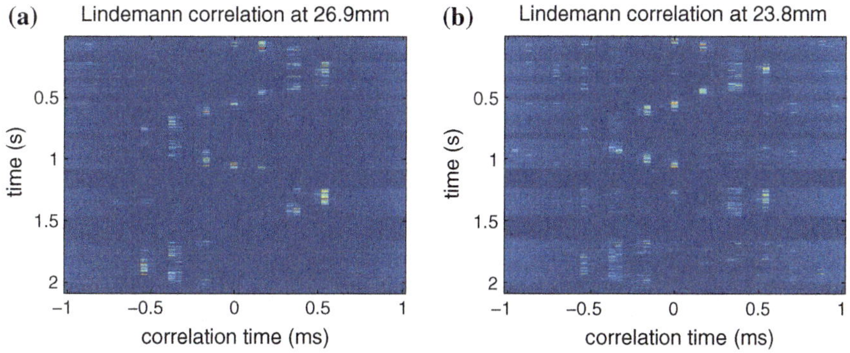


**Fig. 13** Lindemann cross-correlation with ANF-spike-count input of two locations. Location (**a**) is directly at electrode #2 at 26.9 mm from base, and (**b**) between electrodes #3 and 4 at 23.8 mm from the base. The speech-sample input was processed by the FS4-coding strategy. From that data the circling source movement can be identified, but the Lindemann cross-correlation is much more distorted than the NH-listener example from Fig. 11a. The channel crosstalk occurring in electrical stimulation—see Sect. 2.2—is one of the reasons for deterioration of the Lindemann cross-correlation

26.9 mm from the base and the second between electrodes #3 and 4 at 23.8 mm from the base. Compared to the NH case and to electrode-stimulation patterns, results are worse, but the main shape is still observable—which is indeed a major breakthrough for a fully-featured coding strategy for speech input. In less ideal cases, the location cues deteriorate. This is visible at the more basal location, which lies between two electrodes and also gets more input from the CIS electrodes due to electrical crosstalk. If left and right electrodes were inserted at different depths, position mismatch could further reduce correlations. However, as yet it not clear how much mismatch can be counterbalanced by the brain's ability to adapt to unusual cues as long as they are consistent.

## 4 Discussion and Conclusion

This chapter describes a framework to evaluate the extent as to which features required for sound localization are preserved by cochlear-implant coding strategies. Where a correlation model similar to the one proposed by Jeffress [38] is most likely implemented in the barn owl, investigations of the mammalian neuronal sound-localization pathway indicate that humans probably have two systems that extract ILDs and ITDs separately, and probably not with coincidence neurons to estimate the interaural cross correlation—for a review see [31]. Nevertheless, even if the neuronal processing schemes to extract cues for sound localization are still not yet completely understood, it is quite clear that ILDs and ITDs are extracted somehow. This investigation focused on the evaluation of ITD cues and used the model proposed by Lindemann. Thus, there is little doubt that the fundamental findings derived from this procedure hold true even if actual neuronal systems process localization cues somewhat differently.

The Lindemann model was adopted in such a way so that it can process electrical pulse-trains and neuronal spike trains. The analysis was limited to the low-frequency range where neuronal responses exhibit strong phase-locking. At low frequencies, level differences are usually small and ITD processing is assumed to be often dominant in human sound localization [3]. The results presented here show that ITD coding works well also for neuronal spike trains despite their probabilistic behavior. For auditory-nerve spike trains, the Lindemann cross-correlation is more variable as compared to its original input—see Fig. 10a—nevertheless, ITDs are clearly coded. This holds true not only for pure tones but also for complex speech sounds—compare Fig. 11a.

When electrical pulse trains delivered from a CIS coding strategy for one cochlear implant channel was analyzed, it was observed that ITD coding breaks down completely—compare Fig. 12a. This is not surprising, because the pulse train delivered to a single channel, here 1,600 Hz, codes the temporal envelope of the filtered sound signal and was never intended to provide ITD cues with sufficient precision. Given that the left and right processors are not synchronized, the time difference between left and right pulse train is arbitrary and is likely to change over time due to small deviations of the internal clock frequencies.

However, it is known that CI users are indeed able to localize sound sources, albeit less precisely than normal-hearing subjects [39, 61]. CI users almost exclusively use ILDs [70, 80], which was excluded in this investigation. The model results coincide partly with these findings. The Lindemann model was not able to predict sound localization based on ILD cues for CIS strategies, because the cross-correlation mechanism locked on the temporal structure of the signal—see Fig. 12. The temporal precision of the pulse trains from a CIS strategy is not sufficient for, and may even be detrimental to, sound localization. Therefore, ITDs must be ignored by CI users with CIS strategies, if the neuronal system is able to extract them at all at the high stimulation rates used in contemporary CIS strategies.

Nevertheless, FS coding strategies might indeed be able to transmit ITDs with sufficient temporal precision. For instance, the FS4 strategy tested here was found to provide useful ITDs coding, at least at the level of the pulse train of a single electrode—compare Fig. 12. The temporal precision is $\approx$0.2 ms, which is considerable higher than for the CIS strategy due to the higher sampling rate dedicated to the FS channels. When the responses at the level of neuronal spike trains were analyzed—compare Fig. 13—a large degradation of the ITD coding caused by channel crosstalk was found. Therefore, channel crosstalk does not only lead to a spectral smearing of the information but also affects the precision of temporal coding. Where this model predicts that, at least in the best case scenario, there is at least some ITD information left in the neuronal excitation pattern of the auditory nerve, it is unclear if and to what extent this information can actually be extracted by the auditory system.

In summing up, the model proposed in this chapter generates spiking auditory nerve responses and provides a quantitative evaluation of temporal cues for sound localization. The ability of the sound-localization model to process neuronal spike trains makes the model very versatile. It is possible to evaluate not only responses of the intact ear but also of the deaf inner ear provided with a cochlear implant. The model delivers quantitative data and therefore enables comparisons between different cochlear implant coding strategies. As the model of electric excitation of the auditory nerve also includes effects such as channel crosstalk, neuronal adaptation and mismatch of electrode positions between left and right ear,[2] its predictive power goes far beyond pure analysis of the output patterns of implants, which is how contemporary coding strategies were developed. Nevertheless, up to now, this model only extends up to the level of the auditory nerve and can, thus, not answer the question of whether ITDs can still be processed by higher levels of the auditory pathway. Where this final evaluation always has to be done with CI users, this framework provides important answers to the question of how well binaural cues are coded at the first neuronal level, and it allows the design and even the emulation of the required listening experiments. Given the long development cycles including design, fabrication, approval, implantation, and finally extensive measurements in a large group of CI users to yield statistically significant results, the benefit of this approach cannot be overestimated.

**Acknowledgments** This work was supported by the German Federal Ministry of Education and Research within the Munich Bernstein Center of Computational Neuroscience, ref.# 01GQ1004B and 01GQ1004D, and MED-EL Innsbruck. The authors thank V. Hohmann, F.-M. Hoffmann, P. Nopp, J. Blauert and two anonymous reviewers for helpful comments.

[2] Data not shown.

## References


1. P. J. Basser. Cable equation for a myelinated axon derived from its microstructure. *Med. Biol. Eng. Comput.*, 31 Suppl:S87–S92, 1993.
2. P. Blamey. Are spiral ganglion cell numbers important for speech perception with a cochlear implant? *Am. J. Otol.*, 18:S11–S12, 1997.
3. J. Blauert. *Spatial hearing: The psychophysics of human sound localization*. 2nd, revised ed. MIT Press, Berlin-Heidelberg-New York NY, 1997.
4. J. Breebaart, S. van de Par, and A. Kohlrausch. Binaural processing model based on contralateral inhibition. I. model structure. *J. Acoust. Soc. Am.*, 110:1074–1088, 2001.
5. J. Breebaart, S. van de Par, and A. Kohlrausch. Binaural processing model based on contralateral inhibition. II. dependence on spectral parameters. *J. Acoust. Soc. Am.*, 110:1089–1104, 2001.
6. J. Breebaart, S. van de Par, and A. Kohlrausch. Binaural processing model based on contralateral inhibition. III. dependence on temporal parameters. *J. Acoust. Soc. Am.*, 110:1105–1117, 2001.
7. J. J. Briaire and J. H. Frijns. Field patterns in a 3d tapered spiral model of the electrically stimulated cochlea. *Hear. Res.*, 148:18–30, 2000.
8. I. C. Bruce, M. W. White, L. S. Irlicht, S. J. O'Leary, and G. M. Clark. The effects of stochastic neural activity in a model predicting intensity perception with cochlear implants: low-rate stimulation. IEEE Trans. Biomed. Engr. 46(12):1393–1404, 1999.
9. I. C. Bruce, M. W. White, L. S. Irlicht, S. J. O'Leary, S. Dynes, E. Javel, G. M. Clark. A stochastic model of the electrically stimulated auditory nerve: single-pulse, response. IEEE Trans. Biomed. Engr. 46:617–629, 1999.
10. J. Certaine. The solution of ordinary differential equations with large time constants. *Mathematical methods for digital computers*, pages 128–132, 1960.
11. C. Cherry and B. M. Sayers. Experiments upon the total inhibition of stammering by external control, and some clinical results. *J. Psychosom. Res.*, 1:233–246, 1956.
12. G. Clark. *Cochlear implants: Fundamentals and applications*. New York: Springer, 2003.
13. H. S. Colburn. Theory of binaural interaction based on auditory-nerve data. I. general strategy and preliminary results on interaural discrimination. *J. Acoust. Soc. Am.*, 54:1458–1470, 1973.
14. H. S. Colburn. Theory of binaural interaction based on auditory-nerve data. II. detection of tones in noise. *J. Acoust. Soc. Am.*, 61:525–533, 1977.
15. H. S. Colburn and J. S. Latimer. Theory of binaural interaction based on auditory-nerve data. III. joint dependence on interaural time and amplitude differences in discrimination and detection. *J. Acoust. Soc. Am.*, 64:95–106, 1978.
16. R. Cole, Y. Muthusamy, and M. Fanty. The isolet spoken letter database, 1990.
17. J. Colombo and C. W. Parkins. A model of electrical excitation of the mammalian auditory-nerve neuron. *Hear. Res.*, 31:287–311, 1987.
18. M. Dorman, K. Dankowski, G. McCandless, and L. Smith. Consonant recognition as a function of the number of channels of stimulation by patients who use the symbion cochlear implant. *Ear Hear.*, 10:288–291, 1989.
19. N. I. Durlach. Equalization and cancellation theory of binaural masking-level differences. *J. Acoust. Soc. Am.*, 35:1206–1218, 1963.
20. S. B. C. Dynes. *Discharge characteristics of auditory nerve fibers for pulsatile electrical stimuli*. PhD thesis, Massachusetts Institute of Technology, 1996.
21. E. Erixon, H. Högstorp, K. Wadin, and H. Rask-Andersen. Variational anatomy of the human cochlea: implications for cochlear implantation. *Otol. Neurotol.*, 30:14–22, 2009.
22. E. Felder and A. Schrott-Fischer. Quantitative evaluation of myelinated nerve fibres and hair cells in cochleae of humans with age-related high-tone hearing loss. *Hear. Res.*, 91:19–32, 1995.
23. S. Fredelake and V. Hohmann. Factors affecting predicted speech intelligibility with cochlear implants in an auditory model for electrical stimulation. *Hear. Res.*, 2012.
24. L. M. Friesen, R. V. Shannon, D. Baskent, and X. Wang. Speech recognition in noise as a function of the number of spectral channels: comparison of acoustic hearing and cochlear implants. *J. Acoust. Soc. Am.*, 110:1150–1163, 2001.

25. J. H. Frijns, J. J. Briaire, and J. J. Grote. The importance of human cochlear anatomy for the results of modiolus-hugging multichannel cochlear implants. *Otol. Neurotol.*, 22:340–349, 2001.
26. J. H. Frijns, S. L. de Snoo, and R. Schoonhoven. Potential distributions and neural excitation patterns in a rotationally symmetric model of the electrically stimulated cochlea. *Hear. Res.*, 87:170–186, 1995.
27. J. H. Frijns, S. L. de Snoo, and J. H. ten Kate. Spatial selectivity in a rotationally symmetric model of the electrically stimulated cochlea. *Hear. Res.*, 95:33–48, 1996.
28. Q.-J. Fu. Loudness growth in cochlear implants: effect of stimulation rate and electrode configuration. *Hear. Res.*, 202:55–62, 2005.
29. W. Gerstner and W. M. Kistler. *Spiking neuron models*. Cambridge University Press, 2002
30. J. H. Goldwyn, S. M. Bierer, and J. A. Bierer. Modeling the electrode-neuron interface of cochlear implants: effects of neural survival, electrode placement, and the partial tripolar configuration. *Hear. Res.*, 268:93–104, 2010.
31. B. Grothe, M. Pecka, and D. McAlpine. Mechanisms of sound localization in mammals. *Physiol. Rev.*, 90:983–1012, 2010.
32. V. Hamacher. *Signalverarbeitungsmodelle des elektrisch stimulierten Gehörs - Signal-processing models of the electrically-stimulated auditory system*. PhD thesis, IND, RWTH Aachen, 2004.
33. R. Hartmann and R. Klinke. Impulse patterns of auditory nerve fibres to extra- and intracochlear electrical stimulation. *Acta Otolaryngol Suppl*, 469:128–134, 1990.
34. I. Hochmair, P. Nopp, C. Jolly, M. Schmidt, H. Schösser, C. Garnham, and I. Anderson. MED-EL cochlear implants: state of the art and a glimpse into the future. *Trends Amplif*, 10:201–219, 2006.
35. I. J. Hochmair-Desoyer, E. S. Hochmair, H. Motz, and F. Rattay. A model for the electrostimulation of the nervus acusticus. *Neuroscience*, 13:553–562, 1984.
36. A. E. Holmes, F. J. Kemker, and G. E. Merwin. The effects of varying the number of cochlear implant electrodes on speech perception. *American Journal of Otology*, 8:240–246, 1987.
37. N. S. Imennov and J. T. Rubinstein. Stochastic population model for electrical stimulation of the auditory nerve. IEEE Trans Biomed Eng. 56:2493–2501, 2009.
38. L. A. Jeffress. A place theory of sound localization. *J. Comp. Physiol. Psychol.*, 41:35–39, 1948.
39. S. Kerber and B. U. Seeber. Sound localization in noise by normal-hearing listeners and cochlear implant users. *Ear Hear.*, 33:445–457, 2012.
40. A. M. Khan, O. Handzel, B. J. Burgess, D. Damian, D. K. Eddington, and J. B. Nadol, Jr. Is word recognition correlated with the number of surviving spiral ganglion cells and electrode insertion depth in human subjects with cochlear implants? *Laryngoscope*, 115:672–677, 2005.
41. N. Y.-S. Kiang. Discharge patterns of single fibers in the cat's auditory nerve. Special technical report, 166, Massachusetts Institute of Technology, 1965.
42. W. E. Kock. Binaural localization and masking. *J. Acoust. Soc. Am.*, 22:801, 1950.
43. A. Kohlrausch, J. Braasch, D. Kolossa, and J. Blauert. An introduction to binaural processing. In J. Blauert, editor, *The technology of binaural listening, chapter 1*. Springer, Berlin-Heidelberg-New York NY, 2013.
44. A. Kral, R. Hartmann, D. Mortazavi, and R. Klinke. Spatial resolution of cochlear implants: the electrical field and excitation of auditory afferents. *Hear. Res.*, 121:11–28, 1998.
45. M. C. Liberman and M. E. Oliver. Morphometry of intracellularly labeled neurons of the auditory nerve: correlations with functional properties. *J. Comp. Neurol.*, 223:163–176, 1984.
46. W. Lindemann. Extension of a binaural cross-correlation model by contralateral inhibition. I. simulation of lateralization for stationary signals. *J. Acoust. Soc. Am.*, 80:1608–1622, 1986.
47. W. Lindemann. Extension of a binaural cross-correlation model by contralateral inhibition. II. the law of the first wave front. *J. Acoust. Soc. Am.*, 80:1623–1630, 1986.
48. P. C. Loizou. Signal-processing techniques for cochlear implants. 18(3):34–46, 1999.
49. J. Malmivuo and R. Plonsey. *Bioelectromagnetism: principles and applications of bioelectric and biomagnetic fields*. Oxford University Press, USA, 1995.

50. H. J. McDermott. Music perception with cochlear implants: a review. *Trends Amplif*, 8:49–82, 2004.
51. D. R. McNeal. Analysis of a model for excitation of myelinated nerve. BME-23:329–337, 1976.
52. H. Mino, J. T. Rubinstein, C. A. Miller, and P. J. Abbas. Effects of electrode-to-fiber distance on temporal neural response with electrical stimulation. 51:13–20, 2004.
53. H. Mino, J. T. Rubinstein, and J. A. White. Comparison of algorithms for the simulation of action potentials with stochastic sodium channels. *Ann. Biomed. Eng.*, 30:578–587, 2002.
54. H. Motz and F. Rattay. A study of the application of the hodgkin-huxley and the frankenhaeuser-huxley model for electrostimulation of the acoustic nerve. *Neuroscience*, 18:699–712, 1986.
55. H. Motz and F. Rattay. Signal processing strategies for electrostimulated ear prostheses based on simulated nerve response. *Perception*, 16:777–784, 1987.
56. M. H. Negm and I. C. Bruce. Effects of i(h) and i(klt) on the response of the auditory nerve to electrical stimulation in a stochastic hodgkin-huxley model. *Conf Proc IEEE Eng Med Biol Soc*, 2008:5539–5542, 2008.
57. D. A. Nelson, G. S. Donaldson, and H. Kreft. Forward-masked spatial tuning curves in cochlear implant users. *J. Acoust. Soc. Am.*, 123:1522–1543, 2008.
58. M. Nicoletti, P. Bade, M. Rudnicki, and W. Hemmert. Coding of sound into neuronal spike trains in cochlear implant users. In *13th Ann. Meetg. German Soc. Audiol., (DGA)*, 2010
59. M. Nicoletti, M. Isik, and W. Hemmert. Model-based validation framework for coding strategies in cochlear implants. In *Conference on Implantable Auditory Prostheses (CIAP)*, 2011.
60. NIH Publication No. 11–4798. Cochlear implants, March 2011.
61. P. Nopp, P. Schleich, and P. D'Haese. Sound localization in bilateral users of MED-EL combi 40/40+ cochlear implants. *Ear Hear.*, 25:205–214, 2004.
62. S. J. O'Leary, R. C. Black, and G. M. Clark. Current distributions in the cat cochlea: a modelling and electrophysiological study. *Hear. Res.*, 18:273–281, 1985.
63. C. W. Parkins and J. Colombo. Auditory-nerve single-neuron thresholds to electrical stimulation from scala tympani electrodes. *Hear. Res.*, 31:267–285, 1987.
64. F. Rattay. Analysis of models for external stimulation of axons. 33:974–977, 1986.
65. F. Rattay. The basic mechanism for the electrical stimulation of the nervous system. *Neuroscience*, 89:335–346, 1999.
66. F. Rattay, R. N. Leao, and H. Felix. A model of the electrically excited human cochlear neuron. II. influence of the three-dimensional cochlear structure on neural excitability. *Hear. Res.*, 153:64–79, 2001.
67. F. Rattay, P. Lutter, and H. Felix. A model of the electrically excited human cochlear neuron. I. contribution of neural substructures to the generation and propagation of spikes. *Hear. Res.*, 153:43–63, 2001.
68. J. S. Rothman and P. B. Manis. The roles potassium currents play in regulating the electrical activity of ventral cochlear nucleus neurons. *J. Neurophysiol.*, 89:3097–3113, 2003.
69. J. T. Rubinstein, B. S. Wilson, C. C. Finley, and P. J. Abbas. Pseudospontaneous activity: stochastic independence of auditory nerve fibers with electrical stimulation. *Hear. Res.*, 127:108–118, 1999.
70. B. U. Seeber and H. Fastl. Localization cues with bilateral cochlear implants. *J. Acoust. Soc. Am.*, 123:1030–1042, 2008.
71. R. V. Shannon. Multichannel electrical stimulation of the auditory nerve in man. II. Channel interaction. *Hear. Res.*, 12:1–16, 1983.
72. R. K. Shepherd and E. Javel. Electrical stimulation of the auditory nerve. I. Correlation of physiological responses with cochlear status. *Hear. Res.*, 108:112–144, 1997.
73. R. K. Shepherd and E. Javel. Electrical stimulation of the auditory nerve: II. Effect of stimulus waveshape on single fibre response properties. *Hear. Res.*, 130:171–188, 1999.
74. J. E. Smit, T. Hanekom, A. van Wieringen, J. Wouters, and J. J. Hanekom. Threshold predictions of different pulse shapes using a human auditory nerve fibre model containing persistent sodium and slow potassium currents. *Hear. Res.*, 269:12–22, 2010.

75. H. Spoendlin and A. Schrott. The spiral ganglion and the innervation of the human organ of corti. *Acta Otolaryngol. (Stockh.)*, 105:403–410, 1988.
76. R. Stern, Jr and H. S. Colburn. Theory of binaural interaction based in auditory-nerve data. IV. A model for subjective lateral position. *J. Acoust. Soc. Am.*, 64:127–140, 1978.
77. R. M. Stern and C. Trahiotis. Models of binaural interaction. In B. C. J. Moore, editor, *Hearing*, Handbook of perception and cognition, chapter 10, pages 347–387. Academic Press, second edition, 1995.
78. B. Stöbich, C. M. Zierhofer, and E. S. Hochmair. Influence of automatic gain control parameter settings on speech understanding of cochlear implant users employing the continuous interleaved sampling strategy. *Ear Hear.*, 20:104–116, 1999.
79. I. Tasaki. New measurements of the capacity and the resistance of the myelin sheath and the nodal membrane of the isolated frog nerve fiber. *Am. J. Physiol.*, 181:639–650, 1955.
80. R. van Hoesel, M. Böhm, J. Pesch, A. Vandali, R. D. Battmer, and T. Lenarz. Binaural speech unmasking and localization in noise with bilateral cochlear implants using envelope and fine-timing based strategies. *J. Acoust. Soc. Am.*, 123:2249–2263, 2008.
81. B. N. W. Schwartz and R. Stämpli. Longitudinal resistance of axoplasm in myelinated nerve fibers of the frog. *Pflügers Archiv European Journal of Physiology*, 379:R41, 1979.
82. H. Wang. *Speech coding and information processing in the peripheral human auditory system.* PhD thesis, Technische Universität München, 2009.
83. D. Whiten. *Electro-anatomical models of the cochlear implant.* PhD thesis, Massachusetts Institute of Technology, 2007.
84. B. S. Wilson, C. C. Finley, D. T. Lawson, R. D. Wolford, D. K. Eddington, and W. M. Rabinowitz. Better speech recognition with cochlear implants. *Nature*, 352:236–238, 1991.
85. J. Woo, C. A. Miller, and P. J. Abbas. The dependence of auditory nerve rate adaptation on electric stimulus parameters, electrode position, and fiber diameter: a computer model study. *J. Assoc. Res. Otolaryngol.*, 11:283–296, 2010.
86. F.-G. Zeng, G. Grant, J. Niparko, J. Galvin, R. Shannon, J. Opie, and P. Segel. Speech dynamic range and its effect on cochlear implant performance. *J. Acoust. Soc. Am.*, 111:377–386, 2002.
87. F.-G. Zeng, S. Rebscher, W. Harrison, X. Sun, and H. Feng. Cochlear implants: System design, integration, and, evaluation. 1:115–142, 2008.
88. C. Zierhofer. Electrical nerve stimulation based on channel-specific sequences. *European Patent, Office*, WO/2001/013991, 2001.

# Binaural Assessment of Parametrically Coded Spatial Audio Signals

M. Takanen, O. Santala and V. Pulkki

## 1 Introduction

The target of spatial audio reproduction is to deliver to the listener the aspects of spatial sound audible to a human listener with or without modifications. Different methods to reproduce a sound scene over a multichannel loudspeaker setup have been developed over the years [7, 24, 43]. Some of the methods aim at physically accurate reproduction at all audible frequencies. The sound field reproduced by the loudspeaker setup should thus equal the sound field in the recording position. Unfortunately, due to the vast frequency range audible of human hearing covering frequencies approximately from 20 Hz to 20 kHz, such techniques lead into solutions where hundreds or thousands of microphones and loudspeakers are needed [15], which is in most cases not feasible due to the cost, size, and difficulty of use of such ensembles.

When such a target for reconstruction is set, it is implicitly assumed that humans can perceive the spatial characteristics of a sound field in great detail. As humans have only two ears, and as it is known that spatial hearing has many limitations depending on time, frequency, and signal content [8], it can be assumed that such accuracy in the sound-field reproduction is not necessary for transparent audio reproduction. This assumption has been taken into account in recent spatial sound reproduction techniques that parametrize the spatial properties of the sound in time-frequency domain and use that information for compression and for enhancing the quality of the reproduction [12, 27, 39]. Although the aforementioned processing yields in prominently better sound reproduction quality than the traditional methods [51], in some cases these techniques may introduce new types of artifacts in the reproduction. Namely, the spatial image may become too wide or too narrow [34, 35], or the spatial image may not remain stationary in some signals [35]. Further, musical-noise

M. Takanen · O. Santala · V. Pulkki (✉)
Department of Signal Processing and Acoustics, Aalto University, Espoo, Finland
e-mail: ville.pulkki@aalto.fi

J. Blauert (ed.), *The Technology of Binaural Listening*, Modern Acoustics and Signal Processing, DOI: 10.1007/978-3-642-37762-4_13,

effects, also called *spatial bubbling*, where short segments of sound appear at random directions, may be introduced [39].

As the techniques have been developed by using some assumptions taken from psychoacoustics, the only currently valid method for measuring the audibility of such artifacts is the use of human listeners in formal psychoacoustic tests. An appealing alternative for this time-consuming process would be the use of a computational signal-driven binaural auditory model [13], and this chapter shows some successful results in assessment of artifacts caused by spatial audio reproduction techniques using such an auditory model.

This chapter describes some main techniques for time-frequency processing of spatial audio, having focus on the reproduction of spatial sound recorded in a single position. In addition, the related time-frequency-domain multichannel audio-coding techniques are discussed briefly, and the typical artifacts that may occur with such techniques are described. Finally, the binaural model that is used in the study is described and applied in the analysis of the artifacts.

## 2 Parametric Time-Frequency-Domain Spatial Audio Techniques

The basic assumption of all techniques considered here is that the directional resolution of spatial hearing is limited within each auditory frequency band [8]. In principle, all sound within one critical band can only be perceived as a single auditory event with broader or narrower extent. In some special cases, a binaural narrow-band sound stimulus can be perceived as two distinct auditory objects, but the perception of three or more concurrent objects is generally not possible [8]. This is different from visual perception, where the detection of the directions of a large number of visual objects sharing the same color can be performed even by using only one of the eyes.

The limitations of auditory perception imply that such spatial realism needed in visual reproduction is not needed in audio. In other words, the spatial accuracy in the reproduction of an acoustical wave field can be compromised without decreasing the perceived quality. The common schematic of all techniques exploiting this assumption is presented in Fig. 1. The microphone signals from a real sound scene recording, or the loudspeaker signals in an audio file, are transferred into time-frequency domain,

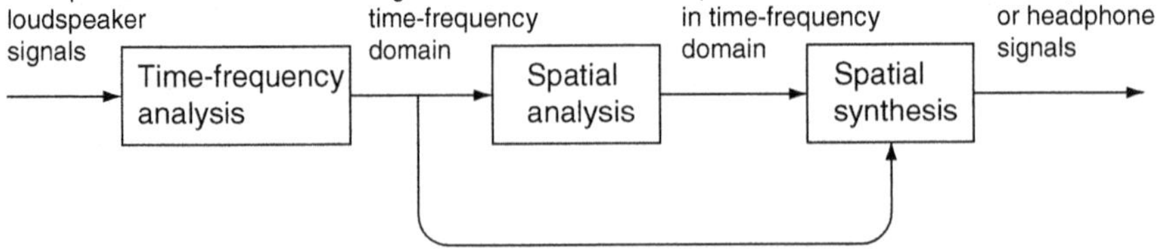


**Fig. 1** Schematic of a parametric time-frequency-domain spatial audio reproduction technique

where a spatial analysis is conducted. The resulting spatial metadata is then utilized in the sound reproduction. The synthesis may target to reproduce the sound scene as such, or to perform modifications, such as spatial filtering [18] or other effects [37], to the reproduced scene.

The techniques have some similarities with the human hearing mechanisms, namely, the input signals are divided into time-frequency domain and the spatial properties of the input are analyzed. However, often the input is not binaural but consists of more than two audio channels. Moreover, the synthesis of audio signals has no direct analog in human hearing. The techniques can thus be thought to have a model of hearing embedded inside them, although they are not actual applications of functional models of hearing mechanisms.

The techniques can be divided into two classes, one consisting of methods that take audio files generated for reproduction over a loudspeaker setup as input, whereas the methods belonging to the other class use microphone signals as input. The major difference is that the first class deals with some differences between the audio channels, while the second class typically analyses some properties of the sound field, such as direction and diffuseness. The classes are described in different subsections below.

## *2.1 Processing of Multichannel Loudspeaker Audio Signals*

This subsection considers the techniques that are designed to process files containing audio signals to be reproduced over loudspeakers. Typically, the files are either two-channel stereophonic audio content, or 5.1-surround audio content.

### Conversion of Spatial Audio Content for Different Loudspeaker-Listening Setups

Early attempts for parametric spatial audio reproduction in general were made in [2], where a stereophonic two-channel audio content was upmixed to a 5.1-surround loudspeaker setup. More specifically, the two signals of stereophonic content are transferred into time-frequency domain, and the ambience and direct components are extracted and synthesized to the 5.1-system with different techniques. The ambient components are applied to all loudspeakers after loudspeaker-specific decorrelation, which targets at reproducing ambient sounds in such a manner that the listener would perceive them to arrive from all directions. The direct components are applied to the loudspeaker setup using amplitude panning [9], which targets at producing a point-like perception of the direct components. A number of upmixing solutions have been suggested after the first attempts—see, for example, [3, 4, 17].

The first solution was targeted to process two-channel stereophonic recordings, after which different approaches have been proposed to transfer audio material between different multichannel loudspeaker layouts. All of the approaches process the input from a file in time-frequency domain to be listened with a certain multichannel loudspeaker setup, and analyze some spatial parameters that are used in the generation of the output for another multichannel loudspeaker setup—see, for example, [27].

#### Compressing Multichannel Audio Signals

The coding techniques for spatial audio take multichannel audio signals as input, such as stereophonic two-channel, or 5.1-surround audio signals, and compute some metadata based on the input [16, 21, 26, 31, 44]. The metadata are transmitted with a downmix of the input channels, and the decoder renders the audio signals to a loudspeaker setup identical with the original setup using the metadata. The target of the processing is to reduce the data rate in the transmission, while reproducing the original audio content perceptually similarly in a corresponding loudspeaker listening setup.

The metadata consists of inter-channel time and level differences [16, 31], or of directional vectors computed as energy- or magnitude-weighted sums of the loudspeaker directions [26]. In listening tests, such techniques have produced good quality reproduction of typical sound materials, with high reduction of the data rate in the transmission [31].

### *2.2 Reproduction of Recorded Spatial Sound*

These techniques assume the input to originate from some microphones typically in coincident or near-coincident positioning with known arrangement in a real or simulated sound field. The knowledge of the microphone arrangement makes it possible to estimate some directional properties of the sound field in each frequency band. The properties are then encoded as metadata as in Fig. 1, and used in the reproduction for different applications, targeting either different or similar perception of the spatial properties of sound.

The techniques that compress multichannel audio signals can be used in this way as well. It should be noted that the techniques are not directly applicable as a tool for sound reproduction from microphone signals. However, if such microphones would be used, which have the directional patterns matched with a loudspeaker setup as suggested in [25], the result of the analysis would be very similar with parametric spatial sound reproduction techniques. Some methods have also been published that focus on how to transform the metadata obtained from directional audio coding, defined below, into such metadata consisting of inter-channel differences [23, 30].

## Representation of Spatial Properties with Direction and Diffuseness Parameters in Time-Frequency Domain

Directional audio coding, DirAC, [39] has been proposed recently as a signal processing method for spatial sound. It is applicable for spatial sound reproduction for any multichannel loudspeaker layout, or for headphones. The steps included in DirAC processing are described here in detail due to the fact that in this work the parameters of DirAC are modified to illustrate artifacts caused by suboptimal settings.

In DirAC, it is assumed that at one time instant and at one critical band, the spatial resolution of the auditory system is limited to decoding one cue for direction and another for interaural coherence. It is further assumed that, if the direction and diffuseness of a sound field is measured and reproduced correctly, a human listener will perceive the directional and coherence cues correctly. Thus, a microphone system has to be used in the sound recording, which allows the analysis of direction and diffuseness for each frequency band.

First-order B-format signals can be used for such directional analysis and are typically utilized in DirAC implementations. The sound pressure can be estimated using the omnidirectional signal, $w(t)$, as $P = \sqrt{2}W$, expressed in the short-time Fourier transform, STFT, domain. The figure-of-eight signals, $x(t)$, $y(t)$, and $z(t)$, are grouped in the STFT domain into a vector, $\mathbf{U} = [X, Y, Z]$, which estimates the 3-D sound field velocity vector. The energy, $E$, of the sound field can be computed as

$$E = \frac{\rho_0}{4}||\mathbf{U}||^2 + \frac{1}{4\rho_0 c^2}|P|^2, \tag{1}$$

where $\rho_0$ is the mean density of air and $c$ is the speed of sound. The capturing of the B-format signals can be obtained with either coincident positioning of directional microphones, or with a closely-spaced set of omnidirectional microphones. In some applications, for instance, in simulated acoustics, the microphone signals may be formed in the computational domain. The analysis is repeated as frequently as needed for the application, typically with the update frequency of 100–1000 Hz.

The intensity vector, $\mathbf{I}$, expresses the net flow of sound energy as a 3-D vector, and can be computed as $\mathbf{I} = \overline{P}\mathbf{U}$, where $\overline{(\cdot)}$ denotes complex conjugation. The direction of sound is defined to be the opposite direction of the temporally integrated intensity vector at each frequency band. The direction is denoted as the corresponding angular azimuth and elevation values in the transmitted metadata. The diffuseness of the sound field is computed as

$$\psi = 1 - \frac{||\mathrm{E}_\tau\{\mathbf{I}\}||}{c\mathrm{E}_\tau\{E\}}, \tag{2}$$

where $\mathrm{E}_\tau$ is the expectation operator. Typically, the expectation operator is implemented with temporal integration. This process is also known as *smoothing*. The outcome of (2) is a real-valued number between zero and one, characterizing if the sound energy is arriving from a single direction or from all directions.

In the reproduction, each frequency channel is divided to diffuse and non-diffuse streams by multiplying the original stream with $\psi$ at a corresponding channel. The non-diffuse stream is then reproduced with amplitude panning, and the diffuse stream by applying the signal to all loudspeakers after phase decorrelation. The gain factors for each loudspeakers are computed with vector base amplitude panning, VBAP, [38] and they are stored to a vector $\hat{\mathbf{g}}$. To avoid nonlinear artifacts due to fast changes in the values of gain factors, a temporal smoothing operation is applied to obtain the gain factors,

$$\mathbf{g} = \mathrm{E}_{\tau}(\hat{\mathbf{g}}), \tag{3}$$

to be used in amplitude panning. Here, $\tau$ is the time constant applied in the temporal averaging process. Two variations of DirAC are used in this study. In mono-DirAC, the $W$ signal is used in amplitude panning, and in B-format-DirAC, instead of basic amplitude panning, the gain factors in $\mathbf{g}$ are used to gate the virtual cardioid signals facing the directions of the loudspeakers in the reproduction setup. Mono-DirAC is typically used in teleconferencing, as it minimizes the number of transmitted audio signals to a single channel, however, the generated virtual sources may suffer from slight directional instability [1]. In B-format-DirAC, all B-format channels are transmitted. This increases the transmission rate, but the virtual sources are more stabile [34, 39, 51].

### Systems Utilizing Two or More Directions

High angular resolution planewave expansion, Harpex, which is a method to reproduce B-format signals for binaural listening, has been proposed in [5]. The method is very similar to DirAC, however, differing from the basic assumptions in DirAC, it is not assumed that there is a single plane wave with superposed diffuse sound, but instead, it is assumed that the sound field consists of two plane waves arriving from different directions with different amplitudes. The directions and amplitudes are computed from the B-format signals, and they are used as metadata in the reproduction. It is stated in [5] that the quality obtained in headphone reproduction is much higher than with traditional conversion from B-format to headphone reproduction.

A version of DirAC utilizing estimation of signal parameters via rotational-invariance parameters, ESPRIT, from linear microphone arrays has been also proposed [49]. The use of ESPRIT makes it possible to analyze in principle any number of directions to be used in the DirAC metadata.

### Wiener-Filtering-Based Methods

A method has been proposed to obtain narrow directional patterns from two cardioid microphones facing opposite directions, which is based on the assumption that at a single auditory frequency band the microphone signals consist of a mixture of a single plane wave and ambient sound. The system computes the degree of similarity of sound in frequency channels based on cross-correlation, and removes the signal of the back cardioid from the front cardioid with a controllable level using Wiener filtering [18]. The width of the directional pattern can be controlled efficiently with this approach in the case where only a single plane wave arrives to the microphone system, and much narrower patterns than the cardioid pattern can be obtained in this way. When the captured field consists only of diffuse sound, the processing does not change the directional pattern of the output signal, which thus remains as the cardioid pattern. This approach has been utilized in a commercial shot-gun microphone [22].

The Wiener-filtering-based approach can be used to transform a B-format signal into loudspeaker signals, as such cardioid signals can be derived from a B-format signal to arbitrary directions [19]. A technique to produce multichannel loudspeaker signals out of stereophonic recordings with known microphone directivities has been proposed in [20], which quite largely follows the same principle. It was also shown that directional metadata compatible with Fig. 1 can be produced as well.

## 3 Spatial Artifacts Specific to Parametric Spatial Audio Techniques

The spatial audio techniques described above are all signal-dependent. Typically, the system steers the sound signal to different loudspeakers based on the properties of the input signals. The output signals may also be decorrelated to obtain the level of diffuseness in the resulting sound field that matches with the original sound field. In the development phase of the methods, the parameters are to be selected in such a way that the artifacts are not audible, however, in some cases, they might still be audible with some critical input signals.

The spatial parameters computed in the analysis phase of the methods may vary rapidly with time. Typically, their time functions are smoothed, for instance, by using first-order low-pass filters, the time constants of which are in a range of 10–100 ms. The parameter that controls the direction of sound source, or similarly, the loudspeaker channel to which the sound is applied, is considered first. The selection of this time constant is a trade-off between the accuracy of the parameter and the capability to respond to fast changes in the sound input. The longer is the temporal averaging, the better is the analysis result of steady-state sound input in terms of tolerance to background noise or distracters. However, in some cases, the signal conditions change rapidly, such as in surrounding-applause recording, or in the case where there are many ongoing sound sources active, such as with two simultaneous talkers. The time constants should thus be selected to follow the human spatio-temporal resolution, which is sometimes challenging to achieve, as the resolution may depend on the signal as well [8].

In such spatially-fast-changing-signal cases, some artifacts may occur. With multiple concurrent sources, the auditory images may migrate towards each other, and typically, a narrower auditory scene is perceived [34]. In addition, the perceived direction of the auditory object may move depending on other simultaneously occurring sounds, which is generally perceived to be annoying. Moreover, it has been found in informal listening that experienced listeners may in some cases perceive movement in only certain frequencies of the sources.

As discussed above, some of the spatial audio techniques have a specific method to generate surrounding diffuse sound for a higher number of loudspeaker channels than there are audio signals delivered to the decoding phase—in practice, the loudspeaker signals are decorrelated by, for instance, convolution with a short loudspeaker-channel-specific burst of noise. In real rooms with a relatively long reverberation time, the reverberant tail after the offset of the sound event is perceived to be surrounding the listener. Consequently, the target for the reproduction would be to reproduce such surrounding effect. If there are some shortcomings in the decorrelation process, different spatial artifacts may occur [36, 39]. For instance, in some cases, the auditory image evoked by the reverberant tail is perceived to be decreased in quality, or the reverberant field is perceived as having different width depending on the listening position.

A related artifact occurs with surrounding applause signals, which has been considered to be a major challenge in multichannel audio coding [11, 32, 33]. The characteristic signal condition causing a new type of artifact in such signals is the case where multiple claps originating from different sources are present very closely in time, and the methods analyze the sound to have low coherence or high diffuseness. In the synthesis of sound, the individual claps are decorrelated, which replaces the impulse-like claps with noise bursts. In [33] it was shown that surrounding applause signals can be reproduced using parametric spatial audio techniques, if the time resolution is made fine enough.

The changing of the listening position may introduce some artifacts to the reproduction as well. One of the reasons for developing time-frequency-domain audio techniques is to extend the listening area in which the quality of the reproduction is good, as compared to the traditional audio techniques that typically have a small sweet-spot. The artifacts introduced by off-sweet-spot listening are therefore most prominent in traditional techniques [1]. A typical artifact is that the loudspeakers nearest to the listener dominate the scenario and most of the sound is perceived to be emitted from the directions of those loudspeakers. Overall, the directions of the sounds in the scenario are inconsistent and fluctuate between the intended direction and the directions of the loudspeakers nearest to the listener.

## 4 Monitoring Spatial Artifacts with a Binaural Model

In this section, a binaural auditory model [46] is briefly introduced, and several simulated test scenarios are used to show the abilities of the binaural auditory model in assessing the perceived quality of spatial sound reproduction methods and monitoring the existence of artifacts in the reproductions. The model was applied to signals obtained from reference scenarios as well as simulated scenarios produced with

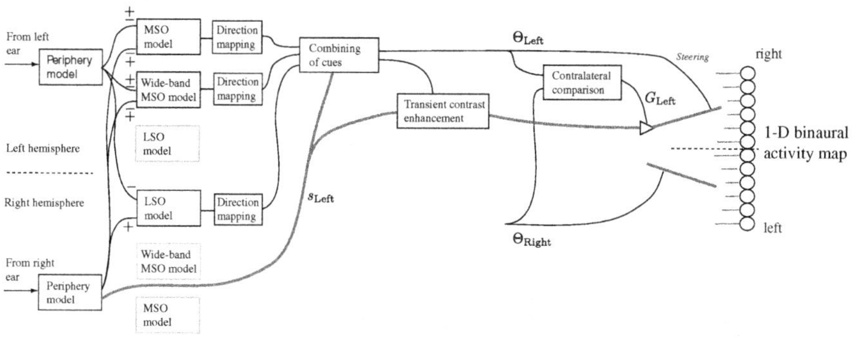


**Fig. 2** Schematic structure of the binaural auditory model that was used in the analysis of the audio-coding techniques. For simplicity, only the pathways leading to the activation projected to the left hemisphere are shown

different spatial sound reproduction methods, and the differences of the model outputs were analyzed. The parametric audio-coding techniques included in these tests are DirAC [39], Harpex [6], and the Wiener-filtering-based Faller method [19]. The first and second-order Ambisonics [24] reproductions were included in one of the tests to illustrate an additional artifact.

## 4.1 Model Structure

The spatial sound scenarios for the simulations explained later in Sect. 4.4 were analyzed with a binaural auditory model based on the count-comparison principle. The functionality of the model used is explained in detail in [41, 46]. The overall structure of the model is illustrated in Fig. 2, and the model parts are briefly introduced and explained in this section.

The binaural input signal is fed to the two models of periphery, one on each hemisphere, containing a nonlinear cochlea model [50] and a model of the cochlear nucleus. Each periphery model feeds the signal to the models of the medial superior olive, MSO, and lateral superior olive, LSO, as depicted in Fig. 2. These model the brainstem nuclei found in the human brain and their functionality is based on neurophysiological studies. They account for the spatial cue decoding. The MSO and LSO models extract spatial cues from separate narrow bandwidths, and, additionally, the wide-band MSO model gathers together spatial information from a wide frequency range. These same MSO and LSO models are included both on the left and the right hemisphere.

After these blocks that model the functionality of the first parts of the auditory pathway in the brains are several functions that are motivated by knowledge obtained from psychoacoustic listening experiments. These functions use the spatial cues from the MSO and LSO models and combine them with spectral cues from the periphery models based on the way the human auditory system analyzes such information. As output, the functions produce a binaural-activity map that is a topographic projection of the binaural input signal on a map. The location of the left/right activation shown in the map indicates the spatial arrangement of the sound scenario, and different colors/grey-shades are used to represent different frequency regions. It should be noted that all the outputs of the model presented in this chapter were obtained by using the model with the same parameters.

## *4.2 Examples of Model Functionality in Typical Listening Scenarios*

In order to illustrate the functionality of the model and the appearance of the binaural-activity map, the outputs of the model for a few example scenarios are presented in Fig. 3. Two types of maps are used in this chapter. In the first type, the binaural-activity map shows the instantaneous activation viewed having left–right dimension in the abscissa, and time in the ordinate. In the second type, the binaural activation in each cell of the map is averaged over time. The map shows the left–right coordinate in the abscissa, and in this case the ordinate denotes the frequency region. The different colors/grey-shades on the map represent the different frequency regions, the division of which is illustrated in Fig. 3e. Positive values on the $x$-axis correspond to activation on the right hemisphere, meaning that in that case, the spatial location of the auditory event is on the left of the listening point, and negative values correspond to activation on the left hemisphere.

All of the samples of this study were simulated using measured head-related transfer functions, HRTFs, of a Cortex–MK2 dummy head. This HRTF database, having a five-degree resolution, was employed in the development of the model and was therefore deemed suitable for the simulations as well. Figure 3a shows the binaural-activity map for short pink noise bursts presented to the model using HRTFs from different azimuthal directions from 0° to 90°. When the sound is emitted from 0°, the activations of all the frequency regions can be seen to be located in the center. As the location of the pink noise burst moves to the right, the activations of all the frequency regions shift towards the left hemisphere, and gradually, with directions from 60° to 90°, the activations become more spread. This reflects the human perception in the sense that the localization accuracy is good for frontal directions and less accurate for directions at the sides [8].

Figure 3b shows the activation of two simultaneous speakers that were presented to the model from directions of ±30° using HRTFs for the simulation. In informal listening, the scenario was perceived to have two spatially separated speakers that were localized close to their correct positions, which is in line with the result from a listening experiment [29] where test subjects were able to localize one of two speech sources with a resolution of ±10° in a similar scenario. Correspondingly, the activation of both speakers can be distinguished separately in the binaural-activity map, one speaker on each hemisphere.

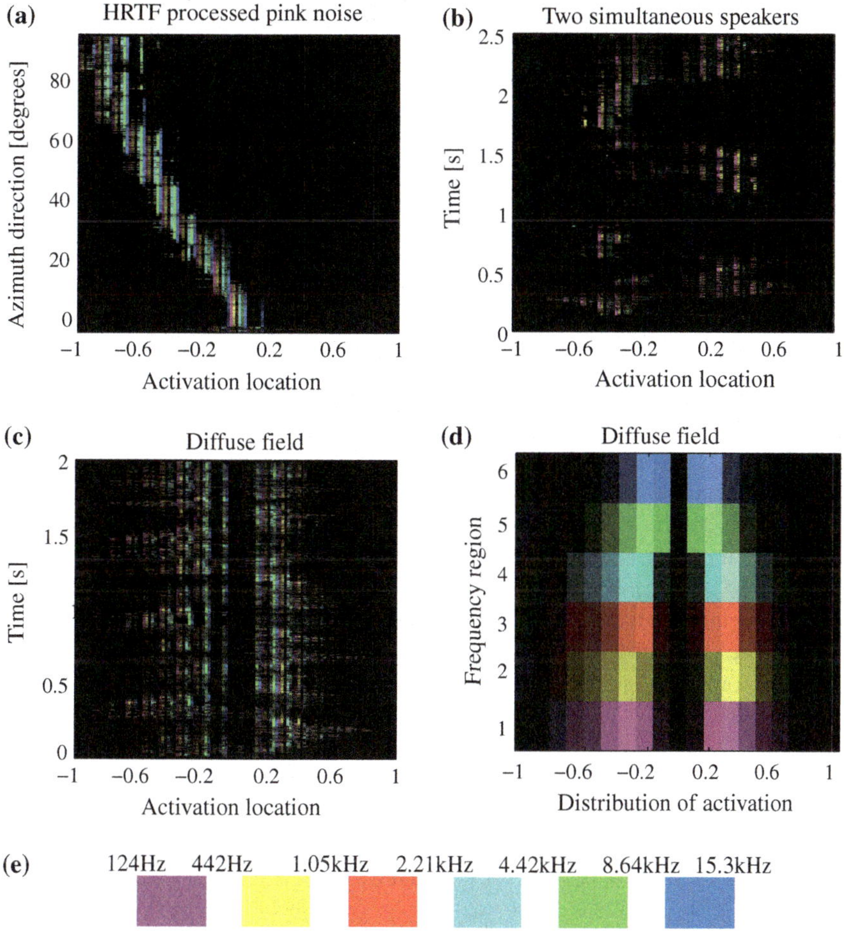


**Fig. 3** Binaural-activity maps obtained with the binaural auditory model [46] for three different scenarios, namely **a** pink noise bursts emitted from different azimuthal directions, **b** two simultaneous speakers positioned at ±30°, and a diffuse field produced with pink noise emitted simultaneously from 12 directions in the horizontal plane. The last-mentioned scenario is shown both with **c** activation location as a function of time and with **d** distribution of activation as a function of frequency region. The *color/grey-shade* codes for different frequency regions are illustrated in **e**

The third example is illustrated in two different ways in Fig. 3c, and d. The figures show the activation caused by a diffuse sound field formed with having pink noise emitted from twelve azimuthal directions at 30° interval all around the listening position simultaneously for a duration of two seconds. The presentation style of Fig. 3c is similar to the above-mentioned examples. It should be noted that the activations evoked by the sound sources behind the listening position are mapped on the same map with no distinction between the front and back. The activation is spread over a wide range, however concentrated on some specific areas, which corresponds to human perception of a spatially distributed auditory event when the head remains stationary. Moreover, it can be seen that different frequency regions cause activation at different locations, the lowest frequencies concentrated on the sides and the highest in the center area. This is in line with observations made in informal listening of high-pass filtered incoherent noise with varying cut-off frequency. The same kind of behavior is illustrated in the distributions of activation as a function of frequency region depicted in Fig. 3d.

## 4.3 Generation of Test Samples

All of the test cases of this study included a reference, that is, the original scenario, as well as several reproductions of the corresponding scenario over a multichannel loudspeaker system in a manner that the signals to the loudspeakers in these reproductions were obtained with different audio-coding techniques. Moreover, a B-format-microphone recording of the original scenario was simulated to obtain compatible signals for the different audio-coding techniques. The signals were then encoded and decoded with the techniques in order to obtain the signals for the reproduction. Finally, the loudspeakers of the multichannel loudspeaker setup were simulated as point sources using the HRTFs of the corresponding directions.

The B-format recording of a given test sample was simulated by computing the components of the 1st-order-B-format-microphone recording following

$$\begin{bmatrix} \mathbf{W} \\ \mathbf{X} \\ \mathbf{Y} \\ \mathbf{Z} \end{bmatrix} = \sum_i \left( \begin{bmatrix} 1/\sqrt{2} \\ \cos(\theta_i)\cos(\gamma_i) \\ \sin(\theta_i)\cos(\gamma_i) \\ \sin(\gamma_i) \end{bmatrix} \times \mathbf{x}_i \right), \tag{4}$$

where $\theta_i$ and $\gamma_i$ denote the azimuth and elevation angles, respectively, from which the given sound event, $x_i$, belonging to the test sample was simulated to be emitted from. It should be noted that a 2nd-order-B-format-microphone recording was simulated to obtain signals for the 2nd-order Ambisonics reproduction. The signals to the loudspeakers for the Ambisonics reproductions were obtained by following the *re-encoding principle* [14], that is, by creating a decoding gain matrix of the B-format components for the different loudspeakers as specified by their azimuth and elevation angles according to (4), and by multiplying signals of the simulated B-format recording with the pseudo-inverse solution of the matrix created. Effectively, in the implementation used, the directivities of the loudspeaker signals of the Ambisonics reproductions match with the ones of hypercardioid microphones. The samples processed with DirAC and the Faller method were obtained with the help of the developers of the methods. Additionally, an online audio conversion service [28] was used to derive the Harpex encoded signals to the loudspeaker setups from the simulated B-format recordings.

## 4.4 Modeling Results of the Tested Scenarios

### Two Simultaneous Speakers

A reference scenario of two simultaneous speakers, a male and a female located at ±30°, was simulated using HRTFs and anechoic samples of speech. The different audio coding techniques were used to derive the signals for a 5.0-loudspeaker system having loudspeakers at directions of 0°, ±30°, and ±90°. The reproduction with such a system was then simulated using HRTFs of the corresponding directions. The directions of speech sources was selected to coincide with directions of two of the loudspeakers in the setup. In this way, the reference condition equals to ideal reproduction of the sources using the loudspeaker setup, which also equals to natural listening of two sources in anechoic conditions. This makes the comparison more straightforward than comparing to reference condition having two virtual sources in corresponding directions.

Figure 4 illustrates the binaural-activity maps obtained for the reference scenario and the reproductions with the different audio coding techniques. All of the activity maps show slightly spread activation on two different areas, one on each hemisphere. The activation changes temporally according to the utterances of the two speakers. The reference scenario in Fig. 4a is the same scenario that was presented in Fig. 3c, and as was observed above, the two speakers and their separate utterances can be distinguished in the activity map. The scenario is perceived relatively similarly with all reproduction methods as compared to the reference, one of the speakers localized on the left side and the other on the right. The similarity of the activity maps reflects this perception. However, some differences between the methods exist, both in the perception of the scenarios and in the activity maps.

The scenario reproduced with mono-DirAC is perceived as having a small but noticeable shift in the speaker locations towards the center [1]. This is due to the averaging of the intensity vector over time in mono-DirAC encoding. Moreover, at some time-frequency bins, the signal contents of the two speakers overlap, resulting in averaged direction angle and increased diffuseness parameter. As a result, the sound of that time-frequency bin is reproduced partly from all loudspeakers after the decorrelation. In the activity map of the mono-DirAC reproduction in Fig. 4b, the activations are shifted slightly towards the center as compared to the reference, reflecting the above-mentioned observations.

The Harpex reproduction of a similar scenario as presented here was evaluated in a listening experiment with a female and a male speaker at ±30° [48]. There, it was found that the perception is very similar to that of the reference scenario. This is in correspondence with the fact that Harpex is designed to be able to accurately detect the directions of arrival of two simultaneous plane waves within each frequency band by paying attention to the phases of the signals [5, 47]. Hence, Harpex performs well in such scenarios as the present one containing two sound sources in anechoic conditions. This is reflected in the activity map in Fig. 4c, which is almost identical to that of the reference scenario.

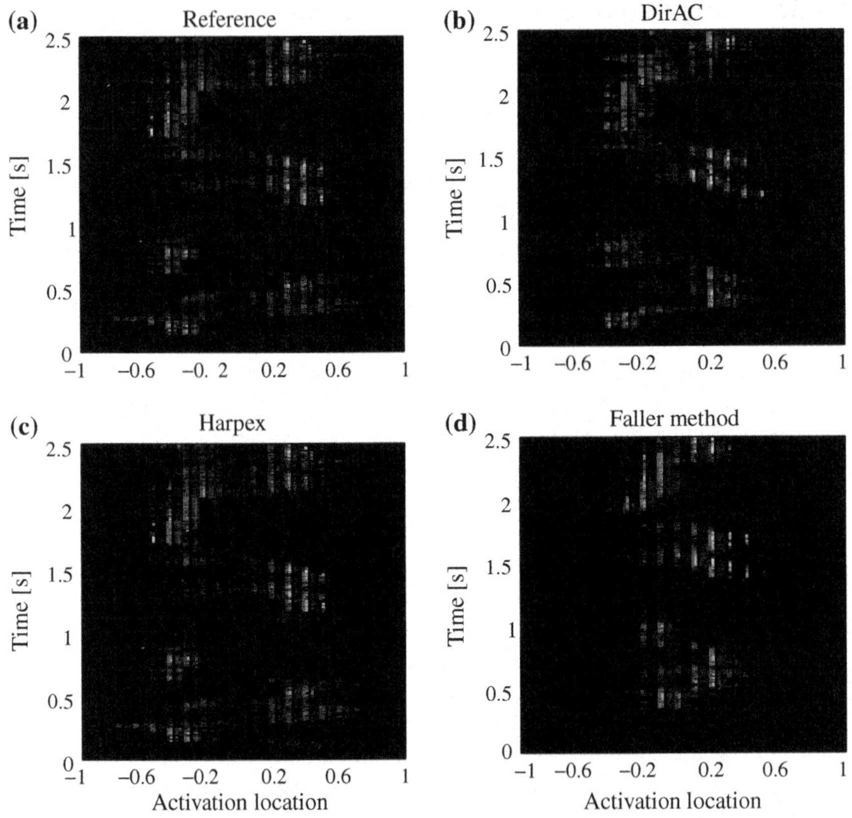


**Fig. 4** Binaural-activity maps obtained with the binaural auditory model for the reference scenario of two simultaneous talkers at ±30° and the reproductions of such a scenario with a 5.0-loudspeaker system employing signals obtained with different audio coding techniques from a simulated B-format recording

The two speakers in the scenario reproduced with the Wiener-filtering based Faller method were perceived in informal listening as point-like but shifted considerably towards the center as compared to the reference scenario. The signals of each loudspeaker were separately listened, and it was noticed that the signal in the loudspeaker at 0° contained both speaker signals with relatively high level—leading into migration of virtual sources towards the center. Figure 4d presents the activity map for the Faller method reproduction. There, temporally changing activation areas in both hemispheres corresponding to the two speakers can be found, albeit having shifted towards the center, following the perceived changes in informal listening. The existence of both speaker signals in the signal for the center loudspeaker at 0° is mainly explained by the look directions of microphones that the Faller method derives for the loudspeaker layout from the B-format microphone input. More precisely, the look directions of the microphones for the loudspeakers at ±30° do not point to the

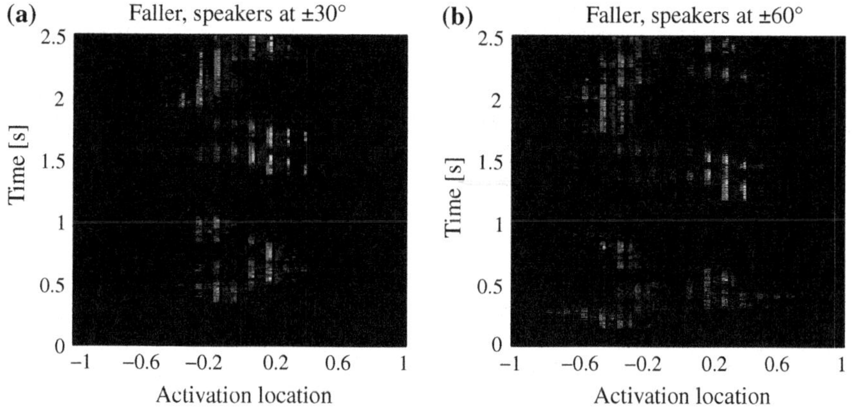


**Fig. 5** Binaural-activity maps obtained with the binaural auditory model for the Faller method reproduction with a 5.0-loudspeaker system. On the *left*, the scenario included two simultaneous speakers at ±30°, while on the *right*, the speakers were at ±60°

directions of the two speakers, instead they point to the directions of ±60°. In order to illustrate this effect caused by the look directions, the reproduction of an additional scenario having the two speakers at ±60° was simulated for the Faller method. The resulting binaural-activity map depicted in Fig. 5b shows that the two activation areas are now separated on different hemispheres, similarly as in the reference shown in Fig. 4a.

### Two Simultaneous Speakers Reproduced with Suboptimal Parameters

In the development of audio coding techniques, there is often a need for tuning various parameter values in order to find the optimal ones for high-quality reproduction. The effects that the tuning introduces to the auditory image may have to be analyzed by comparing the different versions by listening to the reproduction in all cases. This is a time-consuming task and, therefore, binaural auditory models potentially offer an attractive alternative. In order to demonstrate the capability of the binaural auditory model and the binaural-activity map to aid in the development of the audio-coding techniques, the previously used scenario with two speakers was simulated with three different suboptimal parameters in mono-DirAC reproduction as an example. The changing of the parameter values caused visible effects to the binaural-activity maps, and these effects were compared to perceived artifacts in informal listening.

The binaural-activity maps depicted in Fig. 6 illustrate how the quality of the DirAC reproduction decreases when suboptimal parameter values are used in the processing. When the decorrelation of the diffuse stream is switched off, the diffuse stream is reproduced coherently from all of the five loudspeakers. When listening to

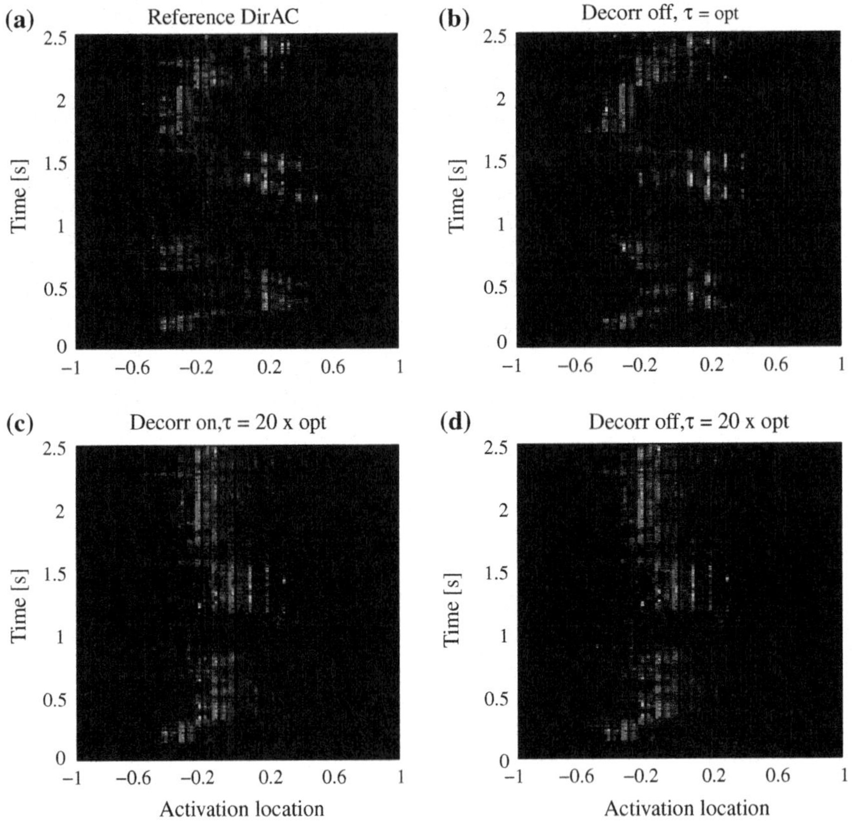


**Fig. 6** Binaural-activity maps obtained with the binaural auditory model for the mono-DirAC reproduction employing suboptimal parameters with a 5.0-loudspeaker system. The scenario is the same as in Fig. 4

the scenario, the increased coherence of the loudspeaker signals results in a shift of the perceived auditory images towards the center area, and the directions of the speakers fluctuate instead of being point-like. Similar changes can be seen in the binaural-activity map of the scenario in Fig. 6b—the activation is spread on a narrower area and there is more activation in the center as compared to the reference.

As mentioned earlier in Sect. 3, the time constant employed in the smoothing of the directional information affects the stability of the auditory images in the reproduction and the ability of the method to respond to sudden changes in the sound input. When the time constant is made 20 times larger than in the normal situation, the perceived locations of the auditory images of the two speakers are affected. The spatial separation is lost and both speakers are perceived to be located close to the direction of the first sound event, that is, the female speaker on the right side. Furthermore, the locations of the speakers are shifted towards the center due to the fact that the directional information used in the positioning of the speakers was computed as an average of the intensity vector containing values corresponding to directions on both sides. The binaural-activity map depicted in Fig. 6c shows activation only on a relatively narrow area in the left hemisphere and the center. This corresponds to the perception of localizing the two speakers close to the original location of the female speaker, the activation of which is on the left hemisphere in the reference scenario.

Moreover, in the case of two simultaneous talkers, the time constant affects the spatial accuracy of the mono-DirAC reproduction more than switching off the decorrelation of the diffuse stream does. This can be seen by comparing the highly similar binaural-activity maps depicted in Fig. 6c, and d, where the time constant is kept the same but the decorrelation is either on or off.

## Off-Sweet-Spot-Listening Scenarios

The size of the listening area within which the different audio coding techniques are able to reproduce the original scenario transparently has been found to vary between the techniques in perceptual studies [1, 10, 42, 45, 51]. The ability of the binaural auditory model to reflect this was evaluated by simulating an off-sweet-spot listening scenario. This was done in a way that the different audio coding techniques were used to derive signals for a reproduction with a 5.0-loudspeaker system depicted in Fig. 7a from the simulated B-format recording in a similar manner as previously, but the simulated scenario was modified so that the listening position was in a point 1.0 m towards the loudspeaker on the right as illustrated in Fig. 7b. The same recording scenario of two simultaneous speakers at $\pm 30°$ as above was used. In practice, the different sound sources of the reproduced scenario were simulated as point sources located in the directions of $-90°$, $-50°$, $-25°$, $0°$, and $90°$, from left to right, due to the five-degree resolution of the HRTF database employed. Moreover, the different signals simulated to be emitted from the different directions were delayed and amplified or attenuated based on the changes in the distances in order to account for the differences in how long it takes for a sound to reach the listener from a given

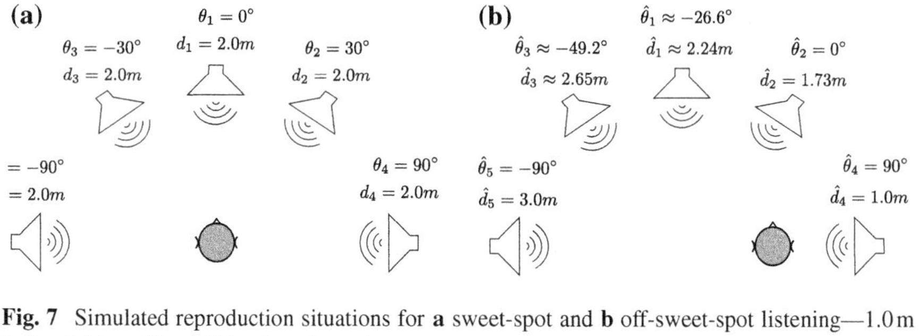


**Fig. 7** Simulated reproduction situations for **a** sweet-spot and **b** off-sweet-spot listening—1.0 m to the right—with a 5.0-loudspeaker setup

location and in the amount the sound emitted from a certain direction gets attenuated before reaching the listener.

Figure 8 shows the binaural-activity maps obtained with the binaural auditory model for the off-sweet-spot listening in the reference scenario and simulated with five different methods. It should be noted that because of the change in the listener position, the two speakers are simulated to be located directly in front and at $-50°$, that is, to the left of the listener.

In the reference scenario, it was found in informal listening that the two speakers were localized close to their actual directions of $0°$ and $-50°$. The binaural-activity map in Fig. 8a shows activations in the center and on the right hemisphere, which correspond to the directions of the two speaker locations and matches with the perception of the scenario. The activations caused by the two speakers are not visually separable in a straightforward way, but rather, the activation areas are fused together. Nevertheless, the temporally separate utterances of one of the two speech samples can be distinguished on the right hemisphere, and the overall activation resembles that of the sweet-spot listening, only clearly shifted towards the right hemisphere.

When the scenario is reproduced with mono-DirAC, the perception changes as compared to the reference scenario in a manner that the female speaker directly at the front is localized slightly to the right while the male speaker remains approximately at $-50°$, corresponding to the observations reported in a listening experiment [1]. The binaural-activity map in Fig. 8b shows the activation in the scenario produced with DirAC. As compared to the reference, there is some excess activation on the left hemisphere, indicating that the loudspeaker nearest to the listening position introduces some diffuseness to the scenario. However, on the right hemisphere, the spreading of the activation is to a large degree similar to that of the reference scenario. The overall impression of the activity map corresponds to the perception of the scenario, as the shifting of the female speaker can be seen in the activity map, and otherwise, a substantial part of the activation is similar to that in the reference scenario.

The Harpex reproduction of the scenario was perceived to be very similar to the reference in terms of spatial impression in informal listening. This is reflected in the activity map in Fig. 8c, which is very close to that of the reference scenario as well. The similarity between the Harpex reproduction and the reference is logical since Harpex has been found to be able to reproduce a similar scenario almost transparently in sweet-spot listening [48], and even in off-sweet-spot listening, the scenario presented is optimal for the design principle of Harpex.

In the case of the Faller method, it was found in informal listening that the speakers were perceived to be closer to each other than in the reference scenario. More precisely, the female speaker was localized directly to the front as is the case in the reference scenario, whereas the male speaker was localized approximately to the direction of the center loudspeaker at $-27°$. No excess sound was perceived from the direction of the nearest loudspeaker and it was possible to perceive the speakers separately. The activity map is presented in Fig. 8d where the activation is concentrated on a narrow area, but the overall shape of the activation is close to that of the reference, illustrating similar observations as in the listening of the scenario.

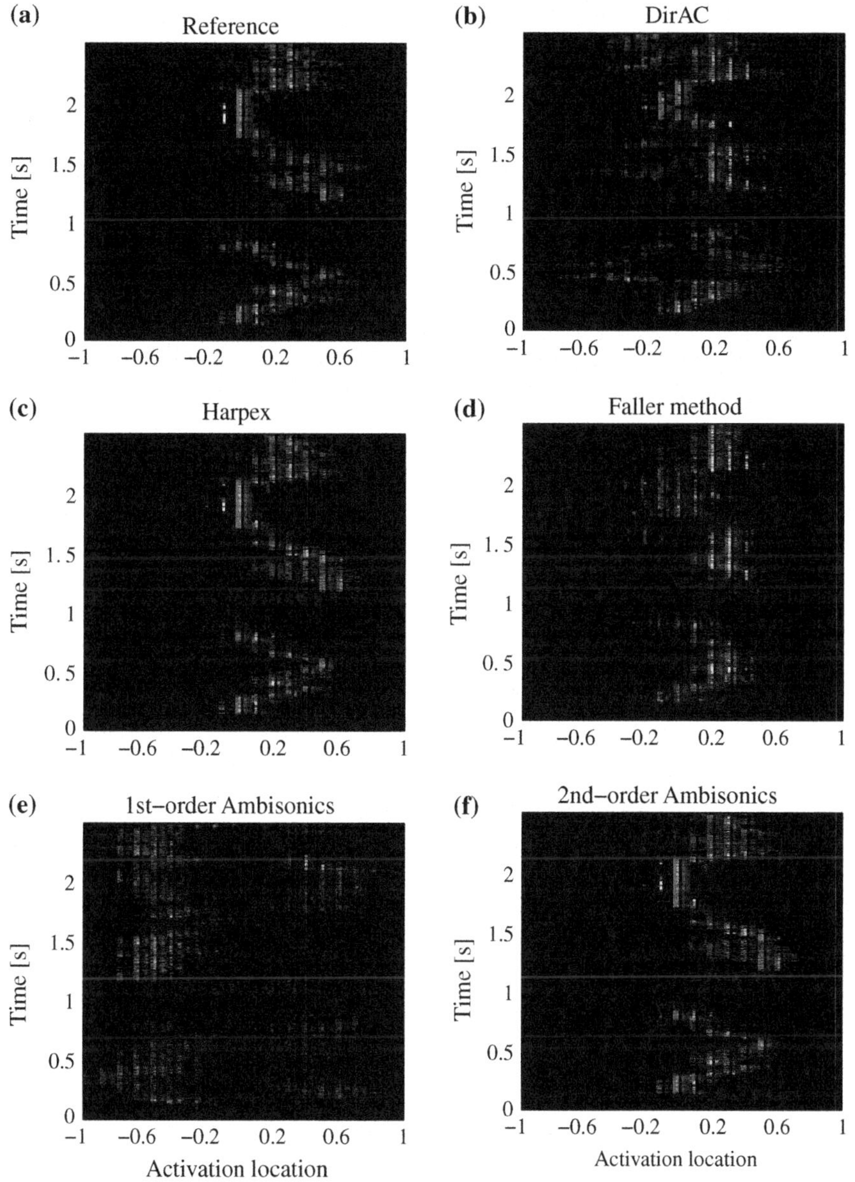


**Fig. 8** Binaural-activity maps obtained with the binaural auditory model for an off-sweet-spot listening scenario with different audio coding techniques as depicted in Fig. 7b

Ambisonics was included in the simulation of this scenario, even though it is not a time-frequency-domain audio technique, in order to illustrate an artifact seen especially in off-sweet-spot listening and detectable with the binaural-activity map. The scenario produced with the first-order Ambisonics reproduction was perceived in informal listening to be emitted mostly from the nearest loudspeaker, that is, the one to the right of the listener. The first-order Ambisonics reproduction has been reported to have a small sweet-spot, because all of the loudspeakers emit sound and the loudspeakers nearest to the listening position dominate the perception [42]. The activity maps for the first-order as well as the second-order Ambisonics are presented in Fig. 8e, and f. The activation in the first-order Ambisonics scenario reflects the perceived effects, as the highest levels of activation are prominently on the left hemisphere as opposed to the reference scenario where the activation is concentrated on the right hemisphere. As opposed to first-order Ambisonics, the perception of the scenario reproduced with second-order Ambisonics was very similar to that of the reference scenario, and this can be seen in the activity map in Fig. 8f as well. This matches with the finding that the second-order Ambisonics is known to have a significantly larger sweet-spot than first-order Ambisonics [45].

### Diffuse Sound Field

Another test was done in order to evaluate the capabilities of the different methods to reproduce a diffuse sound field generated by twelve incoherent pink noise sound sources at azimuth angles of $0°, \pm 30°, \pm 60°, \pm 90°, \pm 120°, \pm 150°$, and $180°$. Again, a B-format recording of the scenario was simulated and the different audio-coding techniques were used to derive signals for a multichannel loudspeaker setup having eight loudspeakers at azimuth angles of $0°, \pm 45°, \pm 90°, \pm 135°$, and $180°$. Both the pink noise sound sources and the loudspeakers were then modeled as point sources and HRTFs of the corresponding directions were used to obtain the binaural input signals to the binaural auditory model. Binaural-activity maps were obtained by using the model, and from them, the distribution of activation as a function of frequency region was analyzed. The activation distributions obtained for the different scenarios are presented in Fig. 9. It should be noted that the current implementation of the model is designed using only left/right cues and, therefore, the activations evoked by sound sources both at the front and the back are mapped by the model to the same left–right axis on the activity map without taking the front-back separation into account.

When listening informally to the reference scenario of the diffuse sound field, the sound is perceived to envelope the listener, forming a surrounding auditory event. However, this perception is aided by small head movements, and when the head is held completely still, some of the sound can be perceived to be concentrated to some specific areas. In the activity map of the reference scenario, illustrated in Fig. 9a, the activation is spread over a wide area on both hemispheres at low frequencies. In contrast, at high frequencies, the activation is concentrated on the center area. The overall distribution of the sound scenario therefore spreads over a wide range and is symmetrically spread on both sides. This corresponds well to the perception of the scenario, having two areas on which the activation is concentrated. The distributions of activation for the scenarios produced with different audio coding techniques show relatively similar patterns as compared to the reference. However, a number of differences can be seen in the maps.

The B-format-DirAC reproduction of a scenario with high diffuseness was perceived to be indistinguishable from the reference [51], which is in line with the high similarity of the activity maps shown in Fig. 9a, and b obtained for the present scenario. At some frequency regions, the activation in the case of DirAC reproduction is spread even over a wider range than it does in the reference. However, at the highest frequencies, the activation is concentrated more on the left hemisphere indicating that the decorrelation has not been fully successful in that frequency region. Furthermore, informal knowledge obtained during the development of such spatial audio techniques has revealed that the selection of the decorrelation filters has an effect on this issue.

In informal listening of the Harpex reproduction of the scenario, a surrounding sound scene was perceived, however, more of the sound was concentrated on certain areas than in the reference scenario. The distribution of the activation shown in Fig. 9c illustrates a similar effect by having a slightly narrower distribution. Furthermore, the highest concentrations of activation shift from one hemisphere to another especially at high frequencies, depending on the frequency region. It is easier to detect the areas where some of the sound is concentrated due to the aspect of the Harpex processing that it aims at decoding the situation in a given time-frequency bin as a combination of two plane waves, which is not optimal for a diffuse field and results in an increased coherence of the loudspeaker signals.

The Faller-method reproduction of the diffuse field was perceived to be surrounding in informal listening, but the areas where the sounds are concentrated were more prominent than in the reference scenario. The activity map in Fig. 9d has the narrowest distribution of activation, which corresponds to the perception. Similarly to the Harpex reproduction, the different frequency regions are concentrated differently on the hemispheres. As mentioned above in Sect. 2.2, in the diffuse field the Faller method does not change the directional pattern of the signal. The narrowing of the reproduced scene is thus simply due to too coherent loudspeaker signals [40].

The parametric audio coding techniques can also take a 5.1-surround signal as input after converting it into B-format. However, it has been found in [35] that the straightforward procedure of simulating a B-format recording of the reproduction of the multichannel signal in anechoic conditions results in suboptimal quality in terms of spatial impression. In order to illustrate this, a B-format recording of a signal consisting of incoherent pink noise samples with a 5.0-loudspeaker system having loudspeakers at azimuth directions of $0^\circ$, $\pm 30^\circ$, and $\pm 110^\circ$ was simulated, and the DirAC was applied to encode and decode the resulting B-format signal for reproduction with a similar loudspeaker layout. Then, the binaural auditory model was used to process a binaural input signal obtained by simulating such a reproduction using HRTFs of the corresponding directions. The binaural-activity map depicted in Fig. 9e illustrates that the sound scene is not reproduced correctly in the sense that the activation is concentrated in the center area. This reflects the perception of the scenario that the virtual sources are biased towards the center loudspeaker in DirAC 5.0 reproduction [35]. This has been found to result from the uneven

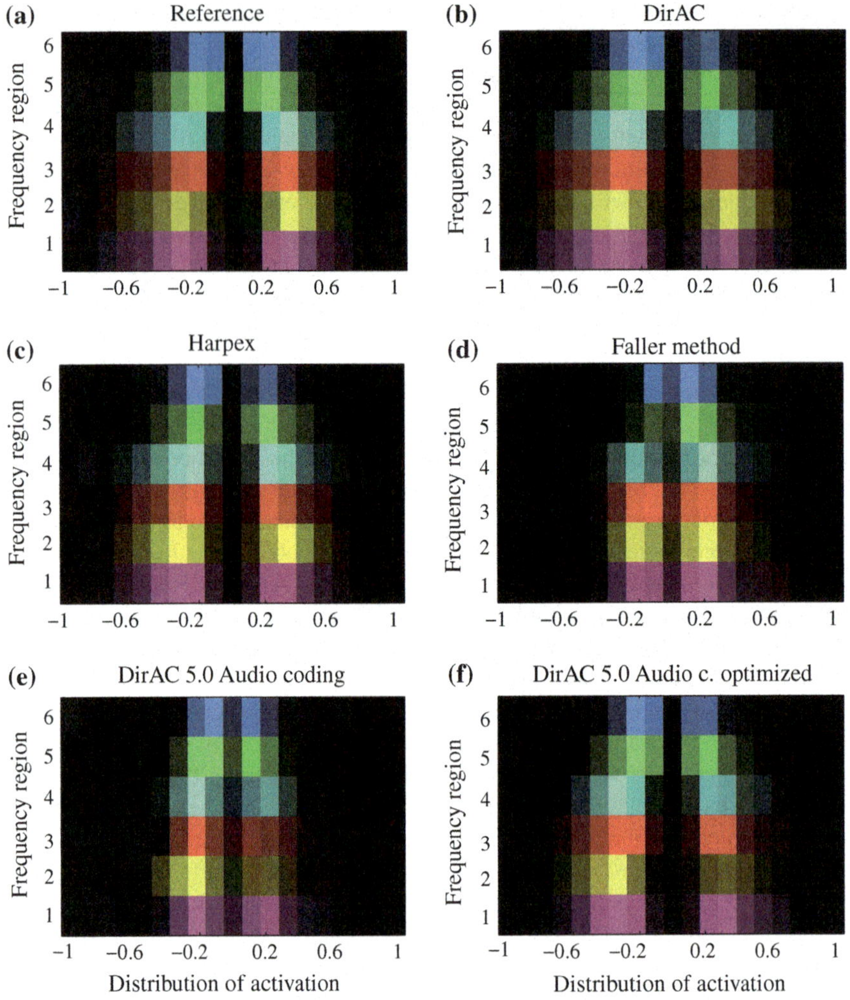


**Fig. 9** Activation distributions obtained with the binaural auditory model for **a** the reference scenario of incoherent pink noise emitted from 12 azimuth directions, and **b–d** the reproductions of such a scenario with an eight-channel loudspeaker system employing signals obtained with different audio-coding techniques. Additionally, the activation distributions when DirAC is used in audio coding of a 5.0-surround signal having pink noise in each channel with **e** the straightforward method, and **f** the even-layout method. Frequency region is on the $y$-axis, the numbers 1–6 corresponding to regions from low to high frequencies as depicted in Fig. 3e

loudspeaker layout of the standardized 5.0-reproduction employed in the simulation of the B-format-microphone recording, and solutions to solve this problem have been previously proposed [35]. When the directions of the loudspeakers in the simulation of the B-format recording are replaced with directions of 0°, ±72°, and ±144°

according to the even-layout method presented in [35], a more surrounding sound scene is perceived [35]. This is reflected in the binaural-activity map shown in Fig. 9f.

## 5 Concluding Remarks

This chapter reviewed a number of recently proposed parametric spatial sound reproduction techniques. The techniques process the sound with the same principle—an analysis of the spatial properties of the sound field is performed based on the microphone signals in time-frequency domain, and the information obtained is subsequently used in the reproduction of spatial sound to enhance the perceptual quality. The constants for the reproduction methods are selected using assumptions of spectral, temporal and spatial resolution of the human hearing mechanisms, and thus, to a large degree, they resemble binaural auditory models.

All of the methods provide some improvement in audio quality as compared to corresponding conventional time-domain methods. However, new types of artifacts may occur in the reproduction that are not necessarily found in traditional spatial sound reproduction, especially if the constants of the methods are selected suboptimally. The most evident artifacts are described. A count-comparison-based binaural auditory model is reviewed and applied to visualize the artifacts.

A number of sound-reproduction scenarios producing spatial artifacts are simulated and the binaural input signals to the model are computed using measured HRTFs of a dummy head. The resulting binaural-activity maps show artifacts as well as various differences between the different techniques. These findings are shown to be in line with results obtained from listening experiments or, in the lack of such experiments, observations found in informal listening. Overall, it can be said that a binaural auditory model can be used in the analysis of the functionality of parametric spatial audio reproduction techniques, and to aid in the development of such techniques.

It should be noted that the results shown in this chapter should not be used to rate the overall quality of the sound reproduction of the techniques presented for several reasons. Most importantly, the aim in this study was to analyze some specific spatial artifacts of the sound reproduction, and perceived timbre among other important factors affecting the overall perception were excluded. Moreover, the techniques were not optimized for best quality specifically for each listening scenario. On the contrary, the selection of the scenarios presented was motivated by their suitability to show a number of spatial artifacts in the reproduction for the purpose of using an auditory model for the assessment of such techniques.

**Acknowledgments** The authors would like to thank S. Verhulst from the Boston University for providing the cochlea model and assisting in its use, C. Faller from Illusonic GmbH for providing the samples processed with the Faller method, V. Sivonen from Cochlear Nordic for providing the head-related transfer functions, and J. Ahonen, M.-V. Laitinen, and T. Pihlajamäki from Aalto University for providing the DirAC-processed samples for the tests. Further, they are indebted to two anonymous reviewers for constructive comments. This work has been supported by The Academy of Finland and by the European Research Council under the European Community's Seventh Framework Programme (FP7/2007-2013)/ERC Grant agreement No. 240453.

## References


1. J. Ahonen. Microphone configurations for teleconference application of Directional Audio Coding and subjective evaluation. In *Proc. 40th Intl. Conf. Audio Eng. Soc.,* Tokyo, Japan, Oct. 8–10 2010. Paper No. P-5.
2. C. Avendano and J.-M. Jot. Frequency domain techniques for stereo to multichannel upmix. In *Proc. 22nd Intl. Conf. Audio Eng. Soc.,* Espoo, Finland, Jun. 15–17 2004. Paper No. 251.
3. G. Barry and D. Kearney. Localization quality assessment in source separation-based upmixing algorithms. In *Proc. 35th Intl. Conf. Audio Eng. Soc.,* London, UK, Feb. 11–13 2009. Paper No. 33.
4. G. Barry, B. Lawlor, and E. Coyle. Sound source separation: Azimuth discrimination and resynthesis. In *Proc. 7th Intl. Conf. Digital Audio Effects,* pages 240–244, Naples, Italy, Oct. 5–8 2004.
5. S. Berge and N. Barrett. A new method for B-format to binaural transcoding. In *Proc. 40th Intl. Conf. Audio Eng. Soc.,* Tokyo, Japan, Oct. 8–10 2010. Paper No. 6–5.
6. S. Berge and N. Barrett. High angular resolution planewave expansion. In *Proc. 2nd Intl. Symp. Ambisonics and Spherical Acoustics,* Paris, France, May 6–7 2010.
7. A. J. Berkhout. A holographic approach to acoustic control. *J. Audio Eng. Soc.,* 36:977–995, 1988.
8. J. Blauert. *Spatial hearing. The psychophysics of human sound localization.* MIT Press, Cambridge, MA, USA, revised edition, 1997.
9. A. D. Blumlein. U.K. Patent 394,325, 1931. Reprinted in Stereophonic Techniques, Audio Eng. Soc., 1986.
10. M. M. Boone, E. N. G. Verheijen, and P. F. van Tol. Spatial sound-field reproduction by wave-field synthesis. *J. Audio Eng. Soc.,* 43:1003–1012, 1995.
11. J. Breebaart, S. Disch, C. Faller, J. Herre, G. Hotho, K. Kjörling, F. Myburg, M. Neusinger, W. Oomen, H. Purnhagen, and J. Rödén. MPEG spatial audio coding / MPEG surround: Overview and current status. In *Proc. 119th Intl. Conv. Audio Eng. Soc.,* New York, NY, USA, Oct. 7–10 2005. Paper No. 6599.
12. J. Breebaart and C. *Faller. Spatial audio processing: MPEG surround and other applications.* John Wiley & Sons, Ltd., Chichester, UK, 2008.
13. H. S. Colburn and N. I. Durlach. Models of binaural interaction. In E. Carrette and M. Friedman, editors, *Handbook of perception,* volume IV, pages 467–518. Academic Press, San Diego, CA, USA, 1978.
14. J. Daniel, S. Moreau, and R. Nicol. Further investigations of high-order Ambisonics and Wave-field synthesis for holophonic sound imaging. In *Proc. 114th Intl. Conv. Audio Eng. Soc.,* Amsterdam, The Netherlands, Mar. 22–25 2003. Paper No. 5788.
15. D. de Vries. *Wave field synthesis.* Audio Eng. Soc. monograph, New York, NY, USA, 2009. 93 pages.
16. C. Faller. Binaural cue coding-Part I: Psychoacoustic fundamentals and design principles. *IEEE Trans. Speech and Audio Processing,* 11:509–519, 2003.
17. C. Faller. Multiple-loudspeaker playback of stereo signals. *J. Audio Eng. Soc.,* 54:1051–1064, 2006.
18. C. Faller. A highly directive 2-capsule based microphone system. In *Proc. 123rd Intl. Conv. Audio Eng. Soc.,* New York, NY, USA, Oct. 5–8 2007.
19. C. Faller. Method to generate multi-channel audio signals from stereo signals. EP Patent 1,761,110, Mar. 2007.
20. C. Faller. Microphone front-ends for spatial audio coders. In *Proc. 125th Intl. Conv. Audio Eng. Soc.,* San Francisco, CA, USA, Oct. 2–5 2008.
21. C. Faller and F. Baumgarte. Efficient representation of spatial audio using perceptual parametrization. In *Proc. IEEE Worksh. Appl. of Signal Processing to Audio and Acoustics,* pages 199–202, New Paltz, New York, Oct. 21–24 2001.

22. C. Faller, A. Favrot, C. Langen, C. Tournery, and H. Wittek. Digitally enhanced shotgun microphone with increased directivity. In *Proc. 129th Intl. Conv. Audio Eng. Soc.*, San Francisco, CA, USA, Nov. 4–7 2010.
23. C. Faller and V. Pulkki. Directional Audio Coding: Filterbank and STFT-based design. In *Proc. 120th Intl. Conv. Audio Eng. Soc.*, Paris, France, May 20–23 2006. Paper No. 6658.
24. M. A. Gerzon. Periphony: With-height sound reproduction. *J. Audio Eng. Soc.*, 21:2–10, 1973.
25. M. M. Goodwin. Enhanced microphone-array beamforming based on frequency-domain spatial analysis-synthesis. In *IEEE Worksh. Appl. Signal Processing to Audio and Acoustics*, pages 6–9, New Paltz, NY, USA, Oct. 21–24 2007.
26. M. M. Goodwin and J.-M. Jot. A frequency-domain framework for spatial audio coding based on universal spatial cues. In *Proc. 120th Intl. Conv. Audio Eng. Soc.*, Paris, France, May 20–23 2006. Paper No. 6751.
27. M. M. Goodwin and J.-M. Jot. Spatial audio scene coding. In *Proc. 125th Intl. Conv. Audio Eng. Soc.*, San Francisco, CA, USA, Oct. 2–5 2008. Paper No. 7507.
28. Harpex Ltd. Online audio conversion service BETA, 2012. (Accessed: Jan. 22, 2013).
29. M. L. Hawley, R. Y. Litovsky, and H. S. Colburn. Speech intelligibility and localization in a multi-source environment. *J. Acoust. Soc. Am.*, 105:3436–3448, 1999.
30. J. Herre, C. Falch, D. Mahne, G. del Galdo, M. Kallinger, and O. Thiergart. Interactive teleconferencing combining spatial audio object coding and DirAC technology. In *Proc. 128th Intl. Conv. Audio Eng. Soc.*, London, UK, May 22–25 2010. Paper No. 8098.
31. J. Herre, K. Kjörling, J. Breebaart, C. Faller, S. Disch, H. Purnhagen, J. Koppens, J. Hilpert, J. Rödén, W. Oomen, K. Linzmeier, and K. S. Chong. MPEG surround-the ISO/MPEG standard for efficient and compatible multichannel audio coding. *J. Audio Eng. Soc.*, 56:932–955, 2008.
32. G. Hotho, S. van de Par, and J. Breebaart. Multichannel coding of applause signals. *EURASIP J. Adv. in Signal Process.*, 2008, 2008. Article No. 10.
33. M.-V. Laitinen, F. Kuech, S. Disch, and V. Pulkki. Reproducing applause-type signals with Directional Audio Coding. *J. Audio Eng. Soc.*, 59:29–43, 2011.
34. M.-V. Laitinen, T. Pihlajamäki, C. Erkut, and V. Pulkki. Parametric time-frequency representation of spatial sound in virtual worlds. *ACM Trans. Appl. Percept.*, 9:1–20, 2012.
35. M.-V. Laitinen and V. Pulkki. Converting 5.1 audio recordings to B-format for Directional Audio Coding reproduction. In *Proc. Intl. Conf. Acoustics, Speech and Signal Processing (ICASSP)*, pages 61–64, Prague, Czech Republic, May 22–27 2011.
36. M.-V. Laitinen and V. Pulkki. Utilizing instantaneous direct-to-reverberant ratio in parametric spatial audio coding. In *Proc. 133rd Intl. Conv. Audio Eng. Soc.*, San Francisco, USA, Oct. 26–29 2012. Paper No. 8804.
37. A. Politis, T. Pihlajamäki, and V. Pulkki. Parametric spatial audio effects. In *Proc. 15th Intl. Conf. Digital Audio Effects*, York, UK, Sept. 17–21 2012. Paper No. 22.
38. V. Pulkki. Virtual sound source positioning using Vector Base Amplitude Panning. *J. Audio Eng. Soc.*, 45(6):456–466, 1997.
39. V. Pulkki. Spatial sound reproduction with Directional Audio Coding. *J. Audio Eng. Soc.*, 55:503–516, 2007.
40. V. Pulkki and C. Faller. The directional effect of cross-talk in multi-channel sound reproduction. In *Proc. 18th Intl. Congr. Acoust.*, pages 3167–3170, Kyoto, Japan, Apr. 4–9 2004.
41. V. Pulkki and T. Hirvonen. Functional count-comparison model for binaural decoding. *Acta Acust./Acustica*, 95:883–900, 2009.
42. V. Pulkki, J. Merimaa, and T. Lokki. Reproduction of reverberation with Spatial Impulse Response Rendering. In *Proc. 116th Intl. Conv. Audio Eng. Soc.*, Berlin, Germany, May 8–11 2004. Paper No. 6057.
43. F. Rumsey. *Spatial audio*. Music Technology. Focal Press, Oxford, UK, 2nd edition, 2001.
44. E. Schuijers, J. Breebaart, H. Purnhagen, and J. Engdegard. Low complexity parametric stereo coding. In *Proc. 116th Intl. Conv. Audio Eng. Soc.*, Berlin, Germany, May 8–11 2004. Paper No. 6073.
45. A. Solvang. Spectral impairment of two-dimensional higher order Ambisonics. *J. Audio Eng. Soc.*, 56:267–279, 2008.

46. M. Takanen, O. Santala, and V. Pulkki. Visualization of functional count-comparison-based binaural auditory model output. Unpublished manuscript, 2013.
47. O. Thiergart and E. A. P. Habets. Robust direction-of-arrival estimation of two simultaneous plane waves from a B-format signal. In *IEEE 27th Conv. Electrical and Electronics Engineers,* pages 1–5, Eilat, Israel, Nov. 14–17 2012.
48. O. Thiergart and E. A. P. Habets. Sound field model violations in parametric spatial sound processing. In *Proc. of IWAENC 2012 Intl. Workshop Acoustic Signal Enhancement,* pages 1–4, Aachen, Germany, Sept. 4–6 2012.
49. O. Thiergart, M. Kratschmer, M. Kallinger, and G. del Galdo. Parameter estimation in Directional Audio Coding using linear microphone arrays. In *Proc. 130th Intl. Conv. Audio Eng. Soc.,* London, UK, May 13–16 2011. Paper No. 8434.
50. S. Verhulst, T. Dau, and C. A. Shera. Nonlinear time-domain cochlear model for transient stimulation and human otoacoustic emission. *J. Acoust. Soc. Am.,* 132:3842–3848, 2012.
51. J. Vilkamo, T. Lokki, and V. Pulkki. Directional Audio Coding: Virtual microphone-based synthesis and subjective evaluation. *J. Audio Eng. Soc.,* 57:709–724, 2009.

# Binaural Dereverberation

**A. Tsilfidis, A. Westermann, J. M. Buchholz, E. Georganti and J. Mourjopoulos**

## 1 Room Reverberation

When a sound is emitted by a source in an enclosed space, a listener will initially receive the direct sound followed by multiple reflections from the walls or objects placed in the room—see Fig. 1.

The energy of the reflected sound will be attenuated according to the frequency-dependent absorption of the reflecting surfaces. Moreover, assuming an omni-directional source, the sound pressure of both the direct and the reflected sounds will decrease according to the $1/r$ distance law [83]. Although the sound energy decays over time, $t$, roughly following an exponential function, the reflection density increases with $t^3$, forming an increasingly diffuse sound field [83]—see also [42]. The required time for the energy in a room to decrease by 60 dB after the sound source has stopped emitting sound is the reverberation time, $T_{60}$, being the most commonly used parameter for specifying the acoustic properties of a given room [83, 110].

Assuming that the room acoustics are modeled as a *linear, time-invariant*, LTI, system, the *room impulse response*, RIR, provides a complete description of the direct and reflective paths in a room from a sound source to the receiver. In a general multichannel scenario with one source and $i$ receivers, the reverberant signal, $x_i(n)$, for each specific source-receiver position in the room can be expressed as the convolution of the anechoic signal, $s(n)$, with the corresponding RIRs, $h_i(n)$, as follows,

$$x_i(n) = \sum_{j=0}^{J_h-1} h_i(j)s(n-j) \ , \tag{1}$$

A. Tsilfidis · E. Georganti · J. Mourjopoulos (✉)
Audio and Acoustic Technology Group, Electrical and Computer Engineering department, University of Patras, Patras, Greece
e-mail: mourjop@upatras.gr

A. Westermann · J. M. Buchholz
National Acoustic Laboratories, Australian Hearing, Macquarie University, Sydney, Australia

J. Blauert (ed.), *The Technology of Binaural Listening*, Modern Acoustics and Signal Processing, DOI: 10.1007/978-3-642-37762-4_14,

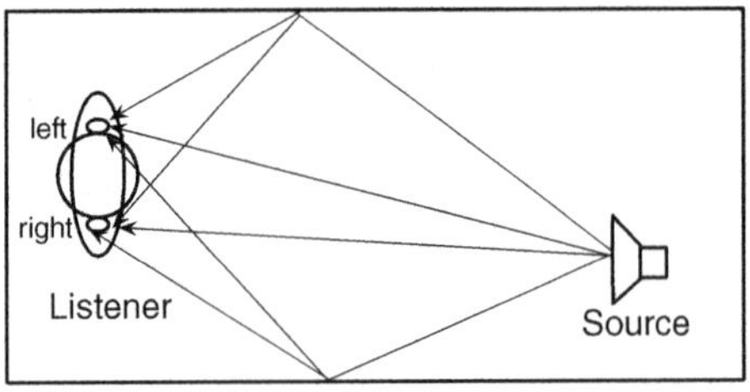


**Fig. 1** A human listener and a sound source in a reverberant room

where $n$ represents the discrete time index and $J_h$ is the length of the impulse response.

In the binaural scenario the room response is combined with the related left- and right-ear *head-related impulse responses*, HRIRs. The latter are measured in anechoic conditions. As a consequence, assuming an ideal omni-directional source, a *binaural room impulse response*, BRIR, for the left-ear channel, $h_L(n)$, can be expressed as

$$h_L(n) = g(r_s)\delta(n - n_s) * h_{HRIR,L,\theta_d,\phi_d}(n) + \sum_{m=0}^{J_{h_m}-1} h_{m,L}(n) * h_{HRIR,L,\theta_m,\phi_m}(n) , \quad (2)$$

where $g(r_s)$ is a gain reduction that depends on the source-receiver distance $r_s$, $\delta(n)$ refers to a Kronecker-delta function, $n_s$ is the delay mainly depending on the source-receiver distance, $r_s$, and the physical characteristics of the propagation medium. $h_{HRIR,L,\theta_d,\phi_d}(n)$ is the left HRIR for the direct sound, corresponding to $\theta_d$ and $\phi_d$, namely, the horizontal and vertical angles between source and receiver. The value $h_m(n)$ denotes the response of the $m$-th reflection. $J_{h_m}$ is the number of individual reflections. $h_{HRIR,L,\theta_m,\phi_m}$ is the HRIR corresponding to such a reflection. Finally, $\theta_m$ and $\phi_m$ are the horizontal and vertical angles between receiver and $m$-th reflection.[1] A similar equation also applies for the BRIR, $h_R(n)$.

Hence, the reverberant signal at the left and right ear of a listener, $x_L(n)$ and $x_R(n)$, can be described as a convolution of the anechoic source signal, $s(n)$, with the left- and right-ear binaural room impulse responses, $h_L(n)$ and $h_R(n)$, that is,

$$x_L(n) = \sum_{j=0}^{J_{h_L}-1} h_L(j)s(n - j) , \quad (3)$$

$$x_R(n) = \sum_{j=0}^{J_{h_R}-1} h_R(j)s(n - j) . \quad (4)$$

Examples of BRIRs measured in a stairway with a reverberation time of approximately 0.86 s are shown in Fig. 2 for the left and right ear [67]. The initial delay before the arrival of the first peak of each RIR depicts the delay, $n_s$, due to the

[1] In some cases $n_s$ is a fractional delay, and the delta function is not well defined.

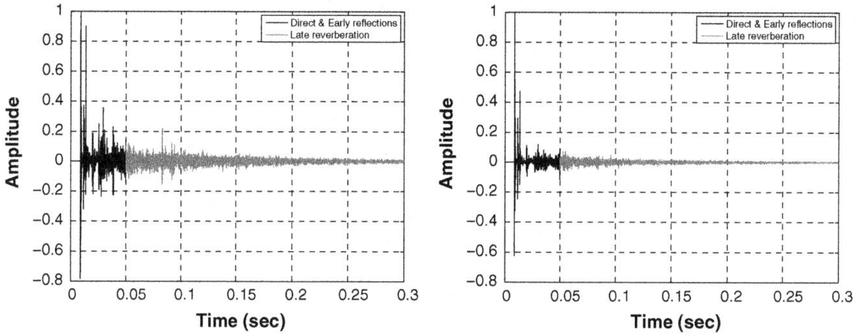


**Fig. 2** Binaural room impulse responses in a stairway having a reverberation time of approximately 0.86 s. *Left* left-ear room impulse response. *Right* right-ear room impulse response

source-receiver distance, $r_s$. After the direct sound, the *early reflections* arrive. The early reflections are considered relatively sparse and span a short time interval of about 50–80 ms after the arrival of the direct sound. The last part of the RIR is called *late reverberation* and results to the reverberant tail of the signal [58, 77]. Due to the interaction of body, head and torso of the listener with the reverberant sound field, the RIRs at the left and right ear exhibit frequency-dependent interaural differences in arrival time and level. These interaural differences are essential parameters for binaural dereverberation algorithms, as is further described in Sects. 3.4 and 4.

In room acoustics and, consequently, in speech- and audio-enhancement applications, RIRs are often modeled as the sum of two components, one denoting the direct path and the early reflections, $h_{i,e}(n)$, and the other one the late reverberation, $h_{i,l}(n)$, as

$$h_i(n) = h_{i,e}(n) + h_{i,l}(n). \tag{5}$$

By combining (1) and (5), each reverberant signal can be written as the sum of a signal part affected by early reflections only, $x_{i,e}(n)$, and a signal part affected by late reverberation, $x_{i,l}(n)$, namely,

$$x_i(n) = \underbrace{\sum_{j=0}^{J_h-1} h_{i,e}(j)s(n-j)}_{x_{i,e}(n)} + \underbrace{\sum_{j=0}^{J_h-1} h_{i,l}(j)s(n-j)}_{x_{l,e}(n)}. \tag{6}$$

As will be further discussed in Sect. 2.2, these two components of room reverberation affect the received signal in a different way and are thus treated separately in most dereverberation applications.

# 2 Speech Signals in Rooms

## *2.1 Auditory Perception in Rooms*

In an anechoic environment, where only the direct sound is present, a normal hearing listener can accurately localize arbitrary sound sources due to the presence of unambiguous interaural-time and -level cues as well as spectral cues that are provided by interaction of pinnae, head and torso with the sound field—see [13]. In particular due to the presence of interaural binaural cues, the auditory system is also able to suppress interfering sounds that arrive from different locations than a target sound and, thereby, for instance, improve speech intelligibility significantly. This phenomenon is commonly referred to as *spatial release from masking* [20].

In the case that a sound is presented in a reverberant environment, the direct sound is accompanied by early reflections and reverberation—see Sect. 1. This results in distortion of the available auditory cues and, typically, leads to reduced auditory performance, for instance, in localization or speech intelligibility. Whereas the early reflections, which arrive within a time window of about 50–80 ms after the direct sound, improve speech intelligibility [4, 17], late reverberation generally has a negative effect on speech intelligibility [61].

Auditory localization in rooms is aided by auditory mechanisms that are associated with the *precedence effect* [13, 88] and may be linked to a cue-selection mechanism that takes advantage of a measure of *interaural coherence* [36]. In particular, early reflections change the timbre of a sound and introduce perception of *coloration* [11, 23], a phenomenon that is significantly suppressed by the binaural auditory system [22, 130]. Also, late reverberation, which is mainly perceived within the temporal gaps inherent in the source signal, is reduced by the binaural system [26]. In [19] it has been shown that familiarization with a reverberant environment can result in enhanced speech intelligibility. Finally, it should be mentioned that auditory masking renders many reflections to be inaudible [24].

Besides the aforementioned detrimental effects of room reverberation on auditory performance, which are partly compensated by different auditory mechanisms, the room also introduces a number of additional cues that are utilized by the auditory system. The direct-to-reverberant energy ratio, for instance, provides a very reliable cue for distance perception [128]. Such mechanisms are described in detail in this volume [42]. Moreover, early lateral reflections extend the apparent width of a sound source [6] and late lateral reverberation energy makes a listener feel enveloped in an auditory scene [18]. The latter two phenomena are highly appreciated when listening to music [46]. The perceived spaciousness introduced by a room has often been related to interaural coherence, whereby, as can be stated as a rule, the lower the interaural-coherence, the higher is the perceived spaciousness [13].

When considering signal processing methods that aim at reducing room reverberation for applications with human listeners, such as in hearing aids or telecommunication devices, it is important that binaural cues are preserved. When binaural cues are distorted, the listener may not be able to correctly localize sound sources any more, although this is obviously very important for the orientation in an auditory scene as well as for perceiving warnings from potential threats. Moreover, with binaural cues being preserved, the binaural system may provide additional benefit by suppressing coloration, reverberation and interfering sound sources. Finally, successful dereverberation methods will increase the direct-to-reverberant energy ratio and, thus, may modify the perceived distance of a sound source and/or also modify the perception of apparent source width and envelopment.

## 2.2 Early- and Late-Reverberation Effects

The typical effects of reverberation in speech spectrograms are presented in Fig. 3, obtained with a fast fourier transform (FFT) length of 23.2 ms. Figure 3a shows

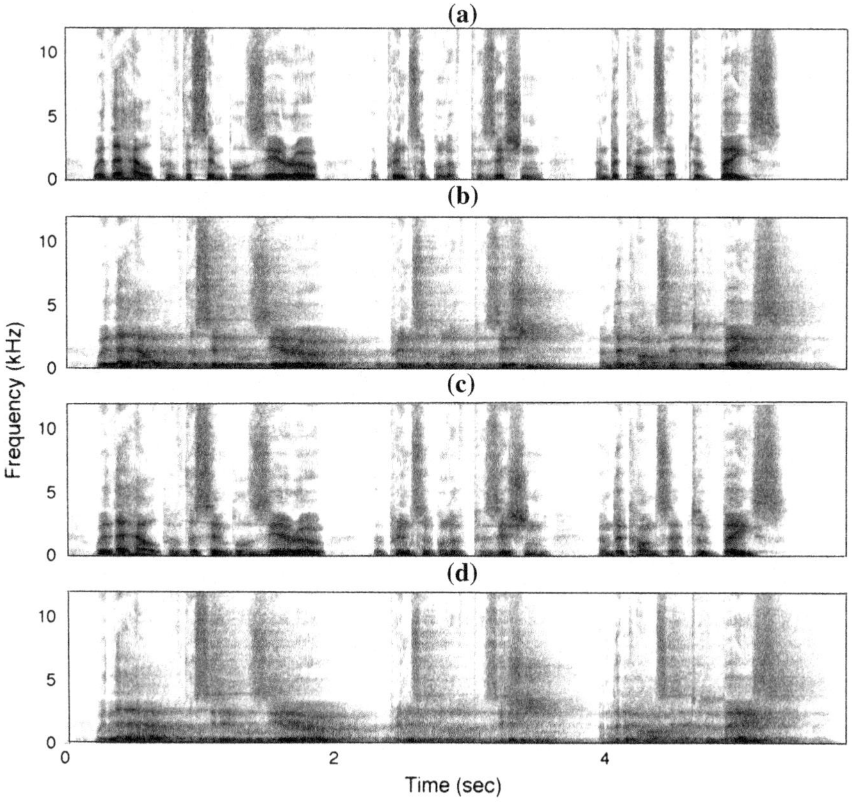


**Fig. 3** Spectrograms illustrating the effects of reverberation on speech. **a** Anechoic input signal. **b** Reverberant signal. **c** Reverberant signal due to early reflections only. **d** Reverberant signal due to late reverberation only

an anechoic speech signal of a male speaker, and Fig. 3b shows the corresponding reverberant signal. The reverberant signal is produced via convolution with an RIR recorded in a lecture hall with a reverberation-time value of $T_{60}$ =1 s at a distance of 4 m from the source. A comparison of Fig. 3a,b reveals that a large number of temporal gaps and spectral dips that can be seen in the anechoic speech are now filled due to the reverberation, that is, reverberation leads to a smearing of both temporal and spectral features. Moreover, the reverberation produced by signal components with high energy may mask later components with lower energy. In Fig. 3c the effect of early reflections on the reverberant signal is shown in isolation, obtained by convolving the anechoic signal with only the first 50 ms of the impulse response, since for speech applications this is considered to be the boundary between early reflections and late reverberation. Obviously, the early reflections alone do not significantly alter the anechoic speech spectrogram, but more careful observation reveals a smearing of the spectral speech profile. In Fig. 3d only the late-reverberant speech is shown, produced by convolving the anechoic signal with an artificially-modified impulse response where the first 50 ms were set to 0. It is evident that late reverberation significantly distorts the spectrogram of the anechoic signal and generates a reverberation tail between temporal speech gaps.

The effects of early and late reflections on the long-term speech spectrum (smoothed in 1/6 octave bands) can be observed in Fig. 4. The FFT length for these illustrations was equal to the signal length, namely, 7.8 s. In Fig. 4a the spectrum of the anechoic signal is compared to the spectrum of the same signal, contaminated by early reflections. In Fig. 4b the long-term spectrum of the anechoic signal is presented along with the spectrum of the same signal, contaminated by late reverberation. It is obvious that the early reflections significantly degrade the long-term speech spectrum, especially in the lower frequencies. This distortion is perceived as coloration of the sounds. In contrast, late reverberation introduces a more flat, white-noise like effect for the same frequency band of the signal.

Finally, Fig. 5 depicts speech spectrograms as obtained via convolution of an anechoic speech excerpt with (a) a left-ear impulse response and (b) a room impulse response obtained from an omni-directional microphone at exactly the same position. Both impulse responses were recorded in a lecture hall with $T_{60} = 0.79$ s at a source-receiver distance of 10.2 m [67]. It clearly appears that the spectrogram of the received speech signal is not significantly different for the binaural scenario. Such observation relates especially to the late reverberation signal components which are usually generated by diffuse reflections and, hence, are less susceptible to binaural cues. These late-reverberation effects can be treated by adapting single-channel dereverberation methods to the binaural scenario.

## 2.3 Interaural Coherence

An often used measure of similarity between two binaural signals or BRIRs is the interaural coherence, IC, defined as

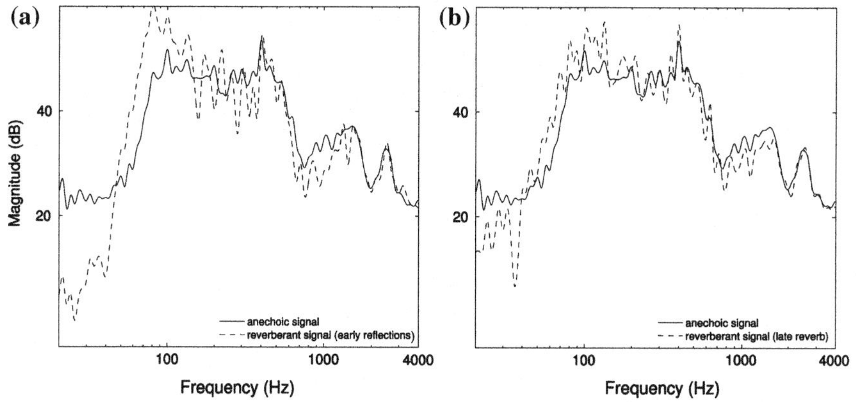


**Fig. 4** Effect of early and late reflections on the long-term speech spectrum, smoothed in 1/6 octave bands. **a** spectrum of anechoic speech signal and of the signal contaminated by early reflections. **b** spectrum of anechoic speech signal and of the signal contaminated by late reverberation

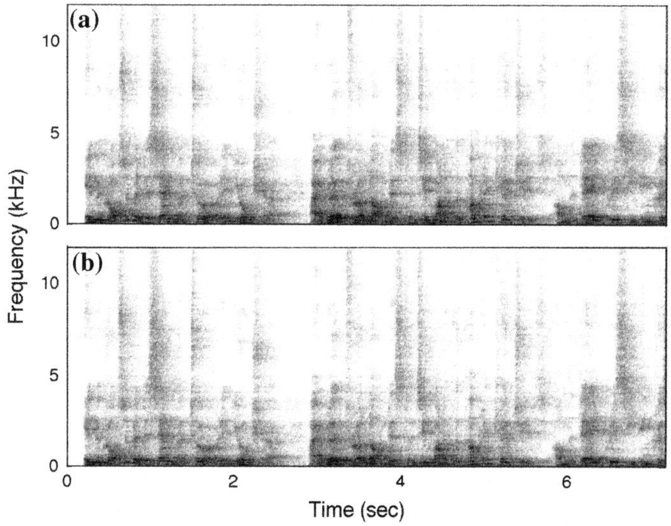


**Fig. 5** Speech spectrograms obtained by convolving an anechoic speech sample. **a** *Left-ear* room impulse response measured in a lecture room with $T_{60} = 0.79$ s at 10.2 m from the source. **b** Room impulse response measured with an omni-directional microphone in the same room at exactly the same position [67]

$$IC_{X_L,X_R}(k) = \frac{|\langle X_L(k) \cdot X_R^*(k)\rangle|}{\sqrt{\langle X_L(k) \cdot X_L^*(k)\rangle \langle X_R(k) \cdot X_R^*(k)\rangle}}, \tag{7}$$

with $k$ being the frequency band. $X_L(k)$ is the Fourier transform of $x_L(n)$. $X_R(k)$ is the Fourier transform of $x_R(n)$. "$*$" denotes the complex conjugate. $\langle s \rangle$ is the expected value of $s$. The magnitude-squared coherence, MSC, is referred to the square of (7). The IC behavior of speech in rooms is highlighted here with four examples—taken from [123] and [124]. The IC is estimated using the method described by (11). Figure 6a shows the IC plot for speech presented in a reverberation chamber, dominated by diffuse reflections and calculated from the binaural recordings of [51]. First, the algorithm defined in Sect. 4.1 was applied to obtain a 6.4 m short-term IC of the binaural representation of an entire sentence spoken by a male talker. From the resulting coherence values, the coherence plots were derived. The gray-graduation scale reflects the number of occurrences in a given frequency channel. As expected for the ideally diffuse sound field, an increased coherence is observed below 1 kHz. Above 1 kHz, most coherence values are between 0.1 and 0.3, whereby the minimum coherence that can be derived is limited by the duration of the time window applied in the coherence estimate.

Figure 6b–d shows examples of coherence plots for 0.5, 5 and 10 m source-receiver distances in an auditorium with $T_{60} = 1$ s. The overall coherence decreases with increasing distance between the source and the receiver. This results from the

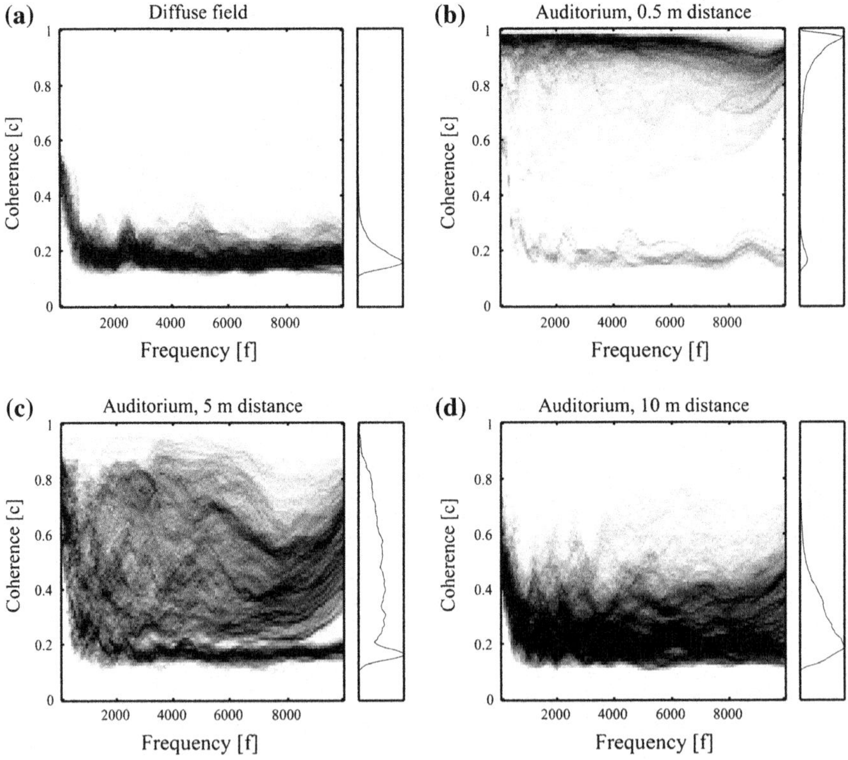


**Fig. 6** **a** Interaural-coherence plots of speech presented in a diffuse field as a function of frequency. **b–d** Interaural-coherence plots in an auditorium at different distances from the sound source. The histograms summed across frequency are shown in the side panels

decreased direct-to-reverberant energy ratio at longer source-receiver distances. At very small distances—Fig. 6b—most coherence values are close to one, indicating that mainly direct-sound energy is present. In addition, coherence values arising from the diffuse field, having values between 0.1 and 0.3, are separated from those arising from the direct sound field. For the 5 m distance, frames with high coherence values are no longer observed. This is because frames containing direct-sound information are now affected by reverberation, and there is no clear separability between frames with direct and diffuse energy. At a distance of 10 m, this trend becomes even more profound as the coherence values drop further and the distribution resembles the one as found in the diffuse field, where very little direct sound information is available.

# 3 Review of Dereverberation-Techniques Literature

Since the early works of Flanagan and Lummis [37], Mitchell and Berkley [95] and Allen, Berkley and Blauert [2], many *blind-* or *non-blind*-dereverberation techniques have been developed, utilizing single or multiple input channels. As was shown in Sect. 2.2, early and late reverberation have different effects on anechoic signals. Hence, most of the research efforts handle early and late-reverberant-signal components separately, enhancing either $x_{i,e}(n)$ or $x_{i,l}(n)$. In the following sections, a summary of the existing literature on devererberation is presented.

## *3.1 Suppression of Early Reflections and Decoloration*

### Inverse Filtering

Inverse filtering of the RIR [97–99, 104] is used to minimize the coloration effect produced by the early reflections. In theory, an ideal RIR inversion will completely remove the effect of reverberation—both early and late reflections. However, the RIR is known to have non-minimum phase characteristics [104] and the non-causal nature of the inverse filter may introduce significant artifacts. In addition, exact measurements of the RIR must be available for the specific source-receiver room position, even if the RIRs are known to present common features in different room positions, as in [50]. The above limitations can be avoided by compensating exclusively for the broad spectral-coloration effect. For this, many single or multichannel techniques have been proposed, such as based on minimum-phase inverse [104], least-squares [100], frequency warping [52, 53, 111], complex smoothing [54, 55], Kautz filters [71, 107], frequency-dependent regularization [75] and filter clustering [10, 99]. Many of them are already incorporated in commercial room-correction systems, which rely on in-situ RIR measurements. However, results from perceptual tests show that some of these techniques do not always achieve the desired perceptual effect [56, 105].

#### Cepstral Techniques

In 1975, Stockham restored old Caruso recordings through cepstral blind deconvolution [106, 113]. The technique was based on homomorphic signal processing, exploring the fact that deconvolution may be represented as a subtraction in the log-frequency domain. Similar dereverberation techniques based on the same principle were later proposed in [8, 100, 108].

#### LP-Residual Enhancement

Using the source-filter production model, the speech can be represented as a convolutive mixture of the *linear-prediction*, LP, coefficients and the LP residual [29]. The fundamental assumption of the LP-residual dereverberation techniques is that the excitation signal is distorted by the room reflections, while the LP coefficients are not significantly affected from reverberation. Hence, the above techniques enhance the LP residual and recover the speech by applying the reverberant LP coefficients [40, 43, 45, 81, 102, 127].

### *3.2 Late-Reverberation Suppression*

#### Temporal-Envelope Filtering

A class of techniques mostly aiming at compensating for late reverberation is based on temporal envelope filtering [5]. They are motivated by the concept of *modulation index* [62]. The modulation index is reduced when the late-reverberation tails fill the low-energy regions of a signal [84]. Mourjopoulos and Hammond [101] have shown that dereverberation of speech can be achieved by temporal envelope deconvolution in frequency sub-bands. Furthermore, the temporal envelope-filtering principle has been found to be advantageous when used in combination with other techniques such as LP-residual enhancement [127] and spectral subtraction [81]. Further, in [119], a sub-band temporal envelope-filtering technique, based on a computational auditory-masking model [21], has been proposed.

#### Spectral Enhancement

A number of dereverberation techniques based on spectral-enhancement techniques have been developed inspired by a multi-microphone reverberation-reducing method proposed by Flanagan and Lummis [37]. The same concept was later explored in the dereverberation method proposed by Allen et al. [2]. Spectral subtraction was mainly explored for denoising applications [9, 16, 29, 33, 89]. The classical technique is implemented in the STFT domain. Its main principle is to subtract an estimate of the noise-power spectrum from the power spectrum of the noisy signal. Usually, a speech-activity detector is involved in order to update the estimation of noise characteristics during the non-speech frames.

The most common processing artifact introduced by spectral enhancement is the so-called *musical noise*. It is generated when spectral bins of the noisy signal are strongly attenuated, because they are close to or below the estimated noise spectrum. As a result, the residual noise contains annoying pure-tone components at random frequencies. Most spectral-enhancement methods are trying to accurately estimate the noise spectra and avoid or reduce the musical noise [25, 89, 118, 120].

As indicated by (3), reverberation is a convolutive distortion. However, late reverberation can be considered as an additive degradation with noise-like characteristics—see (6). Hence, in the dereverberation context spectral subtraction has been adapted for the suppression of late reverberation. The basic principle of spectral-subtraction dereverberation for single-channel signals, originally presented in [86], is estimating the short-time spectrum of the clean signal, $S_e(m,k)$, by subtracting an estimation of the short-time spectrum of late reverberation, $R(m,k)$, from the short-time spectrum of the reverberant signal, $X(m,k)$, that is,

$$S_e(m,k) = X(m,k) - R(m,k) \ , \tag{8}$$

where $k$ and $m$ are the frequency bin and time frame index respectively. Following an alternative formulation, the estimation of the short-time spectrum of the clean signal can be derived by applying appropriate weighting gains, $G(m,k)$, to the short-time spectrum of the reverberant signal, such as

$$S_e(m,k) = G(m,k)X(m,k) \ , \tag{9}$$

where

$$G(m,k) = \frac{X(m,k) - R(m,k)}{X(m,k)} \ . \tag{10}$$

Further examples of spectral enhancement dereverberation methods can be found in [34, 38, 47, 48, 118, 126].

## 3.3 Dereverberation Methods Based on Multiple Inputs

Multichannel dereverberation may be considered as a somewhat easier task than the single-channel dereverberation, since the spatial diversity of the received signals can be further exploited. A set of such multichannel techniques is based on beamforming [121]. They explore the directivity properties of microphone arrays and require some a-priori knowledge of the array configuration. For a given system, the improvement depends on the microphone arrangement and the source-receiver positions, but it is independent of the reverberation time, $T_{60}$, of the room [41]. In simple implementations, the beamforming microphone arrays may present fixed-directivity characteristics such as in fixed-beamforming techniques, however adaptive beamforming setups where the processing parameters are adjusted to the environment also exist. Most beamforming algorithms assume that the noise and the source signal are statistically independent. This assumption does not stand for reverberation, which is a convolutive distortion. Therefore, the performance of such algorithms is poor in the dereverberation context [12].

Some early methods for multichannel dereverberation were presented in [2, 15, 37]. Miyoshi et al. [96] have shown that in non-blind multichannel systems perfect inverse filtering can be achieved when the captured RIRs do not share any common zeros. A technique that performs multiple-point room equalization using adaptive filters has been presented in [31]. Complete reverberation reduction may be theoretically achieved by applying blind deconvolution [59]. However, in order to perform blind deconvolution, the signal and the RIR must be irreducible, that is, they cannot be expressed as the convolution of two other signals [82]. The LTI systems are usually reducible and hence in principle blind deconvolution cannot be applied. In order to overcome the above limitation, single or multichannel blind-deconvolution implementations often involve a very low channel order and the number of reflections in the tested RIRs is unrealistically low—being based on simulations, as in [32, 39, 59, 60]. A set of room impulse response shortening techniques has been also proposed [70, 94, 129].

Further, multichannel blind-deconvolution methods for speech based on the LP analysis have been developed, based on the following principle. When the input of a system is white it can be equalized through multichannel LP. For speech dereverberation, the reverberant speech signal is pre-whitened in order to estimate a dereverberation filter. Then this filter is applied to the reverberant signal [28, 38, 74, 114]. A multichannel combined noise- and reverberation-suppression technique based in matched filtering has been presented in [30].

## 3.4 Binaural Techniques

Dereverberation is particularly important for binaural applications, for example, in digital hearing aids, binaural telephony, hands-free devices, and immersive audio [49, 91, 125]. However, developing models for binaural dereverberation and/or adapting single or multichannel techniques for binaural processing is not a trivial task. Binaural dereverberation cannot be considered as just a subset of the multichannel dereverberation techniques. Apart from the challenging task of reducing reverberation without introducing audible artifacts, binaural dereverberation methods should preserve the interaural-time-difference, ITD, and interaural-level-difference, ILD, cues, because it has been shown that bilateral signal processing can otherwise adversely affect source localization [49].

As was already discussed earlier in this section, such methods can be historically related to Allen et al. [2], who proposed a binaural approach where gain factors are determined by the diffuseness of the sound field between two spatially-separated microphones—see also [14, 15]. The technique involves two methods for calculating gain factors, one of which representing the coherence function of the two channels. However, because of a cophase-and-add stage that combines the binaural channels, only a monaural output was provided by this early method. Kollmeier et al. extended the original approach by applying the original coherence-gain factor separately to both channels, thus providing a binaural output [80]. A binaural variant of the original Allen et al. algorithm was also presented in [85].

In [112], a coherence-based Wiener filter was suggested that estimates the reverberation noise from a model of coherence between two points in a diffuse field. The method was further refined in [93] and [69] where acoustic shadow effects from a listener's head and torso were included. Jeub et al. [68] proposed a two-stage dereverberation algorithm that explicitly preserves binaural cues. They demonstrated that synchronized spectral weighting across binaural channels is important for preserving binaural cues. In [91] and also in [68], a binaural version of the single-channel spectra-subtraction technique presented in [86] is employed. In [115], a unified framework for binaural spectral subtraction dereverberation has been discussed. Lee et al. [87] presented a semi-blind method where they estimated a dereverberation filter from a pre-trained whitening filter and a whitened signal. Note that despite the great importance of binaural dereverberation, only few studies have been published up to now in the existing literature.

# 4 Examples of Dereverberation Algorithms

## *4.1 Method Based on Interaural Coherence*

Historically, coherence-based methods (see 3.4) directly apply the coherence estimates as a gain to both binaural channels. Considering the processing as a mapping between coherence and gain, these methods apply a frequency-independent linear coherence-to-gain mapping. However, the strong source-receiver distance dependency observed in Fig. 6 and the inherent variations of the coherence across frequency highlights the necessity for applying acoustic scenario specific coherence-to-gain mapping functions. While for close source-receiver distances—Fig. 6b—a rather shallow mapping function is already able to suppress reverberant components and to preserve direct sound components, a much steeper function is required for larger distances—Fig. 6c,d. Therefore, a method is proposed here, taken from [123], which applies a parameterized mapping function that is controlled by an estimate of the present coherence statistics. The signal-processing steps for this dereverberation method are illustrated in Fig. 7. Two reverberant time signals, recorded at the left and right ear of a person or a dummy head, $x_L(n)$ and $x_R(n)$, are transformed to the time-frequency domain using short-time fourier transform, STFT [3]. This results in complex-valued short-term spectra, $\underline{X}_L(m, k)$ and $\underline{X}_R(m, k)$, where $m$ denotes the

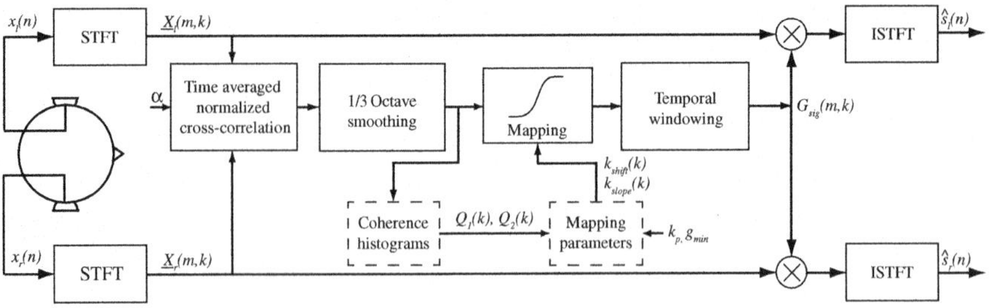


**Fig. 7** Block diagram of a dereverberation method utilizing IC. The signals recorded at the ears, $x_L(n)$ and $x_R(n)$, are transformed via the STFT to the time-frequency domain, resulting in $\underline{X}_L(m,k)$ and $\underline{X}_R(m,k)$. The IC is calculated for each time-frequency bin and third-octave smoothing is applied. Statistical long-term properties of the IC are used to derive parameters of a sigmoidal mapping stage. The mapping is applied to the IC to realize a coherence-to-gain mapping, and subsequent temporal windowing is performed. The derived gains or weights are applied to both channels, $\underline{X}_L(m,k)$ and $\underline{X}_R(m,k)$. The dereverberated signals, $\hat{s_L}(n)$ and $\hat{s_R}(n)$, are reconstructed by applying an inverse SFTF

time frame and $k$ the frequency band. For the STFT, a Hanning window of length $L$, including zero-padding of length $L/2$, and a 75 % overlap between successive windows are used. For each time-frequency bin, the absolute value of the IC, referred also as coherence, is calculated according to (7), which is implemented as follows

$$C_{LR}(m,k) = \frac{|\Phi_{LR}(m,k)|}{\sqrt{\Phi_{LL}(m,k)\Phi_{RR}(m,k)}}, \tag{11}$$

with $\Phi_{LL}(m,k)$, $\Phi_{RR}(m,k)$ and $\Phi_{LR}(m,k)$ representing the exponentially-weighted short-term cross-correlation and auto-correlation functions, namely,

$$\Phi_{ll}(m,k) = \alpha\Phi_{ll}(m,k-1) + \left|\underline{X}_l(m,k)\right|^2 \tag{12}$$

$$\Phi_{rr}(m,k) = \alpha\Phi_{rr}(m,k-1) + \left|\underline{X}_r(m,k)\right|^2 \tag{13}$$

$$\Phi_{lr}(m,k) = \alpha\Phi_{lr}(m,k-1) + \underline{X}_r(m,k)\underline{X}_l^*(m,k) \tag{14}$$

where $\alpha$ is the recursion constant.

The resulting IC estimates are spectrally smoothed using third-octave smoothing [57]. From the long-term statistical properties of the smoothed IC estimates, the parameters of a sigmoidal mapping function are derived. This mapping is subsequently used to transform the coherence estimates to the gain function, $G_{\text{sig}(m,k)}$. In order to suppress potential aliasing artifacts that may be introduced, temporal windowing is applied [72]. This is realized by applying an inverse STFT to the derived filter gains and then truncating the resulting time-domain representation to a length of $L/2{+}1$. The filter response is then zero-padded to a length of $L$ and another STFT is performed. The resulting filter gain is applied to both channels, $\underline{X}_L(m,k)$, and, $\underline{X}_R(m,k)$. The dereverberated signals, $\hat{s_L}(n)$, and, $\hat{s_R}(n)$, are finally reconstructed by applying the inverse STFT and then adding the resulting overlapping signal segments [3].

## Coherence-to-Gain Mapping

In order to cope with the different frequency-dependent distributions of the IC observed in different acoustic scenarios—see Sect. 2.3—a coherence-distribution dependent gain-to-coherence mapping is introduced. This is realized by a sigmoid function which is controlled by an online estimate of the statistical properties of the IC in each frequency channel. The function is derived from a normal sigmoid and given by

$$G_{\text{sig}}(m,k) = \frac{(1 - G_{\text{min}})}{1 + e^{-k_{\text{slope}}(k)(C_{LR}(m,k) - k_{\text{shift}}(k))}} + G_{\text{min}} \ , \tag{15}$$

where $k_{\text{slope}}$ and $k_{\text{shift}}$ control the sigmoidal slope and the position. The minimum gain, $G_{\text{min}}$, is introduced to limit signal-processing artifacts.

In order to calculate the frequency-dependent parameters of the sigmoidal mapping function, coherence samples for a duration defined by $t_{\text{sig}}$ are gathered in a histogram. The method yields best performance with a $t_{\text{sig}}$ in the range of several seconds, assuming that the source-receiver locations are kept constant. For moving sources and varying acoustic environments, the method for updating the sigmoidal mapping function might need revision. The mapping functions are determined as two predefined points, $Q_1$ and $Q_2$, corresponding to the 1st and 2nd quartiles of the estimated IC-histogram distributions. A coherence histogram shown as a Gaussian distribution for illustrative purposes is exemplified in Fig. 8a by a gray curve together with the corresponding 1st and 2nd quartiles. An example sigmoidal coherence-to-gain mapping function is represented by a black solid curve. The linear mapping function as applied by [2] is indicated by the black dashed curve.

The degree of processing is determined by $k_p$, which directly controls the slope of the sigmoidal mapping function. The parameters $k_{\text{slope}}$ and $k_{\text{shift}}$ of the mapping function can be derived from $\varsigma(Q_1) = G_{\text{min}} + k_p$ and $\varsigma(Q_2) = 1 - k_p$ as follows,

$$k_{\text{shift}} = \left(\frac{ln(\varsigma(Q_1)^{-1})}{ln(\varsigma(Q_2)^{-1})} Q_2 + Q_1\right) \cdot \left(1 - \frac{ln(\varsigma(Q_1)^{-1})}{ln(\varsigma(Q_2)^{-1})}\right)^{-1} \tag{16}$$

$$k_{\text{slope}} = \frac{ln(\varsigma(Q_1)) - 1}{Q_1 - k_{\text{shift}}} \ , \tag{17}$$

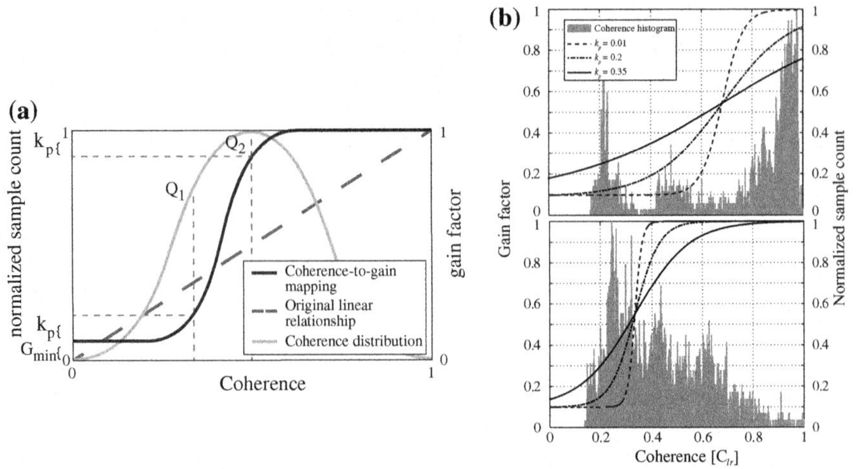


**Fig. 8** **a** Idealized IC histogram distribution in one frequency-channel (*gray curve*). The coherence-to-gain relationship in the specific channel is calculated to intersect $\varsigma(Q_1) = G_{\text{min}} + k_p$ and $\varsigma(Q_2) = 1 - k_p$. Thereby, $G_{\text{min}}$ denotes the maximum attenuation and $k_p$ determines the processing degree. **b** IC histogram distribution of speech presented in an auditorium with 0.5 m source-receiver distance (*top panel*) and 5 m source-receiver distance (*bottom panel*). Sigmoidal coherence-to-gain relationship for three different processing degrees of $k_p$ are shown

whereby $Q_1$ and $Q_2$ are estimated in each frequency channel from the measured coherence histograms and $k_p$ a predetermined parameter—see Fig. 8a. In addition, $G_{\text{min}}$ is introduced to avoid signal artifacts related to applying infinite attenuation.

For speech presented in an auditorium with source-receiver distances of 0.5 m and 5 m—see Sect. 2.3—examples of sigmoidal mapping functions are shown in Fig. 8b for different values of $k_p$ in the 751.7 Hz frequency channel. It can be seen that the coherence-to-gain mapping steepens as $k_p$ increases. In addition, with the distribution broadening, that is, from 5 m to 0.5 m, the slope of the coherence-to-gain mapping decreases. Hence, in contrast to the original coherence-based-dereverberation approach in [2], which considered a linear coherence-to-gain mapping—plotted with dashed line in Fig. 8—the approach presented here provides a mapping function with added flexibility that can be adjusted by the parameter $k_p$ and to any given acoustic condition.

## 4.2 Spectral-Subtraction Framework

As shown in (8), in the spectral-subtraction framework, the dereverberation problem is deduced to an estimation of the late-reverberation short-time spectrum. Several single-channel techniques have been introduced to blindly provide such estimates. Under specific conditions, as discussed later in this section, such blind dereverberation methods may be also adapted for binaural processing. For instance, Lebart et al. [86] proposed a method, referred to in the following as LB, that is based on exponential-decay modeling of the RIR, $h(n)$, as

$$h(n) = b(n)\exp(-3\ln 10/T_{60,\,n}) \;, \tag{18}$$

where $b(n)$ is a zero-mean Gaussian stationary noise. $T_{60}$ is the reverberation time in seconds [83].[2] The short-time spectral magnitude of the reverberation is estimated as

$$|R(m,k)| = \frac{1}{\sqrt{SNR_{pri}(m,k)+1}}|X(m,k)| \;, \tag{19}$$

where $SNR_{pri}(m,k)$ is the a-priori signal-to-noise ratio that can be approximated by a moving average relating to the a-posteriori signal-to-noise ratio, $SNR_{post}(m,k)$, in each frame,

$$SNR_{pri}(m,k) = \beta SNR_{pri}(m-1,k) + (1-\beta)\max(0,(SNR_{post}(m,k)-1)) \;, \tag{20}$$

where $\beta$ is a constant taking values close to one. The a-posteriori SNR is defined as

$$SNR_{post}(m,k) = \frac{|X(m,k)|^2}{E[|X(m,k)|]} \;. \tag{21}$$

Thus, $S_e(m,k)$ is estimated by subtraction and is combined with the phase of the reverberant signal, so that the dereverberated signal in the time domain is finally obtained through an overlap-add process.

An alternative method, as proposed by Wu and Wang [126], referred to hereinafter as WW, is motivated by the observation that the smearing effect of late reflections produces a smoothing of the signal spectrum in the time domain. Hence, similarly to the approach of [86], the late-reverberation power spectrum is considered a smoothed and shifted version of the power spectrum of the reverberant speech, namely,

$$|R(m,k)|^2 = \gamma w(m-\rho) * |X(m,k)|^2 \;, \tag{22}$$

where $\rho$ is a frame delay. $\gamma$ is a scaling factor and "*" denotes convolution. The term $w(j)$ represents an assymetrical smoothing function given by the Rayleigh distribution

$$w(m) = \begin{cases} \frac{m+\alpha}{\alpha^2}\exp\left(\frac{-(m+\alpha)^2}{2\alpha^2}\right) & \text{if } j < -\alpha \;, \\ 0 & \text{otherwise} \;, \end{cases} \tag{23}$$

[2] Note that this model holds when the direct-to-reverberant ratio is smaller than 0 dB [48].

where $\alpha$ represents a constant number of frames. The phase of the reverberant speech is combined with the spectrum of the estimated clean signal and overlap-add is used to extract the time domain estimation.

Alternatively, Furuya and Kataoka [38] proposed a method, referred to hereinafter as FK, where the short-time power spectrum of late reverberation in each frame can be estimated as the sum of filtered versions of the previous frames of the reverberant signal's short time power spectrum, that is,

$$|R(m,k)|^2 = \sum_{l=1}^{M} |a_l(m,k)|^2 |X(m-l,k)|^2 \ , \tag{24}$$

where $M$ is the number of frames that corresponds to an estimation of the $T_{60}$. $a_l(m,k)$ are the coefficients of late reverberation. The FK method assumes that an inverse filtering step, which reduces spectral degradation produced by the early reflections, precedes the spectral subtraction. Hence, in such a case the short-time power spectrum of the reverberant signal is considered to roughly approximate the short-time power spectrum of the anechoic signal. The coefficients of late reverberation are derived from

$$a_l(m,k) = E\left\{ \frac{X(m,k)X^*(m-l,k)}{|X(m-l,k)|^2} \right\} . \tag{25}$$

With these coefficients an estimation of the clean signal in the time domain can be derived through overlap-add from the short-time spectrum of the dereverberated signal, $S_e(m,k)$, as follows,

$$S_e(m,k) = \left\{ \frac{|X(m,k)|^2 - |R(m,k)|^2}{|X(m,k)|^2} \right\} X(m,k) \ . \tag{26}$$

Overlap-add is finally applied in order to estimate the time-domain dereverberated signal.

Although the above methods were originally employed for single-channel dereverberation, they can be adapted for binaural processing. For such case, as discussed in Sect. 3.4, in order to preserve the binaural ITD and ILD cues identical processing should be applied to the left and right signal channels. Similar principles apply to the binaural noise reduction, as in [72]. An effective approach for extending the LB method to a binaural context is to derive a reference signal using a delay-and-sum beamformer, DSB [68], where the time delays are estimated utilizing a method based on the generalized cross-correlation with phase transform as proposed in [76]. The reference signal is then calculated as the average of the time aligned left and right reverberant signals. Using the reference, appropriate weighting gains are derived, and identical processing is applied to both left and right channels. In [115], the DSB approach is also implemented for both the WW and FK methods in order to evaluate the efficiency of different late-reverberation-estimation techniques in a binaural scenario. However, in binaural applications, the time delay between the left and right channels of the speech signal is limited by the width of the human head. Therefore, it can be assumed to be shorter than the length of a typical analysis window used in spectral-subtraction techniques. Hence, in [115], it was shown that the time alignment stage can be omitted.

A different approach in order to adapt single channel spectral subtraction dereverberation in the binaural scenario is to process the left and right-ear channel signals independently. This results in the corresponding weighting gains, $G_L(m,k)$ and $G_R(m,k)$. These two gains can be combined, and different adaptation strategies have been investigated for each algorithm, namely,

(a) The binaural gain can be derived as the maximum of the left and right-channel weighting gains,

$$G(m,k) = \max(G_L(m,k), G_R(m,k)) \ . \tag{27}$$

This approach, maxGain, achieves moderate late-reverberation suppression, but it is also less likely to produce overestimation artifacts.

(b) The binaural gain can be derived as the average of the left and right channel weighting gains,

$$G(m,k) = \frac{(G_L(m,k) + G_R(m,k))}{2} \ . \tag{28}$$

This gain-adaptation strategy, avgGain, compensates equally for the contribution of the left and right channels.

(c) The binaural gain can be derived as the minimum of the left and right channel weighting gains,

$$G(m,k) = \min(G_L(m,k), G_R(m,k)) \ . \tag{29}$$

This adaptation technique, minGain, results in maximum reverberation attenuation, but the final estimation may be susceptible to overestimation artifacts.

After the derivation of the adapted gain, a *gain-magnitude regularization*, GMR, technique can be applied. The purpose of such as step is twofold. Firstly, the GMR has been proved to be a low-complexity approach reducing annoying musical-noise artifacts [79, 117]. Furthermore, the GMR is utilized in order to constrain the suppression and thus, to prevent from overestimation errors. An overestimation of the late reverberation is less likely to happen in spectral regions with a high signal-to-reverberation-ratio, SRR, such as signal steady states [118]. Yet, such problem is more likely to affect the low SRR regions. Therefore a low SRR detector is employed [68] and GMR is applied only on the lower-gain parts. Consequently, the new constrained gain, $G'(m,k)$, is derived as

$$G'(m,k) = \begin{cases} \dfrac{G(m,k)-\theta}{r} + \theta & \text{when } \zeta < \zeta_{th} \text{ and } G(m,k) < \theta\,, \\ G(m,k) & \text{otherwise}\,, \end{cases} \tag{30}$$

and

$$\zeta = \frac{\sum_{k=1}^{K} G(m,k)|Y(m,k)|^2}{\sum_{k=1}^{K} |Y(m,k)|^2}\,, \tag{31}$$

where $\theta$ being the threshold for applying the gain constraints, $r$ is the regularization ratio, $\zeta$ is the power ratio between the enhanced and the reference signal, $\zeta_{th}$ the threshold of the low-SRR detector. $K$ is the total number of frequency bins.

The effect of the GMR is further explained in Fig. 9. In Fig. 9a, a typical illustration of a frequency domain binaural gain is shown. In Fig. 9b–d the effect of the GMR step on the binaural gain is presented for $\theta = 0.2$, $r = 4$, for $\theta = 0.2$, $r = 8$ and for $\theta = 0.6$, $r = 4$ respectively. It can be observed that larger regularization ratios, $r$, result in larger gain values—in other words, the method suppresses less reverberation. Moreover, a constraint of the reverberation reduction is observed for larger thresholds, $\theta$. By comparing Fig. 9b–d, it becomes clear that the effect of the

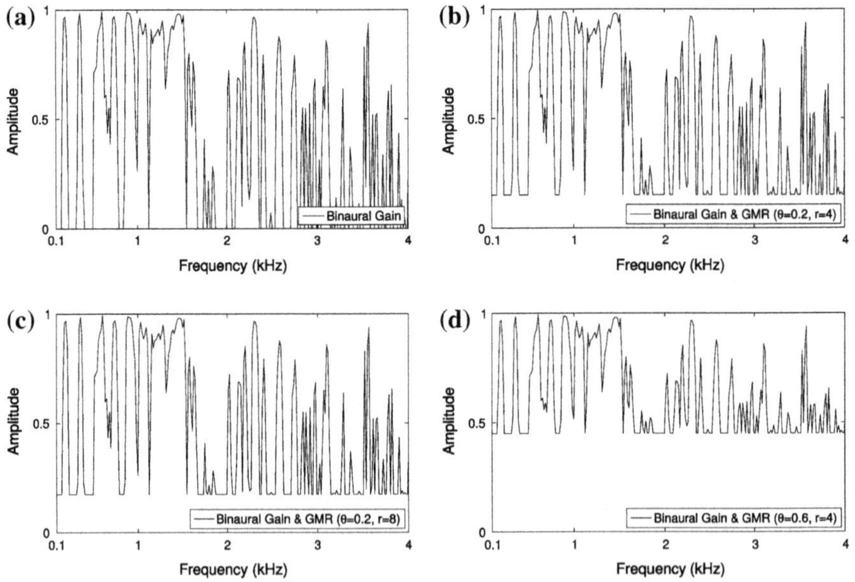


**Fig. 9** **a** Typical binaural gain in the frequency domain. **b** Effect of the GMR step for $\theta = 0.2$ and $r = 4$. **c** Effect of the GMR step for $\theta = 0.2$ and $r = 8$. **d** Effect of the GMR step for $\theta = 0.6$ and $r = 4$

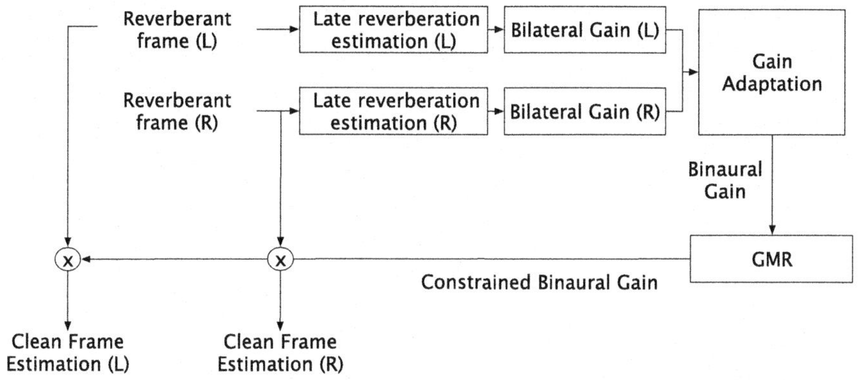


**Fig. 10** Block diagram of the spectral-subtraction binaural-dereverberation approach that preserves the interaural cues. The late-reverberation estimation can be based on either of the techniques described in Sect. 4.2

regularization ratio, $r$, of (30) is more subtle than the effect of the threshold, $\theta$. Therefore, the parameter $r$ can be used for fine-tuning purposes.

To conclude this section, Fig. 10 presents a block diagram of the framework applied for binaural spectral-substraction, as described above.

# 5 Evaluation Methods

## 5.1 *Objective Measures of Dereverberation*

The evaluation of the potential improvement of speech or audio enhancement techniques has proven to be a rather difficult task. Many objective measures have been developed and often they can predict the perceived quality of the enhanced signals accurately enough [89]. However, there are cases where such objective measures fail to correctly evaluate the performance of a reference algorithm [90]. The evaluation of the performance of dereverberation algorithms has proven to be more difficult. This happens for the same reason that dereverberation is generally more demanding than denoising: the reverberation noise is correlated with the anechoic signal. For the above reasons, denoise measures are not always appropriate for dereverberation evaluation.

This difficulty increases further for the evaluation of binaural dereverberation methods. In this case, apart from the assessment of the output signal's temporal and spectral characteristics, the binaural qualities of the processed output must be also taken into account. Unfortunately, the development of objective or perceptual dereverberation evaluation metrics that explicitly take into account the binaural conditions and cues is still an open research issue. Therefore, single-channel measures are adapted in the binaural scenario by combining through simple addition the left and right-channel results. Hence, one must be very careful when interpreting the values of such metrics.

Most dereverberation-evaluation measures require a-priori knowledge of the anechoic signal. In principle they calculate some type of distance between the dereverberated and the anechoic signal. Such metrics are, for example, the *signal-to-reverberation-ratio*, SRR, the *frequency-weighted signal-to-reverberation-ratio*, fwSRR, the *weighted-slope spectral distance*, WSS, the *Itakura-Saito distance*, IS, the *Bark spectral distortion*, BSD, the *cepstral distance*, CD, and the *log-spectral distortion*, LSD [44, 63, 103, 122]. Moreover, metrics based on auditory modeling have been also used for the evaluation of dereverberation algorithms, such as the *perceptual evaluation of speech quality*, PESQ, the *noise-to-mask ratio*, NMR, the *perceptual-similarity measure*, PSM, and the non-intrusive *speech-to-modulation-energy ratio*, SRMR [35]. In order to improve the overall evaluation performance, some researchers have modified and/or combined subsets of the above metrics, as in [27, 78].

## 5.2 Perceptual Measures of Dereverberations

When listening inside reverberant spaces, the auditory system applies several mechanisms assisting both intelligibility and localization. These include monaural/binaural decoloration, binaural auditory dereverberation [13, 22, 130] and the precedence effect [88]. Objective measures of dereverberation processing often do not incorporate or take account of these mechanisms or other features of the auditory system. In addition, these objective measures have shown varying correlation with perceptual measures [78, 122]. If signals processed via dereverberation algorithms are intended for human listeners, such a discrepancy needs to be taken into account.

Until now, the literature has only sporadically used perceptual evaluation for evaluation of dereverberation algorithms. In [38] and [118] a mean opinion score, MOS, of signal quality was used. In [68] a preference comparison task was implemented for the different processed signals. A modified version of the ITU P.835 test has been also employed for perceptual evaluation of dereverberation [66]. The test evaluates (i) the speech signal naturalness, (ii) the reverberation intrusiveness and (iii) the overall signal quality [35, 64, 116]. Similar tests have been extended to cover multiple attributes, such as the amount of reverberation, source width and sound envelopment [92].

The *multiple-stimuli-with-hidden-reference-and-anchor test*, MUSHRA, [109] has been also applied for dereverberation [105, 119]. This test is especially successful at detecting small signal impairments, since stimuli are presented simultaneously and evaluated on a scale. For future evaluation of dereverberation algorithms this test can be extended in order to include attributes such as *amount of reverberation* and *overall quality*. One dilemma faced when designing a MUSHRA test are the anchors. Anchors are an inherent trait of MUSHRA experiments to increase the reproducibility of the results and to prevent contraction bias—see [7]. These are normally made by low-pass filtering the reference signal. To evaluate the quality of speech, the anchor should be implemented by introducing distortions similar to those resulting from the dereverberation processing, for example, by using an *adaptive multi-rate*, AMR, speech coder, available from [1], but other distortion types could be also applied. Anchors for judging the amount of reverberation can be created by applying a temporal half cosine to the BRIRs and thereby artificially reducing the resulting reverberation while keeping direct sound and early reflections. Pilot studies have shown that presenting the unprocessed reference stimulus as a hidden anchor resulted in significant compression bias of the listeners—for further details, see [7]. Therefore, this hidden anchor can be omitted and replaced by a separate *reference-button* which allows listeners to hear the unprocessed signal. This test could be employed and combined with reference-processing methods for more reliable results, as shown later in this chapter.

# 6 Tests and Results

## *6.1 Results for the Coherence-Based Algorithm*

### Signal-to-Reverberation Ratio

In this section, the objective results of the coherence-based algorithm (using the processing parameters in Table 1), described in Sect. 4.1, are presented. The method was compared with the method of Allen et al. [2] and a binaural version of the Lebart et al. [86] spectral subtraction method. Hereinafter the IC-based algorithm of Sect. 4.1 will be referred to as WB, to the Allen et al. method as AB and to the Lebart et al. method as LB. Figure 11 shows the signal-to-reverberation ratio, $\Delta$segSRR, for the different processing schemes. All algorithms show a significant reduction of the amount of reverberation, as all exhibit positive values.

For the 0.5 m distance—left panel—the WB algorithm for $k_p = 0.2$ provides the best performance. For the lowest degrees of processing, $k_p = 0.35$, the performance is slightly below the one attained for the LB algorithm. For the 5 m distance—right panel—the WB method shows a performance that is comparable to the LB method

**Table 1** Processing parameter values for the coherence-based method at $fs = 44.1$ kHz

| Parameter | Value |
|---|---|
| Frame length (L) | 6.4 ms |
| Recursion constant ($\alpha$) | {0.01; 0.2; 0.35} |
| Recursion constant ($\alpha$) | 0.97 |
| Gain threshold ($G_{\min}$) | 0.1 |
| Mapping updating time ($t_{\text{sig}}$) | 3 s |

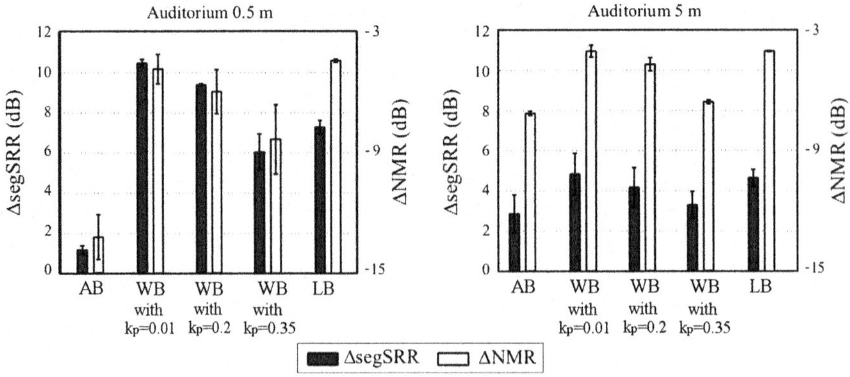


**Fig. 11** Estimates of reverberation suppression, $\Delta$segSRR, and loss of quality, $\Delta$NMR, between the clean signal and the processed reverberant signal for different methods. *Left panel* 0.5 m source-receiver distance. *Right panel* 5 m source-receiver distance

for the highest processing degree, $k_p = 0.01$. As expected, the performance of the WB method generally drops with decreasing processing degree—that is, increasing $k_p$ value. The AB method shows the poorest performance in general and provides essentially no reverberation suppressions in the 0.5 m condition.

## Noise-to-Mask Ratio

The *noise-to-mask ratio*, NMR, is an objective measure that determines the audible non-masked noise components. Lower NMR values denote better signal quality [119]. In Fig. 11, also $\Delta$NMR is shown, whereby smaller values correspond to less audible noise. For the different processing conditions, the AB approach shows the best quality overall for both source-receiver distances. Considering the very small amount of dereverberation provided by this algorithm—see Fig. 11—this observation is not surprising since the algorithm only has a minimal effect on the signal. The NMR performance of the WB method for high degrees of processing, that is, $k_p = 0.01$, is similar or slightly better than that obtained with the LB approach. The sound quality of the WB method increases with decreasing degree of processing, namely, $k_p = 0.2$ and 0.3. However, at the same time, the strength of dereverberation, as indicated by segSRR, also decreases—see the gray bars in Fig. 11. Considering both measures, segSRR and the NMR, the WB method is superior for close sound sources, in our case the 0.5 m condition with $k_p = 0.2$, and exhibits performance similar to the LB method for the 5 m condition.

## Perceptual Evaluation

For the perceptual evaluation of the different dereverberation methods, binaural signals were presented to 10 listeners via headphones. The signals were generated by convolving anechoic sentences with BRIRs, measured in an auditorium with $T_{30} = 1.9$ s. A MUSHRA test was applied to measure (i) strength of dereverberation and (ii) overall loss of quality. As described in Sect. 5.2, an AMR speech coder at 7.95 kbits/sec was used as anchor in the quality measure and a 600 ms long cosine window was applied to the measured BRIRs to generate the anchor for the dereverberation measure. Further details are described in [123]. The results from the perceptual evaluation for each processing method are shown in Fig. 12. For better comparison with the objective results, the measured data were inverted, that is, 100—original score. Considering the strength of dereverberation—indicated by the gray bars—the WB approach exhibited the best performance for $k_p = 0.01$ at both distances. As the degree of processing decreases, that is, for increasing values of $k_p$, the strength of dereverberation decreases. The improvement relative to the LB approach is considerably higher for the 0.5 m distance—left panel—than for the 5 m distance—right panel. The AB approach of [2] produced the lowest strength of dereverberation for both source-receiver distances. The differences in scores between the AB approach and the others were noticeably larger for the 0.5 m distance than for 5 m. This indicates that, for very close sound sources, the other methods are more efficient than the AB approach.

The overall quality loss of the signals processed with the WB method were found to be substantially higher for the 0.5 m condition compared to the 5 m condition. This difference is not as large with the AB approach as well as the LB method,

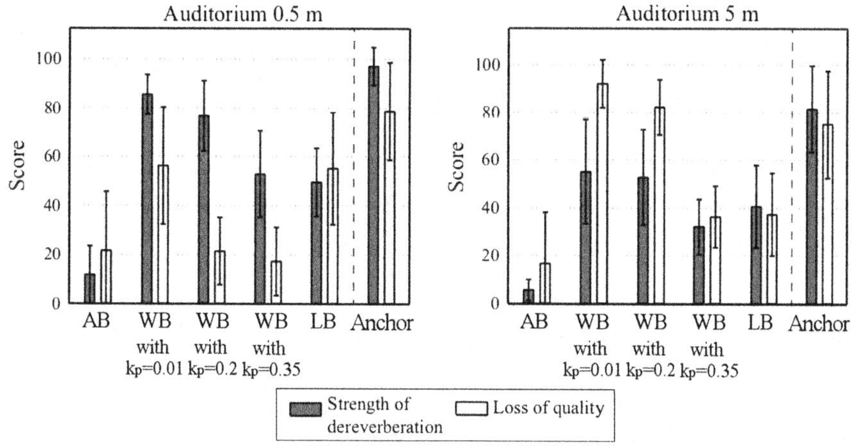


**Fig. 12** The mean and standard deviation of perceptual results judging *Strength of dereverberation* and *Overall loss of quality* for the 0.5 m source-receiver distance (*left panel*) and 5 m source-receiver distance (*right panel*)

indicating that the WB is particularly successful for very close sound sources. As in the objective quality evaluation, increasing the degree of dereverberation processing, that is, by decreasing $k_p$, results in a drop of the overall quality. However, this effect is not as prominent when decreasing $k_p$ from 0.35 to 0.2 at the 0.5 m distance.

Considering the combination of reverberation reduction and overall quality, the WB method with $k_p = 0.2$ exhibits a clearly superior performance at the 0.5 m distance. Even when applying the highest degree of processing, namely, $k_p = 0.01$, the quality is similar to that obtained with LB, but the strength of dereverberation is substantially higher. For the 5 m distance, increasing the degree of processing has a negligible effect on the strength of dereverberation but is detrimental for the quality. However, for $k_p = 0.35$, the performance of the WB method is comparable to that obtained with the LB approach. An analysis of variance, ANOVA, showed significance for the sample effect at source-receiver distances of 0.5 m, namely, $(F = 97.65,\ p < 0.001)$ and 5 m, $(F = 41.31,\ p < 0.001)$. No significant effect of listeners was found.

## 6.2 Results for the Spectral-Subtraction Framework

In this section, the results of the binaural spectral-subtraction framework are shown—for details see Sect. 4.2. The presented dereverberation methods LB [86], WW [126] and FK [38] are binaural extensions of the original single-channel methods.

A first evaluation has been made for 16 kHz signals [115]. Eight anechoic phrases uttered by both male and female speakers of the TIMIT database were convolved with real BRIRs. Four BRIRs measured in a Stairway Hall with a reverberation time of $T_{60}$=0.69 s at a source-receiver distance of 3 m and azimuth angles of 0, 30, 60 and 90° were chosen from the Aachen database [68]. In addition, three BRIRs measured in a Cafeteria with a $T_{60} = 1.29$ s at source-receiver distances of 1.18, 1 and 1.62 m and azimuth angles of approximately −30, 0 and 90° were chosen from the Oldenburg database [73]. The authors made informal tests to select optimal values for the analysis parameters, that is, 16 kHz, 16 bit—see Table 2. The $\theta$ and $\zeta_{th}$ values of the GMR step, described by (30) in Sect. 4.2, were set to 0.15, the regularization ratio, $r$, was 4. The $T_{60}$ was calculated from the impulse responses. All parameter values that are not detailed here were set according to the values proposed by the authors of the

**Table 2** Processing parameter values for comparing the spectral subtraction based methods at $fs$ =16 kHz

| Parameter | LB | WW | FK |
|---|---|---|---|
| Total frame length | 1024 | 1024 | 2048 |
| Zero padding | 512 | 128 | 128 |
| Frame overlap | 0.125 | 0.25 | 0.25 |

original works. In addition, for the FK and LB techniques, two additional relaxation criteria were imposed [118] as they were previously found by the authors to have advantageous effects on the performance. The WW and FK methods assume that an inverse-filtering stage precedes the spectral subtraction implementation. Here, however, the implementation of an 1/3-octave RIR minimum-phase inverse filtering was not found to notably alter the relative improvement achieved by the tested methods. Therefore, a generalized case where the spectral subtraction is applied directly to the reverberant signals is presented.

The produced signals were evaluated by means of the PESQ variation [65], compared to the reverberant signals. PESQ was not originally developed to assess the dereverberation performance—see Sect. 5—and it implements a perceptual model in order to assess the quality of a processed speech signal. Rating is performed according to the five-grade *mean-opinion-score*, MOS, scale. The results are presented in Table 3 with the bold values denoting optimum performance. For the case of the Stairway Hall the bigger PESQ improvement is achieved utilizing the WW method with the minGain adaptation technique. The same gain adaptation technique seems to be also the optimal choice when used in conjunction with the LB method. It can be assumed that in a scenario where bilateral late-reverberation estimations are successful this technique presents superior performance. However, it is not beneficial when used with the FK method where probably the bilateral processing resulted to inferior results. The FK method produces better results when used with the avgGain technique. In general, the WW method shows a significant PESQ improvement for all tested adaptation techniques. For the Cafeteria, the LB method produces a relatively stable PESQ improvement independent of the employed binaural adaptation. On the other hand, better results are derived with the WW method for all binaural-adaptation schemes—although the best results are achieved with the avgGain approach. The FK method seems to produce processing artifacts despite the utilized binaural-adaptation scheme and decreases the PESQ values in every case. Finally, note that the DSB implementation has the advantage of lower computational cost as it involves calculations in a single channel for the estimation of the weighting-gain functions. This is in contrary to the binaural-adaptation schemes that

**Table 3** PESQ improvement for the binaural spectral subtraction framework, showing results for various binaural gain adaptation options—see Sect. 4.2

| Method | BSD | maxGain | avgGain | minGain |
|---|---|---|---|---|
| *Stairway hall* | | | | |
| LB | 0.153 | 0.142 | 0.147 | 0.158 |
| WW | 0.206 | 0.160 | 0.208 | **0.258** |
| FK | 0.160 | 0.180 | 0.186 | −0.029 |
| *Cafeteria* | | | | |
| LB | 0.133 | 0.136 | 0.135 | 0.133 |
| WW | 0.205 | 0.208 | **0.216** | 0.198 |
| FK | −0.235 | −0.141 | −0.228 | −0.428 |

require dual-channel calculations. On the other hand, the gain-adaptation techniques discussed in Sect. 4.2 involve bilateral processing but do not necessitate the initial time-delay estimation.

For the perceptual test, the methods were applied for broadband signals, sampled at 44100 Hz [116]. A modified version of the ITU P.835 test was used for the perceptual evaluation, as explained in Sect. 5.2. Note that the listeners were not guided to directly rate the binaural qualities of the output signals. However, some listeners reported that they were inherently taken them into account in their assessments. Four phrases from two male and two female speakers along with three BRIRs measured in a Stairway Hall with $T_{60} = 0.69$ s, at a source-receiver distance of 3 m and azimuth angles of 0, 45 and 90° were used [68]. The original single-channel dereverberation methods, LB, WW and FK, were optimized for lower signal resolutions. Here, the optimal values for their application in broadband signals were extracted through informal listening tests. The STFT analysis parameters, that is, total frame length, zero padding and frame overlap, for each tested method are detailed in Table 4, the $\theta$ and $\zeta_{th}$ values of the GMR step according to (30) were set at 0.15 and 0.8, respectively, while the regularization ratio $r$ was 4. For the FK and LB techniques, the two additional relaxation criteria proposed in [118] were also implemented. In order to reduce the experimental conditions the authors conducted informal listening tests to choose the optimum gain-adaptation scheme for each dereverberation method. Hence, the avgGain adaptation has been chosen for the LB and WW methods while the maxGain has been used for the FK method. Twenty self-reported normal-hearing listeners participated in the tests and a training session preceded the formal experiment.

Figure 13 presents the perceptual scores in terms of speech naturalness, reverberation reduction and overall signal quality for the proposed binaural dereverberation techniques. The results were subjected to an ANOVA analysis of variance and a highly significant effect for the tested method was revealed for the speech naturalness, namely, $F(3, 228) = 112.7$, $p < 0.001$, for the reverberation reduction, $F(3, 228) = 62.1$, $p < 0.001$ and for the overall quality, $F(3, 228) = 38.8$, $p < 0.001$. No significant effect was found for the tested azimuth angles. Following the ANOVA multiple Tukey's, HSD tests were made to reveal significant differences between algorithms.

In all cases, listeners rated that the unprocessed reverberant signals were significantly more natural than the dereverberated signals—$p < 0.001$. This was due to the artifacts introduced from the dereverberation processing. On the other hand, the FK method performed significantly worse than the other two methods in terms of

**Table 4** Analysis parameter values for the employed methods at $fs = 44.1$ kHz

| Parameter | LB | WW | FK |
|---|---|---|---|
| Total Frame Length | 2048 | 8192 | 8192 |
| Zero padding | 1024 | 4096 | 4096 |
| Frame Overlap | 0.5 | 0.25 | 0.25 |

**Fig. 13** Perceptual results for speech naturalness, reverberation reduction and overall signal quality for unprocessed signals and the three dereverberation methods tested

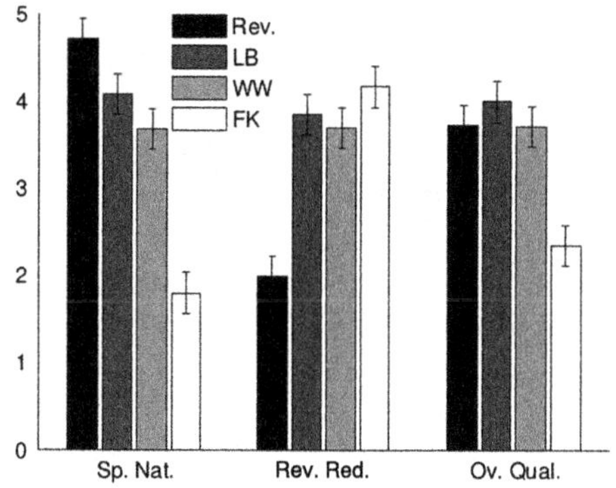


speech naturalness. No significant difference was noticed between the LB and WW methods—$p > 0.05$.

Furthermore, the three dereverberation methods have significantly reduced the reverberation—$p < 0.001$. The FK method performed significantly better than the WW method in terms of perceived reverberation suppression—$p < 0.05$. However, no significant difference between the FK and the LB or the LB and WW methods was found—$p > 0.05$. Finally, the LB, the WW methods and the reverberant signals were rated significantly better in terms of overall quality than the FK method—$p < 0.001$—, but no significant quality difference was found between the LB method, the WW method and the reverberant signals—$p > 0.05$.

From the objective and perceptual results it appears that all methods suppress reverberation significantly, but the introduced processing artifacts reduce the naturalness of the speech signals. The FK method achieves greater reverberation suppression than the LB and WW methods; however, it appears that the produced signals are more degraded. On the other hand, the LB and WW dereverberation methods moderately reduce the reverberation, but they preserve the perceptual signal quality.

## 7 Conclusions

Binaural dereverberation is rapidly evolving as a significant and unique research field having assimilated signal-processing methods and algorithms introduced previously into the broader areas of acoustic signal enhancement and noise suppression but also from more specialized areas such as speech dereverberation, correction of room acoustics, auditory-scene analysis, and from a wealth of perceptual studies and models related to auditory mechanisms. In contrast to some of the mostly single-input-channel methods as were introduced in the signal-enhancement-research fields and were targeted towards machine listening and automatic speech recognition, or were geared towards speech-intelligibility improvements, binaural dereverberation attempts to utilize concepts and optimize processing specifically tailored to binaurally-received signals by human listeners. As is well known, during everyday life, the human auditory system has an impressive ability to analyze, process and select individual source signals from complex acoustic environments, significantly so from signals contaminated by room reflections and reverberation. It has been well established by earlier research that this ability is to a large extend due to auditory and cognitive mechanisms which rely on the binaural signals as these allow the listeners to analyse auditory scenes and suppress unwanted signal components.

Recently, technological and other developments dictate the ever expanding use of portable devices for receiving auditory information via headphones or earpieces and in many such applications binaural dereverberation is an essential preprocessing step in order to ensure reception comparable or better to that of normal listening. Such applications are currently mostly driven by the digital-hearing-aid sector where the problem of reverberation intrusion is prohibitive to speech intelligibility and auditory-scene interpretation. At the same time, the processing capabilities of the commercially available hearing aids allow the real-time implementation of the emerging methods. However, other applications such as binaural telephony, teleconferencing, hands-free devices and interfaces, immersive-audio rendering, and so on, seem to rapidly adopt such binaural dereverberation processing.

Given the extensive capabilities of the hearing system, any binaural dereverberation method is facing significant challenges. Traditionally, most acoustic-signal enhancement methods aim at suppressing some unwanted interference and potentially improve speech intelligibility, and/or audio signal quality. However, the perception of room reverberation is inherently linked to many cues useful to the listener that are generated by room reflections and relate to source position, listener orientation, room size, and further properties that must be retained after processing. Hence, binaural dereverberation methods appear to aim at retaining some useful auditory cues and signal qualitative features, thus compromising their ability to suppress reverberation. From this discussion it is becoming clear that the prominent aims of the binaural methods appear as follows: improving the ratio of direct-to-reverberant signal energy, removing unwanted timbral coloration due to room reflections, removing late-reverberant effects and energy during temporal signal gaps, improving source localization and separation, and suppressing unwanted sound sources and noise. However, after processing, it is imperative that binaural cues necessary for source localization and auditory-scene analysis must be fully preserved or even enhanced. Further, it is sometimes desirable that important cues such as the precedence effect, spectral qualities, as well as room size and envelopment, are retained.

In the previous sections of this chapter it is shown that binaural dereverberation is largely relying on a linear-system model of the room, described via the binaural impulse responses between source(s) and the listener's ears and that often this function is separated into the direct path, early reflection and late-reverberation components. Given that different physical and perceptual effects can be identified due to each of these response components, many dereverberation methods attempt to compensate specifically for effects due to those parts. Another common theme during the evolution of the binaural-dereverberation methods is that some of those methods were based on dual-input-channel processing and others on bilateral adaptation of single channel dereverberation methods, those typically relying on spectral subtraction, a technique which has been widely used for noise suppression. Significantly, all binaural dereverberation must be implemented without any prior measurement of the room response, thus being *blind*, or at least semi-blind, when some broad parameters related to the acoustic environment have to be known.

Starting from those earlier dual-channel dereverberation methods—for example, [2], the significance of the interchannel/interaural coherence function as an indicator of the direct-path-signal contribution in the received signals has been established, leading to techniques that can effectively enhance this signal component in the short-term spectral domain, improving thus the direct-to-reverberant ratio in the processed signal. In this chapter, a detailed presentation of a recent IC based method is given [123, 124]. In contrast to the earlier method [2], where a linear coherence-to-gain function for the spectral modification was proposed, this recent method has introduced a flexible parametric sigmoidal function that can easily be adjusted to the desired filtering-gain form appropriate for specific room-acoustical and source-receiver configurations. Furthermore, the form of the IC estimates and hence the parameters of the filter can be directly obtained from long-term spectral analysis of the received signals. The performance of this IC method was evaluated by testing and comparing the change in the $\Delta$segSRR, the change in the $\Delta$NMR, as well as via perceptual-evaluation tests. It was found that the method achieved superior overall quality compared to the original IC-based method, as well as to a spectral-subtraction based method. The method was also found to perform better for dereverberating signals from closely located sound sources, for instance, at 0.5 m, instead of those derived from distant sources, for instance, at 5 m. This illustrates that dereverberation gains via IC-based methods depend largely on the degree of the direct-signal energy within the received signals, typically this happening for shorter source-listener distances. For such cases, the overall quality of the processed signal is superior to the results obtained via other processing methods. When this condition is not satisfied, then processing may result in signal-quality loss.

Apart from enhancing the direct signal components, many further single channel dereverberation methods have been developed, based on the concept of spectral subtraction. Although reverberation is a convolutive distortion, late reverberation has usually exponentially decaying white-noise-like properties and, hence, it may be effectively modeled as additive-noise distortion. Traditionally, spectral-subtraction methods suppress such additive noise by subtracting in the short-term spectral domain its estimate and, following this line of thinking, they were also adopted for late-reverberation suppression. For the blind estimation of the late reverberation, essential for deriving the short-term spectral filter gains, a number of established methods exist, originally proposed for speech applications. These methods have either employed an exponentially-decaying noise-like estimate function—compare the LB method, [86]—or an estimate based on smoothed-shifted version of the reverberant speech spectrum—WW method, [126]—or an estimate based on a sum of filtered versions of earlier reverberant speech spectra—FK method, [38].

For adopting such methods for the binaural case, their bilateral application to each signal channel may be followed. Given that the relevant ITD and ILD cues must be preserved in the processed dereverberated signals, identical processing must be applied on the left- and right-ear signals by using appropriately adapted common gains derived from the received signals in each channel, since it is likely that filter-gain estimates may vary for the two different paths to the ears. Such alternative gain-adaptation strategies were studied in [115] and were presented in detail in Sect. 4.2 of this chapter—along with an optimized GMR stage [79, 116] that restricts via a parametric function the extend of spectral modifications during processing. For the perceptual performance tests, described in detail in [115, 116], the PESQ was employed along with perceptual tests based on the MOS and a modified version of the ITU P.835 test. The results indicate that a trade-off between the degree of dereverberation and the perceptual quality of the processed signal, with the estimators derived via the LB and WW methods, achieve the best performance. For these two methods it was also found that average-gain weighting of the individually estimated gains for the left- and right-ear paths, was the best way for adapting these functions to the binaural processing scenario.

In this chapter, an analysis of the concepts involved in human sound reception and perception inside reverberant rooms has been presented, along with a literature review concerning past attempts on the open problems of signal dereverberation. A focus was put on more detailed presentation of two recent blind binaural-dereverberation methods, the first one based on interaural coherence to enhance the direct-to-reverberant ratio and the second one geared towards optimal adaptation of single channel, spectral subtraction based methods for suppressing late reverberation. In both cases, as is the case with most other signal enhancement methods, it was found that reverberation suppression and processed signal quality are two mutually exclusive items that restrict the overall performance of the methods. In comparison with the performance achieved by the human auditory system in similar tasks, such as for the precedence effect, signal decoloration and reverberation suppression, the performance of the current dereverberation methods is clearly inferior, in particular, when being considered under all possible acoustic and source-listener configurations. Furthermore, a unique problem facing all these dereverberation methods is the definition of the desirable results that should be aimed at by such processing. The difficulty of adapting existing objective and perceptual-performance measures and methodologies into the binaural-dereverberation case, especially with respect to retaining the complex cues associated by auditory-scene analysis performed by human listeners inside enclosed spaces, as well as the challenge of dealing with both noise and reverberation likewise, illustrates the complexity and the open issues facing this promising and relatively young research field.

**Acknowledgments** The authors would like to thank their two anomynous reviewers for substantial and constructive suggestions. This research has been co-financed by the European Union through the European Social Fund, ESF, and the Greek national funds through its Operational Program *Education and Lifelong Learning* of the National Strategic Reference Framework, NSRF. Research Funding Programs: *Thalis-University of Athens-Erasitechnis, grant number: MIS 375435.*

# References


1. 3GPP TS26.073. ANSI-C code for the Adaptive Multi Rate (AMR) speech codec. Technical report, 3rd Generation Partnership Project, Valbonne, France, 2008.
2. J. B. Allen, D. A. Berkley, and J. Blauert. Multimicrophone signal-processing technique to remove room reverberation from speech signals. *J. Acoust. Soc. Amer.*, 62:912–915, 1977.
3. J. B. Allen and L. R. Rabiner. A unified approach to short-time fourier analysis and synthesis. In *Proc. of the IEEE*, volume 65, pages 1558–1564, 1977.
4. I. Arweiler and J. M. Buchholz. The influence of spectral and spatial characteristics of early reflections on speech intelligibility. *J. Acoust. Soc. Am.*, 130:996–1005, 2011.
5. C. Avendano and H. Hermansky. Study on the dereverberation of speech based on temporal processing. In *Proc. ICSLP*, 1996.
6. M. Barron and A. Marshall. Spatial impression due to early lateral reflections in concert halls: The derivation of a physical measure. *J. Sound Vibr.*, 77:211–232, 1981.
7. S. Bech and N. Zacharov. *Perceptual Audio Evaluation: Theory, method and application,* pages 39–96. Wiley and Sons Ltd., West Sussex, Great Britain, 2006.
8. D. Bees, M. Blostein, and P. Kabal. Reverberant speech enhancement using cepstral processing. In *Proc. IEEE-ICASSP*, volume 2, pages 977–980, 1991.
9. M. Berouti, R. Schwartz, and J. Makhoul. Enhancement of speech corrupted by acoustic noise. In Proc. *IEEE-ICASSP*, volume 4, pages 208–211, 1979.
10. S. Bharitkar, P. Hilmes, and C. Kyriakakis. Robustness of spatial average equalization: A statistical reverberation model approach. *J. Acoust. Soc. Am.*, 116:3491–3497, 2004.
11. F. A. Bilsen and R. J. Ritsma. Repetition pitch and its implications for hearing theory. *Acustica*, 22:63–73, 1969/70.
12. J. Bitzer, K. U. Simmer, and K.-D. Kammeyer. Theoretical noise reduction limits of the generalized sidelobe canceller (GSC) for speech enhancement. In *Proc. IEEE-ICASSP*, volume 5, pages 2965–2968, 1999.
13. J. Blauert. *Spatial Hearing*. MIT Press, 1997.
14. J. P. Bloom. Evaluation of a dereverberation process by normal and impaired listeners. In *Proc. IEEE-ICASSP*, volume 5, pages 500–503, 1980.
15. P. Bloom and G. Cain. Evaluation of two-input speech dereverberation techniques. In *Proc. IEEE-ICASSP*, volume 7, pages 164–167, 1982.
16. S. Boll. Suppression of acoustic noise in speech using spectral subtraction. *IEEE Trans. Acoust. Speech, Signal Process.*, 27:113–120, 1979.
17. J. S. Bradley, H. Sato, and M. Picard. On the importance of early reflections for speech in rooms. *J. Acoust. Soc. Amer.*, 113:3233–3244, 2003.
18. J. S. Bradley and G. A. Soulodre. The influence of late arriving energy on spatial impression. *J. Acoust. Soc. Am.*, 97:263–271, 1995.
19. E. Brandewie and P. Zahorik. Prior listening in rooms improves speech intelligibility. *J. Acoust. Soc. Am.*, 128:291–299, 2010.
20. A. Bronkhorst and R. Plomp. The effect of head-induced interaural time and level differences on speech intelligility in noise. *J. Acoust. Soc. Am.*, 83:1508–1516, 1988.
21. J. Buchholz and J. Mourjopoulos. A computational auditory masking model based on signal-dependent compression. I. Model description and performance analysis. *Acta Acust./Acustica*, 90:873–886, 2004.
22. J. M. Buchholz. Characterizing the monaural and binaural processes underlying reflection masking. *Hearing Research*, 232:52–66, 2007.
23. J. M. Buchholz. A quantitative evaluation of spectral mechanisms involved in auditory detection of coloration by a single wall reflection. *Hearing Research*, 277:192–203, 2011.
24. J. M. Buchholz, J. Mourjopoulos, and J. Blauert. Room masking: understanding and modelling the masking of room reflections. In *Proc. 110th Conv. Audio Eng. Soc.*, Amsterdam, NL, 2001.
25. O. Cappe. Elimination of the musical noise phenomenon with the Ephraim and Malah noise suppressor. *IEEE Trans. Speech and Audio Process.*, 2:345–349, 1994.

26. L. Danilenko. Binaurales Hören im nichtstationären diffusen Schallfeld (Binaural hearing in a nonstationary, diffuse sound field). *Kybernetik*, 6:50–57, 1969.
27. A. A. de Lima, T. de M. Prego, S. L. Netto, B. Lee, A. Said, R. W. Schafer, T. Kalker, and M. Fozunbal. On the quality-assessment of reverberated speech. *Speech Communication*, 54:393–401, 2012.
28. M. Delcroix, T. Hikichi, and M. Miyoshi. Precise Dereverberation Using Multichannel Linear Prediction. *IEEE Trans. Audio, Speech and Language Process.*, 15:430–440, 2007.
29. J. Deller, J. Hansen, and J. Proakis. *Discrete-time processing of speech signals*. Wiley-IEEE Press, 1999.
30. S. Doclo and M. Moonen. Combined frequency-domain dereverberation and noise reduction technique for multi-microphone speech enhancement. In *Proc. IEEE IWAENC*, pages 31–34, Darmstadt, Germany, 2001.
31. S. Elliott and P. Nelson. Multiple-point equalization in a room using adaptive digital filters. *J. Audio Eng. Soc.*, 37:899–907, 1989.
32. K. Eneman and M. Moonen. Multimicrophone Speech Dereverberation: Experimental Validation. EURASIP *J. Audio Speech and Music Process.*, pages 1–20, 2007.
33. Y. Ephraim and D. Malah. Speech enhancement using a minimum-mean square error short-time spectral amplitude estimator. *IEEE Trans. Acoust., Speech and, Signal Process.*, 32:1109–1121, 1984.
34. J. S. Erkelens and R. Heusdens. Correlation-Based and Model-Based Blind Single-Channel Late-Reverberation Suppression in Noisy Time-Varying Acoustical Environments. *IEEE Trans. Audio, Speech and Language Process.*, 18:1746–1765, 2010.
35. T. Falk, C. Zheng, and W.-Y. Chan. A non-intrusive quality and intelligibility measure of reverberant and dereverberated speech. *IEEE Trans. Audio, Speech and Language Process.*, 18:1766–1774, 2010.
36. C. Faller and J. Merimaa. Source localization in complex listening situations: Selection of binaural cues based on interaural coherence. *J. Acoust. Soc. Am.*, 116:3075–3089, 2004.
37. J. L. Flanagan and R. Lummis. Signal processing to reduce multipath distortion in small rooms. *J. Acoust. Soc. Am.*, 47:1475–1481, 1970.
38. K. Furuya and A. Kataoka. Robust speech dereverberation using multichannel blind deconvolution with spectral subtraction. *IEEE Trans. Audio, Speech and Language Process.*, 15:1571–1579, 2007.
39. S. Gannot and M. Moonen. Subspace Methods for Multimicrophone Speech Dereverberation. EURASIP *J. Advances in, Signal Process.*, pp. 1074–1090, 2003.
40. N. Gaubitch, P. Naylor, and D. Ward. On the use of linear prediction for dereverberation of speech. In *Proc. of the IEEE IWAENC*, pages 99–102, 2003.
41. N. D. Gaubitch and P. A. Naylor. Analysis of the Dereverberation Performance of Microphone Arrays. In *Proc. IEEE-IWAENC*, 2005.
42. E. Georganti, T. May, S. van de Par, and J. Mourjopoulos. Extracting sound-source-distance information from binaural signals. In J. Blauert, editor, *The technology of binaural listening*, chapter 7. Springer, Berlin-Heidelberg-New York NY, 2013.
43. B. W. Gillespie, H. S. Malvar, and D. A. F. Florencio. Speech dereverberation via maximum-kurtosis subband adaptive filtering. In *Proc. IEEE-ICASSP*, volume 6, pages 3701–3704, 2001.
44. S. Goetze, E. Albertin, M. Kallinger, A. Mertins, and K.-D. Kammeyer. Quality assessment for listening-room compensation algorithms. In *Proc. IEEE-ICASSP*, pages 2450–2453, 2010.
45. S. Griebel and M. Brandstein. Wavelet Transform Extrema Clustering For Multi-Channel Speech Dereverberation. In *Proc. of the IEEE-IWAENC*, pages 27–30, 1999.
46. D. Griesinger. The psychoacoustics of apparent source width, spaciousness and envelopment in performance spaces. *Acta Acust./Acustica*, 83:721–731, 1997.
47. E. Habets. *Single- and multi-microphone speech dereverberation using spectral enhancement*. PhD thesis, Technische Univ. Eindhoven, 2007.
48. E. Habets, S. Gannot, and I. Cohen. Late reverberant spectral variance estimation based on a statistical model. *Signal Process. Letters, IEEE*, 16(9):770–773, 2009.

49. V. Hamacher, J. Chalupper, J. Eggers, E. Fischer, U. Kornagel, H. Puder, U. Rass. Signal Processing in High-End Hearing Aids: State of the Art, Challenges, and Future Trends. EURASIP *J. Applied, Signal Process.*, pp. 2915–2929, 2005.
50. Y. Haneda, S. Makino, and Y. Kaneda. Common acoustical pole and zero modeling of Room Transfer Functions. *IEEE Trans. Speech and Audio Process.*, 2:320–328, 1994.
51. V. Hansen and G. Munch. Making recordings for simulation tests in the Archimedes project. *J. Audio Eng. Soc.*, 39:768–774, 1991.
52. A. Härmä, M. Karjalainen, L. Savioja, V. Valimaki, U. Laine, and J. Huopaniemi. Frequency-warped signal processing for audio applications. *J. Audio Eng. Soc.*, 48:1011–1031, 2000.
53. A. Härmä and U. K. Laine. A comparison of warped and conventional linear predictive coding. *IEEE Trans. Speech and Audio Process.*, 9:579–588, 2001.
54. P. Hatziantoniou and J. Mourjopoulos. Generalized fractional-octave smoothing of audio and acoustic responses. *J. Audio Eng. Soc.*, 48:259–280, 2000.
55. P. Hatziantoniou and J. Mourjopoulos. Errors in real-time room acoustics dereverberation. *J. Audio Eng. Soc.*, 52:883–899, 2004.
56. P. Hatziantoniou, J. Mourjopoulos, and J. Worley. Subjective assessments of real-time room dereverberation and loudspeaker equalization. In *Proc. 118th Conv. Audio Eng. Soc.*, 2005.
57. P. D. Hatziantoniou and J. N. Mourjopoulos. Generalized fractional-octave smoothing of audio and acoustic responses. *J. Audio Eng. Soc.*, 48:259–280, 2000.
58. T. Hidaka, Y. Yamada, and T. Nakagawa. A new definition of boundary point between early reflections and late reverberation in room impulse responses. *J. Acoust. Soc. Am.*, 122(1):326–332, 2007.
59. J. Hopgood. *Nonstationary Signal Processing with Application to Reverberation Cancellation in Acoustic Environments*. PhD thesis, University of Cambridge, 2001.
60. J. Hopgood, C. Evers, and J. Bell. Bayesian single channel blind speech dereverberation using Monte Carlo methods. *J. Acoust. Soc. Am.*, 123:3586, 2008.
61. T. Houtgast, H. Steeneken, and R. Plomp. Predicting speech intelligibility in rooms from the modulation transfer function. i. general room acoustics. *Acustica*, 61:60–72, 1980.
62. T. Houtgast and H. J. M. Steeneken. A review of the MTF concept in room acoustics and its use for estimating speech intelligibility in auditoria. *J. Acoust. Soc. Am.*, 77:1069–1077, 1984.
63. Y. Hu and P. Loizou. Evaluation of objective quality measures for speech enhancement. *IEEE Trans. Audio, Speech and Language Processing*, 16, 2008.
64. Y. Hu and P. C. Loizou. Subjective comparison and evaluation of speech enhancement algorithms. *Speech Communication*, 49:588–601, 2007.
65. International Telecommunications Union, Geneva, Switzerland. *Perceptual evaluation of speech quality (PESQ), and objective method for end-to-end speech quality assessment of narrowband telephone networks and speech codecs*, 2000.
66. International Telecommunications Union (ITU-T, P.835), Geneva, Switzerland. *Subjective test methodology for evaluating speech communication systems that include noise suppression algorithm*, 2003.
67. M. Jeub, M. Schaefer, and P. Vary. A Binaural Room Impulse Response Database for the Evaluation of Dereverberation Algorithms. In *Proc. 17th Digital Signal Process. Conf.*, DSP, Santorini, Greece, 2009.
68. M. Jeub, M. Schafer, T. Esch, and P. Vary. Model-based dereverberation preserving binaural cues. *IEEE Trans. Audio, Speech, and Lang. Process.*, 18(7):1732–1745, 2010.
69. M. Jeub and P. Vary. Binaural dereverberation based on a dual-channel Wiener filter with optimized noise field coherence. In *Proc. IEEE-ICASSP*, pages 4710–4713, 2010.
70. M. Kallinger and A. Mertins. Impulse response shortening for acoustic listening room compensation. In *Proc. IEEE-IWAENC*, 2005.
71. M. Karjalainen and T. Paatero. Equalization of loudspeaker and room responses using Kautz filters: direct least squares design. EURASIP *J. Appl. Signal Process.*, 2007.
72. J. M. Kates. *Digital Hearing Aids*, pages 221–262. Plural Publishing, San Diego, CA, USA, 2008.

73. H. Kayser, S. D. Ewert, J. Anemuller, T. Rohdenburg, V. Hohmann, and B. Kollmeier. Database of Multichannel In-Ear and Behind-the-Ear Head-Related and Binaural Room Impulse Responses. EURASIP *J. Appl. Signal Process.*, 2009:1–10, 2009.
74. K. Kinoshita, M. Delcroix, T. Nakatani, and M. Miyoshi. Suppression of Late Reverberation Effect on Speech Signal Using Long-Term Multiple-step Linear Prediction. *IEEE Trans. Audio, Speech and Language Process.*, 17:534–545, 2009.
75. O. Kirkeby and P. Nelson. Digital filter design for inversion problems in sound reproduction. *J. Audio Eng. Soc.*, 47:583–595, 1999.
76. C. Knapp and G. Carter. The generalized correlation method for estimation of time delay. *IEEE Trans. Acoust., Speech and, Signal Process.*, ASSP-24:320–327, 1976.
77. A. Koening, J. Allen, D. Berkley, and C. T. Determination of masking level differences in a reverberant environment. *J. Acoust. Soc. Am.*, 61:1374–1376, 1977.
78. K. Kokkinakis and P. C. Loizou. Evaluation of objective measures for quality assessment of reverberant speech. In *Proc. IEEE-ICASSP*, pages 2420–2423, 2011.
79. E. Kokkinis, A. Tsilfidis, E. Georganti, and J. Mourjopoulos. Joint noise and reverberation suppression for speech applications. In *Proc. 130th Conv. Audio Eng. Soc.*, London, UK, 2011.
80. B. Köllmeier, J. Peissig, and V. Hohmann. Binaural noise-reduction hearing aid scheme with real-time processing in the frequency domain. *Scandinavian Audiol. Suppl.*, 38:28–38, 1993.
81. P. Krishnamoorthy and S. R. Mahadeva Prasanna. Reverberant speech enhancement by temporal and spectral processing. *IEEE Trans. Audio, Speech and Language Process.*, 17:253–266, 2009.
82. D. Kundur and D. Hatzinakos. Blind image deconvolution. *IEEE Signal Process. Mag.*, 13:43–64, 1996.
83. H. Kuttruff. *Room acoustics, 4th Edition*. Taylor & Francis, 2000.
84. T. Langhans and H. Strube. Speech enhancement by nonlinear multiband envelope filtering. In *Proc. IEEE-ICASSP*, 1982.
85. K. Lebart, J. Boucher, and P. Denbigh. A binaural system for the suppression of late reverberation. In *Proc. European Signal Process. Conf.*, pages 1–4, 1998.
86. K. Lebart, J. Boucher, and P. Denbigh. A new method based on spectral subtraction for speech dereverberation. *Acta Acust./Acustica*, 87:359–366, 2001.
87. J. H. Lee, S. H. Oh, and S. Y. Lee. Binaural semi-blind dereverberation of noisy convoluted speech signals. *Neurocomputing*, 72:636–642, 2008.
88. R. Y. Litovsky, S. H. Colburn, W. A. Yost, and S. J. Guzman. The precedence effect. *J. Acoust. Soc. Amer.*, 106:1633–1654, 1999.
89. P. Loizou. *Speech enhancement: theory and practice*. CRC Press, 1st edition, 2007.
90. P. Loizou and G. Kim. Reasons why current speech-enhancement algorithms do not improve speech intelligibility and suggested solutions. *IEEE Trans. Audio, Speech, and Lang. Process.*, 19:47–56, 2011.
91. H. W. Löllmann and P. Vary. Low delay noise reduction and dereverberation for hearing aids. EURASIP *J. Advances in Signal Process.*, 2009:1–9, 2009.
92. G. Lorho. *Perceived Quality Evaluation - An application to Sound reproduction over headphones*. PhD thesis, Aalto University, School of Science and Technology, Department of Signal Processing and Acoustics, Finland, 2010.
93. I. McCowan and H. Bourlard. Microphone array post-filter based on noise field coherence. *IEEE Trans. Speech and Audio Process.*, 11:709–716, 2003.
94. A. Mertins, T. Mei, M. Kallinger. Room impulse response shortening/reshaping with infinity- and p-norm optimization. *IEEE Trans. in Acoust., Speech and, Signal Process.*, 18:249–259, 2010.
95. O. M. M. Mitchell and D. A. Berkley. Reduction of long time reverberation by a center clipping process. *J. Acoust. Soc. Am.*, 47:84, 1970.
96. M. Miyoshi and Y. Kaneda. Inverse filtering of room acoustics. *IEEE Trans. Acoustics, Speech and, Signal Processing*, 36:145–152, 1988.

97. J. Mourjopoulos. On the variation and invertibility of room impulse response functions. *J. Sound and Vibr.*, 102:217–228, 1985.
98. J. Mourjopoulos. Digital equalization methods for audio systems. In *Proc. 84th Conv. Audio Eng. Soc.*, 1988.
99. J. Mourjopoulos. Digital equalization of room acoustics. *J. Audio Eng. Soc.*, 42:884–900, 1994.
100. J. Mourjopoulos, P. Clarkson, and J. Hammond. A comparative study of least-squares and homomorphic techniques for the inversion of mixed phase signals. In *Proc. IEEE-ICASSP*, 1982.
101. J. Mourjopoulos and J. K. Hammond. Modelling and enhancement of reverberant speech using an envelope convolution method. In *Proc. IEEE-ICASSP*, 1983.
102. P. A. Naylor and N. D. Gaubitch. *Speech Dereverberation*, pages 57–387. Springer, London, Great Britain, 2010.
103. P. A. Naylor, N. D. Gaubitch, and E. A. P. Habets. Signal-based performance evaluation of dereverberation algorithms. *J. Electrical and Computer Eng.*, 2010:1–5, 2010.
104. S. Neely and J. B. Allen. Invertibility of room impulse response. *J. Acoust. Soc. Am.*, 66:165–169, 1979.
105. S. Norcross, G. Soulodre, and M. Lavoie. Subjective investigations of inverse filtering. *J. Audio Eng. Soc.*, 52:1003–1028, 2004.
106. A. V. Oppenheim. *Applications of digital signal processing*. Prentice-Hall, 1978.
107. T. Paatero and M. Karjalainen. Kautz Filters and Generalized Frequency Resolution: Theory and Audio Applications. *J. Audio Eng. Soc.*, 51:27–44, 2003.
108. A. P. Petropulu and S. Subramaniam. Cepstrum based deconvolution for speech dereverberation. In *Proc. IEEE-ICASSP*, pages I/9-I12, 1994.
109. I. RBS.1534-2001: Method for the subjective assessment of intermediate quality levels of coding systems. 2003.
110. W. C. Sabine. *Collected Papers on Acoustics*. Peninsula Publishing, Los Altos, 1993.
111. L. Savioja and V. Valimaki. Multiwarping for enhancing the frequency accuracy of digital waveguide mesh simulations. *IEEE Signal Process. Letters*, 8:134–136, 2001.
112. K. Simmer, S. Fischer, and A. Wasiljeff. Suppression of coherent and incoherent noise using a microphone array. Ann. *Telecommunications*, 49:439–446, 1994.
113. T. Stockham, T. M. Cannon, and R. B. Ingebretsen. Blind deconvolution through digital signal processing. In *Proc. IEEE*, volume 63, pages 678–692, 1975.
114. M. Triki and D. T. M. Slock. Delay and Predict Equalization for Blind Speech Dereverberation. In *Proc. IEEE-ICASSP*, volume 5, 2006.
115. A. Tsilfidis, E. Georganti, and J. Mourjopoulos. Binaural extension and performance of single-channel spectral subtraction dereverberation algorithms. In *Proc. IEEE-ICASSP*, Prague, Czech Republic, 2011.
116. A. Tsilfidis, E. Georganti, and J. Mourjopoulos. A binaural framework for spectral subtraction dereverberation. In *Forum Acusticum 2011*, Aalborg, Denmark, 2011.
117. A. Tsilfidis, K. E. Kokkinis, and J. Mourjopoulos. Suppression of late reverberation at multiple speaker positions utilizing a single impulse response measurement. In *Forum Acusticum 2011*, Aalborg, Denmark, 2011.
118. A. Tsilfidis and J. Mourjopoulos. Signal-dependent constraints for perceptually motivated suppression of late reverberation. *Signal Processing*, 90:959–965, 2010.
119. A. Tsilfidis and J. Mourjopoulos. Blind single-channel suppression of late reverberation based on perceptual reverberation modeling. *J. Acoust. Soc. Am.*, 129:1439–1451, 2011.
120. D. Tsoukalas, J. Mourjopoulos, and G. Kokkinakis. Speech enhancement based on audible noise suppression. *IEEE Trans. Speech and Audio Process.*, 5:497–513, 1997.
121. H. L. Van Trees. *Optimum array processing*, volume 4. Wiley-Interscience. New York, NY, USA, 2002.
122. J. Wen, N. Gaubitch, E. Habets, T. Myatt, and P. Naylor. Evaluation of speech dereverberation algorithms using the MARDY database. In *Proc. IEEE-IWAENC*, 2006.

123. A. Westermann, J. Buchholz, and T. Dau. Using long-term coherence estimates for binaural dereverberation. In *Forum Acusticum 2011*, Aalborg, Denmark, 2011.
124. A. Westermann, J. Buchholz, and T. Dau. Binaural dereverberation based on interaural coherence histograms. *J. Acoust. Soc. Am.*, xxx:in press, 2013.
125. T. Wittkop and V. Hohmann. Strategy-selective noise reduction for binaural digital hearing aids. *Speech Communication*, 39:111–138, 2003.
126. M. Wu and D. Wang. A two-stage algorithm for one-microphone reverberant speech enhancement. *IEEE Trans. Audio, Speech and Language Process.*, 14:774–784, 2006.
127. B. Yegnanarayana and P. S. Murthy. Enhancement of reverberant speech using LP residual signal. *IEEE Trans. Audio, Speech and Language Process.*, 8:267–281, 2000.
128. P. Zahorik. Auditory distance perception in humans: A summary of past and present research. *Acta Acust./Acustica*, 91:409–420, 2005.
129. W. Zhang, E. Habets, and P. Naylor. On the use of channel shortening in multichannel acoustic system equalization. In *Proc. of the IEEE IWAENC*, 2010.
130. P. M. Zurek. Measurements of binaural echo suppression. *J. Acoust. Soc. Amer.*, 66:1750–1757, 1979.

# Binaural Localization and Detection of Speakers in Complex Acoustic Scenes

T. May, S. van de Par and A. Kohlrausch

## 1 Introduction

The robust localization of speakers is a very important building block that is required for many applications, such as hearing aids, hands-free telephony, voice-controlled devices and teleconferencing systems. Despite decades of research, the task of robustly determining the position of multiple active speakers in adverse acoustic scenarios has remained a major problem for machines. One of the most decisive factors that influence the localization performance of algorithms is the number of microphones. When several pairs of microphones are available, beamforming techniques such as the *steered-response power*, SRP, approach [25] or the *multi-channel cross-correlation coefficient*, MCCC, method [7] can be applied to disambiguate the localization information by exploiting correlation among multiple pairs of microphones. Furthermore, high-resolution subspace techniques such as the *multiple signal classification*, MUSIC, algorithm [66] and the *estimation of signal parameters via rotational-invariance techniques*, ESPRIT, approach [64] generally require that the number of sensors is greater than the number of sound sources. *Blind source separation* approaches, such as the *degenerate unmixing estimation technique*, DUET, attempt to blindly localize and recover the signals of $N$ sound sources from $M$ microphone signals [38, 81]. Although the DUET system is able to deal with

T. May
Centre for Applied Hearing Research, Department of Electrical Engineering, Technical University of Denmark, Kgs. Lyngby, Denmark

S. van de Par
University of Oldenburg, Oldenburg, Germany

A. Kohlrausch (✉)
Eindhoven University of Technology, Eindhoven, The Netherlands
e-mail: a.kohlrausch@tue.nl

A. Kohlrausch
Philips Research Europe, Eindhoven, The Netherlands

J. Blauert (ed.), *The Technology of Binaural Listening*, Modern Acoustics and Signal Processing, DOI: 10.1007/978-3-642-37762-4_15,

underdetermined mixtures, that is, $N > M$, in anechoic conditions, performance deteriorates in reverberant environments.

In contrast to machines, the human auditory system is remarkably robust in complex multi-source scenarios. It can localize and recognize up to six competing talkers [12], in spite of the fact that it is provided with only two signals reaching the left and the right ears. Moreover, listening with two ears substantially contributes to the ability to understand speech in multi-source scenarios [11, 19]. Unlike blind source separation algorithms that aim at separating the sources in such a way that they are fully reconstructed, the human auditory system does not need to perform such a reconstruction of the original signals. It only needs to extract those properties of the signal of interest that are needed for a particular task, such as estimating the direction of a sound source, the identity of a speaker, or the words that are being pronounced. Thus, when particular parts of the target signal are not available, that is, *missing*, due to the presence of other interfering sources, there may still be enough information, in other words, perceptual cues, available to extract the properties of interest, for example, the identity of a speaker. This ability of the human auditory system to handle complex multi-source scenarios and to segregate the contributions of individual sound sources is commonly summarized by the term *auditory scene analysis*, ASA. As described by Bregman [10], the underlying principles that facilitate ASA can be divided into two stages, namely, segmentation and grouping. First, the acoustic input is decomposed into spectro-temporal units, where each individual unit is assumed to be dominated by one particular source. Secondly, in the grouping stage, a set of primitive grouping rules, termed *Gestalt principles*, are employed by the auditory system in order to integrate the information that is associated with a single sound source. These Gestalt principles can be considered as data-driven mechanisms that are related to physical properties of sound generation, leading to certain structures in auditory signals. Common onsets across frequency, common amplitude and frequency modulation, and common spatial location are examples of such Gestalt principles [10, 23, 75]. Apart from data-driven processing—also known as *bottom-up processing*—the auditory system is able to focus the attention on a particular target source and interpret the underlying source, for instance, in order to understand speech. This involves schema-driven processing—also referred to as *top-down processing*—and requires *a priori* knowledge about different sound sources.

Inspired by the robustness of the human auditory system, a research field termed *computational auditory scene analysis*, CASA, has emerged, which aims at reproducing the capabilities of the human auditory system with machines on the basis of sensory input [75]. As the analysis is restricted to binaural signals, the task of automatically localizing multiple competing sound sources is particularly challenging. In this chapter, only two microphone signals will be considered, corresponding to the left- and the right ear signals of an artificial head, and it is shown how principles of human auditory processing can be used to estimate the azimuth of multiple speakers in the presence of reverberation and interfering noise sources, where the number of active speakers is assumed to be known *a priori*. Note that the intention is to develop a robust computer algorithm that is inspired by auditory mechanisms, rather than building a physiologically-plausible model of the human auditory system. Although this chapter focuses on binaural signals, the presented approach can be extended to microphone arrays with multiple pairs of microphones.

After describing the binaural signals that are used throughout this chapter, an overview of different approaches to binaural sound-source localization, ranging from technical approaches to auditory-inspired systems, will be given in Sect. 3. A thorough analysis of localization performance will then be presented in Sects. 4 and 5, using multiple competing speakers in reverberant environments. An important problem is the influence of noise on speaker-localization performance, which will be discussed in Sect. 6. In particular, it will be shown that the ability to localize speakers is strongly influenced by the spatial diffuseness of the interfering noise. Moreover, it will be seen that the presence of a compact noise source imposes severe challenges for correlation-based approaches. By employing principles of auditory grouping based on common spatial-location and missing data classification techniques, it is possible to make a distinction between source activity that originates from speech- or from noise sources. This distinction can substantially improve the speaker-localization performance in the presence of interfering noise.

## 2 Simulation of Complex Acoustic Scenes

In order to evaluate the localization algorithms that are presented in this chapter, complex acoustic scenes are simulated by mixing various speech and noise sources that are placed at different positions within a room. Binaural signals are obtained by convolving monaural speech files with binaural room impulse responses, BRIRs, corresponding to a particular sound-source direction. These BRIRs are simulated by combining a set of head-related transfer functions, HRTFs, with room impulse responses, RIRs, that are artificially created according to the image-source model [4]. More specifically, the MIT database is used, which contains HRTFs of a KEMAR[1] artificial head that were measured at a distance of 1.4 m in an anechoic chamber [30]. These HRTFs are combined with RIRs, simulated with *ROOMSIM*,[2] a MATLAB toolbox provided by Schimmel et al. [65]. The receiver, KEMAR, was placed at seven different positions in a simulated room of dimensions $6.6 \times 8.6 \times 3$ m. For the experiments conducted in this chapter, a set of BRIRs with the following reverberation times are simulated for each of the seven receiver positions, namely, $T_{60} = \{0.2, 0.36, 0.5, 0.62, 0.81$ and $1.05\,\text{s}\}$. The reverberation time, $T_{60}$, of the simulated BRIRs has been verified by applying the energy-decay-curve method developed by Schroeder [67].

Furthermore, a number of databases with measured BRIRs are publicly available [35, 37, 39], each of them focusing on a particular application. For a systematic

[1] Knowles electronic manikin for acoustic research, KEMAR.

[2] Although the problem of moving sources is not covered in this chapter, the MATLAB toolbox *ROOMSIMOVE* for simulating RIRs for moving sources can be found at http://www.irisa.fr/metiss/members/evincent/software.

analysis of localization performance, the measurements provided by the University of Surrey [35] were selected, since they offer BRIRs recorded in four different rooms with an azimuthal resolution of 5°. The following set of measured BRIRs is used for evaluation, $T_{60} = \{0.32, 0.47, 0.68 \text{ and } 0.89\,\text{s}\}$. Note that the BRIRs of the Surrey database are recorded with a Cortex–MK.2 head-and-torso simulator, HATS, which is different from the KEMAR artificial head that was used to create the simulated BRIRs. This allows the investigation of the impact on localization performance that is induced by a mismatch between BRIRs that are used for training and those which are used for testing. The results will be reported in Sect. 5.

Multi-source mixtures are created by randomly positioning sound sources within the azimuth range of $[-90, 90°]$ while having an angular distance of at least 10° between neighboring sources. For the experiments presented in Sects. 4 and 5, speech files are randomly selected from the speech-separation challenge, SSC, database [22]. Signals are either trimmed or concatenated to match an overall duration of 2 s. The level of multiple competing speech sources was always set equal. In addition, the impact of interfering noise on localization performance is systematically investigated in Sect. 6 by using three different types of noise signals, namely, babble noise and factory noise from the NOISEX database [74] and speech-shaped noise that is based on the long-term average spectrum, LTAS, of 300 randomly-selected speech files. Interfering noise sources are simulated by randomly selecting different time segments of the corresponding type of background noise. In contrast to the speech files, there is no constraint on the angular distance between multiple noise sources. The signal-to-noise ratio, SNR, is adjusted by comparing the energy of all speech sources to the energy of the noise. Note that the energies of the left and the right signals are added prior to SNR calculation. The resulting binaural multi-source signals are sampled at a sampling frequency of $f_s = 16\,\text{kHz}$.

## 3 Binaural Sound-Source Localization

The two major physical cues that enable human sound-source localization in the horizontal plane are *interaural time differences*, ITDs, and *interaural level differences*, ILDs, between the two ears [60]. Both cues are complementary in their effectiveness. As already formulated by Lord Rayleigh more than 100 years ago, the ITD cue is most reliable at low frequencies, whereas the ILD cue is more salient at higher frequencies [60]. The spectral modifications provided by the complex shape of the external ears are particularly important for the perception of elevation and help to resolve front-back confusions [68]. In this chapter, the localization of sound sources is restricted to the frontal horizontal plane within the area of $[-90, 90°]$. In the following sections, a short review of popular sound-source localization approaches will be given with the special application to binaural signals.

### 3.1 Broadband Approaches

One of the most frequently-used approaches to sound-source localization is to estimate the time difference of arrival, TDOA, between a pair of two spatially separated microphones. This approach usually consists of the following two steps. First, the relative delay between the microphones is estimated. Secondly, the estimated delay is used to infer the actual angle of the sound source by employing knowledge about the microphone-array geometry.

The *generalized cross-correlation*, GCC, framework presented by Knapp and Carter [41] is the most popular approach to perform time-delay estimation. The TDOA estimate, $\hat{\tau}$, in samples, is obtained as the time lag, $\tau$, that maximizes the cross-correlation function between the two filtered microphone signals, that is,

$$\hat{\tau} = \arg\max_{\tau} \frac{1}{2\pi} \int_{\omega} W(\omega)\, X_{\mathrm{L}}(\omega)\, X_{\mathrm{R}}^{*}(\omega)\, e^{j2\pi\omega\tau} \mathrm{d}\omega \,, \tag{1}$$

where $X_{\mathrm{L}}(\omega)$ and $X_{\mathrm{R}}(\omega)$ indicate the short-time Fourier transforms of the microphone signals, $x_{\mathrm{L}}(n)$ and $x_{\mathrm{R}}(n)$, received at the left and the right ears, and $W(\omega)$ denotes a frequency-dependent weighting function. The classical cross-correlation, CC, method uniformly weights all frequency components by setting $W_{\mathrm{CC}}(\omega) = 1$. To increase the resolution of GCC-based time delay estimation, it is useful to interpolate the GCC function by an oversampled inverse fast Fourier transform, IFFT [27]. Hence, an oversampling factor of four is considered in this chapter, resulting in a $\tau$-step size of 16 $\mu$s.

In ideal acoustic conditions, in which the signals captured by the two microphones are simply time-shifted versions of each other, the most prominent peak in the GCC function reveals the true TDOA between both microphones and can be reliably detected. However, in more realistic scenarios with reverberation and environmental noise, the identification of peaks in the GCC function becomes less accurate, which, in turn, reduces the localization performance. Therefore, a variety of different weighting functions have been proposed in order to sharpen the peak that corresponds to the true TDOA and to improve its detectability [15, 41]. Among them, the so-called *phase transform*, PHAT, is the most frequently-used weighting function, which whitens the cross-spectrum between the two microphone signals, $x_{\mathrm{L}}(n)$ and $x_{\mathrm{R}}(n)$, prior to cross-correlation by choosing the weighting as $W_{\mathrm{PHAT}}(\omega) = |X_{\mathrm{L}}(\omega)\, X_{\mathrm{R}}^{*}(\omega)|^{-1}$. When ignoring the impact of noise, the PHAT weighting eliminates the influence of the source signal on localization and exhibits a clearly visible peak at the true TDOA. One apparent drawback of the PHAT weighting is that is gives equal weight to all frequencies, regardless of their signal-to-noise ratio, SNR. Nevertheless, if all interferences can be attributed to reverberation, the PHAT weighting has been shown to achieve robust localization performance [32, 83], as long as the level of noise is low [83].

Another approach that attempts to improve the robustness of the GCC function in noisy and reverberant environments is to perform *linear prediction*, LP, analysis to extract the excitation source information[3] [59]. The conventional GCC function is then computed based on the Hilbert envelope of the LP-residual signal, which was reported to form a more prominent main peak at the true TDOA in comparison to the conventional CC weighting.

Alternatively, the delay can also be derived from the *average magnitude-difference function*, AMDF, and its variations [17, 36]. For an comprehensive overview of different time-delay-estimation techniques the reader is referred to Chen et al. [18].

Once an estimation of the time delay between the left and the right ears is available, the second step of TDOA estimation requires conversion of the measured time delay to its corresponding *direction of arrival*, DOA. This is commonly achieved by a table-look-up procedure that can roughly account for the diffraction effects of the human head. Therefore, the estimated delay, $\hat{\tau}$, of a particular TDOA method is monitored in response to white noise filtered with HRTFs that are systematically varied between $-90°$ and $90°$ [8, 57]. The resulting mapping function establishes a monotonic relation between time delay, $\hat{\tau}$, and sound-source azimuth, $\varphi$, at an angular resolution of $1°$.

## 3.2 Auditory-Inspired Approaches

It is an important property of the human auditory system to be able to segregate the individual contributions of competing sound sources. In an attempt to incorporate aspects of peripheral auditory processing, the cross-correlation analysis can be performed separately for different frequency channels [8, 46, 50, 57, 63]. The frequency selectivity of the basilar membrane is commonly emulated by a *Gammatone filterbank*, GTFB, that decomposes the acoustic input into individual frequency channels with center frequencies equally spaced on the *equivalent-rectangular-bandwidth-rate* scale, ERB scale, [31]. It is advantageous to use phase-compensated Gammatone filters by accounting for the frequency-dependent group delay of the filters at their nominal center frequencies, $c_f$. This time-alignment can be achieved by introducing a channel-dependent time lead and a phase-correction term [14], allowing for a synchronized analysis at a common instance of time. Further processing stages crudely approximate the neural-transduction process in the inner hair cells by applying half-wave rectification and square-root compression to the output of each individual Gammatone filter [63]. Although not considered in this chapter, more elaborate models of the neural-transduction process might be applied at this stage [53, 54, 72]. Then, on the basis of these auditory signals, denoted as $h_{\mathrm{L},f}$ and $h_{\mathrm{R},f}$, the normalized cross-correlation, $C$, can be computed over a window of $B$ samples as a function of time lag, $\tau$, frame number, $t$, and frequency channel, $f$, as follows,

[3] The corresponding MATLAB code can be found at http://www.umiacs.umd.edu/labs/cvl/pirl/vikas/Current_research/time_delay_estimation/time_delay_estimation.html

$$C\left(t, f, \tau\right) = \frac{\sum\limits_{i=0}^{B-1} \left(h_{\mathrm{L},f}\left(t \cdot B/2 - i\right) - \bar{h}_{\mathrm{L},f}\right)\left(h_{\mathrm{R},f}\left(t \cdot B/2 - i - \tau\right) - \bar{h}_{\mathrm{R},f}\right)}{\sqrt{\sum\limits_{i=0}^{B-1} \left(h_{\mathrm{L},f}\left(t \cdot B/2 - i\right) - \bar{h}_{\mathrm{L},f}\right)^2} \sqrt{\sum\limits_{i=0}^{B-1} \left(h_{\mathrm{R},f}\left(t \cdot B/2 - i - \tau\right) - \bar{h}_{\mathrm{R},f}\right)^2}} \,. \quad (2)$$

$\bar{h}_{\mathrm{L},f}$ and $\bar{h}_{\mathrm{R},f}$ denote the mean values of the left and right auditory signals estimated over frame $t$. The normalized cross-correlation function is evaluated for time lags within a range of $[-1, 1\,\mathrm{ms}]$, and the lag that corresponds to its maximum is used to reflect the interaural time difference, ITD,

$$\widehat{\mathrm{itd}}\left(t, f\right) = \arg\max_{\tau} C\left(t, f, \tau\right)/f_{\mathrm{s}} \,. \quad (3)$$

Instead of using the integer time lag directly for ITD estimation, it is possible to refine the fractional peak position by applying parabolic [36] or exponential [84] interpolation strategies. It has been found that the exponential interpolation performs better than the parabolic one, which is in line with results reported by Tervo and Lokki [73].

The frequency-selective processing allows the frequency-dependent diffraction effects introduced by the shape of the human head [8, 57, 63] to be accounted for. More specifically, the cross-correlation pattern, $C\left(t, f, \tau\right)$, which is usually a function of the time lag, $\tau$, is warped onto an azimuth grid, $S\left(t, f, \varphi\right)$ [57]. This warping is accomplished by a frequency-dependent table look-up, which is obtained in a similar way as the one described in Sect. 3.1 and translates time delay to its corresponding azimuth. The frame-based source position can then be obtained by integrating the warped cross-correlation patterns across frequency and locating the most prominent peak in the summary cross-correlation function, that is,

$$\hat{\varphi}_{\mathrm{GFB}}\left(t\right) = \arg\max_{\varphi} \sum_{f} S\left(t, f, \varphi\right) \,. \quad (4)$$

This across-frequency integration is an implementation of the *straightness approach* where sound-source directions with synchronous activity across multiple frequency channels are emphasized [69, 71].

If more than one sound source should be resolved on a frame-by-frame basis, it might be beneficial to compute a *skeleton* cross-correlation function [57, 63]. The general concept is that each local peak in the cross-correlation function is replaced by a Gaussian function where the corresponding standard deviation is varied linearly as a function of the frequency channel. This processing aims at sharpening the response of the summary cross-correlation function.

Although the computational approaches to binaural sound-source localization discussed so far have been focusing on exploiting the ITD cue, there are some attempts to also consider the information that is supplied by the interaural level differences, ILDs. The aforementioned skeleton cross-correlation function has some similarities with the concept of *contralateral inhibition*, where the ILD information is incorporated into the cross-correlation framework to predict phenomena related to the *precedence effect* [44, 45]. A comprehensive review of the recent development of binaural models can be found in [9]. Moreover, the model presented by Palomäki et al. [57] uses an azimuth-specific ILD template to verify if the estimated ILD is consistent with the template ILD that is expected for the ITD-based azimuth estimate. The ILD cue can be derived by comparing the energy of the left- and the right-ear signals, $h_{\mathrm{L},f}$ and $h_{\mathrm{R},f}$, over a window of $B$ samples, namely,

$$\widehat{\mathrm{ild}}\,(t, f) = 10\log_{10}\left(\frac{\sum\limits_{i=0}^{B-1} h_{\mathrm{R},f}\,(t \cdot B/2 - i)^2}{\sum\limits_{i=0}^{B-1} h_{\mathrm{L},f}\,(t \cdot B/2 - i)^2}\right). \tag{5}$$

## 3.3 Supervised-Learning Approaches

In many realistic environments the observed binaural cues will be affected by the presence of reverberation and noise sources. Although the binaural cues are noisy, there still is a certain degree of predictability associated with these binaural cues, depending on the azimuth of the sound source. Recently, supervised-learning strategies have been employed in order to optimally infer the location of a source on the basis of binaural cues [24, 34, 50, 56, 77–79] where the interdependence between interaural time and level differences can be jointly considered as a function of frequency channel, $f$, and sound-source direction, $\varphi$. Note that supervised-learning approaches based on binaural cues have also been applied in the context of sound-source segregation [33, 63].

In this chapter, a *Gaussian mixture model*, GMM, classifier to approximate the two-dimensional feature distribution of ITDs and ILDs will be described. For the extraction of ITDs and ILDs, auditory front-ends as described in the previous section are commonly employed. In contrast to utilizing a mapping function—see Sects. 3.1 and 3.2—that translates the obtained interaural differences to their corresponding sound-source directions, supervised-learning approaches offer the considerable advantage of providing a probabilistic framework where multiple layers of information can be jointly analyzed. This combined analysis of ITDs and ILDs has been shown to be superior to exclusively relying on the ITD cue [50]. The localization framework based on GMMs is very flexible and can be readily extended to incorporate additional features that depend on the sound-source direction. Likewise, the GMM framework is applicable to array geometries with more than two microphones from which binaural features for multiple microphone pairs could be extracted. To extend the working range of the GMM-based localization model to the dimension of elevation, the additional integration of monaural cues might be beneficial [43, 82]. Further details about vertical sound-source localization can be found in [6], this volume.

During the supervised training process, *a priori* knowledge is available to create training data, namely, binaural features, and the corresponding class labels that categorize the training data according to different sound-source directions, $\varphi$. As analyzed by Roman at el. [63], the joint distribution of ITDs and ILDs is influenced by the presence of a competing source and its strength relative to the target source. This can be accounted for by training the localization model with binaural cues extracted for mixtures with a target and an interfering source at various SNRs. The resulting model was reported to yield substantial SNR improvements [63], however, its application is restricted to anechoic scenarios.

## Multi-Conditional Training of Binaural Cues

Localization models are commonly based on the assumption of single-path wave propagation. To overcome this fundamental limitation, a multi-conditional training stage can be applied in order to incorporate possible variations of ITDs and ILDs that are caused by the presence of competing sound sources, room reverberation and changes in the source-receiver configuration [50]. During the multi-conditional training stage, a variety of different acoustic conditions are simulated, and the frequency-dependent distributions of binaural features are approximated by a Gaussian mixture model classifier. The reverberation characteristic is intentionally simplified by assuming a frequency-independent reverberation time of $T_{60} = 0.5$ s. In this way, the same amount of uncertainty is encoded in each Gammatone channel. To ensure that the model is not trained for a particular room position, the multi-conditional training also involves various receiver positions and radial distances between the source and the receiver. Note that these positions are different from the ones that are used for evaluation—see Sect. 2 for details. More specifically, the following parameters are varied for each sound-source direction, $\varphi$,

- Competing speaker at $\pm 40$, $\pm 30$, $\pm 20$, $\pm 10$ and $\pm\,5°$ relative to the azimuth $\varphi$ of the target source
- Three SNRs between the target and the competing source, 20, 10 and 0 dB
- Three radial distances between the target source and the receiver, 0.5, 1 and 2 m
- Eight positions within the simulated room of dimensions, $6.6 \times 8.6 \times 3$ m

To visualize the influence of reverberation and the presence of multiple sound sources on ITDs and ILDs, the binaural feature space created by the multi-conditional training stage is presented in Fig. 1. Each dot represents a joint ITD-ILD estimate obtained for time frames of 20 ms. Note that the black and the gray distributions correspond to binaural cues associated with a target source at $\varphi = -50°$ and $\varphi = 50°$, respectively. When analyzing the general shape of the joint ITD-ILD feature distributions, it can be seen that the interdependency of both binaural cues results in complex multi-modal patterns. Due to spatial aliasing, the cross-correlation function leads to ambiguous ITD estimates at higher frequencies at which the wavelength is smaller than the diameter of the head. Consequently, the ambiguous ITD information results in multi-modal distributions where the number of individual clusters systematically

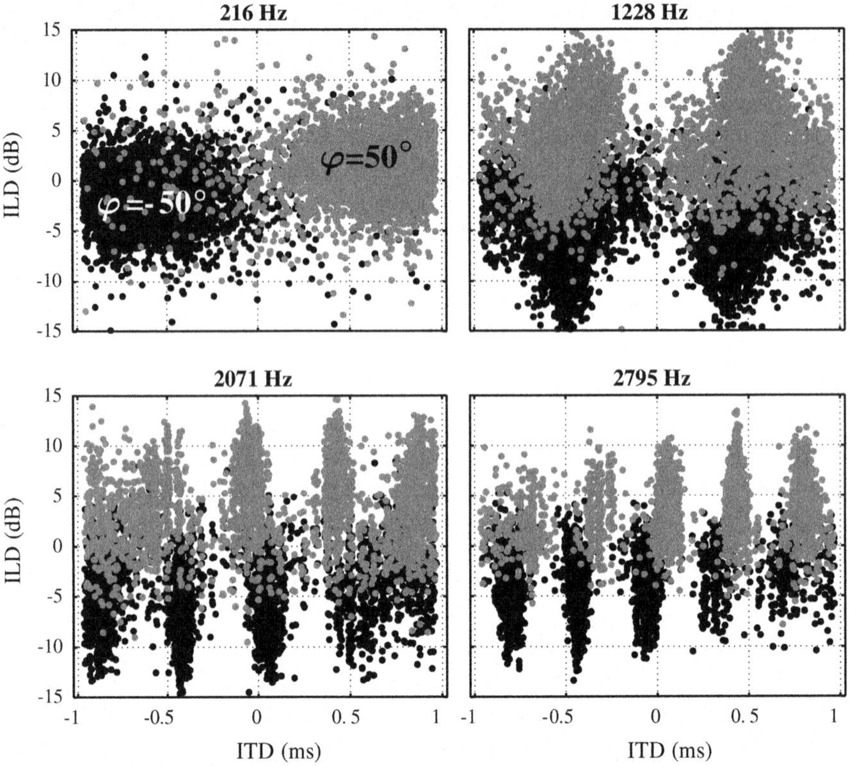


**Fig. 1** Frequency-dependent distributions of interaural time and level differences, ITDs and ILDs, created by the multi-conditional training stage. Each dot represents a frame-based observation of the joint ITD and ILT feature space. The *black* and *gray* distributions correspond to two different sound-source directions, namely, $\varphi = -50°$ and $\varphi = 50°$ respectively. See text for more details

increases with frequency. This ITD fine structure at higher frequencies is deliberately maintained, because experiments showed that a more detailed hair-cell model that simulates the inability of the human auditory system to analyze the temporal fine structure at frequencies above 1.5 kHz performed substantially worse in terms of localization accuracy [50]. This comparison suggests that the fine-structure information of the ITD can be effectively exploited by the GMM classifier for improved localization performance. This is a distinguishing feature from other localization models that attempt to build a physiologically-plausible model of human sound-source localization [26]. Another practical advantage of exploiting ITDs at higher frequencies is that the reverberation energy usually decays towards higher frequencies. As a result, the binaural cues associated with higher frequencies are less affected by reverberation, and thus convey more reliable contributions to overall localization. The spread of the individual clusters can be attributed to the impact of reverberation and the presence of a competing source. Furthermore, when comparing the binaural features for $\varphi = -50°$ and $\varphi = 50°$, it can be seen that the complex structure systematically shifts with sound-source direction.

### GMM-Based Localization

The distinct change of ITDs and ILDs as a function of sound-source direction, which is illustrated in Fig. 1, can now be systematically learned by a GMM classifier. Thus, the multi-conditional training is performed for a set of $K = 37$ sound-source directions, $\{\varphi_1, \ldots, \varphi_K\}$ spaced by 5°, within the range of [−90, 90°]. After training, a set of frequency- and azimuth-dependent diagonal GMMs, $\{\lambda_{f,\varphi_1}, \ldots, \lambda_{f,\varphi_K}\}$, is available. Given an observed binaural feature vector consisting of estimated ITDs and ILDs, $\mathbf{x}_{t,f} = \{\widehat{\text{itd}}(t, f), \widehat{\text{ild}}(t, f)\}$, the three-dimensional spatial log-likelihood can be computed for the $k$th sound-source direction being active at time frame $t$ and frequency channel $f$ as

$$\mathcal{L}(t, f, k) = \log p(\mathbf{x}_{t,f} | \lambda_{f,\varphi_k}) \ . \tag{6}$$

To obtain a robust estimation of sound-source direction, the log-likelihoods are accumulated across all frequency channels and the most probable direction reflects the estimated source location on a frame-by-frame basis, that is,

$$\hat{\varphi}_{\text{GMM}}(t) = \arg\max_{1 \le k \le K} \sum_{f=1}^{F} \mathcal{L}(t, f, k) \ . \tag{7}$$

Note that, in contrast to integrating the cross-correlation pattern across frequency, see (4), the log-likelihoods are accumulated, taking into account the uncertainty of binaural cues in individual frequency channels. This probabilistic integration of binaural cues has been also suggested by Nix and Hohmann [56]. As a result, the model does not require additional selection mechanisms, such as the coherence-based selection of reliable binaural cues [29], because this weighting is already implicitly incorporated into the model by the multi-conditional training stage. In other words, the multi-conditional training considers possible variations of interaural time and level differences resulting from competing sound sources and room reverberation, thus improving the robustness of the localization model in adverse acoustic scenarios.

## 4 Frame-Based Localization of a Single Source

In this section a comparison is performed of the ability of different approaches to localize the real position of one speaker in the presence of reverberation, based on 20 ms time frames. Therefore, binaural mixtures are created by using the simulated BRIRs with different reverberation times, $T_{60}$. A set of 185 binaural mixtures is created for each reverberation time. For evaluation, the following methods are considered,

- Generalized cross-correlation, GCC, function according to (1) with two different weighting functions, $W_{\mathrm{CC}}$ and $W_{\mathrm{PHAT}}$
- GCC function according to (1) with $W_{\mathrm{CC}}$ based on the LP residual
- Gammatone-based cross-correlation, GCC–GTFB, according to (4)
- GMM-based localization according to (7) with multi-conditional training

The GCC-based algorithms used a 20 ms Hamming window and a fast Fourier transform of 1,024 samples. The resolution of the resulting TDOA estimate was improved by applying an IFFT-based interpolation with an oversampling factor of four. The LP residual is created on the basis of 20 ms frames by using ten LP-filter coefficients. The Gammatone-based processing is based on 32 auditory filters that were equally distributed on the ERB-rate scale between 80 Hz and 5 kHz. All mapping functions are derived from anechoic BRIRs based on the KEMAR HRTFs. In general, the number of active target speakers is assumed to be known *a priori*. The blind estimation of the number of active speakers is currently being investigated [48].

The percentage of correctly localized frames is shown in Fig. 2 as a function of the absolute error threshold. Different panels represent different reverberation times, ranging from $T_{60} = 0.2$ up to $T_{60} = 1.05$ s. Apart from the conventional GCC approach, all algorithms reach ceiling performance for a moderate reverberation time of $T_{60} = 0.2$ s. But with increasing reverberation time, performance of all GCC-based methods substantially deteriorates. Due to the fact that these approaches are based on the assumption of single-path wave propagation, the presence of strong reflections causes spurious peaks in the GCC function that are erroneously selected as source positions. Thus, localization performance of the GCC-based approaches will inevitably decrease in more challenging acoustic conditions. While the LP-based preprocessing improves the performance of the conventional GCC approach, the PHAT-weighting produces the overall most reliable estimates of all GCC-based approaches, which supports the findings of previous studies [32, 83]. The GMM-based localization model shows superior performance, especially in conditions with strong reverberation, suggesting that the multi-conditional training stage can account for the distortions of ITDs and ILDs due to reverberation. Furthermore, unlike the other approaches, the GMM-based localization model is able to jointly analyze ITD and ILD information.

# 5 Localization of Multiple Sound Sources

In more complex acoustic scenarios, a variety of sound sources might be active at the same time. As demonstrated in the previous section, the performance of localizing only one speaker on a frame-by-frame basis noticeably degrades with increasing reverberation time. Therefore, an important question is how to integrate localization information across time in order to reliably resolve the position of multiple competing sound sources in reverberant environments.

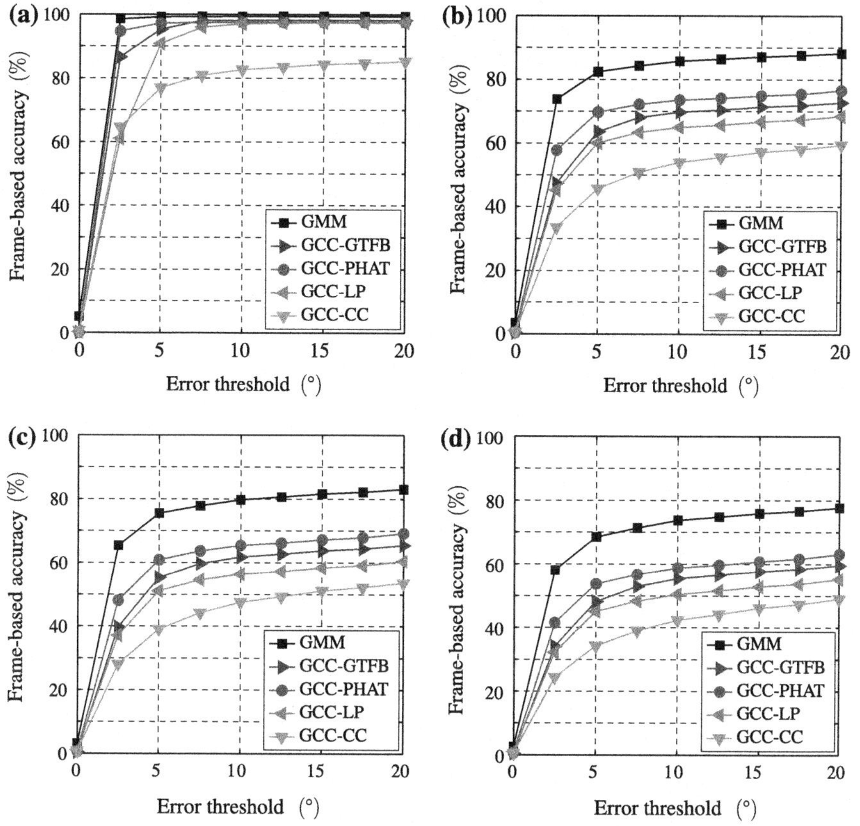


**Fig. 2** Frame-based accuracy in % of localizing one speaker in a reverberant room as a function of the absolute error threshold in °. Results are shown for different reverberation. **a** $T_{60} = 0.2$ s. **b** $T_{60} = 0.5$ s. **c** $T_{60} = 0.81$ s. **d** $T_{60} = 1.05$ s

## Temporal Integration

Recursive smoothing techniques could be considered as a way to calculate a running average of localization information. Regarding the class of GCC-based approaches that require a short-time estimate of the cross-spectrum, a first-order recursive smoothing can be applied [47]. While this approach might help to improve localization performance in scenarios with one target source, recursive smoothing reduces the ability of the localization algorithm to quickly respond to changes in source activity, which is particularly important for complex multi-source scenarios. Furthermore, the optimal smoothing constant might depend on a variety of different factors, such as the number of active sources, the level of noise and the reverberation time. Thus, exponential smoothing is not considered in this chapter.

One possibility of accumulating evidence about the location of sound sources is to average the GCC function across time frames [1, 57]. This approach, which will be referred to as *AVG*, has the potential advantage that activity corresponding to multiple sound sources can be considered per frame. Regarding the GMM-based localization model, the probability of sound-source activity is averaged over all frames.

Alternatively, the most likely source location can be estimated on a frame-by-frame basis, and all resulting short-time estimates can be pooled into a histogram [1, 3, 50]. Assuming that each of the active sound sources is most dominant across a reasonable number of time frames, the histogram will approximate the probability density function, PDF, of the true location of all active sound sources [3]. In addition, variations of time-frequency-based, T–F, histograms, might be considered where competing sources with different spectral contributions can be separated [2]. However, this implies that *a priori* knowledge about the spectral content of active sources is available. Moreover, as this chapter focuses on the localization of multiple speakers that show activity in a similar frequency range, the frame-based histogram technique, denoted as HIST, will be considered.

As discussed in [58], deciding what number of bins is used for the histogram analysis is a difficult task. While a high histogram resolution might be beneficial in scenarios with moderate reverberation, a higher variance of the TDOA estimates due to strong reverberation and noise can cause the histogram to have bimodal peaks, which will be erroneously interpreted as two active sources. Thus, the choice is a trade-off between between resolution and robustness. In accordance with [79], it has been decided to use 37 histogram bins to cover the azimuth range of $[-90, 90°]$ in steps of $5°$, where each individual bin is chosen to represent the time delay of the corresponding anechoic HRTF. To increase the resolution of the final azimuth estimate, exponential interpolation is applied to refine the maximum peak position of the histogram analysis [84].

Recently, a maximum likelihood, ML, framework for localization has been presented by Woodruff and Wang [78, 79], which jointly performs segregation and localization. Although small improvements were reported in comparison to the histogram approach [79], the computational complexity of the resulting search space is only feasible if the number of target sources is low, for instance three. Yet, because acoustic scenes with up to six competing speakers are used for evaluation in this chapter, the ML approach is not considered.

In order to address the problem of moving sources, other approaches aim at tracking the sound-source positions across time by employing *statistical particle filtering*, PF, techniques and *hidden Markov models*, HMMs [26, 61, 62, 76, 80]. But since the position of sound sources is assumed to be stationary throughout the time interval over which the localization information is integrated, these methods are not considered in this chapter. For the application of binaural analysis in combination with particle filtering, see [70], this volume.

The impact of temporal integration on sound-source localization is exemplified in Fig. 3 for a binaural mixture with three competing speakers in a reverberant environment with $T_{60} = 0.5$ s. More specifically, a comparison of two temporal integration strategies, namely, *averaging* versus *histogram*, is shown for two GCC-based methods, $W_{\text{CC}}$ and $W_{\text{PHAT}}$ weighting, and the GMM-based approach. In contrast to the conventional GCC–CC pattern shown in panel (a), the PHAT weighting in panel

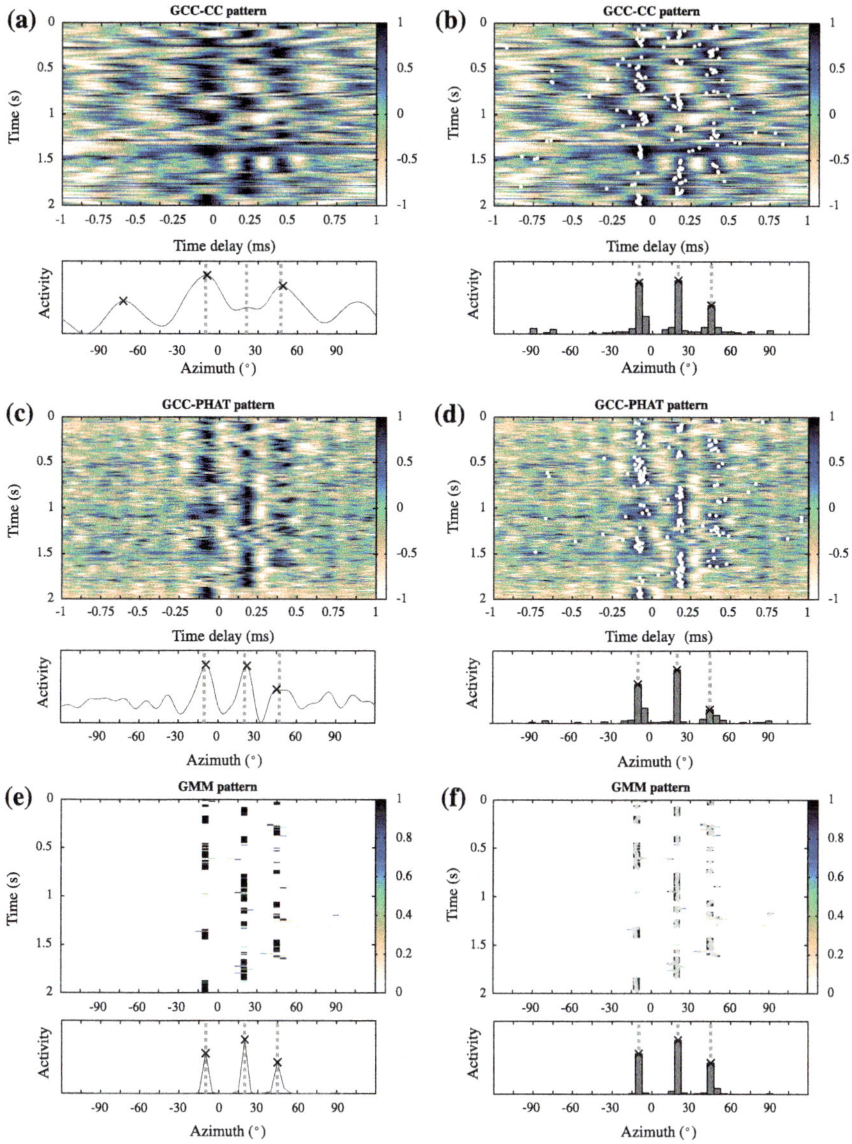


**Fig. 3** Influence of two temporal integration strategies on localizing three competing talkers positioned at $-10$, 20 and 35° in a reverberant room with $T_{60} = 0.5$ s. (**a, c**) Averaging of the GCC function across time. (**e**) GMM-based approach where the frame-based probability of sound-source direction is averaged over time. (**b, d, f**) Histogram-based integration. Dots represent the short-time localization estimates on a frame-by-frame basis. The estimated azimuth of the three speakers is marked by the *black crosses*, whereas their true position is indicated by *dashed vertical lines*

(c) produces sharper peaks, therefore, is able to resolve the positions of all three speakers. When using the histogram-based integration, both GCC–CC and GCC–PHAT achieve accurate predictions of the true speaker locations that are indicated by the vertical lines. The GMM-based approach shows the most prominent peaks of all methods at the true positions of the speakers for both integration strategies, where hardly any secondary peaks are visible.

To systematically compare the impact of these two temporal-integration strategies on localization performance, binaural mixtures of 2 s duration with up to six competing talkers are created, and the ability of various methods to predict the azimuth of all active speakers within $\pm 5°$ accuracy is evaluated. The following acoustical parameters were varied,

- Number of competing speakers, ranging from one to six
- Randomized azimuth within $[-90, 90°]$ with a minimum separation of 10°
- Simulated BRIRs ranging from $T_{60} = 0.2$ to $T_{60} = 1.05$ s

The experimental results are shown in Fig. 4 as a function of the reverberation time. Results are averaged over the number of competing speakers. In comparison to the frame-based localization accuracy reported in Sect. 4, the temporal integration significantly reduces the impact of reverberation on localization performance. In general, averaging the GCC pattern across time—dashed lines—is less robust than the histogram-based approach—solid lines—where short-time localization estimates are pooled across time. This is especially evident for the conventional GCC–CC method where the broad peaks in the accumulated GCC response prevent the detection of spatially close speakers—as seen in Fig. 3. Furthermore the averaging will integrate spurious peaks caused by reverberation, which might be erroneously considered as sound-source activity. In contrast, the histogram approach only considers the

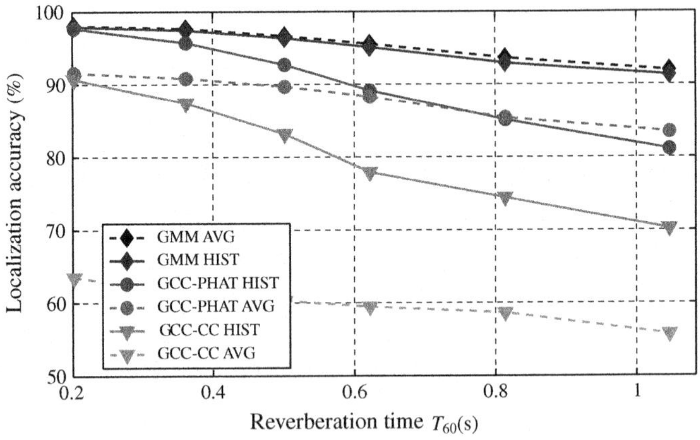


**Fig. 4** Average performance of localizing up to six competing speakers with an accuracy of $\pm 5°$ as a function of the reverberation time, $T_{60}$, for various approaches. The *dashed line* and the *solid line* indicate the two temporal integration strategies, namely, averaging and histogram-based integration

most salient source location on a frame-by-frame basis, thus focusing on the most reliable information. Consequently, the histogram-based integration of short-time localization estimates is very effective and considerably improves the robustness against the detrimental effect of reverberation. Regarding the GMM approach, a marginal benefit over the histogram-based integration is achieved when the probability of sound-source activity is averaged over time. This can be explained by the observation that the azimuth-dependent probability of sound-source activity is almost binary on a frame-by-frame basis—see Fig. 3—suggesting that each frame is approximately dominated by one individual sound source.

Overall, the PHAT weighting is substantially more robust than the conventional GCC–CC. Because the PHAT weighting already provides a sharp representation of the estimated time delay with strongly reduced secondary peaks—see Fig. 3—the additional improvement provided by the histogram integration is smaller than for the GCC–CC, most noticeably at short reverberation times. In anechoic conditions, GCC–PHAT HIST performs as well as the GMM approach. But with increasing reverberation time, the multi-conditional training and the joint analysis of ITDs and ILDs enable the GMM-based localization method to be more robust in reverberant multi-source scenarios.

This benefit of the GMM-based approach over the GCC–PHAT HIST system is presented in more detail in Fig. 5, where the localization performance is individually shown as a function of the number of competing talkers and the reverberation time. With an increasing number of speakers, the amount of reverberation has a stronger impact on the localization performance of the GCC–PHAT HIST approach, as seen in panel (a). This dependency of the localization performance on the reverberation time is substantially reduced in panel (b), showing the robustness of the GMM-based approach.

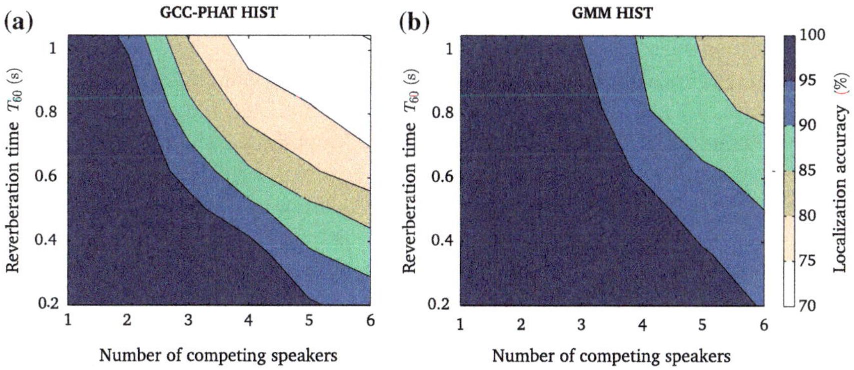


**Fig. 5** Sound-source localization accuracy in % as a function of the number of competing speakers and the reverberation time for two approaches. **a** GCC–PHAT HIST. **b** GMM HIST

**Table 1** Average localization accuracy in % for two sets of BRIRs. **a** Simulated BRIRs based on the KEMAR database [30]. **b** Measured BRIRs based on the HATS taken from the Surrey database [35]

| BRIRs | Methods | # competing speakers | | | | | | |
|---|---|---|---|---|---|---|---|---|
| | | One | Two | Three | Four | Five | Six | Mean |
| | GMM AVG | 100 | 99.5 | 98.0 | 94.9 | 91.6 | 88.7 | 95.4 |
| | GMM HIST | 100 | 99.4 | 97.6 | 94.5 | 90.9 | 87.9 | 95.0 |
| (a) Simulated, KEMAR | GCC–PHAT HIST | 100 | 97.8 | 92.2 | 86.7 | 82.1 | 79.3 | 89.7 |
| $T_{60} = \{0.36, 0.5, 0.62, 1.05\}$ s | GCC–PHAT AVG | 100 | 96.6 | 90.2 | 85.7 | 80.1 | 75.9 | 88.1 |
| | GCC–CC HIST | 96.8 | 85.2 | 79.5 | 74.3 | 72.1 | 70.3 | 79.7 |
| | GCC–CC AVG | 96.2 | 65.4 | 52.3 | 49.3 | 48.7 | 46.4 | 59.2 |
| | GMM AVG | 100 | 99.0 | 97.0 | 92.7 | 89.8 | 86.7 | 94.2 |
| | GMM HIST | 100 | 98.9 | 96.4 | 92.3 | 89.4 | 86.2 | 93.9 |
| (b) Measured, HATS | GCC–PHAT HIST | 99.9 | 98.3 | 95.7 | 89.7 | 84.8 | 80.7 | 91.5 |
| $T_{60} = \{0.32, 0.47, 0.68, 0.89\}$ s | GCC–PHAT AVG | 99.3 | 96.7 | 90.9 | 84.2 | 80.5 | 75.4 | 87.8 |
| | GCC–CC HIST | 93.9 | 82.7 | 76.4 | 71.0 | 69.1 | 67.3 | 76.7 |
| | GCC–CC AVG | 90.4 | 54.5 | 46.4 | 41.9 | 40.8 | 38.8 | 52.1 |

## Generalization to Real Recordings

An important question is to what extent the results obtained with simulated BRIRs can be compared to recorded BRIRs. Therefore, the localization performance of simulated BRIRs is compared with a set of measured BRIRs. To allow for a fair comparison, a subset of the simulated BRIRs, $T_{60} = \{0.36, 0.5, 0.62 \text{ and } 1.05\ \text{s}\}$, was selected such that the reverberation times are as close as possible to the measured BRIRs, $T_{60} = \{0.32, 0.47, 0.68 \text{ and } 0.89\ \text{s}\}$—see Sect. 2 for details. Furthermore, this comparison allows for assessing of how well the localization methods, which have all been trained on one particular artificial head, KEMAR, are able to generalize to the recorded BRIRs, which are based on a different artificial head, HATS.

The analysis involves binaural multi-source mixtures containing between one and six competing speakers that are created using both simulated and measured BRIRs. In Table 1, the localization accuracy of all tested methods is shown separately for (a), the simulated BRIRs based on the KEMAR artificial head and (b), the measured BRIRs based on the HATS artificial head. Results are averaged across all reverberation times. By comparing the mean values for different conditions, it can be seen that the overall performance for the measured BRIRs is fairly well reproduced by the set of simulated BRIRs. Thus, the differences in localization performance evaluated with simulated BRIRs are also valid for real life BRIRs. This is an important statement, justifying the usage of simulation tools for the development of localization algorithms. Furthermore, although the binaural-localization models are calibrated for one particular artificial head, localization performance does not degrade substantially when they are applied in the context of a different binaural-recording setup. Nevertheless, to minimize the sensitivity of the GMM-based localization model to a specific artificial head, the multi-conditional training stage can be readily adopted to include various sets of different HRTFs. Alternatively, it is possible to employ generic head models if only the coarse characteristics of the human head should be captured [13, 28]. Moreover, supervised learning of binaural cues can also be applied in the field of robotics, which is discussed in [5], this volume.

# 6 Localization of Speakers in the Presence of Interfering Noise

When all active sound sources are assumed to be speakers, it is reasonable to cluster the localization information across time and to treat the most significant peaks as estimated source positions. However, if speech activity is corrupted by environmental noise, the task becomes much more difficult and a prominent peak might as well correspond to the position of a noise source. Therefore, a distinction between speech and noise sources is required in order to reliably select sound-source activity that originates from active speakers. In the following, the application of binaural cues to the problem of sound-source segregation is considered.

## *6.1 Segregation of Individual Sound Sources*

In order to distinguish between speech and noise sources, the time-frequency, T–F, representation of multi-source mixtures will be segmented according to the estimated azimuth of sound sources. Assuming that sound sources are spatially separated, all T–F units that belong to one particular sound-source direction will be assumed to belong to the same acoustic source. This source segregation can subsequently be used to control a missing data classifier.

The GMM-based approach to binaural sound-source localization described in Sect. 3.3 was shown to accurately predict the location of up to six competing speakers in reverberation. Instead of using the most prominent peaks in the azimuth histogram as estimated sound-source positions, each local peak in the azimuth histogram will now be considered as a speech-source candidate. The corresponding histogram-bin indices are used to form a set of $M$ candidate positions, $L = \{\ell_1, \ldots, \ell_M\}$. Because the GMM-based approach extracts the likelihood of sound-source activity in individual frequency channels, the resulting spatial log-likelihood function, $\mathscr{L}(t, f, k)$, can be used to determine the contribution of all $M$ candidate positions on a time-frequency, T–F, basis as

$$\mathscr{M}_m(t, f) = \begin{cases} 1 \text{ if } m = \arg\max\limits_{k \in L} \mathscr{L}(t, f, k) \\ 0 \text{ otherwise} . \end{cases} \quad (8)$$

The resulting estimated *binary mask*, $\mathscr{M}_m(t, f)$, is a binary decision whether the $m$-th candidate has been the most dominant source in a particular T–F unit. The binary mask has a wide variety of different application areas, among them automatic speech and speaker recognition [21, 51, 52] as well as speech enhancement [40]. Due to the promising results that were obtained with the *ideal binary mask*, IBM, where the optimal segregation is known *a priori*, the estimation of the ideal binary mask has been proposed as the main goal of computational auditory scene analysis [75].

## Speech-Detection Module

In the following, it will be discussed how the estimated binary mask according to (8) can be used to select the most-likely speech sources among a set of candidate positions. The estimated binary mask can be used to perform *missing data*, MD, classification, where only a subset—indicated by the binary mask—of all time-frequency, T–F, units are evaluated by the classifier, namely those that are assumed to contain reliable information about the target source [21]. In this way, it is possible to selectively analyze and classify individual properties of one particular target source in the presence of other competing sources. Note that the concept of missing data is closely related to the auditory phenomenon of masking, where parts of the target source might be obscured and are, therefore, *missing* in the presence of other interfering sources [55, 75]. To distinguish between speech and noise sources, the amount of spectral fluctuation in individual Gammatone channels is a good descriptor that can be used to exploit the distinct spectral characteristic between speech and noise signals [49]. Based on a smoothed envelope, $e_f$, obtained by low-pass filtering the half-wave rectified output of the $f$th Gammatone channel with a time constant of 10 ms, the mean absolute deviation of the envelope over $B$ samples is calculated as

$$\mathcal{F}(t, f) = \frac{1}{B} \sum_{i=0}^{B-1} |e_f(t \cdot B/2 - i) - \bar{e}_f| \,, \tag{9}$$

where $\bar{e}_f$ reflects the mean envelope of the $t$-th frame. Note that the left and the right ear signals are averaged prior to envelope extraction. This feature, $\mathcal{F}(t, f)$, is subsequently modeled by two GMMs, denoted as $\lambda_{\text{Speech}}$ and $\lambda_{\text{Noise}}$, reflecting the feature distribution for a large amount of randomly selected speech and noise files [49, 51]. Incorporating this *a priori* knowledge about the spectral characteristics of speech and noise signals can be viewed as an implementation of schema-driven processing. Given the estimated mask, $\mathcal{M}_m$, the two GMMs, $\lambda_{\text{Speech}}$ and $\lambda_{\text{Noise}}$, and the extracted feature space, $\mathcal{F}$, the log-likelihood ratio of speech activity for the $m$-th candidate can be derived as

$$p_m = \log\left(\frac{p\left(\mathcal{F}|\lambda_{\text{Speech}}, \mathcal{M}_m\right)}{p\left(\mathcal{F}|\lambda_{\text{Noise}}, \mathcal{M}_m\right)}\right) . \tag{10}$$

In order to emphasize speech-source candidates that are more frequently active in the acoustic scene, the log-likelihood ratio of speech activity is weighted with the *a priori* probability of sound-sources activity, which is approximated by the normalized azimuth-histogram value of the corresponding candidate [49]. Although other weighting schemes can be considered, it was found that putting equal weight on the log-likelihood ratio obtained from the MD classifier and on the histogram-based localization information leads to good results. Finally, these weighted log-likelihood ratios of all $M$ candidates are ranked in descending order, and the azimuth positions corresponding to the highest values are used to reflect the most likely speech source positions. In this way, the plain peak selection based on the most dominant localization information is supported by evidence about the source characteristic, being either speech-like or noise-like, therefore allowing for a distinction between speech and noise signals. The localization based on this *speech-detection module* will be referred to as *GMM SDM*.

Of course, other unique properties of speech signals might be considered at this stage as well, and a joint analysis of multiple complementary features is conceivable to further improve the ability to distinguish between speech and noise signals. As reported by [40, 42], the *amplitude-modulation spectrogram* is an effective feature that provides a reliable discrimination between speech and noise. Furthermore, is has been shown that also the distribution of reliable T–F units in the estimated binary mask, $\mathscr{M}_m$, contains information about the type of source, where the binary pattern shows a more compact representation for speech sources than for noise signals [48].

## 6.2 Influence of the Spatial Diffuseness of Interfering Noise

The impact of environmental noise on the ability to localize speakers does not only depend on the overall signal-to-noise ratio, but furthermore on its spatial distribution. Nevertheless, the vast majority of studies have investigated the influence of diffuse noise on sound-source localization [1, 16, 17, 79], which complies with the assumption of the GCC-based approach. However, the assumption of a diffuse noise field is not necessarily realistic for a real-life scenario. As recently analyzed by [58], real recordings of noise scenarios show a substantial amount of correlation, where the maximum value of the normalized cross-correlation function has been used as an indication of the amount of spatial correlation between the two microphones. Their experimental results showed that the conventional GCC method was superior to the PHAT weighting for acoustic conditions in which the noise had a high degree of correlation [58].

Therefore, the aim of this section is to investigate the impact of noise diffuseness on speaker-localization accuracy. More specifically, the influence of the noise characteristic is analyzed by systematically varying the amount of correlation of the noise between the left and the right ear signals. Therefore, different realizations of a particular noise type are filtered with BRIRs corresponding to a predefined number of randomly-selected azimuth directions. Note that for a given noise signal, each azimuth direction may be only selected once. By systematically varying the number of azimuth directions that contribute to the overall noise field from 1 to 37, the spatial characteristic of the resulting noise can be gradually changed from a compact noise source located at one particular azimuth direction to a noise field where the energy is uniformly distributed across all 37 sound-source directions, thus approximating a diffuse noise field. The spatial diffuseness of the resulting noise field is specified by relating the number of noise realizations that contribute to the overall noise signal to the total number of discrete sound-source directions, ranging from $100 \cdot \frac{1}{37} = 2.7$ to $100 \cdot \frac{37}{37} = 100\,\%$.

In order to quantify the amount of correlation between the left and the right ear signals, the short-time coherence is estimated for 20 ms frames. The resulting coherence is averaged over time and shown in Fig. 6 as a function of frequency for noise signals consisting of 1, 3, 9, 19 and 37 superimposed realizations of randomly-selected azimuth directions. Whereas the average coherence in panel (a) is based on noise signals in anechoic conditions, panel (b) shows the additional influence of reverberation, namely, $T_{60} = 0.36$ s. It can be observed that the coherence functions systematically decrease with increasing number of noise realizations that contribute to the overall noise field. Furthermore, when comparing panel (a) and (b), it can be seen that in addition to the number of noise realizations, reverberation has a decorrelating effect, decreasing the correlation between the left and the right ear signals.

Now, the influence of interfering noise on localization performance is analyzed for binaural multi-talker mixtures with up to four competing talkers. Speech is corrupted with noise with the spatial distribution being gradually changed from compact noise to spatially diffuse noise. The following acoustic conditions are varied,

- Number of concurrent speakers ranging from one to four
- Number of interfering noise sources, 1, 3, 9, 19 and 37

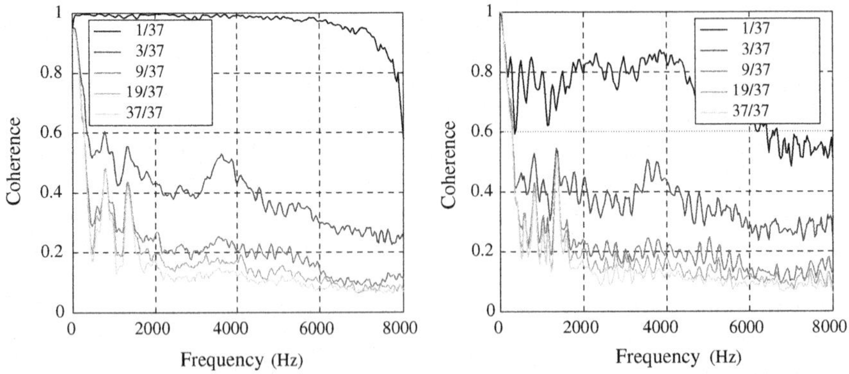


**Fig. 6** Average short-time coherence estimates between the *left* and the *right* ear signals in response to various simulated noise signals. The individual noise signals consist of 1, 3, 9, 19 and 37 superimposed realizations of factory noise excerpts and are filtered with BRIRs corresponding to randomly selected azimuth directions. **a** Results for $T_{60} = 0$ s. **b** Results for $T_{60} = 0.36$ s. See Sect. 6.2 for details

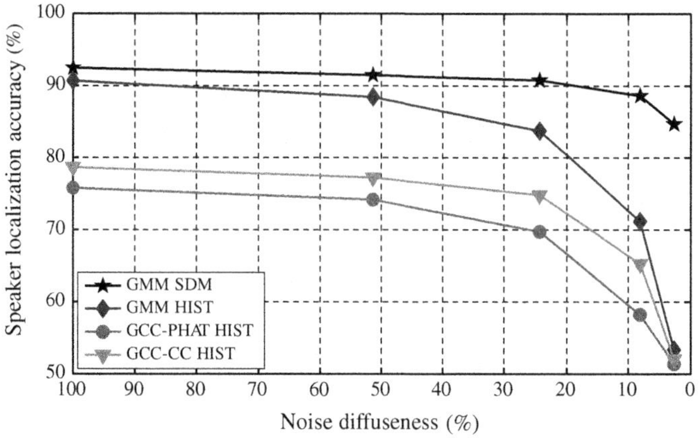


**Fig. 7** Accuracy of localizing up to four competing speakers in a reverberant room, $T_{60} = 0.36$ s, as a function of the spatial diffuseness of the interfering noise. Performance is averaged over three SNRs, that is, 10, 5 and 0 dB, and three types of background noise, namely, factory, babble and speech-shaped noise

- Three noise types, namely, factory noise, babble noise and speech-shaped noise
- SNR between speech and noise, that is, 10, 5 and 0 dB

The performance of localizing up to four competing talkers within $\pm 5°$ accuracy is presented in Fig. 7 as a function of the spatial diffuseness of the noise. Results are averaged over the number of competing talkers, the three noise types and the three SNRs. In general, the presence of noise imposes serious problems for the GCC-based approaches using either the $W_{\text{CC}}$ or the $W_{\text{PHAT}}$ weighting. In contrast to the results presented in Sect. 5, the PHAT weighting performs worse than the classical GCC–CC. This may be attributed to the whitening process, which equally weights all frequency components, thereby also amplifying the noise components. These results are in line with the observation of Perez-Lorenzo et al. [58], where the classical GCC–CC was reported to perform more robustly than the PHAT weighting for scenarios with correlated noise. Although GMM HIST appears to be more robust, the limiting factor that is shared by all of the aforementioned methods is that they solely exploit localization information. However, the most energetic components of speech are sparsely distributed in the presence of noise [20], thereby only a limited set of spectro-temporal units will be dominated by the sound-source direction of the speakers. Thus, as soon as the noise gets more directional, the noise energy is more compactly associated with a particular sound-source direction. As a result, the most dominant localization information will at a certain SNR inevitably correspond to the position of the interfering noise, which in turn reduces the overall speaker-localization accuracy. This observation corroborates the need for a distinction between speech and noise sources, especially for scenarios where the interfering noise has strong directional components. Such a distinction can be realized by using the speech-detection module described in Sect. 6.1, which effectively combines the localization analysis with a

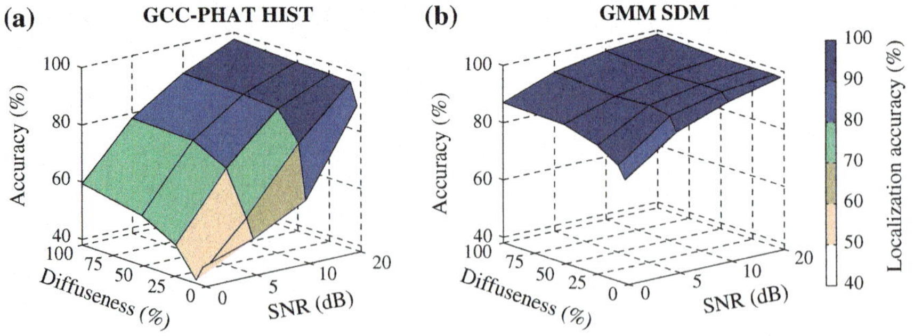


**Fig. 8** Accuracy of localizing up to four competing speakers in a reverberant room, $T_{60} = 0.36$ s, as a function of the spatial diffuseness of the interfering noise and the signal-to-noise ratio

classification stage for selecting the most-likely speech sources according to (10). The experimental results shown in Fig. 7 demonstrate that the GMM SDM allows for a robust localization of up to four competing speakers, where the impact of directional noise is drastically reduced.

In Fig. 8, the localization performance of the two approaches GCC–PHAT HIST and GMM SDM is shown as a function of the SNR and the noise diffuseness. It can be seen that the performance of the PHAT approach systematically decreases with decreasing SNR, quite notably already at SNRs of 5–10 dB. Furthermore, the PHAT approach clearly suffers from interfering noise that is less diffuse, but correlated between the left and the right ears. In contrast, the GMM SDM approach that attempts to separate the contribution of individuals sources on the basis of common spatial location in combination with employing a speech-detection module achieves robust localization performance over a wide range of experimental conditions. In summing up, it can be stated that interfering noise signal with a high degree of directional components will trigger the correlation-based approaches to localize the position of the noise components. Therefore, a distinction between speech and noise signals is required in order to enable a high speaker-localization accuracy in noisy environments.

## 7 Conclusions

This chapter presented an overview of binaural approaches to localizing multiple competing speakers in adverse acoustic scenarios. A fundamental limitation of many methods is that they assume single-path wave propagation, whereby performance inevitably decreases in the presence of reverberation and multiple competing sources. It was demonstrated that it is possible to incorporate the uncertainty of binaural cues in response to complex acoustic scenarios into a probabilistic model for robust sound-source localization, thus significantly improving the localization performance in the presence of reverberation. To reliably estimate the location of multiple competing sound sources in reverberant environments, a histogram analysis of short-time localization estimates can substantially reduce the severe effect of reverberation. Furthermore, a comparison between simulated and recorded BRIRs has confirmed that the presented model produces accurate localization estimates for real-life scenarios and is able to generalize to an unseen artificial head, for which the system was not trained for. In general, considering both the impact of reverberation and noise imposes serious challenges for localization algorithms. A thorough analysis highlighted that in particular the spatial distribution of the noise field is a very important factor that strongly influences the performance of correlation-based localization algorithms, being most detrimental for GCC-based approaches if the interfering noise has a high degree of correlation between the left- and the right-ear signals. This problem can be overcome by separating the contribution of individual sound sources by means of estimating the binary mask. This binary mask can subsequently be used to control a missing data classifier, which is able to distinguish between sound-source activity emerging from speech and noise sources. It was shown that this joint analysis of localization information and source characteristic can be effectively used to achieve robust sound-source localization in very challenging acoustic scenarios.

**Acknowledgments** The authors are indebted to two anonymous reviewers for their constructive suggestions.

## References

1. P. Aarabi. Self-localizing dynamic microphone arrays. *IEEE Trans. Sys., Man, Cybern., C*, 32(4):474–484, Nov. 2002.
2. P. Aarabi and S. Mavandadi. Robust sound localization using conditional time-frequency histograms. *Inf. Fusion*, 4(2):111–122, Sep. 2003.
3. P. Aarabi and S. Zaky. Iterative spatial probability based sound localization. In *Proceedings of the 4th World Multi-conference on Circuits, Systems, Computers and Communications*, Athens, Greece, Jul. 2000.
4. J. B. Allen and D. A. Berkley. Image method for efficiently simulating small-room acoustics. *J. Acoust. Soc. Am.*, 65(4):943–950, Apr. 1979.
5. S. Argentieri, A. Portello, M. Bernard, P. Danès, and B. Gas. Binaural systems in robotics. In J. Blauert, editor, *The technology of binaural listening*, chapter 9. Springer, Berlin-Heidelberg-New York NY, 2013.
6. R. Baumgartner, P. Majdak, and B. Laback. Assessment of sagittal-plane sound-localization performance in spatial-audio applications, chapter 4. In J. Blauert, editor, *The technology of binaural listening*. Springer–Berlin–Heidelberg–New York NY, 2013.
7. J. Benesty, J. Chen, and Y. Huang. Time-delay estimation via linear interpolation and cross correlation. *IEEE Trans. Speech Audio Process*. 12(5):509–519, 2004.
8. M. Bodden. Modeling human sound-source localization and the cocktail-party-effect. Acta *Acust./Acustica*, 1(1):43–55, 1993.
9. J. Braasch. Modelling of binaural hearing. In J. Blauert, editor, *Communication acoustics*, chapter 4, pages 75–108. Springer, Berlin, Germany, 2005.
10. A. S. Bregman. *Auditory scene analysis: The perceptual organization of sound*. The MIT Press, Cambridge, MA, USA, 1990.

11. A. W. Bronkhorst. The cocktail party phenomenon: A review of research on speech intelligibility in multiple-talker conditions. *Acustica*, 86:117–128, 2000.
12. A. W. Bronkhorst and R. Plomp. Effect of multiple speechlike maskers on binaural speech recognition in normal and impaired hearing. *J. Acoust. Soc. Am.*, 92(6):3132–3139, Dec. 1992.
13. C. P. Brown and R. O. Duda. A structural model for binaural sound synthesis. *IEEE Trans. Speech Audio Process.*, 6(5):476–488, Sep. 1998.
14. G. J. Brown and M. Cooke. Computational auditory scene analysis. *Comput. Speech Lang.*, 8(4):297–336, Oct. 1994.
15. G. C. Carter, A. H. Nuttall, and P. G. Cable. The smoothed coherence transform. *Proceedings of the IEEE*, 61(10):1497–1498, Oct. 1973.
16. J. Chen, J. Benesty, and Y. Huang. Robust time delay estimation exploiting redundancy among multiple microphones. *IEEE Trans. Acoust., Speech, Signal Process.*, 11(6):549–557, 2003.
17. J. Chen, J. Benesty, and Y. A. Huang. Performance of GCC- and AMDF-based time-delay estimation in practical reverberant environments. *J. Appl. Signal Process.*, 1:25–36, 2005.
18. J. Chen, J. Benesty, and Y. A. Huang. Time delay estimation in room acoustic environments: An overview. *J. Appl. Signal Process.*, 2006:1–19, 2006.
19. E. C. Cherry. Some experiments on the recognition of speech, with one and two ears. *J. Acoust. Soc. Am.*, 25(5):975–979, Sep. 1953.
20. M. Cooke. A glimpsing model of speech perception in noise. *J. Acoust. Soc. Am.*, 199(3):1562–1573, Mar. 2006.
21. M. Cooke, P. Green, L. Josifovski, and A. Vizinho. Robust automatic speech recognition with missing and unreliable acoustic data. *Speech Commun.*, 34:267–285, 2001.
22. M. Cooke and T.-W. Lee. Speech separation and recognition competition. URL http://staffwww.dcs.shef.ac.uk/people/M. Cooke/SpeechSeparationChallenge.htm, accessed on 15th January 2013, 2006.
23. C. J. Darwin. Auditory grouping. *Trends Cogn. Sci.*, 1(1):327–333, Dec. 1997.
24. M. S. Datum, F. Palmieri, and A. Moiseff. An artificial neural network for sound localization using binaural cues. *J. Acoust. Soc. Am.*, 100(1):372–383, Jul. 1996.
25. J. DiBiase, H. Silverman, and M. Brandstein. Robust localization in reverberant rooms. In M. Brandstein and D. Ward, editors, *Microphone arrays: Signal processing techniques and applications*, chapter 8, pages 157–180. Springer, Berlin, Germany, 2001.
26. M. Dietz, S. D. Ewert, and V. Hohmann. Auditory model based direction estimation of concurrent speakers from binaural signals. *Speech Commun.*, 53(5):592–605, 2011.
27. G. Doblinger. Localization and tracking of acoustical sources. In E. Haensler and G. Schmidt, editors, *Topics in acoustic echo and noise control*, chapter 4, pages 91–124. Springer, Berlin, Germany, 2006.
28. R. O. Duda and W. L. Martens. Range dependence of the response of a spherical head model. *J. Acoust. Soc. Am.*, 104(5):3048–3058, Nov. 1998.
29. C. Faller and J. Merimaa. Source localization in complex listening situations: Selection of binaural cues based on interaural coherence. *J. Acoust. Soc. Am.*, 116(5):3075–3089, Nov. 2004.
30. W. G. Gardner and K. D. Martin. HRTF measurements of a KEMAR dummy-head microphone. Technical report, # 280, MIT Media Lab, Perceptual Computing, Cambridge, MA, USA, 1994.
31. B. R. Glasberg and B. C. J. Moore. Derivation of auditory filter shapes from notched-noise data. *Hear. Res.*, 47(1–2):103–138, Aug. 1990.
32. T. Gustafsson, B. D. Rao, and M. Trivedi. Analysis of time-delay estimation in reverberant environments. In *Proc. ICASSP*, pages 2097–2100, Orlando, Florida, USA, May 2002.
33. S. Harding, J. Barker, and G. Brown. Mask estimation for missing data speech recognition based on statistics of binaural interaction. *IEEE Trans. Audio, Speech, Lang. Process.*, 14(1):58–67, Jan. 2006.
34. J.-S. Hu and W.-H. Liu. Location classification of nonstationary sound sources using binaural room distribution patterns. *IEEE Trans. Audio, Speech, Lang. Process.*, 17(4):682–692, May 2009.

35. C. Hummersone, R. Mason, and T. Brookes. Dynamic precedence effect modelling for source separation in reverberant environments. *IEEE Trans. Audio, Speech, Lang. Process.*, 18(7):1867–1871, Sep. 2010.
36. G. Jacovitti and G. Scarano. Discrete time techniques for time delay estimation. *IEEE Trans. Signal Process.*, 41(2):525–533, Feb. 1993.
37. M. Jeub, M. Schäfer, and P. Vary. A binaural room impulse response database for the evaluation of dereverberation algorithms. *Proc. Intl. Conf. Digital Signal Process.* (DSP), pages 1–5, Jul. 2009.
38. A. Jourjine, S. Rickard, and Yilmaz. Blind separation of disjoint orthogonal signals: Demixing N sources from 2 mixtures. In *Proc. ICASSP*, pages 2985–2988, Istanbul, Turkey, Jun. 2000.
39. H. Kayser, S. D. Ewert, T. Rohdenburg, V. Hohmann, and B. Kollmeier. Database of multichannel in-ear and behind-the-ear head-related and binaural room impulse responses. *EURASIP J. Adv. Sig. Proc.*, 2009.
40. G. Kim, Y. Lu, Y. Hu, and P. C. Loizou. An algorithm that improves speech intelligibility in noise for normal-hearing listeners. *J. Acoust. Soc. Am.*, 126(3):1486–1494, Sep. 2009.
41. C. H. Knapp and G. C. Carter. The generalized correlation method for estimation of time delay. *IEEE Trans. Acoust., Speech, Signal Process.*, ASSP-24(4):320–327, Aug. 1976.
42. B. Kollmeier and R. Koch. Speech enhancement based on physiological and psychoacoustical models of modulation perception and binaural interaction. *J. Acoust. Soc. Am.*, 95(3):1593–1602, Mar. 1994.
43. E. H. A. Langendijk and A. W. Bronkhorst. Contribution of spectral cues to human sound localization. *J. Acoust. Soc. Am.*, 112(4):1583–1596, Oct. 2002.
44. W. Lindemann. Extension of a binaural cross-correlation model by contralateral inhibition. I. Simulation of lateralization for stationary signals. *J. Acoust. Soc. Am.*, 80(6):1608–1622, Dec. 1986.
45. W. Lindemann. Extension of a binaural cross-correlation model by contralateral inhibition. II. The law of the first wave front. *J. Acoust. Soc. Am.*, 80(6):1623–1630, Dec. 1986.
46. R. F. Lyon. A computational model of binaural localization and separation. In *Proc. ICASSP*, pages 1148–1151, Boston, Massachusetts, USA, Apr. 1983.
47. N. Madhu and R. Martin. Acoustic source localization with microphone arrays. In R. Martin, U. Heute, and C. Antweiler, editors, *Advances in Digital Speech Transmission*, chapter 6, pages 135–170. Wiley, 2008.
48. T. May and S. van de Par. Blind estimation of the number of speech sources in reverberant multisource scenarios based on binaural signals. *in Proc. IWAENC*, Aachen, Germany, Sep. 2012.
49. T. May, S. van de Par, and A. Kohlrausch. Binaural detection of speech sources in complex acoustic scenes. In *Proc. WASPAA*, pages 241–244, New Paltz, NY, USA, Oct. 2011.
50. T. May, S. van de Par, and A. Kohlrausch. A probabilistic model for robust localization based on a binaural auditory front-end. *IEEE Trans. Audio, Speech, Lang. Process.*, 19(1):1–13, Jan. 2011.
51. T. May, S. van de Par, and A. Kohlrausch. A binaural scene analyzer for joint localization and recognition of speakers in the presence of interfering noise sources and reverberation. *IEEE Trans. Audio, Speech, Lang. Process.*, 20(7):2016–2030, Sep. 2012.
52. T. May, S. van de Par, and A. Kohlrausch. Noise-robust speaker recognition combining missing data techniques and universal background modeling. *IEEE Trans. Audio, Speech, Lang. Process.*, 20(1):108–121, Jan. 2012.
53. R. Meddis, M. J. Hewitt, and T. M. Shackleton. Implementation details of a computation model of the inner hair-cell auditory-nerve synapse. *J. Acoust. Soc. Am.*, 87(4):1813–1816, Apr. 1990.
54. R. Meddis and E. A. Lopez-Poveda. Auditory periphery: From pinna to auditory nerve. In R. Meddis, E. A. Lopez-Poveda, R. R. Fay, and A. N. Popper, editors, *Computational models of the auditory system*, volume 35, chapter 2, pages 7–38. Springer, New York, 2010.
55. B. C. J. Moore. *An introduction to the psychology of hearing*. Academic Press, San Diego, California, USA, 5th edition, 2003.

56. J. Nix and V. Hohmann. Sound source localization in real sound fields based on empirical statistics of interaural parameters. *J. Acoust. Soc. Am.*, 119(1):463–479, Jan. 2006.
57. K. J. Palomäki, G. J. Brown, and D. L. Wang. A binaural processor for missing data speech recognition in the presence of noise and small-room reverberation. *Speech Commun.*, 43(4):361–378, 2004.
58. J. Perez-Lorenzo, R. Viciana-Abad, P. Reche-Lopez, F. Rivas, and J. Escolano. Evaluation of generalized cross-correlation methods for direction of arrival estimation using two microphones in real environments. *Appl. Acoust.*, 73(8):698–712, Aug. 2012.
59. V. C. Raykar, B. Yegnanarayana, S. R. M. Prasanna, and R. Duraiswami. Speaker localization using excitation source information in speech. *IEEE Trans. Speech Audio Process.*, 13(5):751–761, Sep. 2005.
60. L. Rayleigh. On our perception of sound direction. *Philos. Mag.*, 13:214–232, 1907.
61. N. Roman and D. L. Wang. Binaural tracking of multiple moving sources. In *Proc. ICASSP*, volume 5, pages 149–152, Hong Kong, China, Apr. 2003.
62. N. Roman and D. L. Wang. Binaural tracking of multiple moving sources. *IEEE Trans. Audio, Speech, Lang. Process.*, 16(4):728–739, 2008.
63. N. Roman, D. L. Wang, and G. J. Brown. Speech segregation based on sound localization. *J. Acoust. Soc. Am.*, 114(4):2236–2252, Oct. 2003.
64. R. Roy and T. Kailath. ESPRIT - estimation of signal parameters via rotational invariance techniques. *IEEE Trans. Acoust., Speech, Signal Process.*, 37(7):984–995, Jul. 1989.
65. S. M. Schimmel, M. F. Müller, and N. Dillier. A fast and accurate "shoebox" room acoustics simulator. In *Proc. ICASSP*, pages 241–244, Taipei, Taiwan, Apr. 2009.
66. R. O. Schmidt. Multiple emitter location and signal parameter estimation. *IEEE Trans. Antennas Propagat.*, AP-34(3):276–280, Mar. 1986.
67. M. R. Schroeder. New method for measuring reverberation time. *J. Acoust. Soc. Am.*, 37(3):409–412, 1965.
68. C. L. Searle, L. D. Braida, D. R. Cuddy, and M. F. Davis. Binaural pinna disparity: another auditory localization cue. *J. Acoust. Soc. Am.*, 57(2):448–455, Feb. 1975.
69. T. M. Shackleton, R. Meddis, and M. J. Hewitt. Across frequency integration in a model of lateralization. *J. Acoust. Soc. Am.*, 91(4):2276–2279, Apr. 1992.
70. C. Spille, B. Meyer, M. Dietz, and V. Hohmann. Binaural scene analysis with multi-dimensional statistical filters, chapter 6. In J. Blauert, editor, *The technology of binaural listening*. Springer, Berlin-Heidelberg-New York NY, 2013.
71. R. M. Stern, A. S. Zeiberg, and C. Trahiotis. Lateralization of complex binaural stimuli: A weighted-image model. *J. Acoust. Soc. Am.*, 84(1):156–165, Jul. 1988.
72. C. J. Sumner, E. A. Lopez-Poveda, L. P. O'Mard, and R. Meddis. A revised model of the inner-hair cell and auditory-nerve complex. *J. Acoust. Soc. Am.*, 111(5):2178–2188, May 2002.
73. S. Tervo and T. Lokki. Interpolation methods for the SRP-PHAT algorithm. In *Proc. IWAENC*, Seattle, Washington, USA, Sep. 2008.
74. A. P. Varga, H. J. M. Steeneken, M. Tomlinson, and D. Jones. The NOISEX-92 study on the effect of additive noise on automatic speaker recognition. Technical report, Speech Research Unit, Defence Research Agency, Malvern, UK, 1992.
75. D. L. Wang and G. Brown, editors. *Computational auditory scene analysis: Principles, algorithms and applications*. John Wiley & Sons, Hoboken, NJ, USA, 2006.
76. D. B. Ward, E. A. Lehmann, and R. C. Williamson. Particle filtering algorithms for tracking an acoustic source in a reverberant environment. *IEEE Trans. Speech Audio Process.*, 11(6):826-836, Nov. 2003.
77. V. Willert, J. Eggert, J. Adamy, R. Stahl, and E. Körner. A probabilistic model for binaural sound localization. *IEEE Trans. Sys., Man, Cybern.*, B, 36(5):982–994, Oct. 2006.
78. J. Woodruff and D. L. Wang. Sequential organization of speech in reverberant environments by integrating monaural grouping and binaural localization. *IEEE Trans. Audio, Speech, Lang. Process.*, 18(7):1856–1866, Sep. 2010.
79. J. Woodruff and D. L. Wang. Binaural localization of multiple sources in reverberant and noisy environments. *IEEE Trans. Audio, Speech, Lang. Process.*, 20(5):1503–1512, Jul. 2012.

80. J. Woodruff and D. L. Wang. Binaural detection, localization, and segregation in reverberant environments based on joint pitch and azimuth cues. *IEEE Trans. Audio, Speech, Lang. Process.*, 21(4):806–815, Apr. 2013.
81. O. Yilmaz and S. Rickard. Blind separation of speech mixtures via time-frequency masking. *IEEE Signal Process.* Lett., 52(7):1830–1847, Jul. 2004.
82. P. Zakarauskas and M. S. Cynader. A computational theory of spectral cue localization. *J. Acoust. Soc. Am.*, 94(3):1323–1331, Sep. 1993.
83. C. Zhang, D. Florêncio, and Z. Zhang. Why does PHAT work well in low noise, reverberative environments? In *Proc. ICASSP*, pages 2565–2568, 2008.
84. L. Zhang and X. Wu. On cross correlation based discrete time delay estimation. In *Proc. ICASSP*, volume 4, pages 981–984, Philadelphia, Pennsylvania, USA, 2005.

# Predicting Binaural Speech Intelligibility in Architectural Acoustics

J. F. Culling, M. Lavandier and S. Jelfs

## 1 Introduction

### 1.1 Measures of Acoustic Quality

Speech intelligibility can be impaired by poor room acoustics. This may happen as a result of distortion of the speech signal itself, because many delayed versions of the speech are summed at the ear, causing both spectral coloration and temporal smearing. Reverberation may also exacerbate the effects of background masking noise by impeding the processes by which the auditory system can overcome such masking. The relative importance of these effects depends on the type of listening situation. However, when listening to speech and noise from equidistant sources, it has been shown that the effects of reverberation on noise masking occur at lower levels of reverberation, and thus occur more readily, than the distorting effects on the speech [1].

In the planning and regulation of buildings, the acoustic quality of a room is generally summarized using statistics such as the reverberation time, $T_{60}$, and noise level. For instance, in the U.K., *Building Bulletin 93*, BB93, specifies upper limits for unoccupied ambient noise levels and for $T_{60}$ in different types of classrooms. Ambient noise may vary across the space, in which case an average measure is needed, but the $T_{60}$ should, at least in principle, be independent of measurement position. While single-value indices are convenient, they may not always accurately

J. F. Culling (✉)
School of Psychology, Cardiff University, Cardiff, UK
e-mail: cullingj@cf.ac.uk

M. Lavandier
Laboratoire Génie Civil et Bâtiment, Université de Lyon, Lyons, France

S. Jelfs
Philips Research Europe, Eindhoven, The Netherlands

J. Blauert (ed.), *The Technology of Binaural Listening*, Modern Acoustics and Signal Processing, DOI: 10.1007/978-3-642-37762-4_16,

reflect the speech intelligibility that will result. As will be demonstrated below, the $T_{60}$ in particular, can be misleading.

In some circumstances, the *speech transmission index* [2], STI, or the useful-to-detrimental ratio [3] may be considered. The STI evaluates the degree to which amplitude modulation of speech survives the temporally smearing effect of reverberation. It is dependent upon the positions of the speech source and the receiver within the room. This makes it appropriate for lecture theatres and public address systems, for instance, where one individual communicates from a fixed location to an audience. The STI can be evaluated for each location in the listening space in order to ensure that adequate intelligibility is achieved in all listening locations. It can also produce predictions for intelligibility in noise, provided that noise is continuous and totally diffuse. However, these methods fail to produce accurate results where noise sources are nearby, such as in a busy social environment, where noise sources, such as background voices, may not be diffuse.

## 1.2 Binaural Speech Intelligibility

When speech and noise sources are spatially separated, speech intelligibility always improves compared to a situation in which they are co-located. This effect is known as *spatial release from masking*, SRM, and is likely related to a combination of at least two binaural processes, binaural unmasking and better-ear listening [4, 5]. Since speech and noise generally come from different sources, some SRM occurs in almost all natural listening situations. However, SRM is adversely affected by reverberation [6] and by the presence of multiple noise sources [4]. In order to accurately predict intelligibility in noisy rooms, it is therefore essential to take into account SRM and the influence that reverberation has upon it. This task is complicated by the dependence of these effects on the exact spatial layout of the speech and noise sources—it is not possible to characterize a room as facilitating a given level of SRM. However, it has now become possible to predict SRM for any given situation with considerable speed and accuracy.

Two very successful models of SRM have been developed by research groups in Oldenburg [7, 8] and Cardiff [9–11]. The current version of the Oldenburg model is the more comprehensive, because it can accommodate modulated masking noises and also hearing-impaired listeners. However, this chapter will employ the Cardiff model, which is well adapted to the rapid computation needed for many of the analyses below. This model explicitly evaluates the benefit to intelligibility expected from binaural unmasking and better-ear listening and regards their effects on the *speech reception threshold*, SRT, in noise as additive in decibels. The model has been applied to a wide range of data sets from the literature in both anechoic conditions with multiple noise sources [10] and in reverberant situations [9, 11] and generally provides a very high correlation with the empirical data—see Table 1. At present this model is only strictly applicable to continuous random noise sources. In order to apply them to more structured masking noises, such as voices, additional perceptual

**Table 1** Summary of correlations between empirically measured SRTs from different experiments and corresponding model predictions

| Experiment | Room | Number of noise sources | Correlation |
|---|---|---|---|
| Bronkhorst and Plomp [5] | Anechoic | 1 | 0.86 |
| Bronkhorst and Plomp [12] | Anechoic | 1–6 | 0.93 |
| Peissig and Kollmeier [13] | Anechoic | 1–3 | 0.98 |
| Hawley et al. [4] | Anechoic | 1–3 | 0.99 |
| Culling et al. [14] | Anechoic | 3 | 0.94 |
| Lavandier and Culling [9] | Simulated room #1 | 1 | 0.91 |
| | Simulated room #2 | 1 | 0.98 |
| Beutelmann and Brand [7] | Two real rooms | 1 | 0.99 |
| Lavandier et al. [11] | One real room | 1 | 0.98 |
| | Four real rooms | 1 | 0.98 |
| | One real room | 3 | 0.95 |

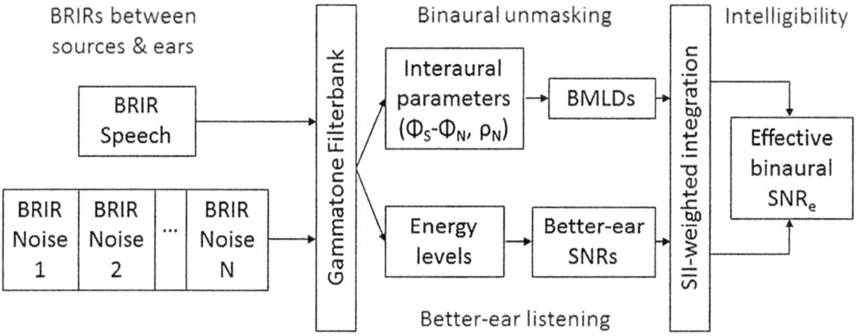


**Fig. 1** Schematic illustration of the binaural intelligibility model. $\Phi_S$ and $\Phi_N$... interaural phase differences of speech and noise, $\rho_N$... interaural coherence of the noise, BMLD ... binaural masking level difference

processes will need to be considered. However, notwithstanding this limitation, the model can make interesting predictions about the effects of room design and layout on communication.

## *1.3 Anatomy of the Binaural-Intelligibility Model*

As noted above, the binaural model is based upon additive contributions from better-ear listening and binaural unmasking—Fig. 1. The model takes as input *binaural room impulse responses*, BRIRs, between the listening location and each of the sound-source locations. Its output is an *effective signal-to-noise ratio*, SNR$_e$, that takes these processes into account. The remainder of this section describes how the

BRIRs are to be prepared and processed in order to generate the $SNR_e$ and may not be of interest to the non-technical reader, who can skip to Sect. 1.4.

BRIRs may be generated by an acoustic model of a virtual room using suitable acoustic modeling software[1] or they may be recorded in a real room using an acoustic manikin. Where multiple noise sources are present, the impulse responses for all these sources are concatenated into one long impulse response. Concatenation has the effect of summing the frequency-dependent energy of each contributing impulse response, and generating an averaged cross-correlation function. It may seem intuitively reasonable to add together the BRIRs, just as one would add together different masking noises. However, summing directly the BRIRs would result in spectral distortion due to mutual interference, which does not occur when summing statistically independent interfering noises that have been convolved with those BRIRs. Only in the particular case of different sound sources, such as loudspeakers, driven by the same acoustic waveform, should the BRIRs be summed, to take into account the interference between these correlated sound sources at the ears.

The impulse responses for speech and noise(s) are separately filtered into different frequency channels, which are processed independently. The two contributions to intelligibility from binaural hearing are then modeled, namely, *better-ear listening* and *binaural unmasking*.

Better-ear listening simply reflects listeners' ability to pick up sound from the ear with the better signal-to-noise ratio. Interaural differences in SNR can occur as a result of head shadow, where the masking noise is occluded at one ear by the head, and also of room coloration, where frequency-dependent room absorption and complex interference between multiple room reflections creates different spectral distortions at each ear. Within each frequency channel the SNRs in dB at each ear are derived from the relative total energies in the filtered noise and speech BRIRs at that ear. The higher SNR of the two is selected as the better-ear SNR for that frequency.

Binaural unmasking is a psychoacoustic phenomenon in which the brain exploits the differences in interaural phase between signal and noise sources in order to improve detection or identification of the signal. These differences in phase are caused by differences in path distance to each ear. The size of the improvement is known as the *binaural masking level difference*, BMLD. The predicted BMLD is calculated within each frequency channel, of center frequency, $\omega_0$. In order to predict speech intelligibility, the filtered BRIRs for speech and noise are separately cross-correlated. The speech and noise interaural phases, $\Phi_S$ and $\Phi_N$, and the noise interaural coherence, $\rho_N$, are extracted from the resulting cross-correlation functions. These values are then used in the following equation, based on equalization-cancellation theory [15].

$$BMLD = 10\log_{10}\left[\frac{k - \cos(\phi_S - \phi_N)}{k - \rho_N}\right] \tag{1}$$

where, $k = (1 + \sigma_\varepsilon^2)\exp(\omega_0^2\sigma_\delta^2)$, $\sigma_\varepsilon = 0.25$, and $\sigma_\delta = 105\ \mu s$.

[1] For example, Odeon or Catt Acoustic.

Following the principle that binaural processing can only improve performance over what is possible based on listening with one ear, the BMLD is reset to zero if it has a negative value.

The better-ear listening and binaural unmasking components are each frequency weighted by the importance function for different frequencies in the *speech intelligibility index* [16], SII, and are assumed to make additive contributions to the *effective signal-to-noise ratio*, $SNR_e$, in decibels. This value is not intelligibility *per se*, because this would depend upon the nature of the speech materials and the integrity of the listeners' auditory systems, but making assumptions about these, one can go on to derive an intelligibility prediction through the SII [15]. The $SNR_e$ can be used to predict differences in speech reception threshold across different listening situations; any resulting increase in $SNR_e$ should give rise to an improvement (decrease) in SRT of equal magnitude. The $SNR_e$ incorporates both the physical signal-to-noise ratio at that location and the benefits of binaural listening.

## *1.4 Suitability of the Binaural Model to Architectural Acoustics*

In architectural acoustics, the effect of a room is fully described by the impulse responses between the positions of sound sources and receivers, for example, stage and seating area. Because the binaural model described above works directly with binaural room impulse responses as inputs, it can very easily be used in connection with room simulation software producing such impulse responses as output, or with acoustical measurements of impulse responses in real rooms. The only requirement is that these impulse responses should be binaural.

Because the model manipulates short impulse responses rather than the long source signals used by other models [7, 9], it produces fast and non-stochastic predictions, avoiding the averaging of predictions over several source signals. Thanks to its resulting computational efficiency, it can be used to draw intelligibility maps of rooms. Such maps were obtained by simulating the listener at different positions in a room containing a speaker and multiple noise sources [11]. The resulting spatial representations offer visualization of the space accessible to a listener who would wish to maintain a given level of intelligibility while moving within the room. Other types of representation can be computed—as illustrated later in this chapter.

Another advantage of the model is its modularity. The contributions of better-ear listening and binaural unmasking are computed independently in each frequency channel. The two contributions of binaural hearing can be considered separately, monaural listening can be simulated, and some frequency regions can be "deactivated". This would allow for specific forms of hearing impairment to be taken into account to guide technical applications directed towards the listener, such as by using directional microphones on hearing aids, or environmental policies concerning room design. For example, as of today, binaurally implanted cochlear implantees benefit from better-ear listening but not binaural unmasking [17], because current implants usually encode the temporal envelope of incoming sounds but not the temporal

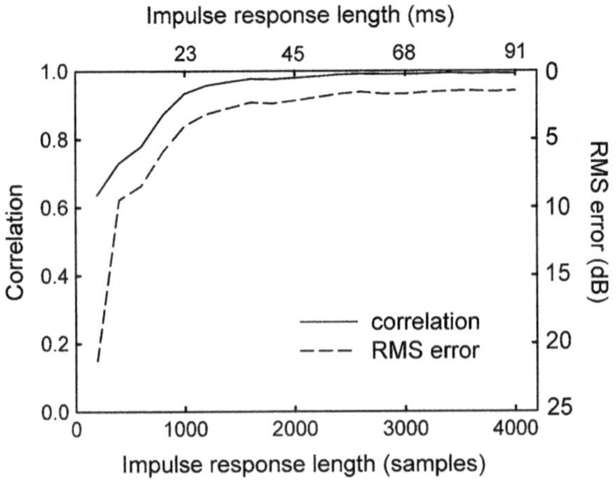


**Fig. 2** Correlation between observed and predicted SRTs, and the RMS error of the prediction, plotted as a function of impulse-response length for the set of conditions examined by Beutelmann and Brand [7]

fine structure. Room intelligibility maps involving monaural or binaural listening, and without binaural unmasking, indicated where listeners can stand without losing understanding [11]. This might prove a useful tool towards predicting *room accessibility* for hearing-impaired listeners—see Sect. 2.6.

A key issue for practical implementation is the length of impulse response necessary to obtain an accurate prediction. Because the predictions of binaural-intelligibility by the model depend on the exact spatial configuration, it may be necessary to make many predictions for different listening positions and for different potential configurations of speech and masking-noise sources. Each prediction would require generation of BRIRs between each of the sources and the listening position. Many BRIRs may therefore be required. The calculation time for predicted BRIRs grows exponentially with the length of the BRIR, so the potential computational explosion may be contained by using the shortest BRIRs necessary for an accurate result.

To examine this, the effect of impulse response length on the accuracy of prediction was evaluated, using real-room impulse responses and corresponding SRT data collected by Beutelmann and Brand [7]. These 1.5-s impulse responses, originally 65,536 samples long, were collected from two different rooms, an office and a large cafeteria. As noted above, the model predicted the SRTs measured by Beutelmann and Brand quite accurately using their impulse responses. The correlation between observed and predicted SRTs was 0.99. In order to examine the effect of impulse response length, their data were modeled with those impulse responses truncated to lengths, between 200 and 4000 samples, that is, between 4.5 and 91 ms.

Figure 2 shows the correlations between observed and predicted SRTs as a function of impulse-response length, as well as the RMS error. It can be seen that long impulse responses are not necessary, with performance reaching asymptote at around 3000 samples, that is, 70 ms. The cafeteria and office rooms in question have reverberation times, $T_{60}$, of 1.3 and 0.6 s, respectively, yet only the first 70 ms of reverberation is needed for an accurate prediction of intelligibility. This may be explained by the fact that, in each case, 96 % of the energy in each of the impulse responses occurred within the first 70 ms.

## 2 Applications to Architectural Design

The binaural model is suitable for answering a number of questions about the acoustic design of spaces in which listeners contend with background noise, such as classrooms, restaurants, cafeterias, railway stations and foyer areas. For the purposes of this chapter an acoustic model of a virtual restaurant is used as an example case, and the predictions of the binaural model will be explored for some simple design choices.

The room-acoustic model employed here was an image-source model [18] restricted to simple rectangular boxes. As noted above, commercial software could produce more accurate modeling of the room acoustics. Consequently, the acoustic model contained no representation of the furniture or the occupants and all sound sources were omnidirectional. On the other hand, the receiver characteristics of the listener are quite accurately modeled, because the acoustic wave fronts arriving at the listeners' heads are represented in the BRIRs by suitably delayed and scaled head-related impulse responses. Each head-related impulse response is selected from a database for the azimuth and elevation of that acoustic ray at the head position. The head-related impulse responses used were recordings made at the Massachusetts Institute of Technology [19] from a KEMAR manikin [20]. Although a more sophisticated room simulation would be preferable for practical applications, the present implementation has the advantage that all the resources for the simulation are in the public domain and the simplicity of the layout allows direct assessment of the principal room parameters. The aim was to demonstrate how the binaural-intelligibility model can be useful in architectural acoustics and to draw out some preliminary conclusions on the influence of these room parameters.

In order to examine these parameters a simple restaurant layout was developed, which included most of the critical factors one might expect to encounter in real life—see Fig. 3. The simulated restaurant contained nine *tables for two* in a regular $3 \times 3$ grid. Each table served to define two potential source/receiver locations, each being 1.2 m above the floor. The restaurant thus included pairs of source/receiver locations that had walls to the side, that is, tables #2 and 8, and others that had walls at one end, namely, tables #1, 3, 4, 6, 7, and 9, and also a pair that was surrounded by other sources—table #5. The room was 6.4 m square, the default ceiling height 2.5 m. The table positions were distributed evenly at 1.6-m intervals, with the source/receiver pairs separated by 0.7 m. The tables all had the common orientation shown in Fig. 3. Walls, ceiling and floor had controllable frequency-independent

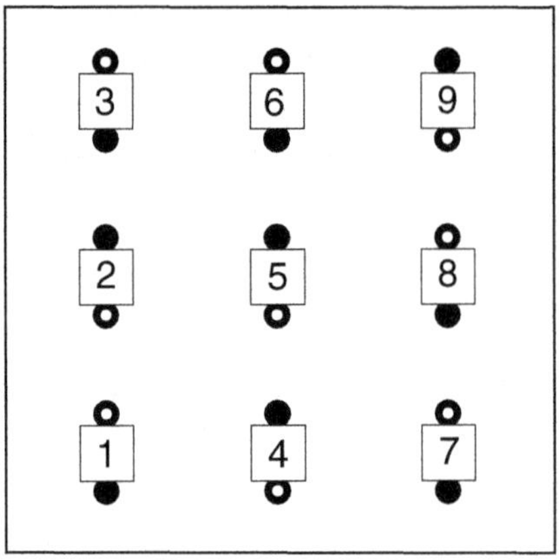


**Fig. 3** Restaurant layout: Each rectangle represents a notional table, across which two diners (*black circles*) may wish to talk. In the model, one diner at each table (*with a white spot*) is nominated as the default location of a noise source

absorption coefficients. Across all simulations, there was one masking source at each table, which was selected at random, indicated by black spots with white dots in Fig. 3.

## *2.1 Effect of Seating Position in a Restaurant*

How should one pick a good seat in a restaurant? Ideally one should be able to hear other individuals at the same table clearly. The model can make predictions of the variations in speech intelligibility across different tables and also within a given table. If one were able to answer such a question in a real restaurant, it would be possible to advise those who require better listening conditions, such as hearing-impaired listeners, to use particular seats. It may also be possible to tailor the acoustic treatment of the room to iron out such variations and provide a consistent acoustic experience across the entire space.

This question was addressed by looking at $SNR_e$ across the different seats in the virtual restaurant. The absorption coefficients of the walls were set to 0.7, that of the floor to 0.1 and that of the ceiling to 0.9. Figure 4 shows the predicted $SNR_e$ for each diner in the room, represented by the size of the corresponding black spot. The size of the spot is related to the $SNR_e$ in dB. One can see that tables in the corner of the room are more favorable than those elsewhere, and that those placed between other tables, namely, the three middle tables, fare worse than those which are aligned with the wall. There are some local modifications to this pattern caused by the particular configuration of noise sources. For instance, one of the diners at table #8 has an interfering noise source immediately behind, that is, on table #7. This decreases the local $SNR_e$—see Fig. 4.

Using suitable acoustic modeling software, similar evaluations could be made in more complex acoustic spaces, such as alcoves, balconies, etc. The effects of different

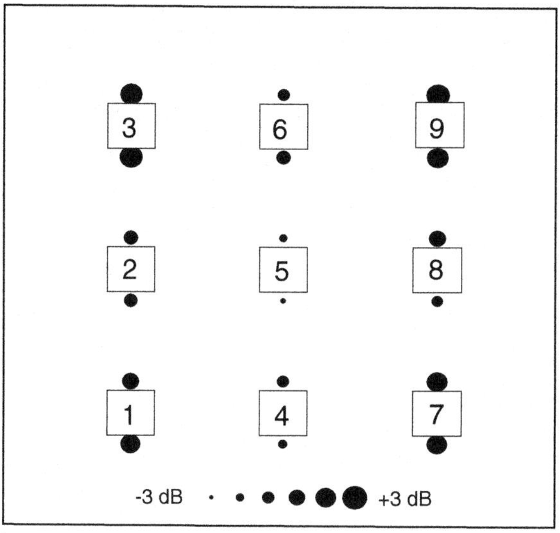


**Fig. 4** $SNR_e$ at each seating position in the virtual restaurant. The diameter of the *black spots* is proportional to $SNR_e$ in dB

assumed configurations of masking noises could be addressed by averaging over a number of different random selections.

## *2.2 Effect of Head Rotation*

As the head is turned horizontally, the relative directions of all sound sources are rotated around the head. Although SRM relies on *differences* in the directions of the target speech and the masking noise(s), the model indicates that changing all source directions together in this way can change the benefit to intelligibility. It is most often assumed that listeners directly face their interlocutor during a conversation, but this is not necessarily the case. In fact, observation of any busy social event will reveal that many people engaged in a conversation have their heads at an angle to each other. It is not currently clear whether this behavior is deliberate or whether it is related at all to optimizing speech intelligibility. Nonetheless, it is instructive to examine the potential impact.

If listeners do orient their heads in order to improve intelligibility, there must clearly be some limit to this behavior. It would be rude to turn one's back, eye contact may occasionally be required and lip-reading, which most listeners use unconsciously to improve intelligibility in noise [21], requires sight of the speaker's face. Counter-rotation of the eyes can be used to some degree in order to maintain sight of one's interlocutor, but it seems unlikely that such a sidelong posture would be practical beyond a head-turn of about 30°. Research on gaze control [22] indicates that, when fixating a target, observers make an initial eye turn of up to about 40°. Some observers will follow this movement with a head turn, which reduces the eye displacement down

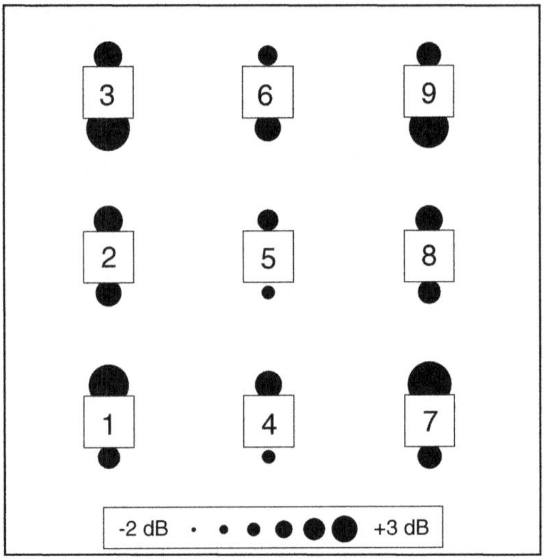


**Fig. 5** As Fig. 4, but assuming that listeners have oriented their heads to the optimum angle for speech intelligibility within a range of ±30°

to 20° or so, while others will maintain a 40° eye displacement. The effect on the situation used in Sect. 2.1, of an optimized head turn of up to 30° was therefore evaluated.

Figure 5 shows revised values of $SNR_e$ after the listeners have made optimal head turns. It can be seen that $SNR_e$ has improved substantially in all cases. The mean improvement is 2.5 dB with values for individual listening positions ranging up to 5.3 dB. In addition to this general improvement, one can see a change in the pattern of results compared to Fig. 4, where no head rotation was assumed. Once head orientation is taken into account, the seats facing the wall at the four corner tables have a clear advantage over other locations. In each case, the optimal head orientation is to turn away from the side wall, such that the interlocutor on the other side of the table is to one side of the head and other sources in the room are on the other side of the head. The ear that is turned towards the interlocutor is thus maximally isolated by head shadow from the sources of masking noise and enjoys an improved $SNR_e$. It is also noticeable that local variations due to the configuration of masking-noise sources are also less evident; for the most part, $SNR_e$ for each seat is similar to that for mirror-image locations across the room.

### *2.3 Effect of Ceiling Height*

Many people have an intuitive sense that high ceilings contribute to a poor acoustic. However, there are good reasons to believe that this intuition is false and that high ceilings are actually beneficial. Their benefits may come from two acoustical factors. First a higher ceiling will increase the total absorbent area of the room, due

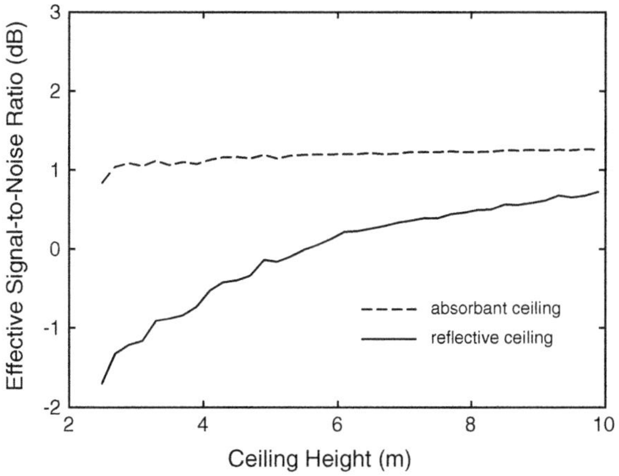


**Fig. 6** Effective signal-to-noise ratio averaged across tables as a function of ceiling height for reflective and absorbent ceilings. Absorption coefficients were 0.9 and 0.1, respectively

to absorbent surfaces provided by the additional wall height. Such an increase in absorption reduces the total amount of reverberant energy within the space. Second, a high ceiling increases the volume of the room. This means that the reverberant energy spreads throughout a larger space, thus reducing the energy density. The prediction model can be used to simulate a range of different ceiling heights and determine the overall effect of these different processes on intelligibility.

To this end, a dining couple on each table in the restaurant was modeled. There were eight masking-noise sources, as in the configuration from Fig. 3. The $SNR_e$ without head rotation was then evaluated as a function of ceiling height between 2.5 and 10 m, in 0.2-m steps. Other parameters were again similar to those of Fig. 4.

As can be seen from the solid line in Fig. 6, the $SNR_e$ increases with the height of this reflective ceiling, indicating that a high ceiling provides easier communication to people in a noisy room. Once the ceiling was raised by 5 m, there was a 2.4 dB mean improvement in $SNR_e$. Across different tables, improvements ranged from 1.7 to 3.1 dB. For comparison, a similar level of benefit could be obtained with an acoustic ceiling that increases the ceiling absorption coefficient from 0.1 to 0.9, but the dashed line shows that only 0.4 dB of improvement would occur if the height of an absorbent ceiling was raised by 5 m, with half of this change occurring in the first 20 cm. To place both these effects in context, a totally anechoic room would only increase the signal-to-noise ratio by a further 2.5 dB.

It can be seen that intuitive impressions of ceiling height as a negative factor in room design are misleading. High ceilings are good. However, intuition is not the only false friend, here. $T_{60}$ is generally used as a measure of how reverberant a room is; a larger $T_{60}$ is usually considered a measure of a "more reverberant" room, which is generally assumed to result in lower intelligibility. Consistent with this association, when the absorption coefficients of the room boundaries are lowered, the corresponding increase in $T_{60}$ is accompanied by a *decrease* in predicted intelligibility [6]. However, the Sabine equation, which can be used to estimate $T_{60}$ from only the volume, $V$, and the effective absorbent area of a room, $\overline{\alpha}S$,—that is, the total surface area, $S$, times the mean absorption coefficient, $\overline{\alpha}$—shows that $T_{60}$ is proportional to the volume, $V$, but inversely proportional to the surface area, $S$, that is,

$$T_{60} \approx 0.163 \frac{V}{\overline{\alpha}S} \text{ [s/m]} . \tag{2}$$

Now, since the volume to surface area ratio of any object or space increases with its dimensions, if the average absorption is held constant, $T_{60}$ will increase with the room dimensions, including the ceiling height. Volume to surface area ratio will increase even if only the ceiling height is changed. Consequently, $T_{60}$ can also be associated with an *increase* in speech intelligibility when ceiling height alone, or room volume in general, is manipulated. This fact is well illustrated by Beutelmann and Brand's data [7], which show consistently lower SRTs in their cafeteria environment with a $T_{60}$ of 1.3 s, than in the office environment with a $T_{60}$ of 0.6 s. In isolation, $T_{60}$ is, therefore, a fairly useless measure of room quality for speech intelligibility unless room volume is factored out in some way. In BB93, there is little cognizance of room volume in the recommended $T_{60}$ targets; particularly, a spacious classroom with a high ceiling would be over-treated in order to meet the specification, while a smaller than average classroom with a low ceiling would be under-treated.

## 2.4 Effect of Absorber Placement

It is most common to provide acoustic treatment to a ceiling. However, the benefits of binaural hearing depend upon the interaural differences produced by spatial separation of different sound sources. Since the ears are usually on the same horizontal plane, these interaural differences tend to be reduced by lateral reflections. Consequently, one might expect that designs which selectively reduce lateral reflections would generally provide greater benefit. Moreover, first-order ceiling reflections tend to reinforce interaural differences, because they come from the same azimuth. Thus, it may be better to place acoustic absorbers on the wall rather than the ceiling.

In order to quantify the potential benefit of laterally placed absorbers, two versions of the restaurant have been created with different absorber placements but the same overall $T_{60}$ of 385 ms, as determined by the Sabine equation. These two rooms had identical floors with an absorption coefficient of 0.07. For the room with a reflective ceiling, the walls had an absorption coefficient of 0.6 and the ceiling an absorption coefficient of 0.06. For the room with an absorptive ceiling, these numbers were 0.05 and 0.9, respectively.

These configurations were tested by calculating the $SNR_e$ for each diner, assuming that they were listening to the diner across the table with their heads fixed and that masking-noise sources were present at all other default locations for masking noise.

The mean benefit of absorptive walls compared to an absorptive ceiling was 0.7 dB. This benefit was entirely driven by better-ear listening. In contrast, the benefit of binaural unmasking fluctuated erratically around the value of 0.5 dB from one table to the next.

The advantage of wall absorbers should not, however, mislead one to thinking that ceiling treatment is ineffective. As shown in Fig. 6, ceiling treatment is always better than no ceiling treatment when the ceiling is high, but the benefit, here, is only 0.35 dB. For a high ceiling therefore, it would be particularly important to consider treating other surfaces. An equivalent change to the floor, for instance, perhaps by adding carpeting, would improve $SNR_e$ by 1.75 dB.

## 2.5 *Effect of Table Orientation*

As noted above, optimum head orientation can substantially assist listeners in background noise, but such orientation is limited to, perhaps, $\pm 30°$ by the need to maintain visual contact with one's interlocutor. This limitation leaves open the possibility that diners may be assisted in reaching beneficial head positions/orientations by turning the whole table by 90°. In other words, might it be possible to use the model to derive an optimal table layout?

In order to investigate this possibility, the restaurant scenario described in Sect. 2.2 has been re-evaluated including optimal head rotations of up to 30°, but with some tables in different orientations. In each simulation, $SNR_e$ was calculated for each pair of diners with eight masking-noise sources randomly distributed across the remaining tables. The results from twenty different random distributions were averaged. Due to the number of seats, head orientations and masker distributions considered, this analysis was quite time consuming. There are $2^8$ unique permutations of table rotations to be considered, so it was necessary to concentrate on just a few interesting alternatives to the regular layout used above. Two layout strategies rotated the tables that were found to be most difficult in the analysis of Sect. 2.2. In one case, only the central table was rotated. In a second case all three of the tables down the centre of the room were rotated. In a third strategy, the case of rotating every second table throughout the room was considered.

The results showed that all three alternative strategies showed some benefit over a regular layout, but the benefits were fairly small. Rotating only the middle table, #5, or rotating every second table, that is, #2, 4, 6 and 8, improved the mean $SNR_e$ by only 0.07 dB. Rotating the three middle tables, #4, 5 and 6—see Fig. 7, yielded a more noticeable mean improvement of 0.3 dB. Moreover, it is noteworthy that this option reduced somewhat the variability in $SNR_e$ across different tables. In this scenario, large improvements of >2 dB were predicted for the diners on tables #4 and 6 who previously had their backs to the wall. None of these interventions produced significant benefit for the diners on table #5, however, and the standard deviation in $SNR_e$ across seats was only reduced from 2 to 1.9 dB.

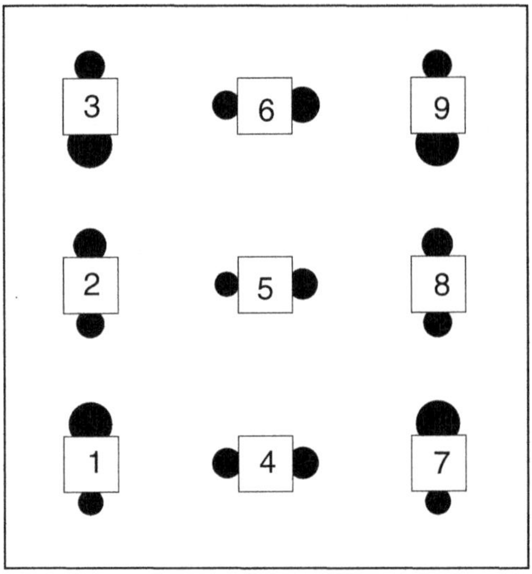


**Fig. 7** Alternative table layout providing improved effective signal-to-noise ratios. The diameters of the *black spots* are proportional to the $SNR_e$ in dB

## 2.6 *Effect of Room Occupancy*

Intelligibility worsens as a room fills up with people. How many people should a room be designed to accept? This has been termed the *acoustical capacity* of the room [23]. One can look at this question using the restaurant simulation. For a couple at each table, a given number of noise sources were distributed at random across the other eight tables. $SNR_e$ of 20 such random distributions was then averaged. No head rotation was assumed.

Figure 8 shows that, unsurprisingly, $SNR_e$ should fall with increasing room occupancy. The critical issue is the level of the $SNR_e$. Even when there is a noise source at every other table, and listeners are making no use of head orientation, the $SNR_e$ falls no lower than $-1.1$ dB. Speech understanding in noise becomes impossible below about $-3$ dB, so this room seems to be acceptable for the assumed table layout.

It should be noted, however, that this analysis takes no account of the Lombard effect [24]. As the level of background noise increases, people instinctively start to raise their voices in order to be heard. As a result, the sound level in a room tends to increase with increased occupancy level more rapidly than would be expected from the number of sources present. Effectively, each doubling in the number of speakers tends to produce an increase of 6 dB in the ambient noise level rather than the expected 3 dB [23, 25]. Because all voices in the room are increasing together, this increase in vocal output and ambient noise level has no effect upon the $SNR_e$. Consequently, the effectiveness of communication is only disrupted to the extent that auditory processing is impaired by elevated sound levels [16, 26]. However, it also has an effect on the experience of the diners. People do not want to be shouting to

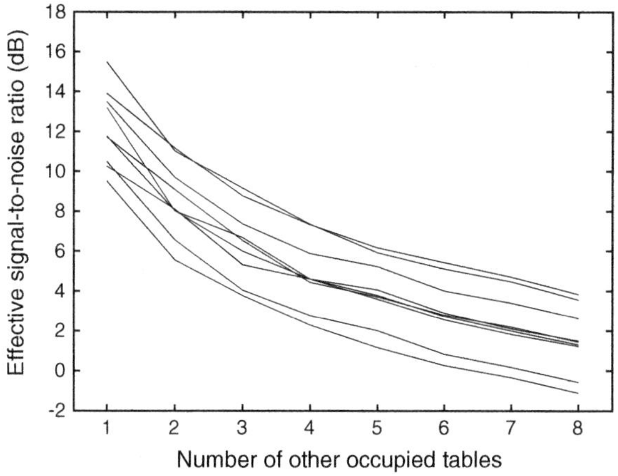


**Fig. 8** Effective signal-to-noise ratio as a function of the number of occupied tables for each of the nine tables in the restaurant

make themselves heard. A separate analysis of the impact of room occupancy on vocal effort would therefore be advised [27].

## *2.7 Towards Predicting Room Accessibility*

Accessibility of public spaces to those with disabilities is an increasingly important aspect of public policy. Architects now need to consider not only whether normally hearing listeners will be able to communicate effectively in a given acoustic space, but also whether the hearing-impaired listeners or non-native listeners will be able to do so. The level of intelligibility corresponding to a given signal-to-noise ratio is dependent on hearing and comprehension abilities. To ensure the same level of understanding, hearing-impaired listeners and cochlear implantees, for example, will require a better ratio than normally-hearing listeners.

The problem is a difficult one to address in a precise way, because hearing loss is a very individual disability. Different listeners will have different patterns of loss across frequency and the different etiologies of hearing losses have different consequences for speech understanding in noise. Moreover, there are currently gaps in our understanding of how a given hearing impairment leads to a given elevation in SRTs, which make it difficult to produce an accurate predictive model. Nonetheless, some notable successes have been achieved. Beutelmann and Brand [7] simulated cochlear hearing loss in their model by assuming that any elevation in pure-tone threshold was equivalent to an increased effective noise floor at that frequency and Culling et al. [17] modeled unilateral cochlear-implant patients simply by running their model in monaural mode, and assuming that each patient had an individually reduced receptive capacity. The same strategy should work for listeners with single-sided deafness, but without the need to vary receptive capacity. Some other maneuvers are possible.

Listeners with cochlear hearing loss tend to have so-called *sloping losses*. This means that their pure-tone detection thresholds increase with frequency. These listeners might be modeled by assuming that they lose information at higher frequencies. This loss in information could be represented in the model by reducing the SII weighting values for high-frequency channels.

Listeners with asymmetric hearing have different SRTs when tested monaurally with each ear. These listeners could be modeled by assuming that their better ear is the ear that has the better signal-to-noise ratio after the difference in monaural SRT has been taken into account. That is, if the left-ear SRT is 3 dB better than the right-ear SRT, the model would assume that in binaural listening situations, the listener uses the left ear until the right ear SNR is at least 3 dB better than the left ear. Culling et al. [17] used this approach in order to model SRT data from bilateral cochlear-implant users [28]. In this instance, taking account of asymmetry in this way did not improve the fit to the data compared to ignoring the asymmetry, but this may be because the asymmetries in these cochlear implant users were fairly small; a minority of cochlear implant users have very large asymmetries, for which this technique might be essential.

The predictions of the model have been explored for the case of an asymmetry in monaural SRT. Such a manipulation does not affect its predictions of binaural unmasking, but only the selection of the better ear within each frequency band. The situation described in Sect. 2.2 was modeled, including the listeners' option to make a head turn of up to 30°, but assuming that each listener's right ear had a monaural SRT that was elevated by 10 dB with respect to their left ear. This has no effect when the better physical SNR is at the ear with the better monaural SRT, but when it is on the other side, it may require the listener to attend to the speech with the ear that has the poorer physical SNR. This inevitably has an impact on $SNR_e$. One would therefore expect only certain seating positions in the restaurant to be affected.

Consistent with this expectation, it turned out that, although the average predicted elevation in SRT was 1.2 dB, the effect was very strongly affected by seating position. Figure 9 shows the uneven distribution of those deficits. Essentially, in those seating positions where a deficit is visible in Fig. 9, it is approximately 3 dB. There are much smaller deficits distributed over the other positions.

The distribution in Fig. 9 can be understood in terms of the spatial distribution of speech and noise sources with respect to each listening position. For instance, for the listener experiencing a problem on table #9, a good right ear would allow them to rotate their head to the left and create a situation in which the target voice is to their right while all the noise sources are on their left. Since their right ear is impaired, they are less successful in following this strategy.

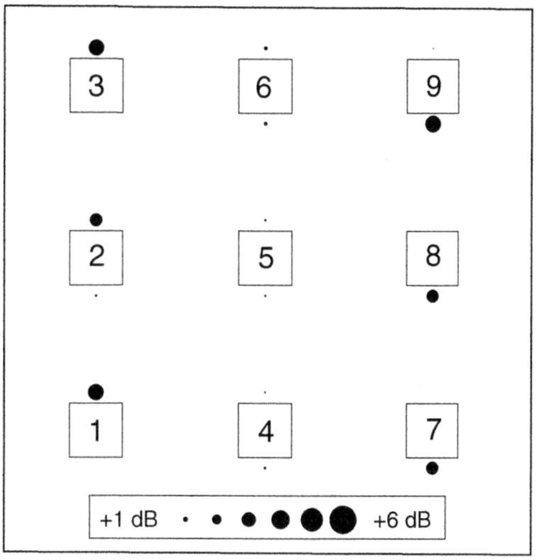


**Fig. 9** Deficit in effective signal-to-noise ratio experienced by a listener, whose right ear has a monaural speech reception threshold that is elevated by 10 dB with respect to that for the left ear. The diameters of the *black spots* are proportional to the decrease in effective signal-to-noise ratio

# 3 Limitations of the Simulations and Further Developments of the Model

The simulations described above illustrate the power of the model to provide insights into the effects of room parameters on speech intelligibility in complex listening situations. A simple room geometry was used. An advantage of this approach is that general principles regarding such things as the effect of ceiling height can be addressed without confounding influences of uncontrolled room parameters. For any practical application, however, one would want to model the specific geometry of a room in order to evaluate the exact effect of making a design change in a specific project. The aim of this chapter was to illustrate the potential applications of the binaural intelligibility model to support the design of social interaction spaces.

## *3.1 Room Simulations*

In order to draw conclusions regarding specific architectural designs, more sophisticated room simulations or real-room measurements need to be used to produce the BRIRs. The room simulations used here only considered the simplest room geometry, without taking into account the strong frequency dependence of room-materials absorption, the diffusion properties of these materials, or the directivity of different sounds sources. The binaural model can be used with any type of BRIRs, and it can only benefit from the use of more realistic BRIRs, be it measured or simulated, that take these acoustic phenomena into account.

The simulations used in this chapter only considered sources with the same sound level and long-term spectrum. The application of the binaural model is not limited to these situations. Sources at different sound levels can be modeled by scaling their respective BRIR to the appropriate relative level; absolute source levels are not relevant, only differences in level between sources are. Sources with different spectra can also be modeled by appropriate filtering of their BRIRs. If the sources all have the same average spectrum, no filtering is required. In the case of multiple masking noises, concatenation of the scaled and/or filtered BRIRs would have the effect of summing the frequency-dependent energy of each contributing BRIR and generating an averaged cross-correlation function, weighted according to the energy in each BRIR.

### 3.2 Model Developments

The binaural model can accurately predict speech intelligibility against any number of stationary noise maskers, in any spatial distribution within a room and for any orientation of the listener. However, because it does not take into account the potential temporal smearing of target speech in very reverberant environments, this model can only predict intelligibility of target speech sufficiently close to the listener, at positions where the direct-to-reverberant ratio is not too low and segregation from sources of masking noise is the overriding factor for intelligibility. It needs to be extended to take into account this direct effect of reverberation on target speech, as has been done in a revision of the Oldenburg model [29]. Because the model works directly with BRIRs, it offers the opportunity to separate the early and late reflections within the BRIRs, so that temporal smearing can be modeled following the concept of useful-to-detrimental ratio [3, 30], in which the early reflections of the speech are regarded as useful because they reinforce the direct sound, while the late reflections are regarded as detrimental and effectively a part of the noise.

A model that intends to completely describe cocktail-party situations in rooms needs to handle competing speech sources and so to predict the segregation mechanisms associated with the temporal envelope modulations and the periodicity of speech. Fundamental frequency, F0, differences facilitate segregation of competing voices [31, 32], but reverberation is detrimental to segregation by F0 differences where F0 is non-stationary [33, 34] as in the case of normal intonated speech. Modulations in the temporal envelope of the masking noise allow one to hear the speech better during the moments when the speech-to-noise ratio is higher [35, 36], so-called *listening in the gaps* or *dip listening*, and this ability is impaired by reverberation which reduces modulations [8, 37], filling in *gaps* of the masker.

Restaurant simulation has been used to test the overall implications of these effects empirically [38]. SRTs were measured as a function of the number of masking sources, where those sources were either speech or continuous speech-shaped noise, and where the room was either reverberant or anechoic. The predictions of the model were accurate for the speech shaped noise, but speech maskers are less effective than noise. That is, SRTs were lower, when there was only a single masking voice, especially in anechoic conditions. On the other hand, SRTs were a few dB higher for speech maskers than for noise when there was more than one masking voice. The advantage for a single masking voice may be attributed to some combination of *F0-difference processing* and *dip listening*, while the disadvantage for multiple masking voices appears to be some form of informational masking. It seems likely that in these multiple-masker cases, both dip listening and F0-difference processing are markedly less effective, although their precise role is, as yet, unclear.

The binaural model has recently been adapted to take dip listening into account, thus providing intelligibility predictions in the presence of speech-modulated noises [39]. However, this modified version of the model does not work directly on BRIRs, it requires the signals produced by the sources in the rooms as inputs. Following an approach proposed by Rhebergen and Versfeld [40] and then Beutelmann et al. [8], it consists of applying the stationary model to short time frames of the speech and noise waveforms, and then averaging the predictions over time. This signal-based approach would need to be adapted to be applied to the model based on BRIRs. For example, signal statistics could be associated with the BRIRs as model inputs, because BRIRs do not contain information about signal modulations. The advantage of having separated inputs for room and source information is that one might be able to simply update signal statistics to make predictions for different speech materials without requiring the actual signals in rooms.

# 4 Conclusions

The modeling presented here has clear limitations, both in terms of the sophistication of the acoustic model employed and the generality of the predictions to more structured masking sources, notably speech. Nonetheless, it captures aspects of the listening task which have hitherto been ignored in the acoustic assessment of rooms. It has been demonstrated in this chapter that the binaural model is markedly better suited to the prediction of intelligibility in rooms than the measures of reverberation time which are generally used. This modeling has also provided novel insights, such as the relative ineffectiveness of acoustically treating a high ceiling, which may well be general.

The limitations of the acoustic model could, for example, be addressed by using impulse responses generated by commercial room-simulation software. Since the binaural model is simple and computationally efficient, it could easily be incorporated into existing software in order to produce maps of the effective signal-to-noise ratio across a room or predictions for particular spatial configurations as in the simulated restaurant. Work continues on gaining sufficient understanding of human hearing to accurately predict the effects of dip listening and exploitation of F0 differences. It is as yet unclear whether they play a significant role in the complex listening situations with multiple maskers, for which the binaural model is designed.

**Acknowledgments** Work supported by U.K. Engineering and Physical Sciences Research Council. The authors thank their two external reviewers for valuable suggestions.

# References

1. M. Lavandier, J. F. Culling (2008) Speech segregation in rooms: monaural, binaural and interacting effects of reverberation on target and interferer. *J. Acoust. Soc. Am.* 113:2237–2248
2. T. Houtgast, H. J. M. Steeneken (1985) A review of the MTF concept in room acoustics and its use for estimating speech intelligibility in auditoria. *J. Acoust. Soc. Am.* 77:1069–1077
3. J. S. Bradley (1986) Predictors of speech intelligibility in rooms. *J. Acoust. Soc. Am.* 80:837–845
4. M. L. Hawley, R. Y. Litovsky, J. F. Culling (2004) The benefit of binaural hearing in a cocktail party: Effect of location and type of masker. *J. Acoust. Soc. Am.* 115:833–843
5. A. W. Bronkhorst, R. Plomp (1988) The effect of head-induced interaural time and level differences on speech intelligibility in noise. *J. Acoust. Soc. Am.* 83:1508–1516
6. R. Plomp (1976) Binaural and monaural speech intelligibility of connected discourse in reverberation as a function of azimuth of a single competing sound source (speech or noise). *Acustica* 34:200–211
7. R. Beutelmann, T. Brand (2006) Prediction of speech intelligibility in spatial noise and reverberation for normal-hearing and hearing-impaired listeners. *J. Acoust. Soc. Am.* 120:31–342
8. R. Beutelmann, T. Brand, B. Kollmeier (2010) Revision, extension and evaluation of a binaural speech intelligibility model. *J. Acoust. Soc. Am.* 127:2479–2497
9. M. Lavandier, J. F. Culling (2010) Prediction of binaural speech intelligibility against noise in rooms. *J. Acoust. Soc. Am.* 127:387–399
10. S. Jelfs, M. Lavandier, J. F. Culling, (2011) Revision and validation of a binaural model for speech intelligibility in noise. *Hear. Res.* 275:96–104
11. M. Lavandier, S. Jelfs, J. F. Culling, A. J. Watkins, A. P. Raimond, S. J. Makin (2012) Binaural prediction of speech intelligibility in reverberant rooms with multiple noise sources. *J. Acoust. Soc. Am.* 131:218–231
12. A. W. Bronkhorst, R. Plomp (1992) Effect of multiple speechlike maskers on binaural speech recognition in normal and impaired hearing. *J. Acoust. Soc. Am.* 92:3132–3139
13. J. Peissig, B. Kollmeier (1997) Directivity of binaural noise reduction in spatial multiple-source arrangements for normal and impaired listeners. *J. Acoust. Soc. Am.* 101:1660–1670
14. J. F. Culling, M. L. Hawley, R. Y. Litovsky (2004) The role of head-induced interaural time and level differences in the speech reception threshold for multiple interfering sound sources. *J. Acoust. Soc. Am.* 116:1057–1065
15. N. I. Durlach Binaural signal detection: Equalization and cancellation theory. In J. Tobias (Ed) Foundations of Modern Auditory Theory, Vol. 2. Academic, New York, pp. 371–462 (1972)
16. ANSI (1997) Methods for calculation of the speech intelligibility index. ANSI S3.5-1997, American National Standards Institute, New York
17. J. F. Culling, S. Jelfs, A. Talbert, J. A. Grange, S. S. Backhouse (2012) The benefit of bilateral versus unilateral cochlear implantation to speech intelligibility in noise. *Ear Hear.* 33:673–682
18. J. B. Allen, D. A. Berkley (1979) Image method for efficiently simulating small-room acoustics. *J. Acoust. Soc. Am.* 65:943–950
19. W. G. Gardner, K. D. Martin (1995) HRTF measurements of a KEMAR. *J. Acoust. Soc. Am.* 97:3907–3908
20. M. D. Burkhard, R. M. Sachs (1975) Anthropometric manikin for acoustic research. *J. Acoust. Soc. Am.* 58:214–222
21. A. MacCleod, Q. Summerfield (1987) Quantifying the contribution of vision to speech perception in noise. Br. *J. Audiol.* 21:131–142
22. J. E. Goldring, M. C. Dorris, B. D. Corneil, P. A. Ballantyne, D. P. Munoz (1996) Combined eye-head gaze shifts to visual and auditory targets in humans. *Exp. Brain. Res.* 111:68–78
23. J. H. Rindel (2012) Acoustical capacity as a means of noise control in eating establishments. Joint Baltic-Nordic Acoustics Meeting, Odense, Denmark
24. H. Lane, B. Tranel (1971) The Lombard sign and the role of hearing in speech. *J. Sp. Hear. Res.* 14:677–709

25. M. B. Gardner (1971) Factors Affecting Individual and Group Levels in Verbal Communication. *J. Audio Eng. Soc.* 19:560–569
26. B. C. J. Moore, B. R. Glasberg (1987) Formulae describing frequency selectivity as a function of frequency and level, and their use in calculating excitation patterns. *Hear. Res.* 28:209–225
27. J. H. Rindel, C. L. Christensen, A. C. Gade (2012) Dynamic sound source for simulating the Lombard effect in room acoustic modeling software. Inter Noise 2012, New York
28. P. C. Loizou, Y. Hu, R. Litovsky, G. Yu, R. Peters, J. Lake, P. Roland (2009) Speech recognition by bilateral cochlear implant users in a cocktail-party setting. *J. Acoust. Soc. Am.* 125:372–383
29. J. Rennies, T. Brand, B. Kollmeier (2011) Prediction of the influence of reverberation on binaural speech intelligibility in noise and in quiet. *J. Acoust. Soc. Am.* 130:2999–3012
30. J. P. A. Lochner, J. F. Burger (1964) The influence of reflections on auditorium acoustics. *J. Sound Vib.* 1:426–454
31. J. P. L. Brokx, S. G. Nooteboom (199 2) Intonation and the perceptual separation of simultaneous voices. *J. Phonetics* 10:23–36
32. J. F. Culling, C. J. Darwin (1993) Perceptual separation of concurrent vowels: within and across formant grouping by F0. *J. Acoust. Soc. Am.* 93:3454–3467
33. J. F. Culling, Q. Summerfield, D. H. Marshall (1994) Effects of simulated reverberation on binaural cues and fundamental frequency differences for separating concurrent vowels. *Speech Comm.* 14:71–96
34. J. F. Culling, K. I. Hodder, C. Y. Toh (2003) Effects of reverberation on perceptual segregation of competing voices. *J. Acoust. Soc. Am.* 114:2871–2876
35. A. W. Bronkhorst, R. Plomp (1992) Effect of multiple speechlike maskers on binaural speech recognition in normal and impaired hearing. *J. Acoust. Soc. Am.* 92:3132–3139
36. J. M. Festen, R. Plomp (1990) Effects of fluctuating noise and interfering speech on the speech-reception SRT for impaired and normal hearing. *J. Acoust. Soc. Am.* 88:1725–1736
37. A. W. Bronkhorst, R. Plomp (1990) A clinical test for the assessment of binaural speech perception in noise. *Audiology* 29:275–285
38. J. F. Culling (in press) Energetic and informational masking in a simulated restaurant environment. in Moore, B C J, Carlyon R P, Gockel H, Patterson R D, Winter I M (eds) Basic Aspects of Hearing: Physiology and Perception (Springer, New York)
39. B. Collin, M. Lavandier (under review) Binaural speech intelligibility in rooms with variations in spatial location of sources and depth of modulation of noise interferers. *J. Acoust. Soc. Am.*
40. K. S. Rhebergen, N. J. Versfeld (2005) A Speech Intelligibility Index-based approach to predict the speech reception threshold for sentences in fluctuating noise for normal-hearing listeners. *J. Acoust. Soc. Am.* 117:2181–2192

# Assessment of Binaural–Proprioceptive Interaction in Human-Machine Interfaces

**M. Stamm and M. E. Altinsoy**

## 1 Introduction

The human auditory system is capable of extracting different types of information from acoustic waves. One type of information which can be assessed from acoustic waves that impinge upon the ears, is the perceived direction of the sound source. For more than 100 years, scientists investigate how the auditory system determines this direction, that is, how we localize an acoustic event. The first knowledge about the auditory localization process stemmed from listening tests and was used to develop localization models trying to mimic human hearing. Pioneering modeling work was done by Jeffress in 1948 and Blauert in 1969. Jeffress proposed a lateralization model [22], which uses interaural differences to explain the localization in lateral direction within the horizontal plane. Blauert proposed a model analyzing monaural cues to explain human localization in the vertical direction within the median plane [5]. Based on these approaches, new and further developed binaural models have been implemented and published over the years—see [8] for a more detailed overview. These models did not only help to understand human auditory localization, they have also been used as a basic requirement for different technical applications, for example, hearing aids, aural virtual environments, assessment of product-sound quality, room acoustics and acoustic surveillance [7].

With the help of binaural models, human localization can be predicted under the assumption that a corresponding localization process is based on acoustic signals, thus, on unimodal information. However, what happens if this localization process is realized in an environment with available bimodal or even multimodal sensory input? This is an important question, because we are encountering such situations in our daily life. For example, we normally look and listen from which direction and at which distance a car is approaching to estimate whether it is safe to cross the street—

M. Stamm · M. E. Altinsoy (✉)
Chair of Communication Acoustics, Technische Universität Dresden,
Dresden, Germany
e-mail: ercan.altinsoy@tu-dresden.de

J. Blauert (ed.), *The Technology of Binaural Listening*, Modern Acoustics and Signal Processing, DOI: 10.1007/978-3-642-37762-4_17,

a case of audio–visual interaction. In close physical proximity to the human body, a further perceptual sense, the haptic sense, becomes important. Everyone has already tried to localize the position of the ringing alarm clock in the morning by listening and simultaneously grabbing after it, that is, audio–proprioceptive interaction. But what role does auditory information play in such situations? Do we consider the auditory modality in the localization process or is it dominated mainly by vision or haptics? These questions have to be investigated within suitable experiments to gain a deeper insight into human intermodal integration. Thus, such experiments also help to understand whether and to which extent binaural models allow to predict human localization in bimodal or multimodal scenes. The gained knowledge could play an important role for further technical developments.

## *1.1 Binaural–Visual Localization*

Many studies have been conducted in the field of audio-visual interaction and perceptual integration. For example, Alais et al. [1] revealed that vision only dominates and captures sound when visual localization functions well. As soon as visual stimuli are blurred, hearing dominates vision. If the visual stimuli are slightly blurred, neither vision nor audition dominates. In this case, the event is localized by determining a mean position based on an intermodal estimation that is more accurate than the corresponding unimodal estimates [1]. Thus, the weighting factor of the binaural signals is chosen indirectly proportional to the perceiving person's confidence in visual input. However, the weighting factor does not only depend on the quality of visual information but also on the corresponding task. For example, if the localization process requires sensitivity to strong temporal variations, binaural sensory input dominates the localization. This is because of the superior temporal resolution of the auditory system over vision [38].

In conclusion, binaural models are definitely required for mimicking human localization even when visual information is available. The sensory input of the auditory and visual modality is integrated in an optimal manner and, thus, the resulting localization error is minimized. These insights can be used for mimicking human audio-visual localization computationally with the help of existing binaural modeling algorithms, for example, to improve technical speaker tracking systems. Equipping such systems with a human-like configuration of acoustical and optical sensors, which feed binaural and visual tracking algorithms, is advantageous because of two reasons. First, the simultaneous analysis of auditory and visual information is superior than the analysis of unimodal information alone. When auditory tracking is weak due to acoustic background noise or acoustic reverberation, the visual modality may contribute to a reliable estimation. The converse is also true. When visual tracking is not reliable due to visual occlusions or varying illumination conditions, the auditory modality may compensate for this confusion. Thus, the integration of auditory and visual information increases the robustness of tracking accuracy. Second, a human-like configuration of sensors means using only two microphones and two cameras simultaneously. Such a configuration is quite effective, because applying only four sensors corresponds to the minimum requirement for spatial auditory and spatial visual localization. Furthermore, such a configuration enables sufficiently high tracking accuracy as we know from our daily life. An approach concerning the application of a human-like configuration of sensors and the computational integration of bimodal sensory input for multiple speaker localization with a robot head is described in [25]. Further details on binaural systems in robotics can be found in [2], this volume.

## 1.2 Binaural–Proprioceptive Localization

There is a further human sense which is capable of localization, namely, the haptic sense. Within the haptic modality, localization is realized by tactile perception and proprioception. Tactile perception enables the localization of a stimulated position on the skin surface. In this chapter, the focus is on proprioception which enables to determine the absolute position of the own arm and to track its movements. Proprioception is realized unconsciously with the help of the body's own sensory signals that are provided by *cutaneous mechanoreceptors* in the skin, by *muscle spindles* in the muscles, by *golgi tendon organs* in the tendons, and by *golgi and ruffini endings* in the joints of the arm.

As mentioned before, binaural models are required for modeling human localization even when visual information is available. However, what happens if the localization process is realized with proprioceptive and auditory signals? Is the auditory modality considered in the localization process or is it dominated mainly by proprioception, that is, by evaluating the body's own sensory signals about the upper limb position? These questions are not only of theoretical interest. If the auditory modality would be considered and could probably even guide the localization process, practical relevance exists especially for virtual environments in which the reproduction of acoustic signals can be directly controlled. Computational models mimicking human binaural-proprioceptive localization could then be used to simulate how an audio reproduction system has to be designed to optimize localization accuracy within a specific workspace size directly in front of the human body. In addition, such simulations may help to reproduce suitable auditory signals to diminish systematically oriented errors in proprioceptive space perception, for example, the radial-tangential illusion [26], and thus to further sharpen human precision. Two questions arise in this context.

- Which applications in virtual environments depend on proprioception?
- Should proprioceptive localization be improved with additional auditory signals?

Proprioceptive localization performance plays an important role for all applications in which a haptic device is involved. Such a device is controlled by the user's hand-arm system within the device-specific workspace and enables haptic interaction in virtual environments. One specific application, for which human interaction with a haptic device is essential, refers to virtual haptic object identification. This field of research will be introduced subsequently. Furthermore, two studies revealing the necessity of improving users' proprioceptive localization precision will be presented in this context.

### Virtual Object Identification

The haptic sense is of increasing importance in virtual environments, because it provides high functionality due to its active and bi-directional nature. The haptic sense can be utilized for solving a variety of tasks in the virtual world. One important task is the exploration and identification of virtual shapes and objects. In particular, blind users, who cannot study graphical illustrations in books, benefit immensely from employing their sense of touch. Creating digital models based on graphical illustrations or models of real physical items allows these users to explore virtual representations and, therefore, to gain information effectively also without the visual sense. A similar idea was followed by the PURE-FORM project [21], which aimed to enable blind users to touch and explore digital models of three-dimensional art forms and sculptures. Providing haptic access to mathematical functions is another exemplary application [42].

The haptic identification of virtual shapes and objects is of great importance for sighted users as well. Studies have shown that memory performance can be increased significantly using multimodal learning methods [37]. Thus, haptic identification has great potential in the field of education, for example, if digitized models or anatomical shapes are explored multimodally rather than solely visually. Another important application refers to medical training or teleoperation in minimally invasive surgery. Because of the poor camera view and the sparsely available visual cues, surgeons must use their long medical instruments to identify anatomical shapes during surgery, for instance, during the removal of a gallbladder [24]. This task is quite challenging, which is why medical students must receive training to perform it [20]. Furthermore, utilizing haptic feedback to identify anatomical shapes is of vital importance for teleoperating surgeons [18, 32]. Finally, another promising application refers to the "haptification" of data, for example, scientific ones [15, 35].

These examples represent only a small portion of the entire spectrum of possible applications, but they demonstrate the importance of haptic virtual shape and object identification for various groups of users.

Enabling the user to touch, explore and identify virtual shapes and objects requires a haptic feedback device that serves as an interface between the user and the application. First of all, this device must be capable of delivering geometrical cues to the user because such cues are of primary importance for creating a mental image of the object. Stamm et al. investigated whether a state-of-the-art haptic force-feedback device providing one point of contact can be successfully applied by test persons in geometry identification experiments [39]. Exploring a virtual geometry with one point of contact means imitating the contour-following exploratory procedure that is intuitively used in daily life to determine specific geometric details of a real object

[27]. Stamm et al. revealed that test persons experience various difficulties during the exploration and recognition process. One such observed difficulty refers to participants' insufficient spatial orientation in the haptic virtual scene. They often reached the boundaries of the force-feedback workspace unconsciously and misinterpreted the mechanical limits as a virtual object. An additional problem occurred if the participants explored the surface of a virtual object and approached an edge or a corner at one of its sides. Often, the haptic interaction point, HIP, which is comparable to the mouse cursor, slipped off of the object and got lost in the virtual space. Thus, participants lost orientation and could not locate their position in relation to the object. They required a considerable amount of time to regain contact with the object and typically continued the exploration process at a completely different position on the surface. This considerable problem was also observed by Colwell et al. [12] and makes it quite difficult and time consuming to explore and identify virtual objects effectively.

In general, studies as those mentioned above have been conducted with blindfolded participants, because the focus is set on investigating the capability of the haptic sense during interaction with the corresponding devices. However, the observed orientation-specific difficulties cannot be solved easily by providing visual information, because the orientation in a workspace directly in front of the human body is not only controlled by vision. In our daily life, the position of the hand is determined by integrating proprioceptive and visual information [14, 41]. Thus, proprioception plays an important role in the localization process. This role may be even more important in virtual workspaces, where the weighting factor of proprioceptive information may often be considerably higher than that of visual sensory input. How can this be explained? Here are five reasons.

1. With a two-dimensional presentation of a three-dimensional scene on a computer monitor, depth cannot be estimated easily using vision. This problem is also often described in medical disciplines [24]
2. Due to varying illumation conditions or often occurring visual occlusions in the virtual scene, visual information cannot be used reliably.
3. The same holds true when the visual perspective is not appropriate or the camera is moving
4. The visual channel is often overloaded during the interaction in virtual environments. This is why information cannot be processed appropriately [9, 31]
5. Visual attention has to be focused on the localization task and, for example, not on specific graphs on the computer monitor, otherwise vision cannot contribute to a reliable estimation

These examples demonstrate that confusion arises even if visual sensory input is available. When the visual information is not used or cannot be used reliably during virtual interaction, the haptic and *auditory* modality has to compensate for this confusion.

### *1.3 Outline of the Chapter*

Motivated by the orientation-specific difficulties observed in the aforementioned object identification experiments, first, the authors of the present chapter developed an experimental design to quantitatively measure proprioceptive localization accuracy within three-dimensional haptic workspaces directly in front of the human body. The experimental design and the results of a corresponding study with test persons are described in Sect. 2. In a second step, the influence of binaural signals on proprioception is investigated to reveal whether synthetically generated spatial sound is considered and can probably even improve human localization performance. In this context, a hapto–audio system was developed to couple the generated aural and the haptic virtual environment. This approach and the corresponding experimental results are described in Sect. 3. On the basis of the results, finally, conclusions will be drawn in Sect. 4 concerning the importance of binaural models for haptic virtual environments.

## 2 Proprioceptive Localization

To describe the accuracy of proprioception quantitatively, studies were conducted by scientists across a range of disciplines, for example, computer science, electrical engineering, mechanical engineering, medicine, neuroscience and psychology. These studies focused primarily on two issues. First, the ability to detect joint rotations was investigated to determine the corresponding absolute threshold of the different joints [10, 11, 19]. Second, the ability to distinguish between two similar joint angles was investigated to determine the corresponding differential threshold of the joints and thus the just noticeable difference, JND, [23, 40]. In addition to the physiological limits, the so-called *haptic illusions* must also be considered. Haptic illusions, and perceptual illusions in general, are systematically occuring errors resulting from an unexpected "discrepancy between a physical stimulus and its corresponding percept" [26]. These illusions enable us to obtain a greater understanding of the higher cognitive processes that people use to mentally represent their environments [26]. A haptic illusion that distorts the proprioceptive space perception is the radial-tangential illusion. This illusion describes the observation that the extent of radial motions away and toward the body is consistently overestimated in comparison to the same extent of motions that are tangentially aligned to a circle around the body. Different explanations have been offered, but the factors that cause this illusion to arise are not yet understood—see [26] for more details.

Proprioceptive accuracy and haptic illusions were investigated within corresponding experiments by restricting the test persons to specific joint rotations, movement directions, movement velocities, and so on. However, how do blindfolded test persons actually perform in a localization task if they are allowed to freely explore the haptic space? To the best of our knowledge, no study has investigated freely exploring test persons' localization performance in a three-dimensional haptic virtual environment. However, this is an important issue for two reasons. First, free exploration is typically used in real world interaction and, thus, should not be restricted in virtual environments. Otherwise, the usability of a haptic system is considerably reduced. Second, the abovementioned observations of the haptic identification experiment indicated that the participants experienced various orientation-specific difficulties. These difficulties should be investigated quantitatively to obtain a cohesive understanding of the proprioceptive orientation capabilities.

## 2.1 Experimental Design and Procedure

The challenge of an experiment that investigates the abovementioned relation is the *guiding* of the test person's index finger, which freely interacts with a haptic device, to a specific target position. If this target position is reached, the test person can be asked to specify the position of the index finger. The difference between the actual and the specified position corresponds to the localization error. The details of this method are explained below.

An impedance-controlled PHANToM–Omni haptic force-feedback device[1] [36] was used in the present experiment. It provides six degrees-of-freedom, 6–DOF, positional sensing and 3–DOF force-feedback. The small and desk-grounded device consists of a robotic arm with three revolute joints. Each of the joints is connected to a computer-controlled electric DC motor. When interacting with the device, the user holds a stylus that is attached to the tip of the robot arm. The current position of the tip of the stylus is measured with an accuracy of approximately 0.06 mm.

For the experiment, a maximally-sized cuboid was integrated into the available physical workspace of the PHANToM. This cuboid defines the virtual workspace and is shown in Fig. 1. Its width is 25 cm, its height is 17 cm, and its depth is 9 cm. The entire cuboid can be constructed with 3825 small cubes whose sides measure 1 cm. The cubes that are positioned on the three spatial axes are shown in Fig. 1. However, the cubes cannot be touched because they are not present as haptic virtual objects in the scene. Rather, they serve to illustrate a specific position inside the virtual workspace.

The test persons were seated on a chair without armrests. At the beginning of each trial of the experiment, the target—a sphere with a diameter of 1 cm—was randomly positioned inside the virtual workspace and thus inside one of the cubes. Then, the experimenter moved the cursor-like HIP—which corresponds to the tip of the stylus—to the reference position, namely, the center of the virtual workspace. Because of a magnetic effect that was implemented directly on the central point, the stylus remained in this position. The blindfolded test persons grasped the stylus and held it parallel to the y-axis in such a manner that the extended index finger corresponded to the tip of the stylus. The test persons were asked to search for the

[1] Manufactured by SensAble Technologies

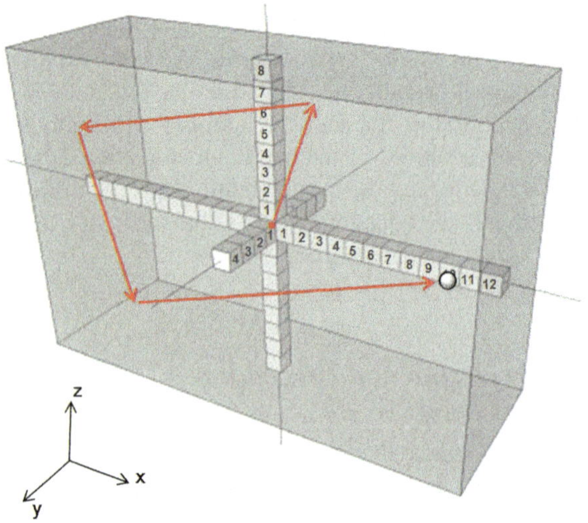


**Fig. 1** The virtual workspace is shaped like a cuboid. The cuboid can be constructed with 3825 small cubes whose sides measure 1 cm. The cubes that are positioned on the three spatial axes are shown here. However, the virtual cubes cannot be touched. Rather, they serve to illustrate a specific position inside the virtual workspace

hidden sphere inside the virtual workspace with random movements starting from the reference position. Once the HIP was in close proximity to the target, the magnetic effect on the surface of the touchable sphere attracted the HIP. Thus, the target position was reached. In the next step, the test persons were asked to specify the current position of the HIP without conducting additional movements. They could use words such as *left/right*, *up/down* and *forward/backward*. In addition, they were asked to specify the exact position relative to the reference position with numbers in centimeters for each axis. The numbers are exemplarily indicated for each axis in Fig. 1.

Participants received training on the entire procedure prior to the test conditions. In the training session, first, the test persons were introduced to the device and the dimensions of the virtual workspace. For this purpose, they used the visualization that is shown in Fig. 2. This visualization helped the participants to imagine what occurs when they move their arms. Because of the arm movement, the index finger respectively the tip of the stylus was displaced. The HIP was displaced in the same direction inside the virtual workspace. Once the HIP left a cube and entered another, the highlighted block moved and visualized the change of position. Furthermore, the test persons were introduced to the mechanical limits of the physical workspace of the device, which were somewhat rounded and slightly outside of the virtual workspace. In the second step, the participants were blindfolded and administered the experimental task in four exemplary trials. Each time they found the hidden sphere and estimated its position, the experimenter gave feedback about their localization error. However, this kind of feedback was only provided in these exemplary trials. After

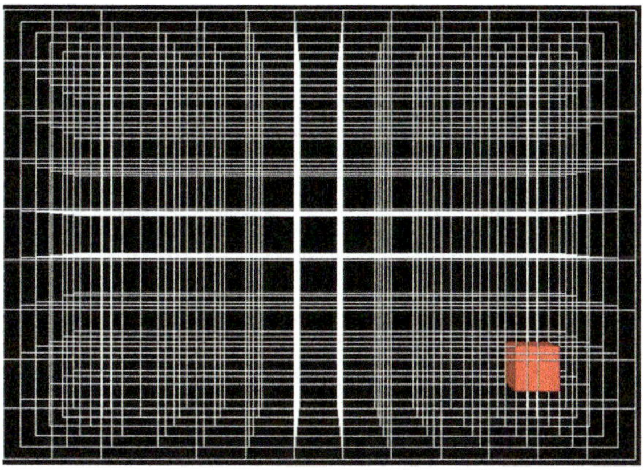

**Fig. 2** Visualization used at the beginning of the training session. It helps the participants to imagine what occurs when they move their arms. Because of the arm movement, the index finger respectively the tip of the stylus is displaced. The HIP is displaced in the same direction inside the virtual workspace. Once the HIP leaves a cube and enters another, the highlighted block moves and visualizes the change of position

the training procedure, subjects indicated no qualitative difficulties in the estimation process. They completed this task in a few seconds.

## *2.2 Condition #1*

To investigate the influence of differently positioned haptic workspaces on localization performance, a construction was built such that the position of the PHANToM could be easily changed. In the first condition, the device was placed at the height of a table—Fig. 3a. This position is quite comfortable and familiar in daily life from writing or typing on a computer keyboard.

Twelve test persons, two female and ten male, voluntarily participated in the first experimental condition. Their ages ranged from 21–30 years with a mean of 24 years. All participants indicated that they had no arm disorders. They were students or employees of Dresden University of Technology and had little to no experience using a haptic force-feedback device. The participants each completed 10 trials.

The results of the proprioceptive measurements are outlined in Fig. 4. The average of the absolute localization error vector in centimeters in the $x$-, $y$- and $z$-direction is shown. The grey bars refer to the overall localization errors, that is, the localization errors averaged over all hidden target spheres. The error increased from 1.4 cm in the forward-backward direction, that is, the $y$-axis, to 1.8 cm in the vertical direction, $z$-axis, and to 2.4 cm in the lateral direction, $x$-axis. The error increased because of the different side lengths of the cuboid—its length in the forward-backward direction is 9 cm, in the vertical direction 17 cm and in the lateral direction 25 cm. This

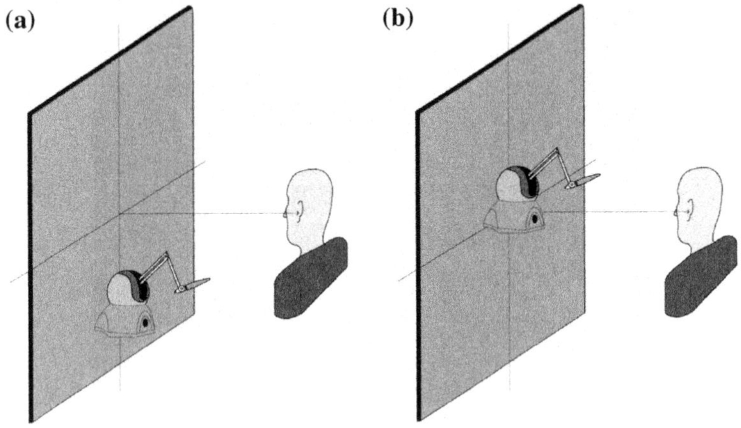


**Fig. 3** A construction was built such that the position of the PHANToM could be easily changed. By varying the position of the device, it is possible to investigate the influence of differently positioned haptic workspaces on localization performance. **a** In the first experimental condition, the device was placed at the height of a table. **b** In the second experimental condition, the device was placed at the height of test persons' head

fact was verified by calculating the localization errors solely for those target spheres that were randomly hidden in the same cuboid within a centered cubic volume with side lengths of 9 cm. In this case, the localization errors were almost the same in all directions. The corresponding error was 1.4 cm in the forward-backward direction and 1.7 cm in the vertical and the lateral directions. Thus, the localization accuracy of the proprioceptive sense did not depend on movement direction within this small workspace, although different types of movements were used, nameley radial movements along the $y$-axis and tangential movements along the $x$- and $z$-axis. Rather, the localization accuracy depended on the distance between the hidden sphere and the reference position. On average, the sphere was hidden further away from the reference point in the lateral direction than in the vertical and forward-backward directions. Therefore, the resulting error was greatest along the $x$-axis. This dependency is verified by the dark bars that are shown in Fig. 4. These bars refer to the localization errors for the spheres that were randomly hidden in the border area of the virtual workspace. The border area is defined as follows.

$$\begin{aligned} x_{pos} &> |6\,\text{cm}|\,\text{OR} \\ y_{pos} &> |2\,\text{cm}|\,\text{OR} \\ z_{pos} &> |4\,\text{cm}|. \end{aligned} \tag{1}$$

The mean localization error increased for all three directions. Therefore, if the sphere was hidden further away from the reference point in a specific direction, the resulting error was greater. An ANOVA for repeated measurements revealed a significant

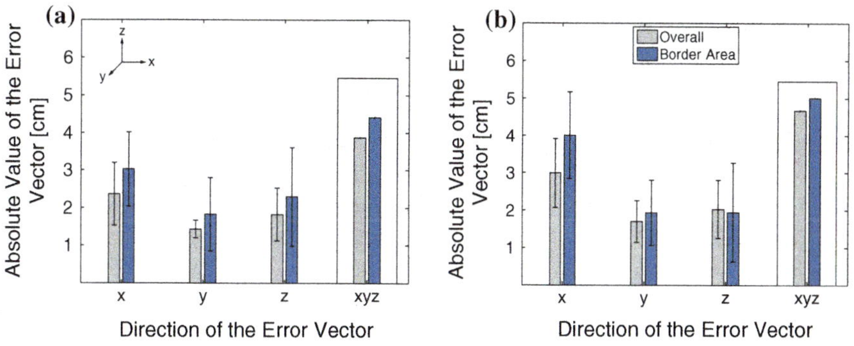


**Fig. 4** Experimental results. **a** Condition #1, PHANToM at table level. **b** Condition #2, PHANToM at head level. The average of the absolute localization error vector ± standard deviations in centimeters in the $x$-, $y$- and $z$-direction is shown. The calculated localization error inside the three-dimensional virtual workspace is also outlined. The *grey bars* refer to the overall localization errors, that is, the localization errors averaged over all hidden spheres. The *dark bars* refer the localization errors for those spheres that were randomly hidden in the border area of the virtual workspace. The border area is defined in (1)

difference between the localization errors indicated by the grey and dark bars for the $x$-axis, namely, $F = 19.164$ and $p < 0.01$. In addition, the standard deviations also increased for all three directions because of the broader spreading of test persons' estimations in these seemingly more complex trials.

However, what is the reason for this increasing error and for the increasing instability of test persons' estimations? Previous works observed a diminished accuracy of the proprioceptive sense once whole arm movements were involved in the exploration process [26]. In the present experiment, participants were required to move their whole arms to reach the border area of the virtual workspace, which may have affected their estimations. As a result, their arms were not in contact with their body; therefore, their body could not be used as a reference point. To investigate whether the described effect can be verified or even strengthened if whole arm movements are provoked, a second experimental condition was conducted.

## 2.3 Condition #2

In the second condition, the PHANToM was placed at the height of the test persons' head—see Fig. 3b. Thus, even if the HIP was located near the reference position, the arm was not in contact with the body anymore.

Twelve different subjects, three female and nine male, voluntarily participated in the second experimental condition. Their ages ranged from 21–49 years with a mean of 29 years. All participants indicated that they had no arm disorders. They were students or employees of Dresden University of Technology and had little to no experience using a haptic force-feedback device. The participants each completed 10 trials.

The results are presented in Fig. 4b. In comparison to condition #1, the overall localization error increased from 1.7 cm in the forward-backward direction, $y$-axis, to 2.0 cm in the vertical direction, $z$-axis, and to 3.0 cm in the lateral direction, $x$-axis. Again, no dependency between the localization accuracy and the movement direction was observed. This was verified as described above in Sect. 2.2. Rather, the error increased because of the different side lengths of the cuboid-shaped virtual workspace. On average, the sphere was hidden further away from the reference point in the lateral direction than in the vertical and forward-backward directions. Therefore, the resulting error was greatest along the $x$-axis. The dependency between the localization performance and the average distance to the target was verified through the calculation of the localization error for the border area of the virtual workspace. This error increased in comparison to the overall localization error for the $x$- and $y$-axis but not for the $z$-axis. An ANOVA for repeated measurements identified significant differences on the $x$-axis with $F = 9.692$ and $p < 0.05$.

When comparing these results with those of condition #1, both the overall localization accuracy and the localization accuracy for selected spheres in the border area decreased, especially for the $x$- and the $y$-axis. Therefore, the resulting overall error in the three-dimensional space increased considerably from 3.9 cm in condition #1 to 4.7 cm in condition #2. The resulting error in the three-dimensional workspace is an important issue concerning virtual haptic interaction. The values of 3.9 and 4.7 cm are considerable amounts, if the length of the virtual objects is limited, for example, to 10 cm. This size was used in the aforementioned identification experiments [39] in which the orientation-specific difficulties were originally observed.

Finally, it is important to note that the outlined results were obtained in a workspace that was slightly smaller than a shoe box. Because it is was found that movement distance directly influences the accuracy of proprioception, the mean-percentage localization errors might be greater in larger workspaces.

## 3 Audio–Proprioceptive Localization

### 3.1 Pre-Study

During the localization experiment described in Sect. 2, the test persons only used their body's own proprioceptive signals and the physical boundaries of the device-specific workspace for orientation. Thus, the experimental conditions were identical to those of the object identification experiments in which the orientation-specific problems were originally observed [12, 39]. This was a crucial requirement for identifying the cause of the difficulties. However, the physical boundaries of the device-specific workspace located slightly outside of the virtual cuboid were somewhat rounded and irregularly shaped due to the construction of the robot arm. This irregular shape could probably impede users' localization performance when interacting with haptic devices. It should be investigated in the pre-study whether such a negative influence actually exists. The experimental procedure was the same as described in Sect. 2.1 except that the test persons utilized their body's own proprioceptive signals and a pink-noise monophonic sound, which was switched off when the HIP touched or moved beyond the boundaries of the cuboid-shaped virtual workspace. These virtual boundaries were straight and thus regularly shaped—Fig. 1.

Two independent groups voluntarily participated in the pre-study. The first group consisted of twelve test personss, two female and ten male, aged from 21–34 years with a mean of 27 years. They took part in the first experimental condition—Fig. 3a. The second group also consisted of twelve test personss, two female and ten male, aged from 21–49 years and a mean of 29 years. They participated in the second experimental condition—Fig. 3b. All participants indicated that they had no arm disorders. They were students or employees of Dresden University of Technology and had little to no experience using a haptic force-feedback device. The participants each completed 20 trials. The trials were divided into two halves whose presentation order was randomized. In one half of the trials, the test persons used proprioception and a monophonic sound. The corresponding results are presented subsequently. The other half of the trials is detailed in the following section.

The results of the pre-study are illustrated with the help of the dark bars in Fig. 5. The grey bars refer to the results of the proprioceptive-only measurements that were already discussed in Sect. 2. They are shown again to aid comparison. Within the first experimental condition, test persons' localization performance in lateral direc-

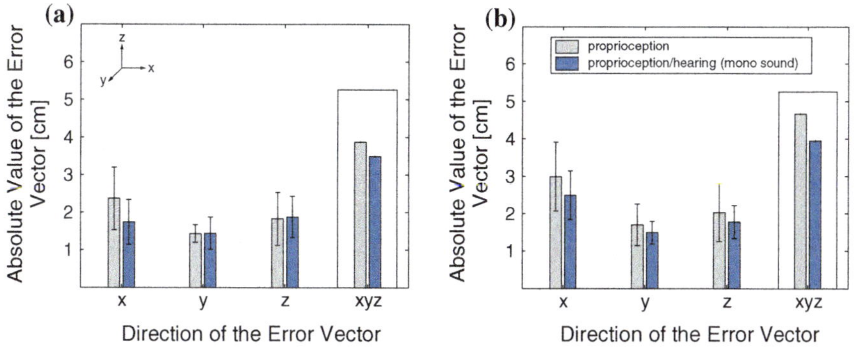


**Fig. 5** Experimental results. **a** Condition #1, PHANToM at table level. **b** Condition #2, PHANToM at head level. The average of the absolute localization error vectors ± standard deviations in centimeters are depicted for the $x$-, $y$- and $z$-direction as well as for the three-dimensional workspace. The *grey bars* refer to the results of the proprioceptive measurements, that is, the test persons only used their body's own proprioceptive signals and the physical boundaries of the device-specific workspace for localization. The *dark bars* refer to the results of the pre-study in which the test persons utilized their bodies' own proprioceptive signals and a monophonic sound that was switched off when the HIP touched or moved beyond the boundaries of the cuboid. The *grey bars* have been previously depicted in Fig. 4; they are shown again for comparison

tion increased due to the acoustically defined, regularly shaped borders of the virtual workspace. An ANOVA for independent samples revealed a statistically significant effect with $F = 4.607$ and $p < 0.05$. However, no effect was observed concerning the localization accuracy in forward-backward and in vertical direction. Within the second experimental condition, test persons' localization performance increased slightly in $x$-, $y$- and $z$-direction, but no statistically significant effect was found.

In conclusion, the pre-study can be summarized as follows. First, the acoustically defined, regularly shaped borders of the virtual workspace helped to prevent users from misinterpreting the mechanical limits of the haptic device as a virtual object. Second, the localization errors and the standard deviations could yet only be partly decreased.

### *3.2 Main Study*

It is investigated within the main study how both proprioception and spatialized sound are used in combination to localize the HIP. Thus, the influence of binaural signals on proprioception is studied to reveal whether synthetically generated spatial sound is considered and might even improve human proprioceptive localization performance. Employing the hearing system to extend proprioceptive perception auditorily seems to be a valuable approach. For example, individuals utilize their hearing system on the streets daily to estimate the direction and distance from which a car approaches without being forced to look at the car. Because spatial audible information is intuitively used in the real world, it should also be incorporated in virtual environments.

The usefulness of auditory localization cues for haptic virtual environments was verified in several studies. Such cues were successfully used in hapto-audio navigational tasks, for example, when users attempted to explore, learn and manage a route in a virtual traffic environment [29]. They were also quite helpful for locating objects in the virtual space [28]. In these studies, the haptic device was used to move the virtual representation of oneself and, thus, one's own ears freely in the virtual scene—that is, as an *avatar*. Therefore, if an object emitted a sound, the user heard this sound from a specific direction depending on the position and the orientation of the hand-controlled avatar. Thus, this method was called the *ears-in-hand* interaction technique [29].

However, the present work aims to investigate how proprioceptive signals and auditory localization cues provided by spatialized sound are used in combination to localize the absolute position of the HIP. The present study's approach is not comparable to the abovementioned approaches. In the current study, the HIP does not correspond to a virtual representation of oneself including one's own ears. Rather, it corresponds to a virtual representation of the fingertip. Furthermore, this study does not aim to localize a sound-emitting object or something similar in the virtual scene. Rather, the study aims to trace the movements of the virtual fingertip auditorily to increase the localization resolution. To investigate whether localization performance can be improved, it is essential to develop a hapto-audio system that auralizes each movement of the user and, thus, the movement of the haptic device. Therefore, each arrow depicted in Fig. 1 must cause a corresponding variation in the reproduced spatial sound, as if the user moves a sound source with his/her hand in the same direction directly in front of the his/her head. The authors of the present chapter call this method the *sound source-in-hand* interaction technique.

## Development of a Hapto–Audio System

There are two main methods for reproducing spatial sound. On the one hand, head-related transfer functions, HRTFs, can be used for binaural reproduction via headphones [6]. In this case, the free-field transfer function from a sound-emitting point in space to the listener's ear canal is used to filter the sound. On the other hand, spatial sound can be reproduced by various techniques using loudspeaker arrays, for example, ambisonics [30], vector-base amplitude panning, VBAP, [34] and wave field synthesis, WFS, [4].

In the present study, generalized HRTFs are used to generate auditory localization cues. An extensive set of HRTF measurements of a KEMAR dummy-head microphone is provided by Gardner et al. [17]. These measurements consist of the left and right ear impulse responses from a total of 710 different positions distributed on a spherical surface with 360° azimuth and −40° to 90° elevation. In a first step, the set of shortened 128-point impulse responses was selected for this investigation. Second, this set was reduced to impulse responses in the range between −30° and +30° in azimuth and elevation. For the real-time binaural simulation of a continuously moving sound source according to the sound source-in-hand interaction technique, an algorithm capable of interpolating between the HRTF measurement points is required. This algorithm should smoothly handle also very fast movements without interruptions. That is the reason why the convincing time-domain-convolution algorithm of Xiang et al. [43] was selected. This algorithm approximates the HRTF at the target position of the current signal block by linearly interpolating the four nearest measurement points. An exemplary case is depicted in Fig. 6. The point $T$ corresponds to the target position which is surrounded by the points $P_1$, $P_2$, $P_3$ and

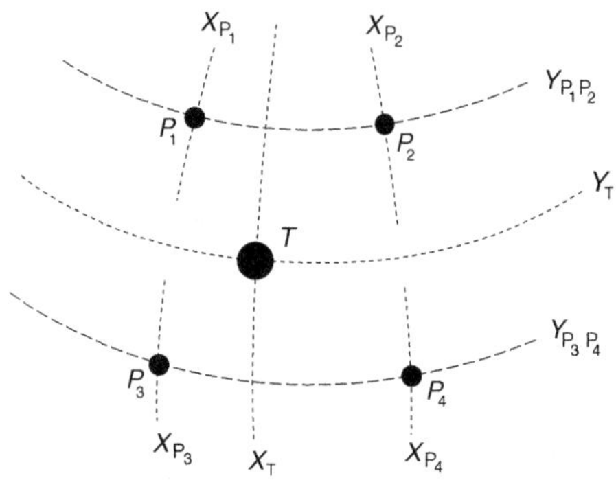


**Fig. 6** The HRTF at target position, $T$, of the current signal block is computed by linearly interpolating the HRTFs at the four nearest measurement points, $P_1$, $P_2$, $P_3$ and $P_4$. This figure is only for demonstration purposes. The four measurement points were chosen arbitrarily—after [43]

$P_4$. The measured HRTFs at these four points, $F_{P_1}$, $F_{P_2}$, $F_{P_3}$ and $F_{P_4}$, can now be used to calculate the HRTF, $F_T$, at the target position, $T$, according to the following equation [43],

$$F_{\mathrm{T}} = \frac{Y_{\mathrm{T}} - Y_{\mathrm{P_1P_2}}}{Y_{\mathrm{P_3P_4}} - Y_{\mathrm{P_1P_2}}} \left( \frac{X_{\mathrm{T}} - X_{\mathrm{P_1}}}{X_{\mathrm{P_2}} - X_{\mathrm{P_1}}} F_{\mathrm{P_1}} - \frac{X_{\mathrm{P_2}} - X_{\mathrm{T}}}{X_{\mathrm{P_2}} - X_{\mathrm{P_1}}} F_{\mathrm{P_2}} \right) + \frac{Y_{\mathrm{P_3P_4}} - Y_{\mathrm{T}}}{Y_{\mathrm{P_3P_4}} - Y_{\mathrm{P_1P_2}}} \left( \frac{X_{\mathrm{T}} - X_{\mathrm{P_3}}}{X_{\mathrm{P_4}} - X_{\mathrm{P_3}}} F_{\mathrm{P_3}} - \frac{X_{\mathrm{P_4}} - X_{\mathrm{T}}}{X_{\mathrm{P_4}} - X_{\mathrm{P_3}}} F_{\mathrm{P_4}} \right) . \tag{2}$$

The next step refers to the transformation of the monophonic signal block, $x$, to the spatialized signal block, $y$. Both blocks can be specified in their lengths by the block-size of $b = 64$ samples. The transformation process is realized by the time-domain convolution of the input signal with a 128-tap filter. This filter is again the result of a linear interpolation between the HRTFs $F_T$ and $F_{T_0}$. $F_T$ was computed for the current signal block and the target position, $T$. $F_{T_0}$ was computed for the previous signal block and corresponding previous target position $T_0$. In conclusion, each sample, $k$, of the output signal is calculated according to the following equation [43],

$$y(k) = \sum_{n=0}^{127} x(k-n) \cdot \left( \frac{k}{b} \cdot F_{\mathrm{T}}(n) + \frac{b-k}{b} \cdot F_{\mathrm{T_0}}(n) \right) , \quad \text{with} \quad k = 0, 1, \ \ldots \ b-1 \ . \tag{3}$$

Using the abovementioned selection of head-related impulse responses, the sound source reproducing broadband noise could be virtually positioned anywhere in the range between $-30°$ and $+30°$ in azimuth and elevation. However, at this point, the generated aural environment was only two-dimensional. To auralize movements in the forward-backward direction, auditory distance cues must be provided. An intuitive way to achieve a high resolution for the localization in the forward-backward direction is to vary the sound pressure level of the corresponding signal. It is well-known from our daily experiences that a distanced sound source is perceived as quieter than a proximal sound source. This fact can be easily utilized in virtual environments. However, because of the small virtual workspace, the sound pressure level variation must be exaggerated in comparison to the real physical world. This exaggeration allows to profit considerably of the high resolution and, thus, the low differential thresholds of the hearing system as concerns level variation, namely, $\Delta L \approx 1\,\mathrm{dB}$, [16]. In the current study, a level range of $L = 60 \pm 12$ dB(A) was applied. Thus, the localization signal became louder when the user moved the HIP forward, and it became quieter when the user moved the HIP backward, that is, away from the body.

Finally, the aural and haptic environments must be linked with each other. The aim is to generate a hapto–audio space by auditorily extending the haptic environment. This extension is performed linearly in depth but also in height and width. Thereby, the height is enlarged more than the width to qualitatively account for the lower

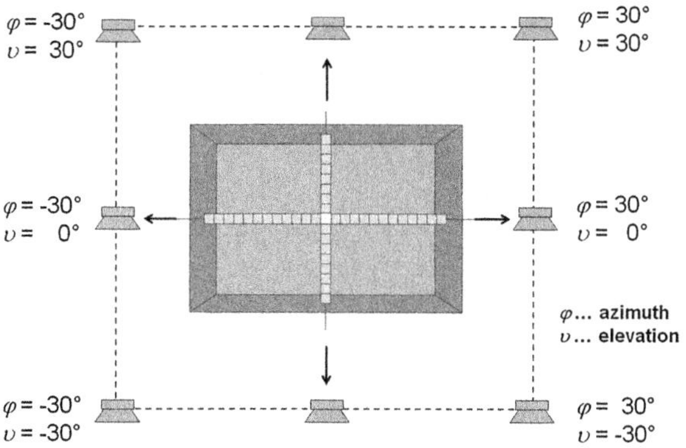


**Fig. 7** The hapto–audio workspace is generated by auditorily extending the haptic environment. This extension is shown here schematically for the height and width

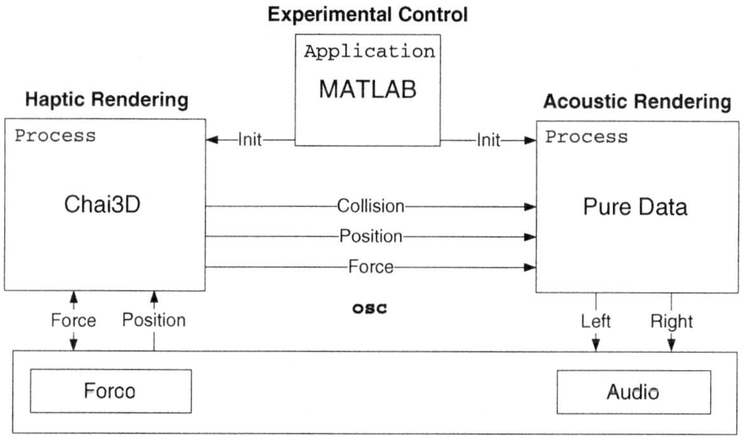


**Fig. 8** Experimental software framework

auditory localization resolution in elevation [6]. The extension is illustrated in Fig. 7 and mathematically detailed in the following section.

The developed software framework incorporating haptic and acoustic rendering processes as well as the experimental control is depicted in Fig. 8. Matlab was used to automate the experimental procedure and question the test subjects. Matlab also initiated the processes for haptic rendering and acoustic rendering.[2] To reproduce the corresponding sound signals, the acoustic renderer requires information from the PHANToM, which communicates with the haptic renderer via a software interface. This information, for example, about the position of the HIP or a detected collision

[2] Chai3D, Stanford University, and Pure Data, Open-Source Project, respectively. For details on the C++ haptic-rendering framework, particularly, the algorithms for collision detection, force control and force response, the reader is referred to [13]. The details of Pure Data are outlined in [33]

with the hidden sphere, is steadily transferred to the acoustic renderer via Open Sound Control, OSC. The information is then processed for subsequent signal output, that is, to reproduce spatialized broadband noise via headphones and to provide force-feedback, once the target position is reached.

### Experimental Procedure

The test persons were seated on a chair without armrests. At the beginning of each trial of the experiment, a sphere with a diameter of 1 cm was randomly positioned inside the virtual workspace, thus, inside one of the 3825 virtual cubes—see Fig. 1. Then, the experimenter moved the cursor-like HIP, which corresponds to the tip of the stylus, to the reference position, that is, the center of the virtual workspace. Because of a magnetic effect that was implemented directly on the central point, the stylus remained in this position. The generated virtual sound source was positioned at 0° in azimuth and elevation. The sound pressure level of the broadband noise was 60 dB(A). The blindfolded test persons grasped the stylus and held it parallel to the y-axis in such a manner that the extended index finger corresponded to the tip of the stylus. The test persons were asked to search for the hidden sphere with random movements starting from the reference position. When moving the stylus of the haptic device, the HIP and the sound source were also displaced. The direction of the sound source, $\varphi$ in azimuth and $\upsilon$ in elevation, depended linearly on the position, $x$ and $z$, of the HIP according to the following equations,

$$\varphi(x) = \frac{\varphi_{max}}{x_{max}} \cdot x \qquad \text{with} \quad x = [-x_{max},\, x_{max}]\,, \tag{4}$$

$$\upsilon(z) = \frac{\upsilon_{max}}{z_{max}} \cdot z \qquad \text{with} \quad z = [-z_{max},\, z_{max}]\,. \tag{5}$$

The right border of the cuboid, $x_{max} = 12.5\,\text{cm}$, and the left border of the cuboid, $-x_{max} = -12.5\,\text{cm}$, corresponded to a maximal displacement of the sound source at $\varphi_{max} = 30°$ and $-\varphi_{max} = -30°$. Similarly, the upper border, $z_{max} = 8.5\,\text{cm}$, and the lower border, $-z_{max} = -8.5\,\text{cm}$, corresponded to a maximal displacement of the sound source at $\upsilon_{max} = 30°$ and $-\upsilon_{max} = -30°$. When the HIP moved beyond the limits of the cuboid-shaped virtual workspace, the sound was immediately switched off. The sound-pressure level of the broadband noise depended on the position of the HIP in $y$-direction. It was calculated according to the following equation with $L_{init} = 60\,\text{dB(A)}$, $\Delta L = 12\,\text{dB(A)}$ and $|y_{max}| = 4.5\,\text{cm}$,

$$L(y) = L_{init} + \frac{\Delta L}{y_{max}} \cdot y \qquad \text{with} \quad y = [-y_{max},\, y_{max}]\,. \tag{6}$$

Once the HIP was in close proximity to the target, the magnetic effect that was implemented on the surface of the touchable sphere attracted the HIP. Thus, the target position was reached. In the next step, the test persons were asked to specify the current position of the HIP without conducting further movements. This method was previously described in Sect. 2.1.

Participants received training on the entire procedure prior to the test conditions. In the training session, first, the test persons were introduced to the device and the hapto-audio workspace. They used the visualization that is depicted in Fig. 2. In the second step, they were blindfolded and listened to consecutively presented sound sources that moved on a specific axis. The corresponding number that specified the actual position of the sound source was provided. This was conducted separately for each axis. Therefore, the test persons experienced the maximal displacements of the sound source and the entire range in between, which is essential for making estimations. Subsequently, the participants conducted the experimental task in four exemplary trials. Each time they found the hidden sphere and estimated its position, the experimenter gave feedback about their localization error. However, this kind of feedback was only provided in these exemplary trials.

**Listening Test**

In a first step, a listening test was conducted to measure test persons' achievable auditory localization accuracy if the aforementioned spatialization technique is applied. Thus, test persons did not interact with the haptic device. Rather, they only listened to a continuously moving virtual sound source. The path of this source was selected automatically in a random manner and also its motion speed varied randomly. After approximately 10 sec the sound source reached the target position, which could then be specified as described above.

Eight test persons, two female and six male, voluntarily participated in the listening test. Their ages ranged from 24–49 years with a mean of 33 years. All participants indicated that they did not have any hearing or spinal damage. They were students or employees of Dresden University of Technology. The participants each completed 10 trials.

The results of the listening test are shown in Fig. 9 with the white bars. These results enable to determine whether it will be at least theoretically possible to further improve localization accuracy with the spatialization technique. The average auditory localization error in azimuth direction was measured to be $4°-5°$. This value corresponds to a localization error of 1.8 cm inside the virtual workspace. In the first experimental condition, test persons already achieved a localization error of 1.8 cm when using the monophonic sound. Thus, they will not be able to further improve their localization performance with spatial auditory cues. However, test persons should be able to clearly decrease the localization error in lateral direction from 2.5 to 1.8 cm in the second experimental condition, in which whole arm movements are provoked. In this case, the spatial auditory cues should be helpful. Furthermore, the localization error in forward-backward direction might even be halved from 1.4 cm in the first condition and 1.7 cm in the second condition to 0.7 cm with available spatial sound. In contrast, no improvement should be achievable in vertical direction, because the

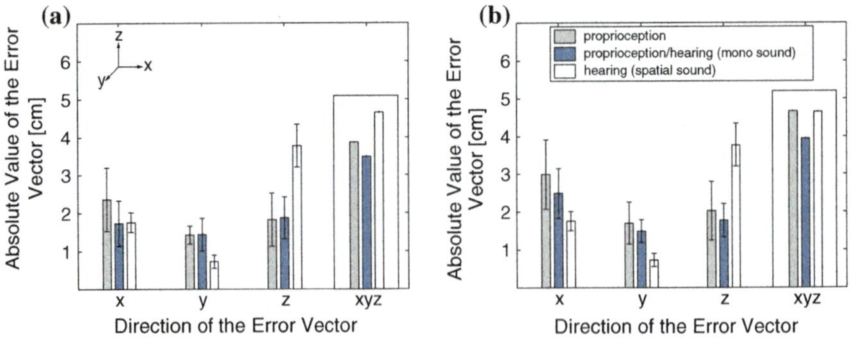


**Fig. 9** Experimental results of the listening test. The average of the absolute localization error vector ± standard deviations in centimeters in *x*-, *y*- and *z*-direction is shown. The calculated localization errors inside the three-dimensional virtual workspace are also outlined. The *grey bars* refer to the results of the proprioceptive measurements, that is, no acoustic signals were available. The *dark bars* refer to the results of the pre-study in which the test persons utilized proprioception and a monophonic sound that was switched off when the HIP touched or moved beyond the boundaries of the cuboid. The *white bars* refer to the results of the listening test. They are depicted in the *left* (**a**) and *right* graphs (**b**) for comparison

average auditory localization error in elevation was measured to be 13°–14°. This value corresponds to a localization error of 3.8 cm inside the virtual workspace. The proprioceptive localization errors of 1.8 cm in conditions #1 and 2 cm in condition #2 are clearly smaller.

Subsequently, results will be presented that reveal how test persons perform if both proprioceptive and spatial auditory signals are used in combination. These results will help to understand whether binaural signals are considered during localization or whether the localization process is mainly dominated by proprioception. In audio–visual interaction, the bimodal sensory input is integrated in an optimal manner. If participants would also integrate proprioceptive and binaural signals optimally, the resulting localization performance could be predicted. According to such a prediction in the context of this investigation, test persons would reject binaural signals for estimating the target position in vertical direction. They would rather trust in proprioception. However, the binaural signals would be considered to estimate the position in forward-backward direction and in lateral direction in condition #2.

## Condition #1

In the first condition, the device was placed at the height of a table, as illustrated in Fig. 3a. The group of participants was already introduced in Sect. 3.1. The participants each completed 20 trials. The trials were divided into two halves, and the presentation order was randomized. During one half of the trials, the test persons used proprioception and a monophonic sound which helped to clearly define the

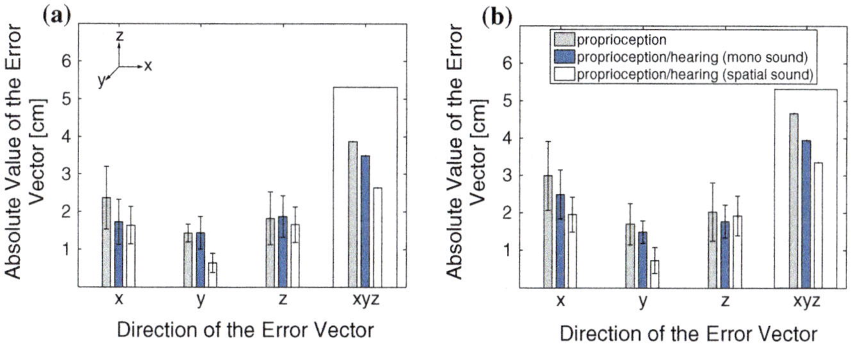


**Fig. 10** Experimental results. **a** Condition #1 – PHANToM at table level. **b** Condition #2—PHANToM at head level. The average of the absolute localization error vector ± standard deviations in centimeters in $x$-, $y$- and $z$-direction is shown. The calculated localization errors inside the three-dimensional virtual workspace are also outlined. The *grey bars* refer to the results of the proprioceptive-only measurements. The *dark bars* refer to the results of the pre-study in which the test persons used their body's own proprioceptive signals and a monophonic sound that helped to clearly define the boundaries of the virtual workspace. The *white bars* refer to the results of the main study in which test persons used both proprioceptive sensory input and spatial auditory cues according to the sound source-in-hand interaction technique

boundaries of the virtual workspace. This part of the experiment was denoted as pre-study. During the other half of the trials, the test persons used proprioception and additionally perceived spatial auditory cues according to the sound source-in-hand interaction technique. This part is analyzed subsequently.

The results are shown in Fig. 10a. The grey bars refer to the results of the proprioceptive measurements that were previously illustrated in Fig. 4. The dark blue bars refer to the results of the pre-study. The white bars show the localization errors that occurred during binaural-proprioceptive interaction according to the sound source-in-hand interaction technique. The results indicate that test persons integrate proprioceptive and binaural signals in such a manner that the resulting localization error is minimized. They achieved a localization accuracy of 1.7 cm in vertical direction. Thus, they rejected binaural signals and rather trusted in proprioception. However, they used binaural signals to estimate the position in forward-backward direction which is why the localization error was significantly reduced from 1.4 cm to approximately 0.6 cm. An ANOVA for repeated measurements rendered $F = 57.863$ and $p < 0.001$. No further improvement was observable concerning the localization performance in lateral direction, as predicted before.

In conclusion, test persons could actually differentiate between the more reliable proprioceptive signals for the localization in vertical direction and the more reliable auditory cues for the localization in forward-backward direction. The resulting error in the three-dimensional workspace was overall reduced by approximately 30 % from 3.9 to 2.8 cm.

**Condition #2**

In the second condition, the PHANToM was placed at the height of the test person's head—Fig. 3b. Thus, whole arm movements were provoked. The group of participants was already introduced in Sect. 3.1. The corresponding results are analyzed below.

In Fig. 10b, the average of the absolute localization error vectors are depicted for the $x$-, $y$- and $z$-direction as well as the three-dimensional workspace. As mentioned before, the grey bars refer to the results of the proprioceptive measurements that were previously illustrated in Fig. 4. The dark bars refer to the results of the pre-study. The white bars show the localization errors that occurred during binaural-proprioceptive interaction according to the sound source-in-hand interaction technique. The results verify that test persons integrate proprioceptive and binaural signals in such a manner that the resulting localization error is minimized. Again, they rejected binaural signals and rather trusted in proprioception when specifying the target position in vertical direction. They achieved a localization accuracy of 1.9 cm. However, they used binaural signals to estimate the position in forward-backward direction which is why the localization error was more than halved from 1.5 to 0.7 cm. An ANOVA for repeated measurements rendered $F = 25.226$ and $p < 0.001$. Furthermore, as predicted before in case of optimal integration, the localization error in lateral direction could be further reduced from 2.5 to 1.9 cm. An ANOVA for repeated measurements revealed a statistically significant effect, namely, $F = 5.532$ and $p < 0.05$.

In conclusion, test persons could again differentiate between the more reliable proprioceptive signals for the localization in vertical direction and the more reliable auditory cues for the localization in forward-backward and lateral direction. The resulting error in the three-dimensional workspace was overall reduced by approximately 30 % from 4.7 to 3.4 cm.

## 4 Summary and Conclusions

With the help of binaural models, human localization can be predicted under the assumption that a corresponding localization process is based on acoustic signals, thus, on unimodal information. However, what happens if this localization process is realized in an environment with available bimodal or even multimodal sensory input? This chapter sets focus on investigating the influence of binaural signals on proprioception to reveal whether synthetically generated spatial auditory signals are considered during localization and might even improve human proprioceptive localization performance in virtual workspaces. Quantitative results were gained with the help of corresponding experiments in which freely exploring test persons' unimodal and bimodal localization performance was measured. To obtain a greater understanding of such complex experiments with actively interacting test persons, the experimental variables, concurrent experimental processes, results and potential errors, it is quite helpful to develop a formalization of the corresponding experimental

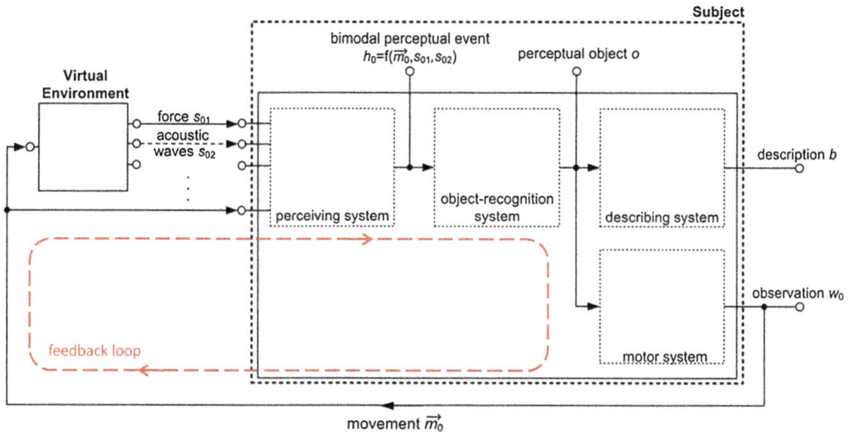


**Fig. 11** Formalization. Basic schematic representation of a test person, (*right block*), that haptically interacts in a virtual environment (*left block*) during a haptic experiment

type. Blauert's system analysis of the auditory experiment serves as an example [6]. However, because of the active and bi-directional nature of the haptic human sense, Blauert's approach must be adapted and extended to describe experiments or situations in which subjects haptically interact in a virtual environment. Therefore, a further developed formalization based on his model is introduced subsequently.

This formalization is depicted in Fig. 11. It contains the basic schematic representation of the active subject—right block—who interacts in the virtual environment—left block—by controlling a specific device. The inputs and outputs of the blocks and the interaction between them can be explained with reference to the main study. During the experiment, the test person actively moved his/her own arm through commands of the motor system. The corresponding movement, $\vec{m}_0$, at the time $t_0$ was transmitted to the haptic device, which was controlled by the test person's index finger. Thus, the device served as an interface and provided input to the virtual environment. According to the sound source-in-hand interaction technique, the subject heard a spatial sound that varied depending on the current movement direction. The corresponding acoustic waves, $s_{02}$, were output ideally without delay at the time $t_0$ by the virtual environment via headphones. At the same time, the subject also directly perceived his/her own joint movements through proprioception. As a result of the proprioceptive signals and auditory cues, the test person perceived a bimodal perceptual event, $h_0$, which can be described. Of course, both unimodal perceptual events, $h_{01}$ and $h_{02}$, can also be described separately. By integrating the preceding bimodal events and the current event, a perceptual object develops. A perceptual object is the result of mental processing. In the present case, it corresponded to a mental image of the virtual workspace and the corresponding position of the HIP in the workspace. A special characteristic and, at the same time, a distinguishing characteristic in comparison to auditory experiments is indicated by the feedback loop depicted in Fig. 11. To find the hidden sphere, the subject explored the virtual workspace continuously and individually by moving the stylus of the PHANToM. Thus, the perceived bimodal events, $h_0 \dots h_n$, differed across subjects because of the individual movements $\vec{m}_0 \dots \vec{m}_n$ and the resulting acoustic signals $s_{02} \dots s_{n2}$. Once the sphere was found at time $t_n$, a collision between the HIP and the sphere was detected. The virtual environment exerts a force, $s_{n1}$, to the test person via the haptic device. The shortly preceding bimodal events and $h_n$, the current event, were integrated to specify the perceptual object, $o$. As a result, the experimenter obtained a description, $b$, of the subject concerning the internal perceptual object, $o$. This description contained the assumed position of the target. With the help of these descriptions, collected from a group of participants, quantitative relations can be obtained, for instance, Figs. 4, 5, 9 and 10. However, the experimenter can also profit from experimental observations, for example, by watching each movement, watching the involved joints and recording the required time. These observations can help in understanding how subjects interact with the virtual environment and in determining possible explanations for their responses.

In the main part of this chapter, first, the localization accuracy of proprioception was investigated. The test persons only perceived their own movements, $\vec{m}_0 \dots \vec{m}_n$, with the help of the proprioceptive receptors. The resulting unimodal perceptual events, $h_0 \dots h_n$, served to develop a perceptual object, that is, a mental representation of the virtual workspace and the HIP in it. The virtual environment provided no additional information. The experimental results help to explain the orientation-specific difficulties that were originally observed in the object identification experiments in Stamm et al. [39] and Colwell et al. [12]. In those experiments, subjects required a substantial amount of time to regain a virtual object after they lost contact with it because they could not easily locate the HIP in relation to the object. Furthermore, they often reached the boundaries of the physical workspace unconsciously and misinterpreted the mechanical borders as an object. These difficulties indicate that their mental representation of the virtual workspace and the position of the HIP in it deviated from reality. The present study found that, indeed, subjects considerably misjudge the actual position of the HIP inside the three-dimensional workspace by approximately 4–5 cm. This is a remarkable amount, for example, if the length of the virtual objects is limited to 10 cm, as it was the case in the aforementioned identification experiments. The current study further found that the localization accuracy of proprioception depends on the distance between the current position of the HIP and a corresponding reference point. If whole arm movements are used to overcome this distance, the localization error and the standard deviations increase considerably. This increase was observed especially for the lateral direction and to a small, non-significant extent for the forward-backward direction.

In the second step, it was investigated whether the localization accuracy improves if proprioception and the hearing sense are used in combination. Thus, the test persons perceived their own movements, $\vec{m}_0 \dots \vec{m}_n$, but also the acoustic signals $s_{02} \dots s_{n2}$. As a result, they perceived the bimodal events $h_0 \dots h_n$. The experimental results of the main study demonstrated that the abovementioned proprioceptive localization inaccuracy was reduced significantly. Proprioception can be guided if additional spatial auditory cues are provided and, thus, the localization performance can be improved. Significant effects were found concerning the localization errors in forward-backward direction and also in lateral direction when whole arm movements were provoked. The resulting errors in the three-dimensional workspace were overall decreased by approximately 30 %.

It is important to mention that the localization accuracy was investigated in a haptic workspace that was slightly smaller than a shoe box. Because it is generally accepted that proprioceptive accuracy depends on the degree to which whole arm movements are involved in the exploration process, proprioceptive localization performance might be worse in larger workspaces. In such workspaces, the positive influence of auditory localization cues may be even stronger. This hypothesis should be investigated in future studies. Furthermore, it would be quite useful to investigate to which extent the localization performance in vertical direction can be improved, for example, if individualized HRTFs are used for the spatialization.

In conclusion, the experimental results clearly show that synthetically generated auditory localization signals are considered in the localization process and can even guide human proprioceptive localization within workspaces directly in front of the human body. The auditory and proprioceptive information is combined in such a way that the resulting localization error is minimized. As described in the introductory part, a similar effect was also observed during audio–visual localization. However, audio–visual interaction involves hearing and vision that both belong to exteroception by which one perceives the outside world. During binaural–proprioceptive localization, hearing and proprioception are involved and, thus, exteroception and interoception are combined. This combination was not investigated before as far as efficient integration of bimodal sensory signals within the given context is concerned.

If auditory localization signals can guide human proprioceptive localization, as it is described in this chapter, then binaural models can also help to model the corresponding bimodal integration process, for example, the model mimicking the localization of an elevated sound source out of [3], this volume. To build a reliable computational model, of course, deep knowledge is required. That is the reason why the complex proprioceptive localization process and bimodal integration have to be studied in more detail. Furthermore, existing binaural models have to be extended, for example, to handle distance cues. New fields of applications will profit of a binaural-proprioceptive localization model. For example, such a model might help to simulate how an audio reproduction system has to be designed to guide proprioception and, thus, optimize bimodal localization accuracy within an arbitrary haptic workspace size directly in front of the human body. Thus, practical relevance exists especially for virtual environments in which the reproduction of acoustic signals can be directly controlled and the localization accuracy can be consciously influenced. Furthermore, simulations of such a model may help to reproduce suitable auditory localization signals to diminish erroneous proprioceptive space perception, for example, the radial-tangential illusion, and thus to further sharpen human precision by auditorily calibrating proprioception.

These examples make obvious, finally, that the field of possible applications of binaural models is not limited to audio-only or audio-visual scenes. The increasingly important field of virtual haptic interaction will also profit from binaural modeling algorithms.

**Acknowledgments** The authors wish to thank the Deutsche Forschungsgemeinschaft for supporting this work under the contract DFG 156/1-1. The authors are indebted to S. Argentieri, P. Majdak, S. Merchel, A. Kohlrausch and two anonymous reviewers for helpful comments on an earlier version of the manuscript.

## References

1. D. Alais and D. Burr. The ventriloquist effect results from near-optimal bimodal integration. *Curr Biol*, 14:257–262, 2004.
2. S. Argentieri, A. Portello, M. Bernard, P. Danès, and B. Gas. Binaural systems in robotics. In J. Blauert, editor, *The technology of binaural listening,* chapter 9. Springer, Berlin-Heidelberg-New York NY, 2013.
3. R. Baumgartner, P. Majdak, and B. Laback. Assessment of sagittal-plane sound-localization performance in spatial-audio applications. In J. Blauert, editor, *The technology of binaural listening,* chapter 4. Springer, Berlin-Heidelberg-New York NY, 2013.
4. A. Berkhout, D. de Vries, and P. Vogel. Acoustic control by wave field synthesis. *J. Audio Eng. Soc.,* 93:2764–2778, 1993.
5. J. Blauert. Sound localization in the median plane. *Acustica,* 22:205–213, 1969.
6. J. Blauert. *Spatial hearing.* Revised Edition. The MIT Press, Cambridge, London, 1997.
7. J. Blauert and J. Braasch. Binaural signal processing. In *Proc. Intl. Conf. Digital Signal Processing,* pages 1–11, 2011.
8. J. Braasch. Modelling of binaural hearing. In J. Blauert, editor, *Communication Acoustics,* chapter 4, pages 75–108. Springer, 2005.
9. S. Brewster. Using non-speech sound to overcome information overload. *Displays,* 17:179–189, 1997.
10. F. Clark and K. Horch. Kinesthesia. In L. K. K. Boff and J. Thomas, editors, *Handbook of Perception and Human Performance,* chapter 13, pages 1–62. Willey-Interscience, 1986.
11. F. J. Clark. How accurately can we perceive the position of our limbs? *Behav. Brain Sci.,* 15:725–726, 1992.
12. C. Colwell, H. Petrie, and D. Kornbrot. Use of a haptic device by blind and sighted people: Perception of virtual textures and objects. In I. Placencia and E. Porrero, editors, *Improving the Quality of Life for the European Citizen: Technology for Inclusive Design and Equality,* pages 243–250. IOS Press, Amsterdam, 1998.
13. F. Conti, F. Barbagli, D. Morris, and C. Sewell. CHAI 3D - Documentation, 2012. last viewed on 12–09-29.
14. M. O. Ernst and M. S. Banks. Humans integrate visual and haptic information in a statistically optimal fashion. *Nature,* 415:429–433, 2002.
15. A. Faeth, M. Oren, and C. Harding. Combining 3-D geovisualization with force feedback driven user interaction. In *Proc. Intl. Conf. Advances in Geographic Information Systems,* pages 1–9, Irvine, California, USA, 2008.
16. E. Fastl and H. Zwicker. *Psychoacoustics - Facts and Models.* Springer, 2007.
17. B. Gardner and K. Martin. HRTF Measurements of a KEMAR Dummy-Head Microphone. *MIT Media Lab Perceptual Computing,* (280):1–7, 1994.
18. T. Haidegger, J. Sándor, and Z. Benyó. Surgery in space: the future of robotic telesurgery. *Surg. Endosc.,* 25:681–690, 2011.
19. L. A. Hall and D. I. McCloskey. Detections of movements imposed on finger, elbow and shoulder joints. *J. Physiol.,* 335:519–533, 1983.
20. K. L. Holland, R. L. Williams II, R. R. Conatser Jr., J. N. Howell, and D. L. Cade. The implementation and evaluation of a virtual haptic back. *Virtual Reality,* 7:94–102, 2004.
21. G. Jansson, M. Bergamasco, and A. Frisoli. A new option for the visually impaired to experience 3D art at museums: Manual exploration of virtual copies. *Vis. Impair. Res.,* 5:1–12, 2003.

22. L. A. Jeffress. A place theory of sound localization. *J. Comp. Physiol. Psychol.,* 41:35–39, 1948.
23. L. Jones and I. Hunter. Differential thresholds for limb movement measured using adaptive techniques. *Percept. Psychophys.,* 52:529–535, 1992.
24. M. Keehner and R. K. Lowe. Seeing with the hands and with the eyes: The contributions of haptic cues to anatomical shape recognition in surgery. In *Proc. Symposium Cognitive Shape Processing,* pages 8–14, 2009.
25. V. Khalidov, F. Forbes, M. Hansard, E. Arnaud, and R. Horaud. Audio-visual clustering for multiple speaker localization. In *Proc. Intl. Worksh. Machine Learning for Multimodal Interaction,* pages 86–97, Utrecht, Netherlands, 2008. Springer-Verlag.
26. S. J. Lederman and L. A. Jones. Tactile and haptic illusions. *IEEE Trans. Haptics,* 4:273–294, 2011.
27. S. J. Lederman and R. L. Klatzky. Haptic identification of common objects: Effects of constraining the manual exploration process. Percept. *Psychophys.,* 66:618–628, 2004.
28. C. Magnusson and K. Rassmus-Grohn. Audio haptic tools for navigation in non visual environments. In *Proc. Intl. Conf. Enactive Interfaces,* pages 17–18, Genoa, Italy, 2005.
29. C. Magnusson and K. Rassmus-Grohn. A virtual traffic environment for people with visual impairment. *Vis. Impair. Res.,* 7:1–12, 2005.
30. D. Malham and A. Myatt. 3D sound spatialization using ambisonic techniques. *Comp. Music J.,* 19:58–70, 1995.
31. I. Oakley, M. R. McGee, S. Brewster, and P. Gray. Putting the feel in look and feel. In *Proc. Intl. Conf. Human Factors in Computing Systems,* pages 415–422, Den Haag, Niederlande, 2000.
32. A. M. Okamura. Methods for haptic feedback in teleoperated robot-assisted surgery. *Industrial Robot,* 31:499–508, 2004.
33. Open Source Project. Pure Data - Documentation, 2012. last viewed on 12–09-29.
34. V. Pulkki. Virtual sound source positioning using vector base amplitude panning. *J. Audio Eng. Soc.,* 45:456–466, 1997.
35. W. Qi. Geometry based haptic interaction with scientific data. In *Proc. Intl. Conf. Virtual Reality Continuum and its Applications,* pages 401–404, Hong Kong, 2006.
36. SensAble Technologies. Specifications for the PHANTOM Omni® haptic device, 2012. last viewed on 12–09-29.
37. G. Sepulveda-Cervantes, V. Parra-Vega, and O. Dominguez-Ramirez. Haptic cues for effective learning in 3d maze navigation. In *Proc. Intl. Worksh. Haptic Audio Visual Environments and Games,* pages 93–98, Ottawa, Canada, 2008.
38. L. Shams, Y. Kamitani, and S. Shimojo. Illusions: What you see is what you hear. *Nature,* 408:788, 2000.
39. M. Stamm, M. Altinsoy, and S. Merchel. Identification accuracy and efficiency of haptic virtual objects using force-feedback. In *Proc. Intl. Worksh. Perceptual Quality of Systems,* Bautzen, Germany, 2010.
40. H. Z. Tan, M. A. Srinivasan, B. Eberman, and B. Cheng. Human factors for the design of force-reflecting haptic interfaces. *Control,* 55:353–359, 1994.
41. R. J. van Beers, D. M. Wolpert, and P. Haggard. When feeling is more important than seeing in sensorimotor adaptation. *Curr Biol,* 12:834–837, 2002.
42. F. L. Van Scoy, T. Kawai, M. Darrah, and C. Rash. Haptic display of mathematical functions for teaching mathematics to students with vision disabilities: Design and proof of concept. In S. Brewster and R. Murray-Smith, editors, Proc. *Intl. Worksh. Haptic Human Computer Interaction,* volume 2058, pages 31–40, Glasgow, UK, 2001. Springer-Verlag.
43. P. Xiang, D. Camargo, and M. Puckette. Experiments on spatial gestures in binaural sound display. In *Proc. Intl. Conf. Auditory Display,* pages 1–5, Limerick, Ireland, 2005.

# Further Challenges and the Road Ahead

**J. Blauert, D. Kolossa, K. Obermayer and K. Adiloğlu**

## 1 Introduction

Auditory modeling has traditionally been understood as a signal-processing task where the model output is derived from the acoustic input signals in a strict bottom-up manner by more or less complex signal-processing algorithms. The model output, then, consists of signal representations that are completely determined by the input signals. In other words, the output is *signal-driven*. It is then taken as a basis for predicting what is aurally perceived. This approach has also been taken for most of the application examples of auditory models reported in this volume [11]. However, notwithstanding the fact that the model output can predict actual aural percepts only in a very limited way, a further fundamental problem is left unsolved, namely, that human beings do not react on what they perceive, but rather on the grounds of what the percepts mean to them in their current action-specific, emotional and cognitive situation.

Inclusion of this aspect requires substantial amendments to the auditory models as they stand today. In addition to perception, assignment of meaning and formation of experience have to be dealt with among many other cognitive functions, for instance, quality judgements. In other words, models of binaural hearing have to be extended to models of binaural listening.

To this end, more advanced models will contain specific, interleaved combinations of signal-driven, that is, *bottom-up* processes and knowledge-based, hypothesis-driven, that is, *top-down* processes. The sub-cortical section of the auditory system has to be seen as an embedded component in a larger system where

J. Blauert (✉) · D. Kolossa
Institute of Communication Acoustics, Ruhr-Universität Bochum,
Bochum, Germany
e-mail: jens.blauert@rub.de

K. Obermayer · K. Adiloğlu
Neural Information Systems, Technische Universität Berlin,
Berlin, Germany

J. Blauert (ed.), *The Technology of Binaural Listening*, Modern Acoustics and Signal Processing, DOI: 10.1007/978-3-642-37762-4_18,

signal-based representations are augmented by symbolic representations at different computational levels. Further, the models will contain explicit knowledge, accessible in a task-specific manner. Additionally, it will be necessary to replace current *static* paradigms of auditory modeling by considering the human being as an intelligent multi-modal agent that interactively explores the world and, in the course of this process, interprets percepts, collects knowledge and develops concepts accordingly. Consequently, models of binaural listening should incorporate means for the exploration of the environment by reflex- as well as cognition-controlled head-and-torso movements. Further, the role of input from other modalities—in particular, proprioceptive and visual input—has to be considered, since human beings are essentially multi-modal agents.

Figure 1 presents the architecture of a model that addresses the demands as described above. It is an architecture as currently discussed in AABBA [12]. The lower part of it schematically depicts the signal-processing, bottom-up processes and modules as can usually be found in today's models—see [37], this volume. The upper part represents modules that perform symbol processing rather than signal processing and are, to a considerable extent, hypothesis-driven, that is top-down controlled. In this upper part, various feedback paths are indicated, which are necessary for building a system that is capable of exploring and developing its world autonomously, thus gaining in knowledge and experience. In old-fashioned terms, one could actually call such a system a *cybernetic* one. It goes without saying that inherent knowledge at different levels of abstraction is required, particularly, when the system is supposed to perform quality judgment on auditory objects and auditory scenes that it has identified and analyzed [14].

## 2 A Framework for Cognitive Aural-Scene Analysis

Conceptualizing a framework for an artificial-listening system starts with the question of what the generic purposes of auditory systems are, or, in other words, why do humans listen at all? There is some consense in the field that three predominant reasons and, consequently, three modes of listening can be identified, namely,

1. Listening to gather up and process information from and about the environment, that is, to identify sound sources with respect to their nature and characteristics, including their positions and states of movement in space. This is also a prerequisit for appropriate action and reaction.
2. Listening for communication purposes. In many species interindividual communication is performed via the auditory pathway. In man, hearing is certainly the prominent social sense. It is, for example, much easier to educate the blind than the deaf.
3. Listening to modify one's own internal state, for instance, listening for pleasure, mood control, cognitive interest, and so on.

The different listening modes determine the strategy that listener persue in given situations and, thus, draw heavily on their cognitive capabilities. This is another case for including cognition in a framework of auditory listening, although it is still not clear as to what extent these modes of operation can be mapped onto a unified architecture. However, it is conceivable that at least for the symbol-processing part in an artificial-listening system some universal strategies can be employed—at least up to the point where meaning is converted into action, or where internal states need to be represented and changed. For the framework introduced in the following, the goal of understanding the external auditory environment, that is, item (1.), provides the main focus. Yet, the authors are aware of the other modes and shall try to accomplish them as well in the further course of the model development.

In the following, the overall function of the model is described in general terms in accordance with the framework depicted in Fig. 1.

The acoustic input to the model is provided at block (a) by a replica of a human head with two realistically formed external ears and two built-in-microphones. This artificial head is connected to a shoulder piece to form, together with the head, a head-and-torso simulator. The head is capable of three-degrees-of-freedom movements with respect to the shoulder piece, namely, rotating, tipping and pivoting. The head-and-shoulder simulator is mounted on a movable cart, which allows for further two degrees of freedom for translatory movements. There are sensors to monitor the positions of head, shoulder piece and cart with respect to each other and to an external reference point. The movements are enabled by actuators which can be remotely controlled. Which of the possible sensors and actuators are actually implemented, depends on the specific tasks that the model is specified for. Depending on the respective tasks, the equipment may further be fitted with sensors for additional sensory modalities, such as visual or tactile sensors.

The audio signals from the two microphones are fed into block (b), which represents major functions as are regarded relevant to be implemented by the human subcortical system up to the midbrain level. The components of this block account for functions that are attributed, for example, to the *middle ears* and the *cochleae*, to the *superior olivary complex*, SO—including *medial superior olive*, MSO, and *lateral superior olive*, LSO—or to the *inferior colliculus*, IC. Those functions as well as their computational implementations are described in more detail in [37], this volume.

The output of block (b) is represented within the computational model through a multidimensional, binaural-activity map including, for example, the dimensions intensity of activity, frequency, lateral position or time—see block (c). This kind of representation is inspired by the existence of activity maps at the midbrain level of animals for coding acoustic features, such as spatial locations of sound sources, fundamental frequencies and spectra and/or envelope characteristics of the acoustic source signals. A specific example of such a computational binaural-activity map is depicted in Fig. 1, namely, a map depicting binaural activity as generated by the binaural impulse responses of a concert hall.

The next step in the model, block (d), has the task of identifying perceptually relevant cues in the activity maps and, based on these cues, organize the activity

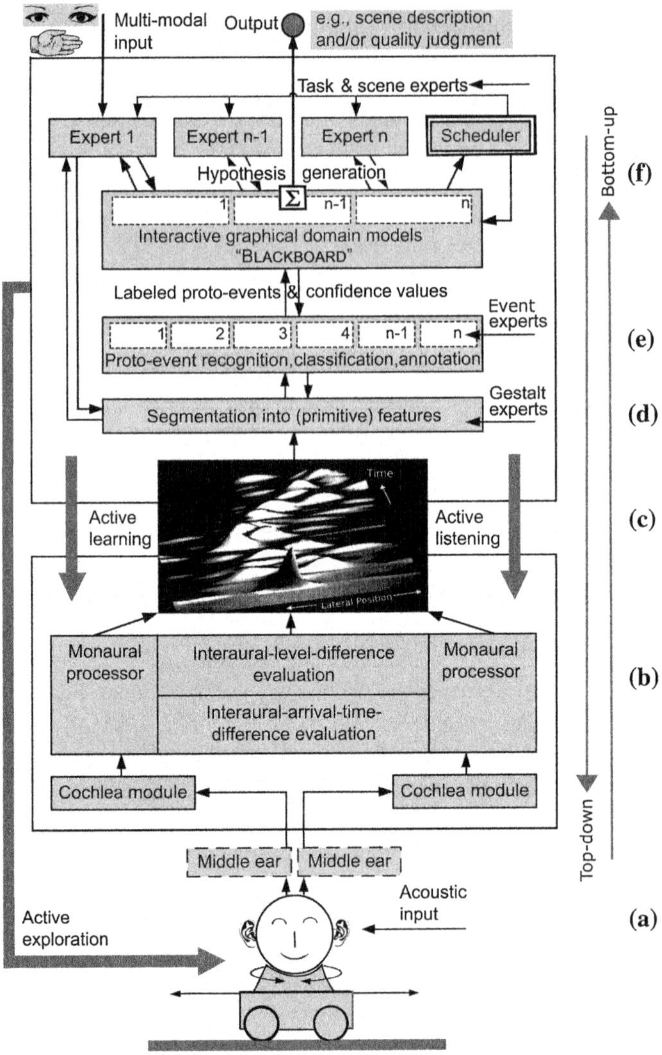


**Fig. 1** Schematic of the architecture of a comprehensive model of binaural listening. **a** Head-and-torso simulator on a mobile platform. **b** Signal processing in the lower auditory system. **c** Internal representation of binaural activity. **d** Rule- and/or data-driven identification and annotation of perceptually-salient (primitive) features. **e** Rule- and/or data-driven recognition, classification and labeling of proto-events. **f** Scene-and-task representation, knowledge-based hypothesis generation, assessment and decision taking, assignment of meaning

maps into segments that represent specific primitive perceptual features. The cues can be signal attributes in the temporal or spectral domain, such as autocorrelation, decrease and centroid, effective duration, energy, interaural arrival-time differences, interaural cross-correlation, interaural-level differences, log-attack time, modulation amplitude, modulation frequency, roll-off frequency, spectral centroid, temporal increase, time frame, zero-crossing rate—just to name a few. Yet, they can also be estimators for *"sensations"* such as pitch, timbre or coloration, loudness, sharpness, roughness, spaciousness, and reverberance, as can be calculated with specialized signal-processing algorithms. The features rendered are typically based on an ensemble of different cues, whereby primitive schemata like the *Gestalt rules* [6, 17] are considered in the formation process. The rendering process can be rule-driven, that is, exploiting prior knowledge, or data-driven, that is, based on statistical procedures—see Sect. 2.1 and 3 for more details of relevant machine learning methods. Multiple processes of this kind may act in parallel, also as a combination of data-driven and rule-driven approaches.

The primitive features derived from the activity map provide the input to the next model stage, block (e), where sets of features are interpreted as indicators for specific auditory events. These indicators are called *proto-events* here, since their actual character is subject to statistical uncertainty. The formation process for proto-events may, for instance, follow a sequence of detection, classification and annotation. Again, rule-driven as well as data-driven approaches may be used. Clearly, the successful extraction of proto-events depends on whether appropriate feature sets have been chosen in the beginning. In the context of an artificial system, feature-selection techniques can be employed, and small sets of informative features can in principle be learned for given combinations of (classes of) auditory tasks and (classes of) auditory environments. However, in the context of bio-inspired processing as for human-listening modeling, this becomes a highly non-trivial task. One possible approach to proceed is to conduct model-driven psychoacoustic experiments and to ask, whether human listeners employ the same features as the artificial system suggest.

By the way, blocks (d) and (e) have been described here as two sequential processing steps. However, for certain statistical procedures, such as being used, for example, in the machine-learning field, the difference between primitive features and proto-events may not always be clear-cut and processing steps (d) and (e) may well be combined into one. It is at this model stage, that a *transition from signal processing to symbol processing* takes place, since the proto-events can be represented by symbols.

The last stage, block (f), represents the world-knowledge of the modeling system—among other functions. At this stage, contextual information is used to build task-related representations of the auditory scene, namely, prior knowledge is integrated, hypotheses about auditory events are generated and validated, and meaning is assigned. The auditory scene is evaluated, decisions are made and signals or commands may be sent back down to the lower processing levels—blocks (a) and (e).

In the framework of Fig. 1, a so-called *blackboard structure* [20, 27, 28] is proposed for this purpose. It works as follows. The input from the lower stages is put on a "blackboard", which is visible to a number of specialized experts, namely, computer programs that try to interpret the entries on the blackboard based on their respective expert knowledge. There can be various different experts, for instance, acoustic experts, psychoacoustic experts, psychologic experts, experts in spatial hearing, experts in cross-modal integration of vision, tactility, proprioception etc., speech-communication experts, music experts, semiotic experts, and so on—depending on the specific task that a listening model is constructed for.

Once an expert finds a reasonable explanation of what is shown on the blackboard, it puts this up as a hypothesis to be tested against the available entries. The hypothesis will then be accepted or rejected based on rules or on statistical grounds. For the control of the activities of the experts, a special program module, the *scheduler*, is provided.

The scheduler acts like the chairman of a meeting. Firstly, it determines the order in which the individual experts intervene, controls the statistical testing and makes a decision regarding the final outcome—which may well be a mixture of various accepted hypotheses—compare [30] as to how to provide such a mixture. Secondly, it will also select groups of experts as well as modify the computations performed by them on the bottom-up data according to the current goals, such as extracting the *"what"* versus the *"where"* of a sound event.

In the light of the main focus of the current chapter, that is, the analysis and assessment of aural objects and scenes, two types of models are of particular relevance, namely, (i) *object models*, to understand single aural perceptual entities known as aural objects, and (ii) *scene models*, to understand interactions between aural objects in arrays of objects, for instance to cover questions like: Which of the objects may be simultaneously aurally present. Hereby, the following definitions may apply.

*Objects* are perceptual entities that are characterized by specific attributes and invariances and by their relations to other objects

*Scenes* are arrays composed of multiple objects, again specified by specific attributes and invariances

For creating models of aural objects and models of aural scenes, a wide range of knowledge can be exploited, such as rules of how object and scenes develop, physical knowledge of type, location, and movement of sound sources, perceptual knowledge such as Gestalt rules, or semantic knowledge with respect to the information that the sound sources may intend to convey. Here, so called *graphical models* are proposed to represent these models and, at the same time, act as the blackboard. *Graphical domain models* allow for invention both from block (e), that is the proto-event level, as well as from the experts' level. For more details see Sect. 2.2.[1]

The outcome from block (f) and, thus, the output of the listening model at large, may be a description of an aural object or the description of an aural scene. Yet, it is further planned to develop the model to such a state, where it can assess aural objects and scenes with regard to their perceptual quality. Judgment on quality requires a further processing stage for the following reason. Quality, in general terms is the amount

[1] Graphical models are convenient when it comes to the implementation of a working artificial-listening system, but whether and—if yes—how it actually maps to the processes which integrate and disambiguate sensory information in the human brain remains a matter of future research. It has been suggested that neural systems implement *Bayesian inference* including even *belief propagation* [24, 25], but there is also evidence, that competition between *neural assemblies* and an *attractor dynamics* [23] may play an important role in sensory processing.

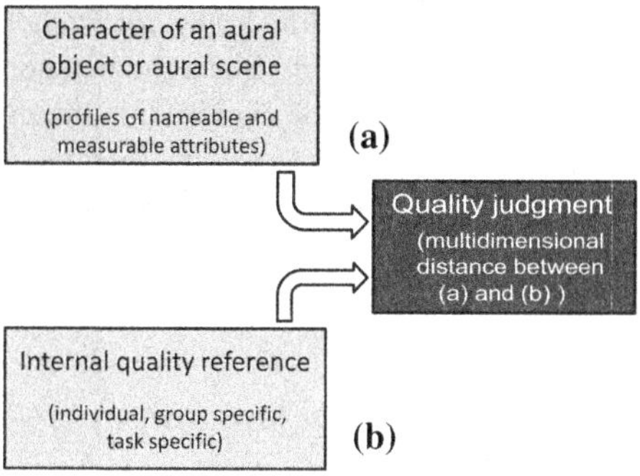


**Fig. 2** Quality judgment as a multi-dimensional distance measure between the character, (a), of an item and a reference, (b), representing expectations regarding this item

as to which an item fulfills expectations with regard to it [26]. A quality judgment thus requires a set of measured and nameable features of the item under consideration, that is, its *character, (a),* and a set of values expected for these features, the *reference, (b).* A quality judgment can then be seen as a multi-dimensional distance measure between those two—see Fig. 2. It follows that the expert must have internal references to apply when judging on quality. These reference are individual and/or group dependent and task specific [10, 13]. To construct them can be impossible, where internal quality references are concerned, and it is tedious even in the best of cases.

To avoid this, it is also possible to perform the quality assignment as the result of machine learning. In this case, the machine-learning algorithm needs the quality judgments assigned to specific situations as a-priori information. Whether this turns out to be less tedious in the end than collecting information about internal references, remains to be explored for each specific case. Also, it is of advantage for many applications to have direct and explicit access to the internal references behind the quality judgments. With machine learning this would require further analysis.

At this point, some remarks on *signal-driven*—bottom-up—and *hypothesis-driven*—top-down—processing are due, since they proceed in an interleaved way in the model framework as proposed here. In purely signal-driven processes the output is completely determined by the input. If the processing requires multiple variables to be combined, with each of them having a number of possible states, this can quickly lead to an immense number of potential output states—combinatorial explosion—which all have to be followed up and evaluated until a final decision has been reached. In top-down processing, in contrast, the number of states to be evaluated is substantially reduced, as the process *knows what to look for*, that is, focuses attention on states which *make sense* in a given specific situation. Of course, such a strategy is limited to known scenarios, unless means for adaption and assimilation are provided.

To avoid this deadlock and for other reasons, Fig. 1 provides various feedback paths, some more specific, others more general. The general ones originate from the concept that the listener model, that is, the *"artificial listener"*, actively explores its aural world and thereby differentiates and develops it further in an autonomous way, very much like human beings do. Following this line of thinking, it is attempted to model listeners according to the *autonomous-agent* paradigm, where agents *actively learn* and *actively listen*. Since the listener model can deliberately move its sensors about in the space to be explored, it can use proprioceptive cues besides aural ones to perform these tasks. Cross-modal cues, like visual and/or tactile ones [4, 38] may be included if appropriate. Active learning and active listening are further discussed in Sect. 3.

## *2.1 Feature Extraction and Proto-Event Detection*

In the framework of Fig. 1, statistical machine-learning techniques are planned for the extraction of primitive features, block (d), and for the formation of proto-events, block (e). As their input, these techniques require signal partitions and/or features that carry information that is relevant for what is finally perceived and considered to be meaningful.

By using such input, machine-learning models can be developed that are able to extract proto-events from autonomous and/or interactive environments. These proto-events can then be used at higher levels for comparison and verification—for example, in graphical models, see Sect. 2.2—as well as for providing feedback to lower-level model stages. The final goal is to arrive at proto-events that make sense to human beings, particularly in the light of previous experience.

In the machine-learning field, agents learn tasks from data that are provided to them by the environment, that is, tasks are learned by induction. Within the processing stages of blocks (d) and (e), in other word, agents will primarily perform pattern recognition tasks. Following the machine-learning paradigm, agents will first have to undergo a learning phase during which informative acoustic features are selected for input, and the agents' parameters are tuned in such a way that the pattern-recognition task can be performed sufficiently well. In the following recognition phase, these agents will then fulfill their "duty", which is to extract the relevant acoustic features from the input signal and to combine them for the detection, classification and annotation of proto-events.

Traditionally, one distinguishes three learning schemes on the basis of what kind of information is available to the agent. In *supervised* learning, the agent is provided with the acoustic input simultaneously with the correct annotation. In *reinforcement* learning, the agent is provided with the acoustic input, but the environment provides only a summary feedback signal that tells the agent whether the annotation was correct or not. In *unsupervised* learning, finally, only the acoustic input signals are available, and the agents' task is to utilize statistical regularities within the acoustic environment in order to generate a new representation that is optimal—given some predefined learning criteria.

Although reinforcement learning is a key learning scheme when it comes to human beings, its machine-learning counterparts are computationally expensive and require extremely long learning periods. Therefore, it is currently not advisable to use it in the framework of Fig. 1. As to the other two paradigms, unsupervised-learning paradigms are better suited for learning feature representations—block (d)—while supervised-learning paradigms are generally better suited for generating symbolic representations—block (e).

Successful learning and recognition strongly depends on the representation of auditory signals and scenes. While many different approaches should be implemented, evaluated and compared, a particularly interesting class of representations are the biologically-inspired spike-based representations [48]. These representations are typically sparse and provide a decomposition of a complex auditory object into a pattern of brief *atomic events*. This decomposition can, for example, be derived from the binaural-activity map generated at block (c), where the atomic events would then be localized in both auditory space and time. Another example is discussed in more detail in the next subsection.

Although many auditory objects are distinctly localized in time and space, they may still occur at variant spatial locations and points in time. Vector-based representations typically have difficulties capturing this and other types of variability, for instance, different durations and/or spectral shifts. Relational representations, where auditory events are described by their similarity to other auditory events rather than by their individual features, are much better suited, because the required invariances can often be built into the similarity measure in a straightforward way. In addition, many kernel-based machine-learning methods have been devised over the last 20 years that naturally operate on relational representations, including, for example, the well known support-vector machines. Although standard kernel methods impose certain constraints on the similarity measure, that is, the *kernel*, extensions have been suggested, such as by [34, 35], that can also be applied to a wide class of similarity measures for spike-based representations. Two examples for the supervised and unsupervised learning paradigms are described in the according subsection herewithin.

### Sparse Event-Based Representation

Spike-based representations [48] provide a decomposition of a given sound signal via a linear combination of normalized basis functions taken from a predefined or learned dictionary. This kind of representation will now be illustrated by a simple example from the auditory domain. Let $x(t)$ be a monaural sound signal and let $\gamma_{f_k,t_k}(t)$ be the basis functions, then

$$x(t) = \sum_{k=1}^{K} a_k \gamma_{f_k,t_k}(t) + \epsilon_{K+1}(t). \qquad (1)$$

Every basis function, $\gamma_{f_k,t_k}(t)$, in (1) corresponds to one atomic event located at time, $t_k$, in the auditory stream. The corresponding coefficient, $a_k$, and "property", $f_k$, characterize this event. The parameter $f_k$ determines the type of basis function taken from the dictionary and can, for example, be the center frequency of a time-frequency localized filter from a given filterbank. $a_k$ then describes how well the filter function locally matches the auditory stream. The most efficient representation of the type defined by (1) is one that achieves a small residual, $\epsilon_{K+1}(t)$, for a small number, $K$, of atomic events. A greedy way of iteratively constructing such a representation employs the matching-pursuit algorithm [40]. During each iteration, $k$, a basis function, $\gamma_{f_k,t_k}$, is selected that maximally correlates with the residual signal, $\epsilon_k$, remaining from this iteration, that is,

$$(f_t, t_k) = \text{argmax}_{f_m,t^*} < \epsilon_k(t), \gamma_{f_m,t^*}(t) > . \tag{2}$$

The total number, $K$, of iterations determines the number of events with which a particular auditory object is described, in other words, the level of sparseness as well as the accuracy of this representation, namely, the magnitude of the remaining residual. There is a trade-off between both. However, since the goal of this representation is *not* to reconstruct the sound at later processing stages, the absolute size of the residual is not so important, and the focus should be on creating a representation that uses a small number of basis functions, that are most informative for later classification and annotation. Different filter functions are suitable for generating an overcomplete dictionary—such as Gabor atoms, Gabor chirps, cosine atoms, or Gammatone filters. Gammatone filters are popular for auditory-adequate filtering, because a subset of them approximates the magnitude characteristics of human auditory filters [43]. Figure 3 shows the spike-based representation for a specific sound

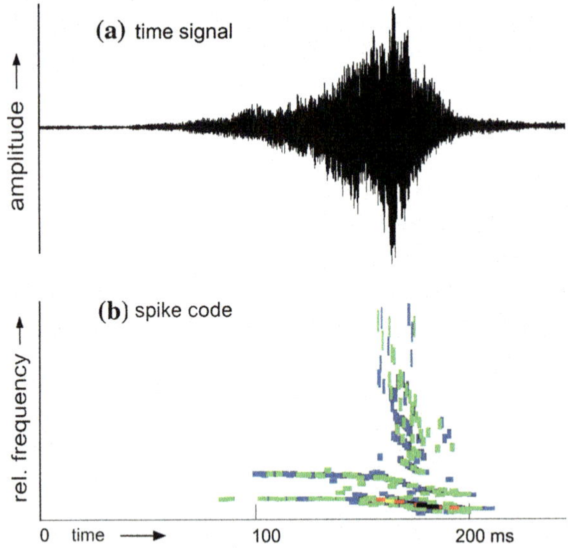


**Fig. 3** **a** Time-domain signal of a *whoosh* sound, **b** corresponding event-based representation using Gammatone filters. Each *rectangle* corresponds to one basis function in the expansion of (1). The magnitude of the coefficient, $a_k$, is represented by the *gray shade* (darker → larger). Time, $t_k$, and center frequency, $f_k$, are represented by the location of the *rectangles*. The size of the *rectangles* indicates the localization of the basis functions in time-frequency space

that has been labeled *"whoosh"*. A Gammatone filter bank of 256 filters is used for generating a representation based on $K = 32$ atomic events. The salient areas in the signal are represented by more events than other areas.

Different auditory events can now be compared via their event-based representations by using appropriate distance or similarity measures, respectively. These distance measures can account for invariances in a straightforward way and can be devised such that they capture the event-structure of a complex auditory event [2]. Take, for example, the representation shown in Fig. 3b. The total distance between two such representations can, for example, be decomposed into a sum of distances between pairs of corresponding atomic events which, in turn, can be calculated as a weighted sum of the distances between the parameters describing them. But how can corresponding events be found? They can be found by minimizing the total distance over all the candidate pairs. The problem of optimally matching two event-based representations can be transformed into the problem of optimally matching the two node-sets of a weighted bipartite graph, which can in turn be solved using the so-called *Hungarian algorithm* [39]—details can be found in [2].

## Supervised Learning: Support-Vector Classification

*Support-vector learning* is an efficient machine-learning method for learning the parameters of perceptrons for classifying patterns but also for assigning real-valued attributes, for example the quality ratings for auditory events event. Consider a simple binary classification problem, where data points represented by feature vectors, $\mathbf{x}$, —for example, some low-level descriptors of sound events-should be assigned to one of two possible classes, $y \in \{-1, +1\}$. Support-vector learning is a supervised learning method, hence it requires a so-called training set of labeled examples, that is, of pairs, $(\mathbf{x}_i, y_i)$, $i = 1, \ldots, p$, during the learning phase. The *support-vector classifier* is a standard perceptron and has the following simple form,

$$y(\mathbf{x}, \mathbf{w}) = \text{sign}\left\{ w_0 + \sum_{k=1}^{M} w_i K(\mathbf{x}_{i(k)}, \mathbf{x}) \right\}. \tag{3}$$

$\mathbf{w}$ is a vector of model parameters, and the data points, $\mathbf{x}_{i(k)}$, $k = 1, \ldots, M$, are data points from the training set.

During the learning phase, model parameters have to be determined and specific data points, $\mathbf{x}_{i(k)}$,—the so-called *support vectors*—have to be chosen from the training set, such that the resulting classifier will perform well during recognition. Details about standard support-vector learning can be found in a number of textbooks, for instance [46]. Important for the following, however, is the function $K(\mathbf{x}_i, \mathbf{x}_j)$, the *kernel*, which is part of the perceptron. It can be interpreted as a similarity measure that quantifies how similar two data points, $\mathbf{x}_i$ and $\mathbf{x}_j$, are. Learning and recognition make both use of similarity values $K(\mathbf{x}_i, \mathbf{x}_j)$ for pairs of data points only, and feature values would only enter these processes through the similarity measure $K$.

Therefore, the function $K(\mathbf{x}_i, \mathbf{x}_j)$, which maps pairs of feature vectors to similarity values, can be replaced by a more general function, $K(i, j)$, that maps pairs $(i, j)$ of patterns directly to similarity values. This allows support-vector learning and perceptron recognition to directly operate on relational rather than on feature-based representations.

Now consider event-based representations and the example of Fig. 3 . For every pair of event-coded sounds, $(i, j)$, the Hungarian algorithm can be used to determine their distance, $d(i, j)$, which can then be transformed into a similarity measure, $K_{i,j}$, for example, by use of the common Gaussian kernel function

$$K(i, j) = \exp\left(-\frac{d(i, j)^2}{2\sigma^2}\right). \tag{4}$$

$\sigma^2$ is a variance parameter to be determined—sometimes called length scale. Unfortunately, this choice may not lead to a valid kernel function, valid in the sense of standard support-vector learning, as the kernel function should be positive semi-definite. Consequently, variants of support-vector learning have to be used that do not require this property [34, 35].

In [2] event-based representations and support-vector classification have been applied to recognize everyday sounds. The dataset contained ten classes of different types of everyday sounds. Figure 4 shows the results for a one-vs-the-rest recognition task, and compares the performance of the event-coded representation with the performance achieved for a number of standard feature-based representations for the same sound. The event-based representation, SPKE, outperforms the other representation schemes, including the popular feature-based representation using

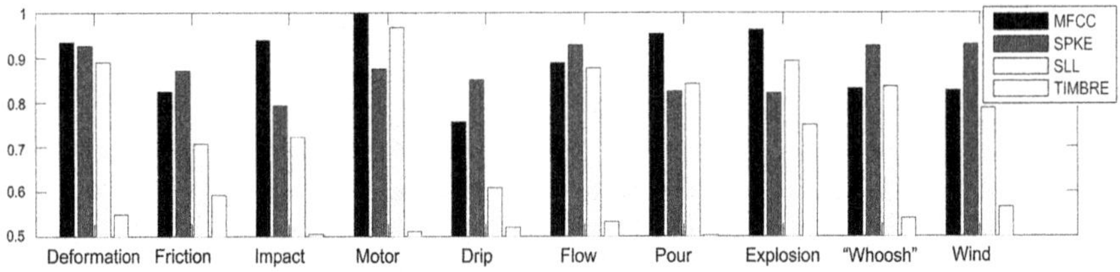


**Fig. 4** Recognition of everyday sounds with perceptrons and support-vector learning. The figure shows the recognition performance for one particular sound class against the rest for a sound-data set consisting of ten sound classes. Four different low-level representations were used, mel-frequency cepstral coefficients, MFCC, an event-based representation using Gammatone filters, SPKE, a set of spectral low-level descriptors, SLL (energy, zero crossing rate, spectral centroid, roll-off frequency and their variances, finite differences, and the variances of the differences) and a set of timbre descriptors, TIMBRE (perceptual spectral centroid, relative specific loudness, sharpness, roughness, signal autocorrelation, zero crossing rate, time frame, log attack time, temporal increase, decrease and centroid, effective duration, energy-modulation frequency, energy-modulation amplitude)

*mel-frequency cepstral coefficients*, MFCCs,[2] in five particular cases. In the other five cases, MFCCs perform best.

As a result, the event-based representation can be considered as a denoising procedure that emphasizes those contours of a given sound that are perceptually important. On the other hand, however, from a certain number of atoms selected, some perceptually non-existent contours (phantom-spikes) can be emphasised. This could degrade the recognition performance. By means of the results of some suitable psycho-acoustical experiments, the perceptually optimal number of spikes can be determined to avoid this effect.

Furthermore, the total distance computed between two spike codes is a weighted sum of the parameters. These weights can be adjusted using prior knowledge. For impact sounds, for example, time differences can be weighted more strongly than differences in amplitude or frequency, which in turn improves recognition performance.

## Unsupervised Learning: Prototype-Based Clustering

The use of relational representations is not limited to supervised learning paradigms and support-vector learning but can be applied to unsupervised learning as well. Although in the framework as laid out in Fig. 1, unsupervised learning is generally better suited for learning feature representations, clustering could also be a promising method for defining proto-objects—provided that the quality of the pre-segmentation is sufficiently high. If applicable, unsupervised learning has the benefit of not requiring annotations of auditory objects during the learning phase, since these are often expensive to obtain.

During clustering, data points are grouped according to a predefined similarity or distance measure. A *cluster* is then formed by data points whose inter-point distances are small compared to the distances to data points that are members of the other clusters. For some of the methods, a prototypical data point is generated for every cluster. These methods are usually called central or prototype-based clustering methods. In the following, it will be illustrated how prototype-based clustering methods can be applied to auditory objects in an event-based representation.

Let $d(i, j)$ be the distance between two auditory objects in their event-based representation as computed, for example, by using the distance measure introduced in the preceding subsection. A particular clustering of a set of auditory events can then be quantified using binary assignment variables, $M_{ib}$, where

$$M_{ib} = \begin{cases} 1 & \text{if the auditory event, } i \text{, is assigned to a cluster, } b \text{, and} \\ 0 & \text{otherwise} . \end{cases} \tag{5}$$

[2] *Mel-frequency cepstral coefficients*, MFCCs, are the DCT coefficients of the logarithm of a mel-scaled signal spectrum. They have been introduced for the purpose of speech recognition [21], but have since proven versatile and found use in many other acoustic classification applications.

Let $b$ denote the prototypical event-based representation of all the sounds assigned to group $b$. Then a good grouping and a good prototype should minimize the cost function,

$$\mathcal{H}^{oc}(\{b\}, \{M_{ib}\}) = \sum_{i=1}^{I} \sum_{b=1}^{B} M_{ib}\, d(i, b), \tag{6}$$

with respect to both the assignment variables, $M_{ib}$, and the parameters of the prototypical event-based representations, $b$. Following [29], the optimization can be performed via an expectation-maximization algorithm. This is an iterative procedure, where each iteration consists of two steps. In the first step, all distances, $d(i, b)$, are calculated and representations are preferably assigned to the group for which the distance to its prototype is smallest. In the second step, the parameters of the prototypical representations are chosen by minimizing (6), keeping the assignment variables fixed. For better convergence, a probabilistic version of this procedure is used in practice, where the binary assignment variables are replaced by assignment probabilities—details can be found in [29] or [1].

Figure 5 shows the prototypical event-based representations for three clusters from a sound class called *whoosh* on the left-hand side. The figure shows that the prototypes well represent the sounds assigned to a particular cluster and that information can be

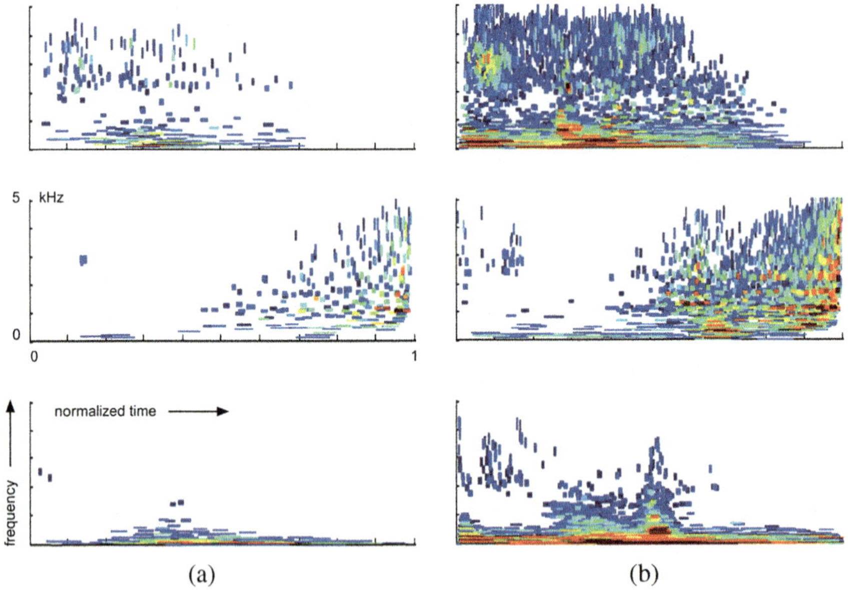


**Fig. 5** **a** Event-based representations generated by prototype-based clustering of everyday sounds. Shown are three different prototypes for clusters of a class of sounds annotated by "whoosh". **b** Overlay of the event-based representations of all sounds as have been assigned to the corresponding prototype by the clustering method. For details of the graphical representation, see the caption of Fig. 3b

derived about the pattern of atomic events that are typical for a particular group. If annotations of sounds become available at this point, meaning can be assigned to the different groups, and further acoustic events can be recognized by matching them to the closest prototype.

## 2.2 *Graphical Models as Dynamic Blackboards*

At the highest level of the model framework depicted in Fig. 1, level (f), it is attempted to make sense of the input from level (e), which consists of sets of annotated proto-events and the confidence levels assigned to them. The task is to find out about the following.

- In how far does the input correspond to any patterns that are known to the system, that is, can any aural objects an/or aural scenes be *recognized*?
- How far can pre-known patterns be adjusted using available input information? In other words, can something be *learned* from the input—for instance, by taking advantage of any sort of understanding that we have of the interaction of aural objects?

In this section, a mathematical tool is introduced that can be used to encode just such knowledge, namely, the so-called *graphical model*. Graphical models are originally models of a statistical nature, and they are typically designed on case-by-case information. They are thus generally ignorant of physical laws and mathematical/logical rules, such as acoustic-wave-propagation theory, the constancy of source identity, physical limitations to source and sensor movements, and other relevant knowledge.

To overcome these limitations of graphical models, it will now be attempted to reconcile the flexibility of graphical models with the precision of physical and mathematical knowledge—without loosing the advantages of either. To illustrate one way of obtaining an appropriate framework, the following three subsections will first present a brief background of graphical models and, consequently, "Auditory-Scene Understanding with Graphical Models" will show how rule-based knowledge can guide the design of graphical models to the end of achieving maximum precision with a task-specific limitation to relevant available knowledge.

### Graphical Models

The dependence relationships within groups of random variables are often described very concisely by denoting their statistical dependence with the help of a graph. Such graphs come in two forms, directed and undirected. Yet, in the following, the discussion will be limited to the directed ones. In this type of graphs, two variables that are statistically dependent upon each other are connected with a line, with the arrow pointing from the independent to the dependent variable.

**Fig. 6** Dependence tree

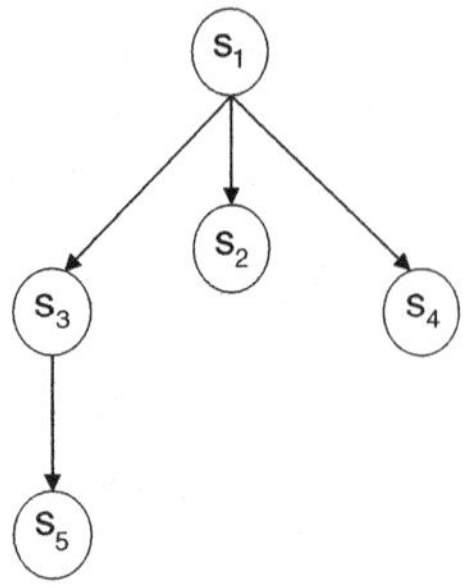


An example of such a dependence graph is given by Fig. 6. In this example, the graph indicates that the two variables, $s_3$ and $s_5$, exhibit some form of statistical dependence, while there is no direct dependence between $s_3$ and $s_2$. Based on this understanding, the graph can be used to simplify the joint probability density of the variables, $s_1$ through $s_5$, as follows,

$$\begin{aligned} \mathrm{p}(s_1, s_2, \ldots s_5) &= \mathrm{p}(s_1)\mathrm{p}(s_2|s_1)\ldots\mathrm{p}(s_5|s_1\ldots s_4) \\ &= \mathrm{p}(s_1)\mathrm{p}(s_2|s_1)\mathrm{p}(s_3|s_1)\mathrm{p}(s_4|s_1)\mathrm{p}(s_5|s_3). \end{aligned} \tag{7}$$

More generally speaking, a tree-shaped dependence graph indicates that the joint-probability-density function, PDF, of all variables, $\mathrm{p}(s_1, \ldots s_n)$, can be factorized in the following manner,

$$\mathrm{p}(s_1, s_2, \ldots s_N) = s_r \prod_{n \in N, n \neq r} p(s_n | s_{\mathrm{A}(n)}), \tag{8}$$

where $r$ denotes the root node of the graph and $\mathrm{A}(n)$ yields the ancestor (or parent) nodes of node $n$. Graphs that are more complex can also be factored according to the same principle, that is, by using the fact that the probability density of statistically independent variables may be factorized into a product of the PDFs of all interdependent subgroups.

## Graphical Models for Non-Stationary Processes

In many contexts, graphical models denote statical-dependence relationships, as is also true in the example shown in Fig. 6. It is, however, in the nature of auditory events that they are temporally evolving, and exhibiting only approximate short-time stationarity. Thus, when a description of auditory scenes is desired, it becomes necessary to extend graphical models to describe temporally evolving variables in addition to stationary ones.

One example of such a temporal graphical model is a *hidden Markov model*, HMM, which is highly popular due to its flexibility and the availability of easily

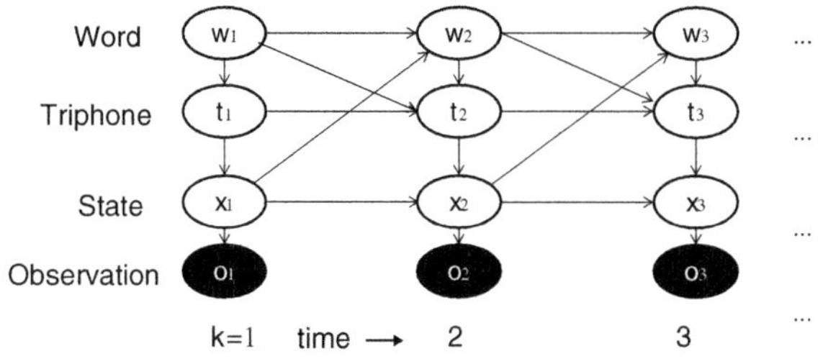


**Fig. 7** Hidden Markov model, shown as an unrolled temporal graphical model

implementable, statistically optimal algorithms. Figure 7 depicts an HMM in the notation of a directed graphical model.

In this figure, variables are shown either as filled ellipses, meaning that they are observable, or as white ellipses, indicating that they are hidden. The exemplary variables here are taken from automatic speech recognition, ASR. They indicate the word identity, $w_k$, the triphone identity, $t_k$, the state index, $x_k$, and the observation vector, $o_k$, for each time frame, $k$.

As above, their dependence relationships are depicted in the notation of graphical models, but now, in addition, their temporal evolution is shown, because in contrast to the stationary variables in Fig. 6, it becomes necessary to model their value at each time frame by a separate node.

This step of adding duplicates of variables for each of the points in time is often described as *unrolling* the graphical model over time. As visible in this example, an originally static dependence graph can be extended by temporal unrolling such as to model the properties of a temporally-evolving, non-stationary statistical process.

## Rule-Bases in Graphical Models

While graphical models, both for stationary and for non-stationary processes, are statistical models by nature, they allow for the incorporation of rule-based knowledge in two distinct ways, that is,

- On the one hand, conditional probabilities in the graph can also collapse to deterministic ones—namely, conditional-probability tables can contain ones and zeros—if physical rules or rules based on other knowledge sources state that certain values in one variable of the model directly imply setting other variables to specific values.
- On the other hand, the structure of the graphical model itself, or of sub-structures in it, can be compiled rather than determined manually, and this compilation process can be carried out by automatically integrating sets of rules. An example of how this may be accomplished in the context of automatic speech recognition by compiling possible sequences of phonemes from a pronunciation dictionary and a task grammar, can be found in [8].

## Auditory-Scene Understanding with Graphical Models

By using graphical models like those described above, very precise source models may be attained, provided that training takes place on sufficient amounts of representative data. Such models can serve as a repertoire of possible source signals for schema-based, rather than primitive grouping. As one success story of graphical models in multi-source scene understanding, Hershey et al., [33], use factorial hidden Markov models as models and apply them to perform inference with variational loopy-belief propagation. The computational effort of this approach scales linearly in the number of sources and leads to super-human ASR rates on multi-speaker single-channel recordings.

This is already a highly promising result for the application of graphical models in aural-scene understanding. However, the current approach as exemplified by this work, still has several drawbacks, such as the following. The model of [33] and other similar graphical-model-based systems like [22] need to exploit very strong source models, that is, top-down or schema-based segregation, in order to come close to or even surpass the recognition rate of human listeners on the tested but highly specialized task. In natural scenes, there is not usually such an ample experience regarding the possible waveforms of sound sources. However, in any realistic cases, a much higher variability of possible sounds, coming with a far greater vocabulary and a much wider range of admissible sources, is the standard.

Thus, dynamical Bayesian networks have been applied and already proven their merit in cases where detailed source models were available. Yet, achieving a general applicability with this approach is still an open issue.

As a step forward, even for quite general tasks where only coarse source models are available, additional information in two specific forms could often be exploited.

- Physical and mathematical knowledge may be used, for instance, regarding the mixing process—which may be supported by input from other modalities.
- Psychoacoustic heuristics for source separation may be applied, as based on an implementation of *primitive-segregation* rules mimicking human auditory streaming in general environments.

How these two information sources, namely, physical models and perceptual rules, should be combined with all available source information, for the purpose of gaining an optimally informed understanding of auditory scenes, is an interesting open issue.

As one option, this should be possible by compiling a combination model from all information sources—similarly to well known model compilations for ASR, where a lexicon, phonetic and linguistic information are used to form a search network. Yet, for the applications envisaged here, the compiled model would not be linear in topology but rather allow for superposition of all acoustic sources according to the internal acoustical wave-propagation sub-model.

Such a compiled graphical model could also possess an interface for higher level processes that might search over variable allocations in the manner of an expert system. Only, in contrast to standard expert systems, this search would operate on the graphical-model variables directly, effectively making the graphical model an *active blackboard*, which could be simulated to measure the goodness-of-fit between all observations and the internal variable occupation probabilities, corresponding to possible internal scene interpretations. Thus, of all scene interpretations, the most fitting one could be selected, whereby the "fit" is assessed, among other information, based on what is known about the source and further physical and mathematical knowledge, and guided by streaming mechanisms as are also active in human perception.

# 3 Active Learning and Active Listening

## *3.1 Active Learning*

Active learning, that is, autonomous inductive learning, is one of the key features of the envisaged artificial listener and plays a prominent role in most of the higher-level processing stages of the framework that has been laid out in Fig. 1. Ideally, the artificial listening system would autonomously explore its environment and use the information gathered through this interaction in the learning process. Strategies for autonomous learning exist in principle, and reinforcement learning is a prominent example of this. However, *pure* autonomous learning is still notoriously slow in complex environments, and one has to resort to supervised learning strategies for many of the subtasks involved. Still, an artificial listener is an excellent testbed for concepts to improve *autonomy*.

Supervised learning suffers from the fact that acoustic events have to be annotated. Given the large amount of data needed by standard learning methods when tasks become complex, the required human interaction can become excessive. For an artificial cognitive system it is therefore important to make best use of the available information. One idea, which has been around for several decades by now, is to replace a passive *scanning* of the environment by strategies where a learning system *actively* sends out requests for training data that are particularly informative. There is a large amount of empirical evidence about the fact that *active data selection* is more efficient in terms of the required number of training examples for reaching a particular level of performance. It follows, that the amount of human interactions can be reduced by supervised-learning paradigms. The next subsection provides an example where active data selection is applied for learning a predictor for perceptual attributes assigned to everyday sounds by humans.

### Active Data Selection

For a binary classification problem, consider a parameterized family of perceptron classifiers,

$$y(\mathbf{x}, \mathbf{w}) = \text{sign}\left\{ w_0 + \sum_{k=1}^{M} w_i K(\mathbf{x}_{i(k)}, \mathbf{x}) \right\}, \tag{9}$$

similar to the classifiers that have already been introduced in the context of support-vector learning in Sect. 2.1. Assume that the reference-data points, $\mathbf{x}_{i(k)}$, for the functions $K$ have been chosen in a sensible way. Then the goal of inductive learning is to find values for the parameters, $w_i$, such that the perceptron predicts the class memberships sufficiently well.

In active data selection, inductive learning is interpreted as a process, where predictors from the set—for example, perceptron classifiers from the parameterized family of (9)—are discarded if their predictions are inconsistent with the training data. Every new data point from the training set splits the current set of classifiers in two sets that differ in their prediction of the class label. Assume that the set of classifiers is endowed with a useful metric, for example, a metric taking into account that two classifiers are similar if their predictions are so too. Then a space of classifiers can be constructed, and volumes and distances can be defined. With those concepts one can then assess how useful a new data point is for training: A new data point is useful, given that the space of classifiers predicting membership of one class has about the same *size* as the other ones with regard to volume or maximum diameter of the corresponding subspace. At any stage during learning, an active-learning agent, when implementing, for example, the perceptron classifier, will select a useful data point and will ask for its class. When the information arrives, the agent will no longer consider the subset of classifiers with more or less disagreeing predictions but rather continue the learning process with those classifiers that have shown to predict correctly [32].

If there is no noise in the problem, and if the two classes can be separated in principle by a perceptron, active learning leads to an exponential decrease in the size of the set of consistent classifiers with the number of training data. For size meaning volume, for example, this follows from the fact that every well chosen new data point cuts the size of the set by half. Given a distance measure between classifiers that is related to their difference in prediction performance, the exponential reduction in volume then carries over to an exponential reduction of the classification error with training-set size. Unfortunately, this assertion may no longer hold if classes cannot be separated without errors. The reduction of classification error may then become polynomial again. Still, empirical evidence is abundant, that shows that active data selection strategies lead to a significant improvement of learning over standard inductive learning strategies.

For illustration, the kernel perceptron (9), has been applied together with active data selection for training an agent to predict the perceptual quality of sounds. Four classes of impact sounds were generated by an acoustic model and played to ten human listeners whose task was to rate them as *glass, metal, plastic* or *wood*. Sound-rating pairs were then used to train ten multiclass predictors based on the kernel perceptron, one for each listener—either using standard methods or active data selection. During standard training procedures, sound-rating pairs were randomly chosen for every new training sound while, during active data selection, the agent requested the most informative data point under the volume-reduction criterion—see Fig. 8a. For this demonstration, sounds were represented by the parameters of the acoustic models that were used for their creation, but a representation based on aural features

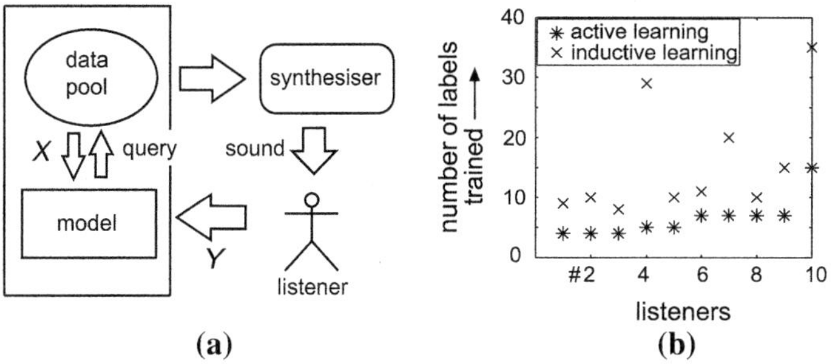


**Fig. 8** Listening test using active learning. **a** A parameter vector, **x**, is used to synthesize a new sound by a physical model, that is, a synthesizer. After listening, the listener labels it with *glass, metal, plastic* or *wood* (label *y*). The learning machine then updates its prediction model and the query algorithm will suggests a new sound. **b** The graph shows the number of labeled training data required to reach a test-error rate of 0.35 for agents implemented through multiclass kernel-perceptron classifiers. The results for active data selection are compared with the results for standard inductive learning separately for every listener

could have been used as well. Figure 8b shows the number of training examples that were required to achieve a performance significantly above the chance level of 35 %. For all ten listeners, active learning led to a significant reduction of the number of training examples. Standard inductive learning required on average 2.5 times the number of labels compared to active learning.

## 3.2 *Active Listening*

The active-listening[3] approach, recently popular in robot audition [5], is based on the concept that perception and action come in couples [3]. While exploring their aural environment actively, listeners recognize sound sources and analyze aural scenes by simultaneously monitoring their auditory sensations and their motoric actions, taking advantage of both the auditory and proprioceptive modalities. Further, during this process, they permanently adjust their auditory system task-specifically. Modeling active listening implies various feedback mechanisms.

### Feedback

Incorporation of multiple feedback loops into a model of sensory perception and cognition reaches out to the edge of current knowledge. In the auditory system, although there is strong evidence for numerous physiological feedback pathways

[3] The term *active listening* in the sense used here is not synonymous with a specific oral-communication technique that requires listeners to feed back to talkers what they hear.

[31, 45], little work has been done so far to incorporate feedback into processing models. In vision there is more experience with this type of modeling, but it is not yet clear to what extent this can be translated to the auditory domain [36, 49]. The aspect of porting domain experience across modalities has thus to be followed up carefully. While, in principle, the model framework of Fig. 1 allows for feedback between any stages, it makes sense at the time being to limit this structure with respect of physiological and operational evidence. In the following, a list of feedback paths is given that appear of particular relevance to the model framework presented here [15].

- Feedback from the binaural-mapping stage, that is, the output of auditory signal processing, to head-position control to keep a tracked sound source in aural focus—compare the so-called *turn-to reflex* as can already be observed in infants [19].
- Feedback from the cognitive stage to head-position in order to control for deliberate exploratory head movements, eventually moving the complete head-and-torso cart [16] this volume, and [7].
- Feedback from the segmentation stage to the signal-processing stage to solve ambiguities by activating additional preprocessing routines, for instance, cocktail-party effect and/or precedence effect processing—compare [50].
- Feedback from the cognitive stage to the signal-processing stage. This is intended to model efferent/reafferent effects of attention by modifying filter characteristics and/or putting a special focus on dominant spectral regions—for example [41, 42, 44, 47].
- Feedback from the cognitive stage to the segmentation stage to request task-specific and/or action-specific information on particular features—for example [18].

In the following, some exemplary feedback ideas along those lines are discussed in more detail.

1. To improve localization accuracy, a movable head-and-torso platform can perform movements, properly controlled by mimicking human strategies when exploring auditory scenes, for example, to derive estimates of distances and to solve front-back ambiguities [16], this volume.
2. Feedback can be used to adjust parameters for bottom-up processing, such as auditory-filter bandwidths, spectral weights in combining information across auditory filters, operating points of the temporal adaptation processes. Further, it can provide additional information supporting auditory-stream segregation, for instance, classifying groups of features of the same auditory stream in the binaural-activity maps [9].
3. At the *cognitive* level of the model framework, feedback from higher levels can make use of the interactive graphical-domain models as an *active blackboard*—as already mentioned in Sect. 2.2. Higher-level processes in an application-specific subsystem, such as an expert system for scene analysis, can set variables according to their specific intentions. Through modeling, it can be monitored how higher-level feedback corresponds with the rules and observations of the system and implications can be tested regarding the interpretation and initiation of new feedback information to lower model stages.

4. An important aspect of feedback is the incorporation of cross-modal information into the auditory processes. It is well known that profound interrelations between auditory and visual cues exist—compare, for example [38]. Visual cues can be introduced to the model system at two stages, namely, the pre-segmentation stage—the turn-to reflex—and the cognitive stage, exploiting prior knowledge about the visual scene. Specific proprioceptive information, such as the current position and movement of the head-and-torso platform, can also be used, particularly, at the pre-segmentation stage or even lower.

# 4 Conclusion

The model architecture as described in this chapter offers a comprehensive approach to modeling aural perception and experience. The listeners are modeled as an intelligent system exploring its surroundings actively and autonomously via an *active exploratory listening* process. It is assumed that, in the course of this process, the perceptual and cognitive world of the modeled listeners evolves and differentiates. This notion of the essence of listening stands in contrast to a widespread view that sound signals impinging on listeners' ears are processed by their auditory systems in a purely bottom-up manner. While it is an advantage of the latter approach that invention of the listeners is not needed, its prognostic power is limited to some primitive perceptual features, such as loudness, roughness or pitch. In a pure bottom-up approach, percepts are solely determined by the given ear-input signals, that is, formed in a signal-driven way. The active-explorative-listening approach, in contrast, requires a more complex model structure in which bottom-up, signal-driven, and top-down, hypothesis-driven, processes interleave in a complex way. The formation of hypotheses, a major feature of such a model structure, requires explicit knowledge inherent to the system. Part of the knowledge is acquired by the system itself in the exploration processes mentioned above, other knowledge has to be imported from external sources—potentially including physical knowledge as well statistical knowledge derived from possibly large datasets—or it may originate from other sensory modalities, such as proprioception, vision and/or tactility.

**Acknowledgments** The authors gratefully acknowledge suggestions of their external reviewers who helped to improve the clarity of presentation. Particular thanks are due to P. A. Cariani, who contributed relevantly by commenting the chapter from the viewpoint of biological cybernetics.

# References

1. K. Adiloğlu, R. Annies, H. Purwins, and K. Obermayer. Deliverable 5.2, visualisation and measurement assisted design. Technical report, Neural Information Processing Group, TU Berlin, 2009.
2. K. Adiloğlu, R. Annies, E. Wahlen, H. Purwins, and K. Obermayer. A graphical representation and dissimilarity measure for basic everyday sound events. IEEE *Transactions Audio, Speech and Language Processing*, 20:1542–1552, 2012.

3. J. Aloimonos. *Active perception*. Lawrence Erlbaum, 1993.
4. M. Altinsoy. The quality of auditory-tactile virtual environments. *J. Audio Engr. Soc.*, 60:38–46, 2012.
5. S. Argentieri, A. Portello, M. Bernard, P. Danés, and B. Gas. Binaural systems in robotics. In J. Blauert, editor, *The technology of binaural listening*, chapter 9. Springer, Berlin-Heidelberg-New York NY, 2013.
6. L. Avant and H. Helson. Theories of perception. In B. Wolman, editor, *Hdb. of General Psychology*, pages 419–448. Prentice Hall, Englewood Cliffs, 1973.
7. M. Bernard, P. Pirim, A. de Cheveign, B. Gas, and IEEE/RSJ. Sensomotoric learning of sound localization from auditory evoked behavior. In: *Proc. Intl. Conf. Robotics and Automation*, ICRA ' 2012. pages 91–96, St. Paul MN, 2012.
8. J. Bilmes and C. Bartels. Graphical model architectures for speech recognition. *Signal Processing Magazine*, IEEE, 22:89–100, 2005.
9. J. Blauert. Analysis and synthesis of auditory scenes. In J. Blauert, editor, *Communication Acoustics*, chapter 1, pages 1–26. Springer, Berlin-Heidelberg-New York, 2005.
10. J. Blauert. Conceptual aspects regarding the qualification of spaces for aural performances. *Act. Acust./Acustica*, 99:1–13, 2013.
11. J. Blauert, ed. *The technology of binaural listening*. Springer, Berlin-Heidelberg-New York NY, 2013.
12. J. Blauert, J. Braasch, J. Buchholz, H. Colburn, U. Jekosch, A. Kohlrausch, J. Mourjopoulos, V. Pulkki, and A. Raake. Aural assessement by means of binaural algorithms - the AABBA project. In J. Buchholz, T. Dau, J. Dalsgaard, and T. Paulsen, editors, *Binaural Processing and Spatial Hearing*, pages 303–343. The Danavox Jubilee Foundation, Ballerup, Denmark, 2009.
13. J. Blauert and U. Jekosch. Concepts behind sound quality, some basic consideration. In *Proc. InterNoise 2003*, pages 72–76. Korean Acoust. Soc., 2003.
14. J. Blauert and U. Jekosch. A layer model of sound quality. *J. Audio-Engr. Soc.*, 60:4–12, 2012.
15. J. Blauert and K. Obermayer. Rückkopplungswege in Modellen der binauralen Signalverarbeitung (feedback paths in models of binaural signal processing). In *Fortschr. Akustik*, DAGA 2012, pages 2015–2016. Deutsche Ges.f. Akustik, DEGA, Berlin, 2012.
16. J. Braasch, S. Clapp, A. P. T. Pastore,, and N. Xiang. Binaural evaluation of auditory scenes using head movements. In J. Blauert, editor, *The technology of binaural listening*, chapter 8. Springer, Berlin-Heidelberg-New York NY, 2013.
17. A. Bregman. *Auditory scene analysis - the perceptual organization of sound*. MIT press, Cambridge MA, 1990.
18. N. Clark, G. Brown, T. Jürgens, and R. Meddis. A frequency-selective feedback model of auditory efferent suppression and its implication for the recognition of speech in noise. *J. Acoust. Soc. Am.*, 132:1535–1541, 2012.
19. R. Clifton, B. Morongiello, J. Kulig, and J. Dowde. Newborn's orientation towards sounds: Possible implication for cortical development. *Child develop.*, 52:883–838, 1981.
20. D. Corkhill. *Collaborating software: blackboard and multi-agent systems and the future*. Proc. Intl. Lisp Conf., New York NY, 2003.
21. S. Davis and P. Mermelstein. Comparison of parametric representations for monosyllabic word recognition in continuously spoken sentences. IEEE *Trans. Acoust., Speech, Signal Processing*, 28:357–366, 1980.
22. M. Delcroix, K. Kinoshita, T. Nakatani, S. Araki, A. Ogawa, T. Hori, S. Watanabe, M. Fujimoto, T. Yoshioka, T. Oba, Y. Kubo, M. Souden, S.-J. Hahm, and A. Nakamura. Speech recognition in the presence of highly non-stationary noise based on spatial, spectral and temporal speech/noise modeling combined with dynamic variance adaptation. In *Intl. Worksh. Machine Listening in Multisource Environments*, CHiME 2011, pages 12–17, 2011.
23. L. Dempere-Marco, D. Melcher, and G. Deco. Effective visual working memory capacity: An emergent effect from the neural dynamics in an attractor network. PLoS ONE, 7:e42719, 2012.
24. S. Deneve. Bayesian spiking neurons I: Inference. *Neural Computation*, 20:91–117, 2008.
25. S. Deneve. Bayesian spiking neurons II: Learning. *Neural Computation*, 20:118–145, 2008.

26. DIN EN ISO 9000. *Qualitätsmanagementsystem, Grundlagen und Begriffe (quality management system, fundamentals and concepts)*. Dtsch. Inst. f. Normung, Berlin, 2005.
27. R. Engelmore and A. Morgan (eds.). *Blackboard systems*. Addison-Wesley, Boston MA, 1988.
28. L. Erman. The Hearsay II speech-understanding system - integrating knowledge to resolve uncertainty. *Computing surveys*, 12:213–253, 1980.
29. S. Gold, A. Rangarajan, C.-P. Lu, and E. Mjolsness. New algorithms for 2d and 3d point matching: Pose estimation and correspondence. *Pattern Recognition*, 31:957–964, 1998.
30. S. Haykin. *Neural networks - a comprehensive foundation*. Macmillan, New York NY, 1994.
31. J. He and Y. Yu. Role of descending control in the auditory pathway. In A. Rees and A. Palmer, editors, *Oxford Hdb. of Auditory Science*, volume 2: The auditory brain. Oxford Univ. press, New York NY, 2009.
32. F.-F. Henrich and K. Obermayer. Active learning by spherical subdivision. *J. Machine Learning Res.*, 9:105–130, 2008.
33. J. R. Hershey, S. J. Rennie, P. A. Olsen, and T. T. Kristjansson. Super-human multi-talker speech recognition: A graphical modeling approach. *Comput. Speech Lang.*, 24:45–66, 2010.
34. S. Hochreiter, T. Knebel, and K. Obermayer. An SMO algorithm for the potential support vector machine. *Neural Computation*, 20:271–287, 2008.
35. S. Hochreiter and K. Obermayer. Support vector machines for dyadic data. *Neural Computation*, 18:1472–1510, 2006.
36. B. Julesz and I. Hirsh. Visual and auditory perception - an essay of comparison. In E. Davis jr and P. Denes, editors, *Human communication* - a unified view, pages 283–340. McGraw Hill, New York NY, 1972.
37. A. Kohlrausch, J. Braasch, D. Kolossa, and J. Blauert. An introduction to binaural processing. In J. Blauert, editor, *The technology of binaural listening*, chapter 1. Springer, Berlin-Heidelberg-NewYork NY, 2013.
38. A. Kohlrausch and S. van de Par. Audio-visual interaction in the context of multi-media applications. In J. Blauert, editor, *Communication Acoustics*, pages 109–134. Springer, Berlin-Heidelberg-New York NY, 2005.
39. H. W. Kuhn. The Hungarian method for the assignment problem. *Naval Research Logistics Quarterly*, 2:83–97, 1955.
40. S. G. Mallat and Z. Zhang. Matching pursuits with time-frequency dictionaries. IEEE *Transactions Signal Processing*, 41:3397–3415, 1993.
41. R. Meddis, R. Ferry, and G. Brown. Speech innoise and the medial olovo-cochlear efferent system. *J. Acoust. Soc. Am.*, 123:3051–3051, 2008.
42. D. Messing, L. Delhorne, E. Bruckert, L. Braida, and O. Ghitza. A non-linear efferent-inspired model of the auditory system - matching human confusion in stationary noise. *Speech Communication*, 51:668–683, 2009.
43. R. D. Patterson and J. Holdsworth. A functional model of neural activity patterns and auditory images. *Advances in Speech*, Hearing and Language Processing, 3:547–563, 1996.
44. B. Scharf. Human hearing without efferent input to the cochlea. *J. Acoust. Soc. Am.*, 95:2813, 1994.
45. B. Schofield. Structural organization of the descending pathway. In A. Rees and A. Palmer, editors, *Oxford Hdb. of Auditory Science*, volume 2: The auditory brain. Oxford Univ. press, New York NY, 2009.
46. B. P. Schölkopf and A. J. S. AJ. *Learning with Kernels: Support Vector Machines, Regularization, Optimization, and Beyond*. MIT Press, Cambridge, 2002.
47. L. Schwabe and K. Obermayer. Learning top-down gain control of feature selectivity in a recurrent network of a visual cortical area. *Vision Research*, 45:3202–3209, 2005.
48. E. Smith and M. S. Lewicki. Efficient coding of time-relative structure using spikes. *Neural Computation*, 17:19–45, 2006.
49. R. Welch and D. Warren. Intersensory interaction. In K.R. Boff, L.Kaufmann, and J. Thomas, editors, *Hdb. of Perception and Human Performance*, chapter 25, pages 1–36. Kluwer Academic, Dordrecht, 1989.
50. S. Wolf. *Lokalisation von Schallquellen in geschlossenen Rumen (Localization of sound sources in enclosed spaces)*. doct. diss., Ruhr-Univ. Bochum, Germany, 1991.

# Index

**A**
ABR wave V latency, 38
Absolute hearing thresholds, 38
Absorbent area, 436, 438
Absorption coefficient, 434, 437, 438
Accessibility, 441
Acoustic
  near-field, 274
  scattering, 277
Acoustic absorber, 438
Acoustical capacity, 440
Acoustic manikin, 430
Acoustic modeling software, 430, 434
Acoustic treatment, 434, 348
Activating function, 314, 317
Active
  audition, 236, 239, 241, 242, 244, 246, 248
  Direction pass filter (ADPF), 231, 239
  Listening, 495, 497
  perception, 240, 241, 248
Active learning, 495, 497
Adaptation, 126
Adverse acoustic condition, 123, 129
Alcoves, 434
Ambient component, 335
Ambiguity of direction, 155
Ambisonics, 344, 351
Amplitude modulation, 130
Amplitude-modulation spectrum, 286
Anechoic, 437, 444
Anechoic conditions, 428
Anechoic environment, 131, 138
Apparent source width (ASW), 216, 219
Architectural acoustics, 433
Artifact, 333
Artificial neural network, 96
Asymmetric hearing, 442
Atomic events, 485
Audibility noise, 127, 128, 135, 138
Audio technology, 6
Audiology, 6
Audio-proprioceptive localization, 451
Audio-visual localization, 450
Auditory
  cortex, 8, 9
  event, 2, 4, 5, 7, 12, 9, 15, 16, 18, 256, 352
  filter, 270
  nerve, 8, 9
  scene, 256
  stream segregation, 4
Auditory brainstem responses, 38
Auditory filter bandwidth roadening, 127
Auditory frequency band, 334
Auditory image, 340
Auditory model, 33, 478
Auditory modeling toolbox (AMToolbox), 96
  experiments, 36
  getting started, 51
  psychoacoustic data, 36
  status, 40
Auditory modeling, 478
Auditory nerve, 8, 9, 44, 311, 315, 317, 318
  ANF, 321, 322
Auditory perception, 334
Auditory scene analysis (ASA), 147, 280, 398
  binaural, 149
  computational, 149
Auditory-scene understanding, 494, 495
Aural objects, 482
Aural scenes, 482
Aural space, 3
Authentic, 257
Automatic gain control, 309
  AGC, 309
Automatic speech recognition (ASR), 163
  classification, 164
  feature extraction, 163
  training and test, 164

J. Blauert (ed.), *The Technology of Binaural Listening*, Modern Acoustics and Signal Processing, DOI: 10.1007/978-3-642-37762-4,

Autonomous agent, 484
Azimuth, 203

**B**
Background noise, 129
Balconies, 434
Basilar membrane (BM), 9, 43, 126
Bayesian classifier, 286, 287, 302
Beamforming filter, 280
Beamformer, 163
  super directive, 163
Better-ear listening, 428, 430, 431, 439
Better-ear-I3, 298
Better-ear-STOI, 298
B-format, 337, 344, 353
Bimodal, 449
Bimodal perceptual event, 471
Binary mask
  estimated, 415
  ideal, 416
Binaural
  decoloration, 4, 6, 7
  dereverberation, 3
  difference
    intelligibility-level, BILD, 5
    masking-level, BMLD, 5, 20–22
  hearing, 2, 3, 6, 7, 10, 19
  noise suppression, 5, 20
  sluggishness, 25
  synthesis, 264, 265, 268
  transmission, 255
  unmasking, 21–23, 26, 428–431, 439
Binaural-activity map, 128, 129, 342, 479
Binaural advantage, 280
Binaural auditory model, 340, 355
Binaural decoloration, 7
Binaural front-end, 286
Binaural interaction, 280
Binaural listening, 477
Binaural manikin, 208, 221
Binaural masking level difference (BMLD), 429, 430, 431
Binaural model, 430
  structure, 154
Binaural processor, 124, 128
Binaural-proprioceptive localization, 451
Binaural recordings, 107
  non-individualized, 107
Binaural room impulse response (BRIR), 429–431, 433, 443, 444
  recorded, 124, 133, 136, 399, 414
  simulated, 399, 414
Binaural room transfer function (BRTF), 177
Binaural sluggishness, 146
Binaural spectral magnitude difference standard deviation (BSMD-STD), 180–182, 191–194
Binaural statistics, 282
Binaural tuning, 155
Binaural unmasking, 428–431, 439
Binaural-visual localization, 450
Binaural weighting, 94, 100
  contralateral ear, 94
  ipsilateral ear, 94
Bivariate distribution, 296
Blackboard architecture, 481, 495
Blind source separation, 11
Bone conduction, 2
Bottom-up processing, 483
Broadening of auditory filters, 127, 136
Building bulletin 93 (BB93), 427, 438

**C**
Cafeteria, 432, 433
Calibration, 103
Carpeting, 439
Ceiling, 438, 443, 445
Central processor, 128
Channel crosstalk, 310, 314
  electrical crosstalk, 313, 321
  electrical field spread, 311, 314, 322
Classroom, 427, 433, 438
Cochlea, 8, 9
Cochlear compression, 126
Cochlear implant, 110
  CI, 309–311, 313, 318
  listener, 319
Cochlear implantees, 431, 441
Cochlear nucleus, 315
Cocktail-party, 444
Cocktail-party effect, 5, 6
Cocktail party problem, 234
Cocktail party scenario, 309
Coding strategy, 312
  CIS, 313, 321–324
  FS, 313
  FS4, 321–324
Cognitive aural-scene analysis, 478
Coincidence detector, 280
Coincidence model, 311
Coloration, 3, 4, 6, 427, 430, 481
Commercial room-simulation software, 445
Comparison process, 97, 98
Complex acoustic scenes, 399

Computational auditory scene analysis (CASA), 226, 236, 238, 398
Computer vision, 226, 228, 231, 235–237, 239, 248
Concurrent talkers, 130
Cones of confusion, 203
Count-comparison principle, 341
Critical band, 38, 334
Critical distance, 174, 176, 184, 195
Cross-correlation, 320, 430, 444
Cross-correlation function (CCF)
  normalized, 42
  summary, 403
Cross-correlation model, 20, 21, 25
Cue-selection mechanism, 124, 126, 131

**D**
Decoder, 336, 344
Decorrelation, 339, 347, 353
Delay lines, 145
Dereverberation, 3, 361, 363, 364, 379, 380
Diffraction, 262, 263
Diffuse sound field, 340, 343, 352
Dip listening, 444, 445
Direct component, 335
Direct sound, 3, 4, 17, 18
Direction of arrival (DOA), 154, 402
Directional audio coding, 337, 345
Directional microphones, 431
Directional pattern, 339
Direct-to-reverberant ratio (DRR) , 444
Disabilities, 441
Disjointness, 282
Distance, 2, 10
Distance perception, 50, 171–173, 195
Distance-to-similarity mapping, 100
Double-pole coordinate system. *See* Interaural-polar coordinate system
Downmix, 336
Driving signal, 260, 261
Dual-resonance nonlinear filterbank (DRNL), 124
Duplex theory, 14, 17
Dynamic range, 309, 313, 317

**E**
Early reflections, 444
Echo, 3
  detector, 6
  threshold, 3
EC theory, 21, 23, 24
Edge waves, 263
Effective signal-to-noise ratio (SNRe), 429, 431, 434, 436, 437, 439, 440, 442
Ego-noise, 228, 236
  cancellation, 229, 232, 234, 236, 239
Electric stimulation, 310
Electrode model, 313
Elevation, 11, 18, 93, 115, 203
Encoder, 336, 344
Envelope, 126
Equalization and cancellation, 20, 22, 26
Equalization-cancellation (EC), 124, 128, 183
  model, 280, 311
Equalization-cancellation theory, 430
Equivalent rectangular bandwidth (ERB), 35, 98, 186, 191
Estimated ILD, 132
Estimated ITD, 132
Excitation signal, 79
Excitation-inhibition (EI), 128
Expectation-maximization, 302
Eye contact, 435

**F**
Fidelity
  spatial, 257, 276
  timbral, 257, 276
Fine structure filter, 155
Filter
  Bayesian, 245
  Kalman, 245
  Particle, 244, 245
Filter bank, 98
  auditory nerve response, 44
  dual-resonance nonlinear, DRNL, 34, 36, 97
  Gammatone, 40, 41, 43, 97
  invertible gammatone, 43
  parabolic-shaped, 97
  spacing, 98
  transmission line, 44
Floor, 438
Force-feedback, 452
Foyer, 433
Frequency channel, 430
Front/back confusions, 203, 204, 212
Front-back discrimination, 93
Fundamental frequency (F0), 444, 445
Fusion, 3

**G**
Gammatone filterbank, 183, 186, 191, 402
Gaussian mixture, 246

Gaussian mixture model (GMM), 190, 195, 301, 404
Gaussian-envelope sinusoid vocoder, 104
Gaze control, 435
Generalized cross-correlation (GCC), 182, 244
  classical cross-correlation (GCC-CC), 401
  framework, 401
  phase transform (GCC-PHAT), 401
  skeleton, 403
Generalized likelihood ratio test (GLRT), 246
Genetic algorithm, 288
Gestalt principles, 398
Gestalt rules, 481
Glimpse, 155
Graphical models, 482, 491, 495, 498
Green's function, 260, 262

**H**

Hair-cell transduction, 126
HA processing, 138, 140
Haptic device, 451
Haptic illusion, 454
Haptics, 450
Haptic space, 454
Hapto-audio system, 454, 462
Head, 435, 439
  tracker, 265, 266
Headphone
  reproduction, 265, 266
Headphone transfer functions, 38
Head-related
  impulse response, HRIR, 265
  transfer function, HRTF, 265, 266, 270, 277
Head-related coordinate system, 209
Head-related impulse response (HRIR), 360, 433
Head-related transfer function (HRTF), 15, 41–43, 94, 184, 202, 231, 237, 244, 283, 342, 399, 463
  common transfer function, 95
  directional transfer function, 95
Head rotation angle, 208
Head rotations, 202, 208
Head shadow, 430, 436
Head shadow effect, 280
Head-and-torso simulators, 64
Hearing aid (HA), 6, 7, 121, 123, 138, 431
Hearing-assist device, 109
  casing, 109
  microphone placement, 110
    behind-the-ear, 110
    in-the-ear, 110
Hearing-impaired (HI), 428, 434, 441
Hearing impairment, 431
Helmholtz equation, 260
High angular resolution planewave expansion, 338, 345
High ceilings, 436
Higher-order ambisonics (HOA), 256
Histogram, 296
Histogram lookup table, 287
Hodgkin–Huxley, 310, 315
Horizontal plane, 203
Horizontal-polar coordinate system. *See* Interaural-polar coordinate system
HRIR
  band-limited spectrum, 78
  continuous, 69
  continuous-azimuth acquisition, 72
    3D, 73, 84
  convolution model, 71
  cues, 57
  definition, 57
  dynamical measurement, 69, 79, 83
  field, 77
  Kalman filter identification, 72
  multichannel NLMS, 74
  NLMS identification, 73
  quality, 75
  quasi continuous, 71
  realtime demo, 76
  spatial spectrum, 78
  state-space model, 72
  storage, 63, 76
  usage for rendering, 76
HRTF
  analytical, 62
  anthropometry, 63
  applications, 76
  club fritz, 62
  confusion, 58
  cross fading, 67
  cues, 57
  databases, 60
  definition, 57
  discrete measurement setup, 64
  dynamical measurement, 79
  equalization, 66
  externalization, 59
  extrapolation, 68
  fast measurement, 67
  field, 78
  individual, 58
  interpolation, 67
  measurement point, 58
  range dependency, 62, 67

reciprocal, 62
sampling, 64
smoothing, 59
spatial transformation, 63
storage, 63
technical properties, 60
Humanoid robots, 227, 232, 234–236, 238, 239
Human-robot interaction, 225–227, 233, 235, 236, 238, 248
Hungarian algorithm, 487
Huygen's principle, 255

I
ILD detection threshold, 129
ILD histogram, 132, 133, 135
Image-source model, 433
Immersion, 255
Inferior colliculus (IC), 8–10
Informal listening, 347
Informational masking, 445
Input/output function, 126
Instrumental measure of speech intelligibility, 286
Intelligibility, 435
Intelligibility maps, 431
Interaural, 231
coherence, 155
cross-correlation, IACC, 12, 16, 17
level difference (ILD), 2, 5, 10–12, 14–17, 19–21, 145, 230, 258, 281, 295, 310, 311, 320, 370, 376, 388, 389, 400, 404, 405
phase difference (IPD), 146, 230
time difference (ITD), 2, 5, 10–12, 14–17, 20, 21, 24, 230, 237, 258, 265, 270, 320, 324, 326, 327, 400, 403, 405
transfer function, 155
vector strength, 155
Interaural axis, 203
Interaural coherence (IC), 303, 362, 364, 366, 371, 372, 381, 390, 430
Interaural cross correlation (ICC), 205, 206
Interaural phase, 430
Interaural phase differences, 280, 295
Interaural-polar coordinate system, 93
lateral angle, 93
polar angle, 93
Interaural spectral differences, 93
Interaural time difference (ITD), 43, 178, 179, 185, 187, 191, 203, 280, 370, 376, 389
ITD, 310, 311, 320, 324, 326, 327
Interaural transfer function, 281
Interlocutor, 435, 439
Internal reference, 257
Internal representation, 97, 98
Internal template, 97
template angle, 98
Inter-spectral differences, 98
Ion channel, 315
Ipatov sequence, 81
IPD model, 154
ITD detection threshold, 129
ITD histogram, 132, 133, 135
ITU P.835, 380, 386, 390

J
Jeffress model, 320

K
Kent distribution, 99
Kirchoff approximation, 261
Kurtosis (KURT), 179, 187

L
Late reflections, 444
Late reverberation, 361–365, 367–369, 374–377, 385, 388–390
Lateral superior olive (LSO), 8, 10, 26, 341
Learning process, 97
Lecture theatre, 428
Likelihood statistics, 101
Lindemann, 320
Linear prediction (LP), 402
Linear time-frequency-analysis toolbox (LTFAT), 35
Lip-reading, 435
Listener envelopment (LEV), 216
Listening area, 259, 263, 269
Listening experiment, 341, 350, 355
Listening position, 340
Localization, 2, 7, 10, 17, 19, 22, 27, 226, 229, 230, 256, 257, 265, 268, 276
accuracy, 268, 276
auditory epipolar geometry, 230
auditory-inspired, 402
blur, 2, 276
broadband, 401
cue, 5, 14, 16, 17
dominance, 3
ease of, 268
elevation, 230
horizontal, 15
HRTF, 231

Localization (*cont.*)
  learning-based approaches, 231, 241–244
  map, 269, 271
  model, 11, 18, 22, 26
  monaural, 18
  multiple sources, 408, 415
  performance, 10, 18
  prediction, 269, 271
  revised auditory epipolar geometry, 230
  scattering theory (ST), 230
  single source, 407
  supervised learning, 404
Localization blur, 132, 133
Localization error, 455
Localization process, 449
Localization task, 454
Local polar error, 102
Logarithmic compression, 126, 128
Lombard effect, 440
Loss of cochlear compression, 124
Loss of sensitivity, 134
Loss of temporal fine structure, 130
Loudness, 256, 257
Loudspeaker
  array, 256, 259, 262, 263, 266, 272, 274
Loudspeaker setup
  multichannel, 352
  stereophonic, 335
  surround, 335
Loudspeaker system, 345

**M**
Machine learning, 483, 484
Masking thresholds, 98
MCL, 313
Mean-ear-Q3, 298
Measurement signal, 65
  distortion, 66
  MLS, 66
  sweep, 66
Medial superior olive (MSO), 341
Median plane, 203
Median-plane sound localization data, 38
Mel frequency cepstral coefficient (MFCC), 164
Metadata, 336
Microphone
  array, 253
Microphone arrangement, 336
Middle-ear, 126
Minimal audible pressures, 38
Minimal-redundancy\maximal-relevance criterion (mRMR), 190, 191, 193
Missing data, 153
Missing data classification, 416
Missing features framework, 232, 234
Model
  adaptation loops, 45
  auditory brainstem responses, 45
  binaural activity, 48
  binaural detection, 46
  continuous-azimuth HRTF, 42
  directional time-of-arrival, 43
  distance perception, 50
  implementation verification, 36, 38
  inner hair cell, 36, 44
  lateral direction, 47, 48
  level conventions, 40
  masking, 46
  median-plane localization, 49
  modulation filter, 45
  sagittal-plane localization, 49
  sound lateralization, 47
  spatial speech unmasking, 51
  speech intelligibility, 50
Modulation filter, 156
Monaural cues, 230, 231, 234
Monaural heairng, 2
Monaural spectral cues, 93
  local spectral features, 94
  macroscopic patterns, 94
Moving speakers, 157
Multichannel
  reproduction, 257
Multichannel audio
  5.1-surround, 336, 353
  stereophonic, 335
Multichannel filter, 280
Multi-channel Wiener filter, 280
Multi-conditional training, 405
Multidimensional scaling, 257
Multimodal, 449
Multimodal input, 478, 499
Multiple stimuli with hidden reference and anchor (MUSHRA), 380
Multivariate, 301

**N**
Neuron models
  multi-compartment model, 310, 317
  point neuron models, 310, 317
  population models, 310, 318
Noise
  compact, 418
  diffuse, 418
Noise to mask ratio (NMR), 380, 382

Non-native listeners, 441
Normalization, 101
Normal hearing listener
 NH, 319, 320, 322
Novelty processing, 148

**O**
Odd-perfect ternary sequence, 80
Odd-PSEQ. *See* Odd-perfect ternary sequence
Office, 432
Off-sweet-spot, 340, 349
Oldenburger sentence test, 157
Optimal detector, 124, 128
Optimization, 287
Ordinary differential equations, 44
Organ of corti, 9
Orientation, 1, 453
Otoacoustic emissions, 44
Outer and middle ear responses, 38

**P**
Parameter-based statistical filtering, 287
Parameter value, 347
 suboptimal, 347
Particle filter, 150, 159
Percentile kurtosis (PKURT), 189
Percentile skewness (PSKEW), 189
Percentile symmetry (PSYM), 188
Percentile width (PWIDTH), 188
Perception, 342, 345
Perceptual evaluation of speech quality (PESQ), 380, 385, 390
Perceptual integration, 450
Perfect
 sequence, 80
 sweep, 43, 81
Peripheral processing, 97, 124, 126
Peripheral processor, 126
Phantom source, 111
Phase lock, 326
Phase locking, 155
Peripheral processing, 97
 basilar membrane, 97
 inner hair cells, 98
Pitch, 256
Place code, 145
Plane wave, 338, 345
Plausible, 257
Point of minimum activity (PMA), 128, 129, 132
Point source, 344, 349
Pool of models, 103
Position-variable model, 16
Power spectral density, 295
Precedence effect, 3, 6, 17, 122, 258, 271
Prediction matrix, 101
Primitive grouping scheme, 286
Prior probability, 287
Probability density function, 282
Probability mass vector, 101
Proprioception, 451
Proprioceptive accuracy, 454
Proprioceptive localization, 452
Proprioceptive orientation, 455
Prototype-based clustering, 489, 490
PSEQ. *See* perfect sequence
Psychoacoustic, 334
Psychoacoustic data, 38
Public address system, 428

**Q**
Quadrant error, 101
Quality, 257
Quality assessment, 477, 483

**R**
Railway station, 433
Rao-blackwellization, 160
Rate code, 145
Recognition
 speaker, 229, 234
 speech, 226, 227, 229, 231, 233, 236, 239
Reduced frequency selectivity, 127, 136
Reduced sensitivity, 134, 135
Reflection, 2–4, 16–19
Relational representations, 485, 489
Remapping, 206
Repertory-grid technique, 257
Reproducible research, 34
Response pattern, 99
Restaurant, 433, 434, 439
Reverberance, 3, 6, 7
Reverberant room, 131
Reverberation, 444, 445
Reverberation time, $T_{60}$, 427, 433, 437, 438
Room acoustics, 427
Room impulse response (RIR), 175, 183, 185, 359, 361, 367, 375, 399
Room-materials, 443
Room reverberation, 129
Room-related coordinate system, 208
Room transfer function (RTF), 176–178, 182

## S

Sabine equation, 438
Sagittal plane, 93
Sagittal-plane localization, 49, 93
  band limitation, 106
  baseline performance, 103
  chance performance, 108
  spectral channels, 104
  spectral warping, 106
Sagittal-plane localization data, 38
Scala tympani, 314
Segregation, 415
Self-noise. *See* Ego-noise
Sensitivity loss, 135
Sensorimotor theory, 239, 241, 242, 244, 248
Sensorineural hearing loss, 122, 124, 135
Sequence
  Ipatov, 81
  odd-perfect ternary, 80
  perfect, 80
Sequential importance resampling, 160
Sequential Monte-Carlo methods, 150
Short-time Fourier transform, 283, 294
Signal-to-noise ratio (SNR), 430
Signal to reverberation ratio (SRR), 377, 378, 380, 382
Similarity index, 99
Simulated scenario, 340
Simultaneous speakers, 342, 345, 349
Single-compartmental model. *See* Point neuron models
Single-layer potential, 259
Situational awareness, 1
Skewness (SKEW), 179, 188
Soft-mask, 295
Sound
  event, 257
  field, 255, 256, 260, 261
  reproduction, 257
  scene, 257
Sound localization, 319
Sound localization model, 311
Sound pressure level, 173, 174
Sound quality, 6
Source extraction. *See* Source separation
Source separation, 227, 229, 231, 233, 234, 236, 239, 240
Spaciousness, 256
Sparse representation, 485
Spatial
  aliasing, 256, 258, 263, 268
  extent, 257
  sampling, 262
  truncation, 262, 263
Spatial artifact, 339, 355
Spatial coherence, 285
Spatial cues, 121
Spatial hearing, 333
Spatial impression, 350
Spatial release from masking (SRM), 428, 435
Spatial sound, 454, 462
Spatial sound reproduction, 333
Spectral envelope, 313
Spectral standard deviation, 176–179, 182
Spectral subtraction, 368, 369, 374
Speech, 427, 430
Speech-detection module (SDM), 416
Speech intelligibility, 279, 427, 443, 444
Speech intelligibility index (SII), 431, 432
Speech processor CC, 285
Speech processor CLP, 284
Speech processor ELT, 285
Speech reception threshold (SRT), 279, 428, 431
Speech recognition, 311
Speech technology, 6
Speech transmission index (STI), 428
Spike-based representations, 485, 487
Spike pattern, 321, 326
Spiral ganglion neuron
  SGN, 318, 319
Spread of excitation, 314
Stapes footplate displacement, 38
Statistical quantities, 176, 179, 187, 196
Stereophony, 255, 257, 258
Stimulus, 103
Straightness weighting, 403
Summing localization, 3, 258
Supervised learning, 296, 404, 484, 487
Support-vector learning, 487, 488
Surround-sound system, 111
  Audyssey DSX, 114
  Auro-3D, 114
Sweep
  perfect, 81
Sweet spot, 340, 352
Synthesis area, 259

## T

Temporal coding, 135
Temporal envelope, 431, 444
Temporal fine structure (TFS), 432
Temporal integration
  averaging, 410
  histogram, 410
Temporal integrator, 129, 132
Temporal resolution, 128, 129

Temporal smearing, 427, 444
Temporal smoothing, 337, 338, 348
THR, 313
Timbre, 256, 257
Time of arrival, 43
Time constant, 348
Time delay estimation (TDE), 182, 244, 246
Time difference of arival (TDOA), 401
Time-frequency domain, 333
Tonotopy, 320
Top-down processing, 483
Torso reflections, 98
Tracking
  dynamic model, 160
  Kalman filter, 160
  measurement model, 161
Training, 297

**U**

Uncertainty parameter, 100
Unimodal, 449
Univariate distribution, 295
Unsupervised learning, 484, 489
Upmixing, 335
Useful-to-detrimental ratio, 428, 444

**V**

Vector base amplitude panning, 111, 338
  desired polar angle, 114
  law of tangents, 114
  panning ratio, 112, 115
Vertical-plane localization. *See* Sagittal-plane localization
Virtual
  sound scene, 256
  sound source, 256, 260, 261, 265
Virtual environment, 6, 7, 451
Virtual scene, 453
Vision, 450. *See also* Computer vision
Vocal effort, 441
Voice activity detection (VAD), 229, 231
Volume, 437

**W**

Wall, 438
Wave-field synthesis (WFS), 256, 257, 259, 264, 265, 268, 276
Welch method, 295
Wide-dynamic-range compression (WDRC), 123, 138
Wiener filtering, 338, 346
Workspace, 451

CPSIA information can be obtained
at www.ICGtesting.com
Printed in the USA
LVOW02*2052140516
488281LV00001B/3/P

9 783642 377617